Lecture Notes in Computer Science 5954

Commenced Publication in 1973
Founding and Former Series Editors:
Gerhard Goos, Juris Hartmanis, and Jan van Leeuwen

Sølvi Ystad   Mitsuko Aramaki
Richard Kronland-Martinet
Kristoffer Jensen (Eds.)

Auditory Display

6th International Symposium, CMMR/ICAD 2009
Copenhagen, Denmark, May 18-22, 2009
Revised Papers

Volume Editors

Sølvi Ystad
Richard Kronland-Martinet
CNRS - LMA
31 Chemin Joseph Aiguier, 13402 Marseille Cedex 20, France
E-mail: {ystad, kronland}@lma.cnrs-mrs.fr

Mitsuko Aramaki
CNRS - INCM
31 Chemin Joseph Aiguier, 13402 Marseille Cedex 20, France
E-mail: aramaki@incm.cnrs-mrs.fr

Kristoffer Jensen
Aalborg University Esbjerg
Niels Bohr Vej 8, 6700 Esbjerg, Denmark
E-mail: krist@aaue.dk

Library of Congress Control Number: 2010924065

CR Subject Classification (1998): H.5, H.3, H.4, C.2, H.2, J.5

LNCS Sublibrary: SL 3 – Information Systems and Application, incl. Internet/Web and HCI

ISSN 0302-9743
ISBN-10 3-642-12438-0 Springer Berlin Heidelberg New York
ISBN-13 978-3-642-12438-9 Springer Berlin Heidelberg New York

springer.com

Printed in Germany

Typesetting: Camera-ready by author, data conversion by Scientific Publishing Services, Chennai, India
Printed on acid-free paper 06/3180

Preface

Computer Music Modeling and Retrieval 2009 was the sixth event of this international conference series that was initiated in 2003. Since the start, this conference has been co-organized by the University of Aalborg, Esbjerg, Denmark (http://www.aaue.dk) and the Laboratoire de Mécanique et d'Acoustique in Marseille, France (http://www.lma.cnrs-mrs.fr) and has taken place in France, Italy and Denmark. The five previous editions of CMMR offered a varied overview of recent years' music information retrieval and sound modeling activities in addition to alternative fields related to human interaction, perception and cognition, as well as philosophical aspects linked to the field. We believe that the strength and the originality of this international conference both lie in its multidisciplinary concept and its ability to permanently evolve and open for new trends and directions within the related fields of interest. This year's CMMR took place in Copenhagen, Denmark, May 18–22, 2009 and was associated with the International Conference on Auditory Display (http://www.icad.org), hereby introducing new topics related to sound design, sonification and augmented reality to the computer music modeling and retrieval community.
Research areas covered by ICAD include:

- Auditory exploration of data via sonification (data-controlled sound) and audification (audible playback of data samples)
- Real-time monitoring of multivariate data
- Sound in immersive interfaces (virtual environments) and teleoperation
- Perceptual issues in auditory display
- Sound in generalized computer interfaces
- Technologies supporting auditory display creation
- Data handling for auditory display systems
- Applications of auditory display

As Derek Brock describes in the short foreword that follows this preface, the auditory display community is a very active one that opens for a large number of applications. Even though musical sounds have not traditionally been part of auditory display topics, the joint CMMR-ICAD conference revealed mutual interests and uncovered new perspectives of applications and fundamental research in both areas.

The proceedings of the previous CMMR conferences were published in the *Lecture Notes in Computer Science* series (LNCS 2771, LNCS 3310, LNCS 3902, LNCS 4969 and LNCS 5493), and the present edition follows the lineage of previous ones, including a collection of 25 papers of which a majority are directly related to auditory display topics. These articles were specially reviewed and corrected for proceedings volume.

The current book is divided into two main parts, one that concerns auditory display which deals with new CMMR topics such as sound design, sonification and audio-augmented reality and another entitled Modeling and Retrieval, which concerns more traditional CMMR topics linked to sound events, perception and cognition as well as music analysis and MIR.

We would like to thank the Program Committee members for their valuable paper reports and thank all the participants who made CMMR - Auditory Display an exciting and original event. Finally, we would like to thank Springer for accepting to publish the CMMR/ICAD 2009 proceedings in their LNCS series.

January 2010

Sølvi Ystad
Mitsuko Aramaki
Richard Kronland-Martinet
Kristoffer Jensen

Foreword

In May 2008, the International Community for Auditory Display (ICAD) was invited to hold its 15th conference in partnership with the 6th International Computer Music Modeling and Retrieval (CMMR) Symposium and the annual Danish re-new digital arts festival. The result of this joint meeting, which convened in Copenhagen one year later, was an exciting mix of research, presentations, and performances and a stimulating cross-fertilization of practice and ideas from three communities with many interests in common. This was the first collaboration between these organizations, so my purpose in this foreword to the latest volume in the CMMR series of selected proceedings papers is to briefly introduce the reader to a bit of the history, scope, and ambitions of the ICAD community.

ICAD's interests and goals have evolved and expanded over the course of 15 plus conferences and meetings, but it continues to be a research community whose primary focus—to paraphrase its founder, Gregory Kramer—is the study and technical application of sound, as it is used or occurs in familiar settings, to convey meaningful information. When ICAD was formed in the early 1990s, this was hardly an undiscovered domain, but it was nevertheless an area of inquiry that was just beginning to achieve a critical body of applied and theoretical work. Thus, the first conference, which took place at the Santa Fe Institute in October 1992, was attended by a comparatively modest, international group of researchers. These individuals hailed from a wide variety of backgrounds and, in many cases, were motivated by very different objectives. However, most were already deeply invested in the enterprise of displaying information with sound, and the range of topics that emerged from this initial gathering have gone on to shape the core interests of the community. These include the sonification of data, the psychology, semiotics, and design of aural information, virtual and spatialized sound, auditory interfaces (both pure and mixed) for human/machine interaction, and the nature of technologies and tools needed for auditory displays. A quick glance at the range of material presented at ICAD 2009 and papers selected for this volume will confirm the ongoing relevance of each of these areas of work in the ICAD community.

By the end of the 1990s, ICAD had grown well beyond its origins and, in the new century, its meetings began to be hosted annually by research groups and institutions in Europe, the Pacific rim, and elsewhere in North America. Through the activities of the community, a wider recognition of how techniques for aurally exploring and transforming data can make meaningful contributions in both scientific and applied settings was achieved, and ICAD was asked to develop a formal research agenda and executive overview of sonification for the U.S. National Science Foundation. There are numerous other developments in the past decade that can be cited, too. ICAD has, for instance, enjoyed an important growth in work and colleagues addressing aural information design research for blind and sight-impaired individuals. It has also worked to broaden the community's distinct take on the study and use of sound as information, both internally and by reaching out to other entities and groups involved in the sound-based arts and sciences. Notable examples of this include ICAD's seminal concert of sonifications of brain activity, "Listening to the Mind Listening," presented at the Sydney Opera House in 2004, and, more recently, of course, its joint/concurrent conference with the CMMR community and re-new digital arts festival leading to the present volume.

Finally, and perhaps most vitally, ICAD has recently launched an online presence for both the ICAD community and all who have an interest in topics, news, and events related to auditory display. Membership in ICAD continues to be free, and, at this new website, one can find an extensive bibliography, read about and hear examples of auditory displays, learn about relevant auditory tools and design techniques, and find and interact with others who are working in the field and/or related areas, as well as much more. If any of the work in the following pages—or the notion of an auditory display—sounds interesting, you are encouraged to go to www.icad.org where you will find yourself welcomed and invited to listen, explore, and join the conversation.

January 2010

Derek Brock
Secretary, ICAD

Organization

The 6th International Symposium on Computer Music Modeling and Retrieval (CMMR2009) was co-organized with the 15th International Conference on Auditory Display (ICAD 2009) by Aalborg University (Esbjerg, Denmark), LMA/INCM-CNRS (Marseille, France) and Re:New - Forum for digital arts.

Symposium Chairs

General Chair

Kristoffer Jensen — Aalborg University Esbjerg, Denmark

Digital Art Chair

Lars Graugaard — Aalborg University Esbjerg, Denmark

Conference Coordination

Anne Bøgh

Program Committee

Paper Chair

Mitsuko Aramaki — Institut de Neurosciences Cognitives de la Méditérranée - CNRS, Marseille, France

ICAD Program Chair

Richard Kronland-Martinet — Laboratoire de Mécanique et d'Acoustique - CNRS, Marseille, France

CMMR Program Chair

Sølvi Ystad — Laboratoire de Mécanique et d'Acoustique - CNRS, Marseille, France

CMMR/ICAD 2009 Referees

Mitsuko Aramaki
Frederico Avanzini
Stephen Barrass
Terri Bonebright
Eoin Brazil
Derek Brock
Douglas Brungart
Antonio Camurri
Laurent Daudet
Benjamin Davison
Kees van den Doel
Milena Droumeva
Alistair Edwards
Mikael Fernström
Cynthia M. Grund
Matti Gröhn
Brian Gygi
Kristoffer Jensen
Richard Kronland Martinet
Rozenn Nicol
Sandra Pauletto
Grégory Pallone
Camille Peres
Stefania Serafin
Xavier Serra
Malcom Slaney
Julius O. Smith
Tony Stockman
Patrick Susini
Vesa Välimäki
Thierry Voinier
Bruce Walker
Sølvi Ystad

Table of Contents

I Auditory Display

I-i Sound Design Theory and Methods

I-ii Sonification

I-iii Audio Augmented Reality and 3D

II Modeling and Retrieval

II-i Sound Events: Perception and Control

II-ii Music Analysis an MIR

Tools for Designing Emotional Auditory Driver-Vehicle Interfaces

Pontus Larsson

Volvo Technology Corporation, Dept. 6310 M1.6, SE-40508 Göteborg, Sweden
pontus.larsson@volvo.com

Abstract. Auditory interfaces are often used in vehicles to inform and alert the driver of various events and hazards. When designed properly, such interfaces can e.g. reduce reaction times and increase the impression of quality of the vehicle. In this paper it is argued that emotional response is an important aspect to consider when designing auditory driver-vehicle interfaces. This paper discusses two applications developed to investigate the emotional dimensions of auditory interfaces. EarconSampler is a tool for designing and modifying earcons. It allows for creating melodic patterns of wav-snippets and adjustment of parameters such as tempo and pitch. It also contains an analysis section where sound quality parameters, urgency and emotional response to the sound is calculated / predicted. SoundMoulder is another tool which offers extended temporal and frequency modifications of earcons. The primary idea with this application is to study how users design sounds given a desired emotional response.

Keywords: Auditory interfaces, sound, emotions, earcons, auditory icons, vehicles.

1 Introduction

The safety and information systems of modern cars and trucks are becoming increasingly complex and these systems must be interfaced to the driver in the most efficient way to ensure safety and usability. Using sound in human-machine interfaces in vehicles may be a good way to relieve some of the visual load and be able to give warning, information and feedback when the driver has to focus on the primary task of safely controlling the vehicle. When properly designed, auditory displays are efficient means of increasing performance of Driver Vehicle Interfaces (DVI's) and may lead to shorter reaction times, improved attention direction, less ambiguities, and an increased quality impression [1]. Designing sounds for DVIs is however not an easy task as there are several design goals which must be fulfilled, including a wide range of urgency levels and related behavioural responses - some which may be crucial for the safety of the driver and the surrounding traffic.

A characteristic which is central in many discussions on auditory Driver- Vehicle Interfaces (DVIs) is the sound's level of urgency. Urgency can be defined as "...an indication from the sound itself as to how rapidly one should react to it." [2]. Too urgent sounds may cause annoyance, unwanted startle effects and even lead to the

S. Ystad et al. (Eds.): CMMR/ICAD 2009, LNCS 5954, pp. 1–11, 2010.

wrong behavior. On the other hand, if the sound is not urgent enough, reaction may be unnecessarily slow or result in that the warning is neglected.

There are several ways to vary urgency in a sound, including repetition speed, number of repeating units, fundamental frequency, inharmonicity [3], and loudness which appears to be one of the stronger cues for urgency [4]. However, the range within which loudness can be varied before the sound becomes un-ergonomic is in practice rather small since the sound should of course be loud enough to be heard over the background noise in the operator's environment and quiet enough not to cause annoyance or hearing impairment.

Although urgency is important to consider, it is not always a sufficient defining characteristic of sound. Designing a sound with low urgency is for example rather easy, just by using a low loudness, slow repetition speed and soft timbre. However, if we say that the sound should indicate a new SMS the urgency measure does not indicate whether the sound elicits the desired understanding, quality impression and behavioral response. And, designing sounds with unambiguous and appropriate meaning is perhaps the most important task in auditory warning design [2]. Using Auditory icons may be one possible way to increase understanding of the auditory display. Auditory icons have the advantage of carrying an inherent meaning and therefore require no or little learning. Still, this meaning may not be the same to all persons. As an example, the sound of a drain clearing may to some mean "Wet road" while others may interpret it the way the designer intended, namely "Low on fuel" [5]. In general it may be difficult to find a suitable match between the function/event to be represented and the sound.

While there is some ongoing research on how to design auditory signals to optimize the degree to which the sound elicits the intended meaning and correct action (the "stimulus-action compatibility" [6]) there is currently a lack of theory that can guide sound design.

We argue that emotional reactions to auditory information may help further guiding the design of auditory DVIs and there are several reasons why emotions are useful in sound design. One basic reason is that emotions are central in our everyday life and inform us about our relationship to the surrounding environment [7]. From an evolutionary perspective emotion can be seen as the human alarm system where positive emotions signal that everything is safe and no specific action is needed to be undertaken to survive, while negative emotions signal a potential threat and need to take quick action. Emotions thus have strong consequences for behavior and information processing.

Moreover, urgency is both a cognitive and emotional sensation with the function of motivating behavior [8]. Basic emotion research suggests that urgency is a form of cognitive preprocessing of information [9]. At its most basic level, a person in a specific environment (i.e. a truck) has a specific goal (i.e. driving the truck to its destination). When an event occurs (i.e. a warning sound is heard indicating something is wrong), the person appraises the event on a number of psychological dimensions designed to evaluate the seriousness of the threat (unexpectedness, familiarity, agency including urgency). This appraisal process automatically leads to an emotion with changes in experience, physiology, cognition, and behavior. The perception of a state of emotion is the internal event and, as a consequence of this experience, the person tries to cope with the situation by taking external or internal actions to improve the relationship between his goals and the environment. The more negative and the more arousal this experience has, the more serious is the event (thus calling for more immediate or urgent action, see [10,11]).

Following this, it may be argued that designing sounds that elicit an emotion is also a way of designing sounds that will elicit a certain level of urgency response and reaction and may therefore be particularly suited for design of sounds for DVI's. Behavioral responses are often difficult and time-consuming to assess but within emotion psychology there is a large number of instruments for measuring emotion which thus may be used as a proxy measure of behavior.

Our previous research has indicated that emotions are indeed useful proxy measures of behavior in auditory DVIs [12]. Moreover, some of the basic parameters of auditory DVIs have been mapped with regards to the emotional response they elicit (e.g [11]). While previous research has aimed at measuring the response of existing sounds, the current paper takes a further step and describes an attempt to use the emotional response as a design parameter. We describe two applications ("tools") where emotions serve as the defining measure of the sound. We believe that by incorporating the emotional response already in the sound design process, this would speed up and simplify the process, create a better understanding for how the sound is will work in reality and eventually lead to more efficient sounds.

2 Emotional Response

It has been shown that the emotional reaction to both natural and product sounds can be efficiently described by two bipolar dimensions, activation and pleasantness-unpleasantness (valence) [13-16]. Taking this approach, it is assumed that any emotional state can be described by the combination of these two orthogonal dimensions [17].

The so-called affect circumplex [17], shown in Figure 1, visualizes the two dimensional approach to describing emotional reactions. As an example, an emotional state such as excitement is the combination of pleasantness and high activation (upper left quadrant in Figure 1). An emotional reaction such as calmness is similar in pleasantness, but low in activation (lower left quadrant). Boredom is the combination of unpleasantness and low activation (lower right quadrant) and distress is the combination of unpleasantness and high activation (upper right quadrant).

2.1 Measurement of Emotional Reactions to Sound

There is a number of different ways to measure emotional reactions, including self report, physiological measures such as EEG and behavioral measures.

Self-report measures rely on that participants accurately can report their felt emotion. One common self-report measure is the Self Assessment Manikin (SAM) scales ([18], see Figure 2) which aims at capturing the activation/valence dimensions described in the previous section. The advantage of the SAM measure is that it can be understood by different populations in different cultures and that it is easy to administrate.

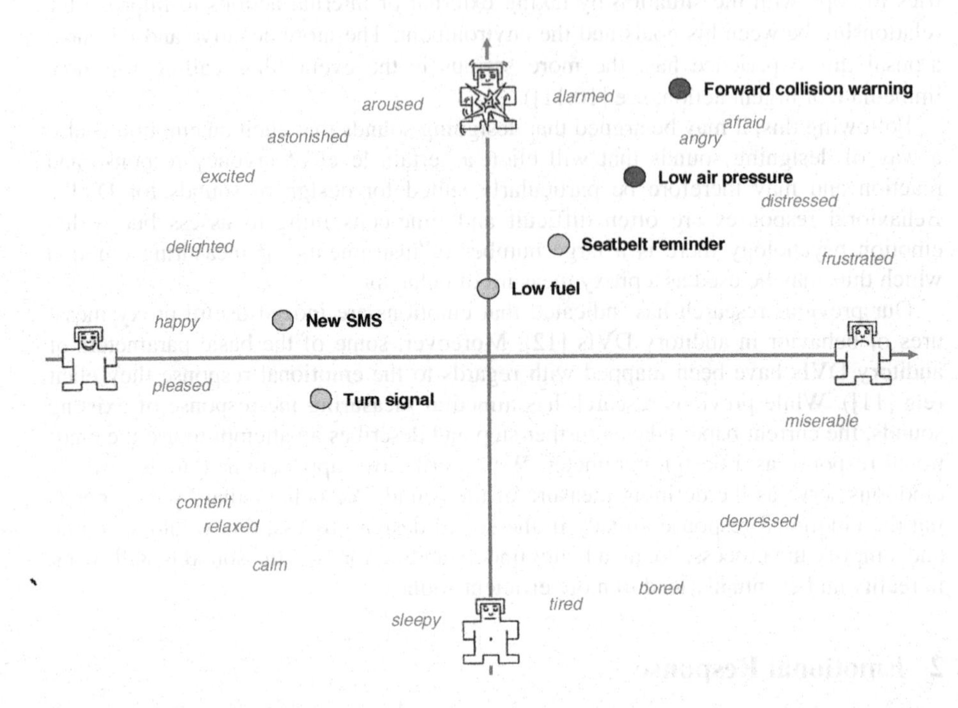

Fig. 1. The affect circumplex (Adapted from Russell, 1980), with some examples of desired responses to different warning- and information sounds. On the ends of each axis are shown the endpoints of the Self-Assessment Manikin (SAM) scale.

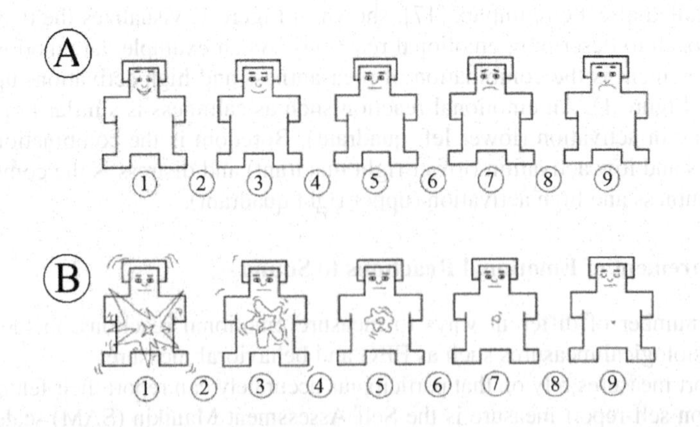

Fig. 2. The Self Assessment Manikin (SAM) scales for valence (*top*) and activation (*bottom*) (Bradley & Lang, 1994)

However, emotional reactions can also be captured via various physiological measures. An example of this is Electromyographical (EMG) measurement of facial muscle contractions which can be used to capture the valence dimension [15].

Facial EMG is typically measured by attaching electrodes in the facial region and measuring muscle micro-movements. Unpleasant emotions (negative valence) are linked to activity in the Corrugator supercilii (which controls eyebrow contraction) whereas pleasant emotions (positive valence) are reflected in activity in the Zygomaticus major (the "smile muscle"). Activation may be measured physiologically using EDA (Electro- Dermal Activity) which can be obtained by measuring the galvanic skin resistance on subjects' fingers or palms [15].

3 EarconSampler

EarconSampler is a simple sample sequenser/player application implemented in Matlab™ which can be used to test and design/modify earcon sound snippets (brief musical sounds that may be used for HMI purposes). The purpose of this software is to provide a tool with which prototype sounds can be easily and rapidly developed and/or modified for further use in user studies or for the final product. Although EarconSampler was designed primarily for automotive applications, it is not by any means limited to these kinds of displays but can be used as a general sound design/analysis tool. It should be stressed that although EarconSampler has been used for actual sound design work and research, it is a work-in-progress and all functionalities have not been extensively validated.

EarconSampler's algorithms calculate objective sound quality values (so far loudness and sharpness has been implemented) of the designed sound and estimate the emotional reaction (activation/valence) and urgency to the sound. It is the intention that with this tool, the designer will be more guided by research results and can easily get an understanding of how the designed sound will work in reality. Moreover, people who are not familiar with sound design may be involved in the total HMI design and may therefore have opinions on how the sound should sound. With the tool, they can easily try different variations to a certain sound, and give suggestions on modifications to the sound designer.

Short wav file segments ('patches') are required as input to the software as well as a 'melody' scheme which resembles a midi-file – e.g. it contains information on how the 'patches' should be played. The patches can be created in any sound synthesis software, recorded or obtained by other means. Different modifications to the melody (i.e. the earcon itself) can be easily done via EarconSampler's GUI: tempo, pitch, the patch's decay time, level and the length of the patch. EarconSampler also has an analysis functionality which visualizes the earcon in temporal and frequency domains.

The envisioned design loop when working with EarconSampler is shown in Figure 3. One starts from wav sound snippets designed in external synthesis software, which is then played and modified until the desired perceptual and emotional responses are achieved. If the desired response is not achieved, a new sound snippet is selected (or created) and run through the loop again. When the desired response is achieved, the sound can be taken to implementation and testing in driving situations, either real or simulated.

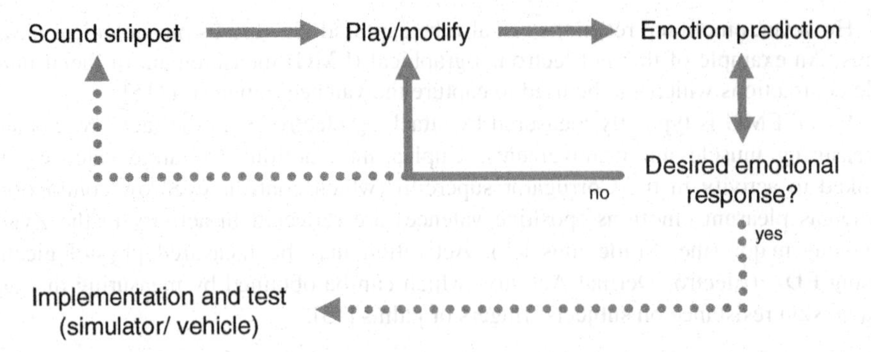

Fig. 3. The envisioned design loop when working with EarconSampler

3.1 The Graphical User Interface

The Graphical User Interface (GUI) of EarconSampler, shown in Figure 4, was implemented using Matlab's GUIDE interface design tool. It contains basically two sections; the upper one containing five sliders, some radio buttons and a drop down menu corresponding to the earcon design/modifcation section, and the middle/lower one containing graphs and push buttons corresponding to the render/analysis section. There is also a menu section which allows loading new patches or melodies and setting some plot options. Via the File menu, one can also save the current earcon to a wavefile.

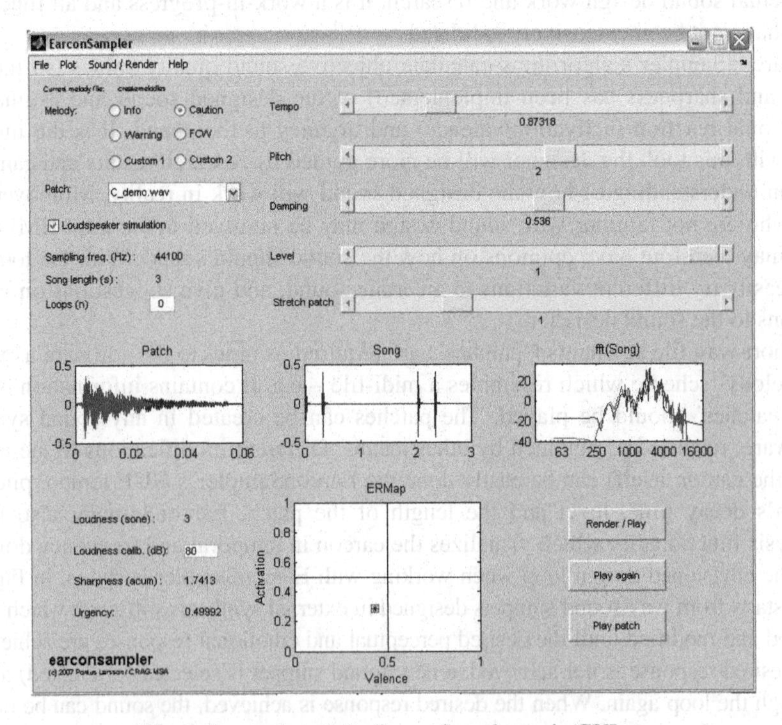

Fig. 4. Screenshot of the EarconSampler main GUI

3.2 The Design/Modification Section

In the upper left corner of the GUI one finds six different radio buttons with which one can set the melody scheme for the earcon (the 'song'). It is possible to change the melodies themselves by editing textfiles containing melody information which can be loaded via the File menu. Below the radiobuttons there is a drop down menu where one selects the desired wavefile which will be used for the 'patch'. A default patch list is always loaded but it is possible to add/change patches by using the File/Add patch menu item. The sampling frequency of the selected patch and the resulting song length (which naturally varies with tempo, see next paragraph) is also shown.

The four sliders found to the right in the design/modification section control Tempo, Pitch, Damping, Level and Patch stretch (the length of the patch) respectively. The current value of each parameter is also shown below each slider.

Tempo adjusts the time intervals between each 'note on' in the melody scheme. A tempo of 1, which is the default value, corresponds to unaltered tempo (e.g. as given in the melody scheme), higher or lower values scale the note onset intervals with the inverse of those values.

The pitch slider adjusts the fundamental pitch of the melody in semitone steps. Consequently, a pitch value of –12 or 12 lowers or raises the pitch by one octave. A pitch value of 0 gives an unaltered pitch. The algorithm which handles these pitch shifts also retains the length of the patch/melody.

Damping is a parameter which acts upon the patch itself. This parameter decreases the decay time of the patch by multiplying the patch with an exponentially decaying window. This can be useful if the patch is too long, making consecutive notes sound "on top" of each other. In figure 1 can be seen the result of such damping; the red curve in the "patch" graph shows the patch after windowing. The damping slider ranges from 0 to 1 where 0 corresponds to no window (no damping) and 1 corresponds to a decay of 60 dB / patch length.

The level slider naturally controls the level of the melody and ranges from 0-1 where 0 means no sound at all and 1 means full amplitude.

Finally, the stretch patch slider alters the length of the patch without changing the pitch, which could be useful for e.g. speech patches.

3.3 The Render/Analysis Section

In the lower half of the GUI one first of all finds three different plots which display the temporal view of the selected patch and the result of the damping operation (plotted in red on top of the original patch), the whole song (middle plot), and the FFT of the whole song (right plot). The plots are updated after each press on the "render/play"-button.

Below these plots and to the left are shown some sound quality metrics (Loudness, Sharpness) and Urgency and as with the plots, these values are updated each time the "render/play" button is pressed. The loudness calculation (and hence the sharpness calculation, which is based on loudness) requires a calibration value to produce correct values, which is entered into the edit box below the loudness. If no calibration value is available and if one simply wants to compare loudness/sharpness of different earcons, any reasonable value could be given.

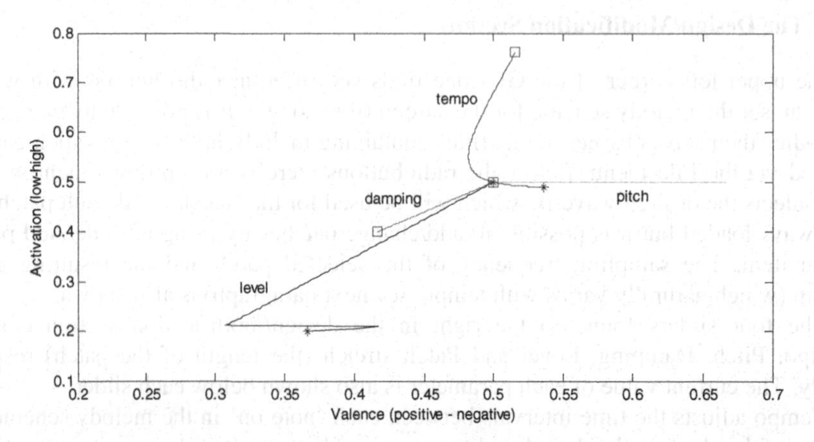

Fig. 5. Activation/valence functions of parameters tempo, pitch, damping, and level. The overall predicted response is a linear combination of these functions. Asterisks marks lowest parameter value and square marks highest parameter value.

In the middle, lower half of the GUI the predicted emotional response from the sound in terms of Valence/Activation for the current earcon is plotted. The Valence axis goes from 0 (positive) to 1 (negative) and the Activation goes from 0 (low activation) to 1 (high activation). The calculation of the V/A values is based on previous measurements using the SAM scales and extrapolations of this research [5-6]. The functions used to calculate the V/A values are shown in Figure 5. As can be seen, the functions are mainly linear apart from the tempo function and it should be noted that these are rather course approximations based on SAM measurement data. In some cases there is an exact correspondence between the current settings (patch/melody/tempo etc) and settings used in previous studies; in these cases, the blue '*' marker is emphasized by a red square marker.

Urgency, shown below Loudness/sharpness is calculated as a linear combination of Valence/Activation values (see next paragraph) – i.e. activating and negative sounds will result in higher urgency values than non-activating, positive sounds.

Finally, to the right one finds three buttons with which one can 1) create and play the song with the current settings ('render/play'), 2) play the most recently rendered song ('play again'), and 3) play only the patch ('play patch') –without damping.

4 SoundMoulder

SoundMoulder is a further development of the EarconSampler which allows also for more extensive shaping, "moulding" of both the temporal and spectral characteristics of an earcon. In addition, a step sequencer panel has been added to enable on-the-fly melody edits. As with EarconSampler it was implemented using Matlab's GUIDE interface design tool.

The purpose of the development of the SoundMoulder software is twofold. First, one intention is simply to provide an integrated fast and easy tool for creating and modifying earcons. Today, much of the existing software for sound design and editing is far too general, complex and not entirely suitable for the design of DVI earcons. SoundMoulder is intended to give the user a simple but efficient and, in many cases, sufficient set of tools for creating, modifying and auditioning earcons.

The second and perhaps the main intention, is that it should serve as an experimenting platform for investigations on how users/designers design sounds given a desired emotional response. In sound quality research in general, one uses a predefined set of sounds assumed to cover the dimensions under investigation and see how respondents rate different target attributes. The approach taken here is rather the opposite – to give the respondents target attributes and let them create the sound which represents this attribute.

4.1 The Graphical User Interface

The GUI of SoundMoulder is divided into seven different panels. In the top panel, the target emotion is displayed along with a text box for optional comments and a button which stores the sound/settings and advances to the next target emotion.

In the panel to the left on the second row, the basic sample snippet to be used as the "tone" for the earcon is selected. A currently inactive option of mixing two samples is also available. The user can in this panel change bitrate/sampling frequency, although this option is not intended for the proposed experiment. In the middle panel on the second row, an amplitude – time plot of the sample is shown. Above the plot, there are four buttons with which one can modify the envelope of the sample by inserting four attenuation points at arbitrary time/amplitude positions in the sample. With this feature, one can for example slow down the attack or decay of the sample or create an echo effect. Below the plot are two buttons with which one can select new start and endpoints in the sample. To the right in the second row one finds two sliders which control the length and pitch of the sample.

On the third row, there are two panels which contain various filtering options. The left panel has a drop down menu in which one can select from various pre-measured playback systems, so one can audition how well the earcons will work in different systems. The right panel contains a magnitude/frequency plot of the earcons and the possibility to apply two different types of filters; one bandpass filter with selectable high/low cutoff frequencies and filter order and one notch/peak filter with selectable frequency, Q-value and gain.

Finally, on the fourth row, one finds a melody edit panel which in essence is a compact one-track step sequencer. Here, one can either select from pre-defined melodies via the drop down menu or create/edit the melody directly in the melody matrix. There are three editable rows in the melody matrix: "Note on", "Velocity" and "Pitch". In the "note on"-row, a "1" means that a note should be played and all other values indicate delays (in ms) between two consecutive note onsets. Several "1"s in a row (no delays between consecutive notes) means you can play intervals and chords as well. On the "Velocity" row, one gives values of the amplitude (0-1) with which each note on should be played. Velocity values in columns with "Note on" delays has no significance. Finally on the "Pitch" row, each "Note on" can be given a pitch offset (in semitones) relative to the original sample pitch. In the melody edit panel, one can also alter the number of repetitions of the melody (the number of loops) and the playback tempo.

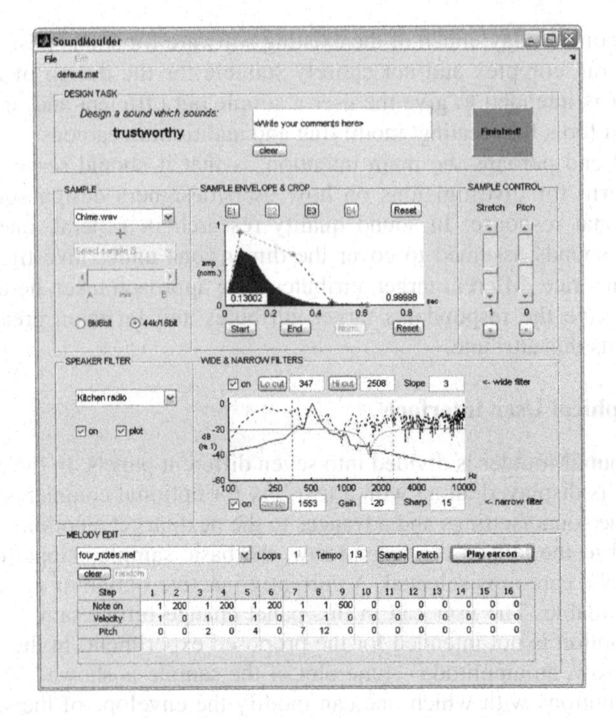

Fig. 6. Screenshot of SoundMoulder's GUI

5 Conclusions

Considering emotions when evaluating auditory DVIs has been shown to be useful as emotional response is linked to behaviour and may give a better understanding of users' response to the sound than measures of e.g. urgency or sound quality metrics. The tools/applications discussed in this paper represent a first approach to involve emotions also in the design process and to gain further understanding of the connection between sound design and emotional response.

The EarconSampler application has so far been successfully used both in focus groups meetings for the purpose of evaluating and improving sound designs, and for actual editing and design of sounds (in conjunction with sound synthesis software). SoundMoulder which is a further development offers more control of the design and the main intention here is to study how designers use different sonic parameters to create various emotional responses.

A useful addition to both these applications/tools would be a visual vehicle interior module where the sound can be played together with the visual information presented by e.g. LEDs, information displays and HUDs and perhaps also some traffic scenarios can be simulated. This would provide an even better impression of whether the sound fits the context and other information and allow for matching the sound to visual impressions.

Acknowledgements

This research was supported by the Swedish Foundation for Strategic Research (SSF).

References

1. Ho, C., Spence, C.: The multimodal driver. Ashgate Publishing, UK (2008)
2. Edworthy, J., Hellier, E.: Complex nonverbal auditory signals and speech warnings. In: Wogalter, M.S. (ed.) Handbook of Warnings, Lawrence Erlbaum, Mahwah (2006)
3. Hellier, E.J., Edworthy, J., Dennis, I.: Improving auditory warning design: quantifying and predicting the effects of different warning parameters on perceived urgency. Human Factors 35(4), 693–706 (1993)
4. Haas, E.C., Casali, J.G.: Perceived urgency and response time to multi-tone and frequency modulated warning signals in broadband noise. Ergonomics 38(11), 2313–2326 (1995)
5. Winters, J.J.: An investigation of auditory icons and brake response times in a commercial truck-cab environment. Virginia Polytechnic Institute and State University, M. Sc. Thesis (1998)
6. Guski, R.: Psychological methods for evaluating sound quality and assessing acoustic information. Acta Acustica 83, 765–774 (1997)
7. Schwarz, N., Clore, G.L.: Feelings and Phenomenal Experiences. In: Higgins, E.T., Kruglanski, A. (eds.) Social Psychology, A Handbook of Basic Principles, 2nd edn., pp. 385–407. Guilford Press, New York (2007)
8. Scherer, K.R.: Appraisal considered as a process of multilevel sequential checking. In: Scherer, K.R., Schorr, A., Johnstone, T. (eds.) Appraisal Processes in Emotion: Theory, Methods, Research, pp. 92–120. Oxford University Press, New York (2001)
9. Smith, C.A., Ellsworth, P.C.: Patterns of cognitive appraisal in emotion. Journal of Personality and Social Psychology 48, 813–838 (1985)
10. Västfjäll, D., Sköld, A., Genell, A., Larsson, P., Tajadura, A., Väljamäe, A., Kleiner, M.: Emotional acoustics: A new framework for designing and measuring the effects of information and warning sounds. Manuscript submitted for publication in Acta Acoustica.
11. Sköld, A.: Integrative Analysis of Perception and Reaction to Information and Warning Sounds in Vehicles, Chalmers University of Technology, Ph D Thesis (2008)
12. Larsson, P., Opperud, A., Västfjäll, D., Fredriksson, K.: Emotional and behavioural response to auditory icons and earcons in driver-vehicle interfaces. In: Proceedings of 21st Enhanced Safety of Vehicles conference, Stuttgart, Germany, June 15-18, Paper No. 09-0104 (2009)
13. Bisping, R.: Emotional effect of car interior sounds: Pleasantness and power and their relation to acoustic key features. SAE paper 951284, 1203–1209 (1995)
14. Bisping, R.: Car interior sound quality: Experimental analysis by synthesis. Acta Acustica 83, 813–818 (1997)
15. Bradley, M.M., Lang, P.J.: Affective reactions to acoustic stimuli. Psychophysiology 37, 204–215 (2000)
16. Västfjäll, D., Kleiner, M., Gärling, T.: Affective reactions to interior aircraft sound quality. Acta Acustica 89, 693–701 (2003)
17. Russell, J.A.: The circumplex model of affect. Journal of Personality and Social Psychology 39, 1161–1178 (1980)
18. Bradley, M.M., Lang, P.J.: Measuring emotion: The self-assessment manikin and the semantic differential. Journal of Behavioral Therapy and ExperimentalPsychiatry 25, 49–59 (1994)

Investigating Narrative and Performative Sound Design Strategies for Interactive Commodities

Daniel Hug

Universities of the Arts Linz & Zurich,
Linz, Austria / Zurich, Switzerland
daniel.hug@zhdk.ch

Abstract. Computing technologies turn everyday artefacts into narrative, procedural objects. This observation suggests that the narrative sound design strategies used in films and many video games could also be applied for the design of interactive commodities. However, it is unknown whether these strategies from immersive media can be applied in physical artefacts of everyday use. In this paper we describe methodological considerations and outline a structure of a revisable, design oriented, participatory research process, which allows to explore narrative sound designs and their possible application in interactive commodities in a systematic yet explorative way. The process, which focused on interpretational aspects, has been applied in two workshops and their results are reported and discussed. The experience of the prototyping and evaluation method, which made use of theatrical strategies, raised important questions about the role of performativity in the emergence of meaning and the possible limitations of a strictly hermeneutic aesthetics, when dealing with sonically enhanced interactive commodities.

Keywords: Sound Design, Interactive Commodities, Design Research, Hermeneutics.

1 Introduction

1.1 The Nature of Interactive Commodities

Computing is increasingly moving out of the desktop into household appliances, entertainment systems, mobile communication devices, clothes and to places we might not even know about, turning objects of everyday use into interactive commodities. One thing such interactive commodities have in common, is the reduced importance of visual display. Also they are often "black boxes", objects who's inner workings are unknown. Such devices may seem to have a will of their own and to act without assistance. Sound can provide useful means to give such devices a "display", a means of expression and communication, and a sonic identity, which is present even in the peripheral awareness of people.

1.2 The Potential of Narrative Sound Design

There is still little knowledge available about how to design sound for interactive commodities. Designers are starting to work on scenarios and prototypes [1] and

S. Ystad et al. (Eds.): CMMR/ICAD 2009, LNCS 5954, pp. 12–40, 2010.

first approaches to a sound design oriented theory for such devices have been made (e.g. [2], [3]). It has been proposed that the sound design of interactive commodities could be informed by narrative and even fictional design strategies [4]. According to Csikszentmihalyi and Rochberg-Halton [5] things embody goals, make skills manifest and shape the identities of their users. Hence they alter patterns of life, they reflect and define the personality, status and social integration of both, producer and owner. In addition, things evoke emotions through interpretations in the context of past experiences, thus becoming signs or symbols of one's attitude. Artefacts thus are socio-cultural components in the experience of our everyday lifes. Vastokas [6] states, that the meaning of artefacts is constituted in the life of the objects themselves, not in words or texts about them, artefacts are not static, "dead" things, but interactive agents in sociocultural life and cognition and theoretical insights about them derive from direct observation and experience of the phenomenal world of nature and culture. Artefacts thus are dynamic agents in social networks and narrative in nature. This is even more true for computerised artefacts which are procedural and interactive, and often even anthropomorphised or interpreted as "magical" [7].

1.3 Strategies for Designing Narrative Sounds

Fictional media, in particular film, have developed an immense body of knowledge and know-how about how to convey meaning through the narrative qualities of sounds. Moreover sound designs in films are often highly elaborated, semantically rich and subtly tuned to the identity of the protagonists, objects and processes depicted. This satisfies several fundamental requirements for sounds of interactive commodities, which are to become a meaningful, enriching and appreciated part of our everyday life. Thus it seems worthwhile to investigate possibilities to leverage this design knowledge. Although documentations and reflections on sound design are relatively scarce, there are several publications that shed light on how meaning is made not only by a sound's functional use in the audio-visual medium but also by its compositional and phenomenological qualities (e.g. [8], [9], [10]). Some texts also provide examples to how such sounds could actually be designed (e.g. [11], [12], [13]). The use of semantic layers, archetypal templates, the play with familiarity, traces of materiality and traces of psychology are examples of commonly applied sound design strategies. These are described in more detail in [4].

But two important issues need to be taken into consideration which prevent the straightforward application of film sound design strategies in the context of interactive commodities.

1.4 Issue 1: Immersive vs. Non-immersive Media

Even if it turns out to be possible to extract and describe the design strategies used for specific narrative and dramaturgic aims in narrative audio-visual media, this does not grant their applicability in the domain of actual physical artefacts and everyday interactions. The reason for this is mainly that films are *immersive* media. They are consumed - not used - and their consumption takes place in a more or less defined spatio-temporal context. Activity and intentionality is present only through a protagonist's audio-visually represented actions and intentions. A mediated perspective on events and protagonists that requires an internal "re-enactment" of the recounted (fictional) experience is enforced on the spectator. We are experiencing a "second order" experience.

The artefacts investigated here are of an altogether different nature. They are used non-immersively, pragmatically and in the periphery of attention. Different interpretative strategies are thus to be expected.

1.5 Issue 2: Interpretation of Schizophonic Artefacts

The possibility to integrate a practically limitless range of sounds into physical objects by means of miniaturized electroacoustic devices and to control them through computer technology makes the relationship between physical object and its sound arbitrary. Not only can the sounds naturally occurring from an interaction with a physical object be extended with signals, but also it is possible to transfer physical qualities, even a virtual presence of other artefacts (and the associated semantic qualities), into artefacts. This creates inherently *schizophonic*[1] artefacts and it is unknown how such sounds will be interpreted and what varieties of synchresis[2] and diegetics[3] are emerging.

1.6 Issue 3: Experience of Sonic Agency

Another important aspect to consider, which cannot be addressed using narrative strategies alone: Sounds of interactions are inherently performative and can be understood and conceived as manifestations of our agency in the world. We have to move our bodies, press our lungs, and interact with objects to produce sound. Chion [15] coined the term "ergo-audition" to denote the experience of hearing oneself acting. According to Chion, ergo-audition is potentially more joyful if the action-sound relationship is not too simple and also always a bit surprising (e.g. through variations). However, as sound is an inherently social phenomenon, very often this expressive sonic agency is a shared phenomenon that may either connect to or disconnect from acoustic communities, a process that has been described by Schafer [14] and, more specifically, by Truax [16]. Therefore, sonic agency is a form of performing a relationship with the environment and also potentially part of our presentation of self in everyday life in a social setting, as described, for instance, by Goffman [17].

[1] Schizophonia is the term coined by R. M. Schafer to denote the separation of sound from their - natural - sources by means of electroacoustics [14].

[2] Synchresis, a combination of the terms "synchronism" and "synthesis", denotes the association of a sound with a synchronous image (usually related to an action or a process), resulting in a irreducible audio-visual percept. The term was introduced by Chion [8].

[3] A term originating in narratology, denoting the (fictional) world in which narrated situations and events (including sounds) occur.

2 Methodical Approach

As described above, the proposition states that narrative sound design strategies from fictional media like film or video games could inform the sound design of interactive commodities. Moreover, the approach should make it possible to investigate the performative aspect of sound making. The research pivots around questions of interpretation in two different experiential settings: immersive and non-immersive, disembodied screen and schizophonic, physical artefacts.

A single methodological framework is not sufficient for dealing with the complex socio-cultural context of this research in a non-reductionist manner. This is the case in most design oriented research, and it has been suggested to take a pragmatist stance towards methodology, selecting and combining methods according to their usefulness for achieving specific goals [18]. This practice of design research finds argumentative support in many methodological writings, particularly from the social and health sciences (e.g. as "multimethod research" [19]). Another challenge was to formulate initial design hypotheses which would not be biased by analytic theories from film sound. For this reason, aspects of Grounded Theory [20] were adopted in the process, in particular for the collection, coding and clustering of interpretations.

2.1 Research through Design

As sounding interactive commodities are supposed to be part of our everyday life, the understanding of our experience of technology as a dynamic, dialogical process [21] forms the philosophical foundation of this work. Hence, the research methodology had to support the collection and evaluation of interpretive strategies and to "grasp lived cultures" [22]. However, as this research deals with possible futures, instances of possible lived experiences had to be created that can be further investigated, which calls for an integration of design as creative component. Designing artefacts which can then serve as cases for further study is a useful practice in research and provides an ideal link between what designers and researchers can do best (c.f. [23], [24]).

The selection and integration of design strategies also makes sense when dealing with complex socio-cultural issues: Following Rittel, Krippendorff [25] argues that design deals with "wicked" problems, as opposed to more or less well structured technological problems and that it has been successful in developing methods to cope with this challenge. "Research through design" thus is a suitable strategy for tackling the issues presented here [26].

2.2 Designing a Revisable Design Process

An essential challenge at the outset was: How can design knowledge be developed in a way that is both structured enough to provide valid, sufficiently generalizable and useful results and at the same time supports a design process flexible enough for ad-hoc inspiration and creative intuition? The design process had to be structured in such a waythat it supported the creation and exploration of ideas while allowing to compare specific aspects of the outcome to an initial design hypothesis. Moreover the research process needed to be accessible and open to the contribution, both in discourse and application, of designers at any moment.

Krippendorff suggests a science for design, a "systematic collection of accounts of successful design practices, design methods and their lessons (...). It also (...) [provides] methods for validating designs." ([25], p. 209) Such a systematic methodological collection can in turn be used to implement designerly practices into research processes. This is particularly useful if the research deals with possible futures rather than with the present or the past. By joining documented design methods in a structured way, artefacts can be created that can be evaluated against an initial design hypothesis [26].

The aim of this study was to compare the interpretations of the sound designs in interactions represented in immersive, fictional media with sound designs for actual interactions with artefacts. Krippendorff [25] proposes several human-centered design methods of which the following were selected as conceptual building blocks to create the framework for this research:

- Reframing: Useful for creating conceptual spaces, similar to brainstorming but following specific cognitive devices
- Narratives of ideal futures: Analyzing fictional scenarios and their depiction in movies and games can help to understand stakeholder's concepts, motivations and visions
- Stakeholder participation in the design process: For this research both analytic as well as creative processes involved group work
- (Re)designing the characters of artefacts: First desirable attributes for the character of an artefact or an interaction are defined. Then related sensory manifestations are designed and evaluated against the initial descriptions
- Designing expressive artefacts, guided by narratives and metaphors: How sound can be used to (re)design the expressivity of an artefact has been outlined above and is described in more detail in [4]. This method also is in line with the underlying concept of applying strategies of narrative sound design. The steps of this method are:
  - Collection or generation of relevant narratives and metaphors
  - Analysis of narratives into scenarios and the semantics of components and the exploration of available technical details
  - Synthesis and realisation (in this case using Wizard-of-Oz - prototyping, see below) and testing of the results in terms of narratives and interpretations supported by the actual prototype

2.3 Experience Prototyping and the OZ Paradigm

As mentioned, part of the research goal was to create instances of possible futures and lived experiences that then can be evaluated and discussed. This is a common practice in design called "experience prototyping". According to Buchenau and Suri, researchers need "to explore and communicate what it will be like to interact with the things we design" ([27] p. 424). They describe experience prototypes as "any kind of representation, in any medium, that is designed to understand, explore or communicate what it might be like to engage with the product, space or system we are designing" ([27] p. 425). For this research a variation of the Wizard-of-Oz prototyping method was used. The method was developed in the eighties by John F. Kelley to simulate natural language computing and he describes it as an "(...) experimental simulation which I call the OZ paradigm, in which experimental participants are given the impression that they are interacting with a program that understands English as well as another human would" ([28], p. 26). More recently a variation of the method has also been used in the context of auditory display research [29]. The presentation of the Wizard-of-Oz prototype can be compared to a theatrical performance. The integration of theatrical performance or of methods inspired by it has a long history in design methodology, in particular in human-computer interaction design [30]. Some examples of this approach include scenario-based design [31], role-playing [32] and body-storming [33]. The general idea is that users play different roles, and the interactive system to be designed is considered in terms of interactions mediated by computers.

2.4 Implementation in a Workshop Setting

Participatory workshop settings have been found to be useful in order to deal with complex, design oriented issues like the one presented here (see e.g. [34] [1], [29], [35]). Therefore a structure for participatory workshops has been conceived and applied. By going through a specific sequence of steps from analysis to evaluation with the group of participants, and by iterating the overall process, the reliability and validity of the findings was increased. So far, over 30 individuals participated in the process, and about 20 more will be added to this number in the future.

Based on the considerations given above, the main components that were developed for the workshops were:

- *A methodical framework and conceptual toolbox:* Before the design researcher can start to work on concepts she must know the tools and methods at her disposal. These tools are primarily of a conceptual nature, a combination of theoretical expertise and practical know-how. In this case two essential domains had to be covered: Methods for designing and prototyping on the one hand, and a solid background in sound theory and design. Part of the toolbox were also handouts with procedural guidelines for analytic and conceptual stages of the process.
- *A structured process:* The second important task was to define the temporal structure of the creative research, to create specific phases that build on and interact with each other and form a coherent whole. This structure must not only be understood by the researcher but by all stakeholders in the process.
- *An evaluation strategy:* As mentioned above, the results of the creative process are only of value for future research if they can be evaluated in some

way. This was achieved by creating a level of comparability between existing designs and the newly designed prototypes.

In the following, the resulting workshop structure is described in detail.

3 The Workshop Structure

So far three workshops have been carried out. The third iteration has not been evaluated yet and thus is not discussed here. Each iteration maintained the basic structure in order to establish comparability of specific aspects. Other aspects were slightly modified in order to accommodate circumstances and insights from previous iterations. Several reasons would motivate and justify the modifications, for example the time available, group size or the background of participants. Some variations were intentionally introduced in order to create richer and less expectable results - this is a necessary trade-off between creative potential and scientific rigor in design oriented research. The thorough documentation of each iteration made it possible to identify differences and their possible impact, thus ensuring accountability. The core components remained unchanged and are applied in every iteration. They are described in the following.

3.1 Step 1: Establishing Fundamental Competencies

For the overall research strategy to yield the desired results it has to be assured that all participants share a comparable and sufficient level of sound related competence. This is necessary for both analytic and creative tasks. Therefore, the following topics are introduced:

- Introduction to Sonic Interaction Design, especially about the characteristics of interactive artefacts and the rationale to use sounds in their design. This is illustrated with examples from industry and design research, e.g. [36] or [37]. Also an overview of the activities of the European COST-Initiative Sonic Interaction Design[4] is provided.
- Competence for dealing with sound conceptually, linguistically and practically. This involves an introduction to sound studies and acoustic communication as well as ways of analyzing and describing sonic objects, sonic events and soundscapes.
- Competence related to the relationship of sound, object and interaction. This includes product sound quality and the analysis of action-sound relationships [38]. This is combined with the creation of a "foley-box", containing all kinds of objects with interesting sounds, to inspire creation and provide material for recordings.
- Competence in a theory of sounding interactive commodities. This entails the considerations about object-sound reconfigurations, which extends the argumentative basis for using sound in interactive commodities (mostly built on [4]). A typology for analysing sounding interactive commodities based on [3] is provided as well.

[4] http://www.cost-sid.org

3.2 Step 2: Analyzing Fictional Sound Designs

A collection of up to 23 short extracts from movies and 5 extracts from video games are provided to the participants. Each clip contains a narratively closed scene, depicting either an interaction with an artefact, or changing internal states of objects or protagonists (provided the change is mediated by an artefact). The sound design should play an important role in the interpretation. The clips are divided in two sets and anonymised.

The clips are provided together with an analytic protocol. According to the protocol, the following analytical steps have to be followed (summary):

- Short description of the narrative content (protagonists, actions and events)
- Interpretation of the narrative content
- Introspection about what lead to this interpretation (visual or acoustic aspects, narrative context, characteristic actions)
- Creation of labels with qualitative keywords for the narratives depicted in the scenes. This abstraction makes it possible to find cross-links between narratives and their judgement. Each label should be composed of an *adverb*, a *verb*, an *adjective* and a *noun* (e.g. "dangerous movement of fragile box")

After this first part of the analysis, the second part focuses on the sonic aspects:

- Description of reason why a particular sound might have been important for the interpretation
- Description of the sound itself and the supposed design strategy to create it
- Investigation of the relationship between action and sound (using the framework provided in step 1)
- Creation of labels with qualitative keywords for the sounds, again using attributes, objects (possible source) and verbs. Also here the labels provide an abstraction that aids cross-linking and comparison.

The labels are then written on post-it notes of two different colours, one for general interpretations and the other for sound specific aspects. After that, the short description and the labels for each scene are grouped and stuck to a big matrix with one clip per column. This process results in an extensive collection of condensed narratives and the related interpretations by each group for each clip (see Figure 1).

Finally, the descriptions and interpretations are discussed together with all participants. The researcher moderates this discussion and everything is recorded to allow a closer evaluation later on. In the last and most important step, so-called *"metatopics"* are formulated. These are abstracted narrative themes and attributes of artefacts and interactions that appear in several clips and seem important recurring elements of storytelling using sound. These metatopics are distilled in a discussion with all participants. To this end, the abstractions provided by the labels for qualities of both general narratives and the associated specific sound designs are used. First, common attributes across cases are highlighted and subsequently grouped along their common denominators. Based on these semantic clusters the moderator proposes and discusses the emerging metatopics

Fig. 1. An analytic matrix. Each post-it colour stands for a class of labels.

with the participants. This process is a "rapid" version of coding, memo-writing, clustering and saturation known from Grounded Theory [20].

3.3 Step 3: Collecting and Analysing Everyday Interactions

After the clip analysis, participants are asked to document everyday interactions and objects using video recordings. These interactions have to be analyzed using the same steps as the movie clips (step 2). This helps to prepare the ground for the next step.

3.4 Step 4: Sonic Redesign of Everyday Object Interactions

The participants are asked to find narrative links between the fictional metatopics identified and the everyday experiences documented, and to redesign the sounds of the everyday experiences, using the available sound designs identified in the fictional scenarios of the clips. This allows the rapid prototyping of design ideas, relying on the elaboration and quality of professional sound design, without having to go through an elaborate sound design process. It also allows a preliminary verification of the usefulness and transferability of the fictional metatopics to everyday artefacts.

3.5 Step 5: Experience Prototyping

This step consists in the conception and development of an experience prototype by each team. The only requirement is, that it has to deal with an interactive artefact and that metatopics should drive the sound design. In order to provide a fruitful work environment, every team receives a package of the basic tools and materials needed for electro-acoustic prototyping including a midi keyboard. This package includes two small loudspeakers of ca. two cm diameter, a small mono-amplifier, cables, connectors and plugs for setting up a small active loudspeaker system and a midi controller keyboard for creating and controlling sounds, especially during the life performance.

After making sure that every team has a goal and there are no obstacles to the design process, the lecturer remains in the background, only intervening when problems with implementation of an idea occur. The sound designs in particular should be kept secret by the teams. This serves to assure a certain level of "ignorance" from the lecturer and the teams, which allows them to participate in more or less equal position in the final interpretation and discussion.

3.6 Step 6: Interpretation of Prototype Performances

In the final step the participants perform their scenario, interacting with the prototypes. These performances are given without initial explanations. After the performance there is a discussion, following a semi-structured protocol. The topics discussed follow the same structure as the analysis of the clips, in order to maintain comparability: what were the interactions? what was the role of the various actions and objects? what could have led to a specific interpretation? to what extend sound was relevant and what sonic aspects would contribute to the interpretation? After the interpretations are formed and discussed, the presenting team can provide their explanations, which helps to verify whether the interpretations of the demonstration corresponded to the intentions of the designers.

Also it is discussed, to which class of sounding interactive commodities the artefact proposed could belong. The classes were: authentic, extended, placeholder and omnivalent commodity (for an explanation of these types see [3]).

4 Pilot Workshop at the University of Helsinki

4.1 Background

The initial workshop was held on October 21st and 22nd 2008 at the Helsinki University of Technology, hosted by lecturers Cumhur Erkut and Inger Ekman. It was held in the "Design Factory"[5], which provided an ideal environment. Thirteen students attended, all of them had a background in audio engineering, some also in music. None of them had a formal education in design, but they had taken courses in which design-oriented methods were taught and used. This ensured a degree of familiarity with design related issues.

The workshop was run to pilot the whole process. The duration was one and a half days, which required a few modifications to the program described

[5] http://www.aaltodesignfactory.fi

Fig. 2. Students doing field recording and adding content to the clip analysis

above: Step 2 and 3 were carried out simultaneously in two separate groups and the number of clips and the number of everyday situations to be analysed was reduced. Moreover step 4 and 5 were merged and step 6 was slightly simplified. All essential elements of the overall process were still contained in the program and it provided sufficient data to contribute to the overall research.

4.2 Metatopics Identified

The following metatopics were identified in the group discussion of the clips:

- Artefact turning evil
- Artefact matching user
- Autonomous or remote control
- Scary or positive magic
- (In)correct or (in)appropriate use
- Life and life-cycle of artefact

4.3 Prototyping Results

Group 1: Large Hadron Collider
Raine Kajastila, Olli Oksa, Matti Pesonen
The group developed a sound design for a short story about a scientist operating the Large Hadron Collider[6]. Several operational steps can be heard however it

[6] The Large Hadron Collider (LHC) is the world's largest and highest-energy particle accelerator. See http://lhc.web.cern.ch/lhc/

is clear something is not working. The scientist attempts to start the collider but fails. He then fixes it with a mechanical device but when the machine is restarted something goes terribly wrong.

The sound design combined several field recordings that were interpreted similarly like the metatopics investigated, e.g. combining pneumatic or glassy sounds with film sounds associated with "magic".

The following *metatopics* and sounds design strategies were used: *Artefact matching user, scary or positive magic, (in)correct or (in)appropriate use, life and life-cycle of artefact.*

Group 2: Ancient Autonomous Steel Tape Measure

Emil Eirola, Tapani Pihlajamäki, Jussi Pekonen

The scenario from this group featured a "living" measuring tape, which would become uncontrollable, opposing the user and turning evil. It was presented in a small performance using a real tape measure, synchronized with a pre-fabricated sound design. First the protagonist extends the tape, however it rapidly becomes uncontrollable. The protagonist fights with the tape, tries to submit it and finally succeeds, forcing the tape to snap back.

The sounds used came from field recordings and were mixed with sound extracts from clips associated with the metatopics of interest. These were enhanced with effects like reverb or distortion, partially to create sonic coherence, partially to enhance the expressive quality of the sounds.

The physical motion sounds were associated with the metatopic of uncontrollability by sonically merging features previously associated with these narratives. The battle to submit the artefact also used sounds used in a clip depicting the same narrative context. These narrative components were merged with sounds of the tape itself. The following *metatopics* and associated sound design strategies were used: *Artefact turning evil, autonomous or remote control, scary or positive magic.*

Group 3: Dangerous Microwave Oven

Sakari Tervo, Juha-Matti Hirvonen, Joonas Jaatinen

This project presented the interaction with a microwave oven that turns into a dangerous device, almost killing its user.

The design was built on a distorted recording of the hum and the familiar beeps of a microwave oven and several sounds from various clips associated with the desired metatopics but also with phenomena like energy discharge. The sounds of the microwave oven were partially reversed or played faster to create familiar yet strange sounds.

The following *metatopics* and associated sound design strategies were used: *Artefact turning evil, life and life-cycle of artefact.*

Group 4: Door to a Restricted Magic Place

Qiwen Shao, Esko Järnfors, Antti Jylhä

In this scenario, a person opens a door, which leads to a magic place. The magic is announced by sounds coming from behind the door. After entering, everything is fine at first, but then the protagonist breaks something. This disturbs an order of the place, which makes it angry. All of a sudden it become hostile and uses its magical powers to attack the protagonist.

The magical door was created using several different squeaking door sounds to establish a degree of familiarity. These everyday sounds were specifically chosen to contain either aggressiveness or eeriness and were mixed with samples from clips associated with the metatopics related to magic energy and sounds that were associated with dangerous characters and eerie magic as well as animal squeaks which were associated with animated artefacts.

The following *metatopics* and associated sound design strategies were used: *Scary or positive magic, (in)correct or (in)appropriate use.*

5 Workshop at Zurich University of the Arts

5.1 Background

The second iteration of workshops was held between December 3rd and 19th 2008 at Zurich University of the Arts (ZHdK) in the Design Department. This version of the workshop lasted for 12 days. The longer time span allowed for more introductory exercises and for more time to be spent on both the production of sound designs and on the actual prototypes. Wizard-of-Oz techniques were used extensively for prototyping and for the theatrical performance (see figure 3). The clip analysis was extended, both in breadth and duration, including also several clips created from video games. The increased duration allowed also to create physical prototypes, either as novel constructions or by "hacking" and modifying existing artefacts. The workshop was attended by 16 students. Their background was Interaction Design (5) Game Design (9), Scenography (1) and Style & Design (1 student).

5.2 Metatopics Identified

The larger corpus found in this workshop, in combination with longer discussions, significantly increased the number of metatopics. However, most of the "new" entries are in fact refinements of the categories identified in the first workshop:

- Atmospheric machine
- Moody / emotional machine
- Presence or infestation with alien power
- Dead matter becoming alive
- Anthropomorphisation of machines
- Moral attitude
- Known/friendly or unknown/evil
- Level of power
- Invisible energy
- Charging / discharging (energy)
- Comprehensible or incomprehensible magic

Fig. 3. The workspace of a "wizard". The keyboard controller is used to trigger and control sounds.

- Metamorphosis
- Compatibility
- Inappropriate use
- Speed
- Pain

5.3 Prototyping Results

Group 1: TakeMeAway (Figure 4)
Balz Rittmeyer, Daniel Lutz, Bruno Meilick
Sometimes we receive a call when we do not want to be disturbed, and have to invent an excuse. The phone service proposed by this group allows a callee to transport herself sonically into an imaginary reality, making such excuses more convincing. Buttons on the phone activate virtual soundscapes and modify the voice of the callee to fit the scenario. For example, the callee might decide to pretend to be in a church, on a busy road or in the middle of a war.

It was proposed that the phone itself could change its character based on the moral decay of the user, reflected in the amount of times the function was used or in the extremity of the chosen fake scenarios.

The following *metatopics* and associated sound design strategies were used: *Moral attitude, level of power, metamorphosis.*

Group 2: Assembly Line (Figure 5)
Philipp Lehmann, Julian Kraan, Didier Bertschinger
According to this group, in the near future assembly lines for car manufacturing will be entirely virtualised. Robotic arms execute the precision work, remotely controlled through touch screens. The fine and expressive manipulative gestures of a car manufacturing expert on the touch screens are mediated by audio-visual representations that translate them into a superhuman robotic system.

Fig. 4. Preparing the demonstration of the phone prototype. On the right is a "wizard", controlling the sound events.

Fig. 5. Demonstration of the audiovisual assembly line

Fig. 6. Trying to wake the pen up

A complex sonic environment gives feedback about the movements and expresses aspects of precision and overall quality in a multilayered sound composition.

The sounds were designed using abstracted synthesised representations of car sounds and musical metaphors of dissonance, consonance and closure. Different basic keynotes were used to represent different areas of the car and provided the context for the sounds that represented the process.

The following *metatopics* and associated sound design strategies were used: *Atmospheric machine, invisible energy, metamorphosis, compatibility.*

Group 3: Franz, the Smart Pen (Figure 6)
Rosario Azzarello, Konradin Kuhn,Patric Schmid
This scenario features a smart, precious pen with a history: It has been given from father to son through many generations and has changed its character over the centuries of its existence.

The pen would react to the users movements. For instance, as it is an old pen, it would start blurring the sounds when becoming dizzy from abrupt movements. The sounds would change also depending on how the paper is touched, how fast it was moved or when the exhausted pen falls asleep and is waken up by shaking. Thus, the sonic expressions of Franz's moods and emotions merge with a feedback related to the way "he" is operated.

The following *metatopics* and associated sound design strategies were used: *Moody / emotional machine, dead matter becoming alive, anthropomorphisation of machines, inappropriate use, speed, pain.*

Fig. 7. Interacting with sofa and fridge in the smart home

Group 4: Smart Home (Figure 7)
Luigi Cassaro, Christoph Böhler, Miriam Kolly, Jeremy Spillmann
This performance featured a flat inhabited by a single man. He owns three different interactive commodities: His sofa modulates the sound of sitting down, depending on the amount of people and the social context; the fridge expresses his content or discontent with the selection of food contained in it, judging the user morally; and the home-office chair reacts friendly when a user sits on it, signaling a successful connection to the internet when the user assumes a healthy seating position.

The following *metatopics* and associated sound design strategies were used: *Atmospheric machine, moral attitude, invisible energy, charging / discharging (energy).*

Group 5: Moody Hat (Figure 8)
Kai Jauslin, Monika Bühner, Simon Broggi
This group demonstrated a hat with an attitude. First it would attract a potential wearer by whispering and making attractive sounds. When somebody would pick it up it would attach itself violently to the person and finally sit on his head, grabbing it firmly. Depending on whether the person would touch it appropriately it would create different sounds and start to comment on the wearer, either insulting or praising him. Also the hat's sounds would have a close link to the gestural interaction with it, e.g. when touching its rim.

Fig. 8. Various interactions with a moody hat

The following *metatopics* and associated sound design strategies were used: *Moody / emotional machine, presence or infestation with alien power, dead matter becoming alive, anthropomorphisation of machines, moral attitude, known/ friendly or unknown/evil, invisible energy, charging / discharging (energy), inappropriate use.*

6 Discussion and Findings

6.1 Evaluating the Design Process

Overall, the process worked very well. The number and originality of the resulting projects shows, that it was well suited for creative design. To approach the complexity of sound design using the "ready-made" sounds from film in the everyday film clip remix exercise (step 4 of the process) made the topic accessible. At the same time this strategy was surprisingly successful in anticipating where a design idea could lead.

The Wizard-of-Oz paradigm for prototyping was successfully used within this design process. Building functional prototypes can be very problematic as it is dependant upon sufficient resources and skilled expertise, dragging attention away from sound and interaction design. The workshop's setting could be improved to help create more possibilities for thorough introspection or protocol analysis by people interacting with the prototypes. The aspect of interpretation-in-action needs further attention. This can be achieved by slightly altering the creative process, for example by establishing team-internal test sessions with protocol analysis. A post-review stage could be added to integrate interpretations from people who did not attend the workshop.

The temporal flexibility of process is an important factor in order to implement it in actual curricula. Despite the time constraints of the first workshop it was possible to follow the essential steps of the process. However, the reduction of discussion time for step 6 created a serious limitation to the evaluative quality. This was addressed in the subsequent workshop, whose duration (12 days) seems optimal for developing elaborate prototypes. But the results from the first short workshop integrate very well with the results of the second one, which suggests, that the overall process is solid and scales well. The third workshop, which was held at the Media Lab of the University of the Arts in Helsinki, Finland, lasted five days. As mentioned, this iteration is not fully evaluated yet, but it can be said that the duration was well suited to the methods employed, however only if the students have already some basic skills and experience in working with sound.

6.2 Narrative Sound Design and Metatopics

The strategy of deriving narrative metatopics from film and game scenes was successful. It allowed the metatopics to be used both as analytical entities as well as design material within the process. The high overlap of metatopics identified in both workshops indicates that they provide a relatively consistent and reliable tool. Finally, the transfer of narrative metatopics to the sound design of interactive commodities worked surprisingly well in most of the cases, their interpretation was usually very close to the intended meaning.

In a few cases, however, the interpretation was biased by the way the artefact was presented in the theatrical performance. For example, the virtual assembly line (group 2 of the second workshop) was initially interpreted as a musical performance by many participants due to the setting chosen. This suggests that a prototyping method refinement is needed, which helps to eliminate potentially undesired interpretive associations at an early stage. The impact of performativity on the findings will be discussed below.

Based on their use in the designs and the discussions during the workshops, the following summary and grouping of the emerging narrative metatopics could be achieved:

- Artefact - user relationship: circles around the narrative of suitability and appropriateness of an artefact (and its characteristics) to a user. The perspective can be either on the artefact or the user. The enunciation of this relationship can follow a notion of "acceptance" (foregrounding psychological enunciations) , or "compatibility" (foregrounding traces of material processes)
- (In)correct use: relates to the way an artefact is handled by a user
- (In)appropriate use: relates to the use of the artefact in a specific sociocultural context
- Quality of energy: Apart from materiality, energetic quality is a fundamental aesthetic element of schizophonic artefacts. Energy, manifested through sound, can transcend its physical framework and be used in higher level semantics, e.g. metaphorically or symbolically, and even for expressing a positive, supportive or negative, dangerous or adversary quality of an artefact. Moreover an artefact's (metaphorical) life cycle can be expressed through energetic sounds.
- Source of control or energy: An artefact, in particular a networked one, may be controlled or "possessed" by a known or obscure agent. This relates to the perception of an energy source, which may be explicitly placed beyond the artefact, e.g. by sounds that are not interpreted as coming from the artefact itself.

- Animated artefact: Here an ensouling or subjectivisation of the device is expressed sonically. This involves also transitions: Dead matter becomes "alive", and vice versa.
- Atmospheric artefact: According to Böhme [39], "Atmosphere" can be defined as constituting the "In-between" between environmental qualities and human sensibilities. Atmospheric sound in particular leads us outside ourselves, between our body and the sound's origin. This definition seems to suit the implicit understanding of the artefacts in question: When speaking about this quality, people were alluding to the effect of the machine's presence, mediated by sound, to the humans sensibilities.

In most of these metatopics, an essential semantic trajectory is established between the two poles of the location of agency in the experience of interactions: artefact or User. The sonic mediation of this location of agency can be ambiguous and may even shift during the course of an interaction.

6.3 Emerging Sound Design Strategies: Fictional Realities - Realistic Fiction

The sound design approaches chosen by the participants usually relied on a mix between sounds from field recordings and narratively significant sounds extracted from the clips. Often the modification of the everyday sounds followed patterns identified in the clip analysis (adding reverberation, distortion or modification of speed or pitch and the like), a strategy which helped unify the aesthetic and semantic merging of sounds from different sources. Quite often participants found strong parallels between narrative components in the fictional clips and their own everyday sound recordings, even if unedited. This was very often the case when the sounds were detached from the recording context (source). Finding narrative aspects in the "natural" sounds helped creating aesthetically fitting mixes of everyday sounds with fictional sounds. And conversely, the transfer from fictional to real life scenarios seemed to work especially well where the sound design in the fictional scenario would rely already on a certain familiarity with a real life experience, usually associated with material qualities. Examples are sounds of unstable structures (wobbling, creaking) or electric discharges.

It was interesting to observe how the identifiability of a specific sound became a tool for design: One strategy was to "defamiliarize" sounds that had an identifiable source, e.g. by cutting or processing. Sometimes at the same time hints to possible sources, e.g. traces of materiality, were left in it. This common strategy from film sound design (see introduction) transferred well into the realm of physical artefacts.

6.4 Hermeneutics as Challenge

Interpretations are hard to do, especially without resorting to structuralist and reductionist strategies. Many participants struggled with expressing their experience and why something was interpreted as it was. Circular statements like "the scenery was oppressive, hence the interpretation" were quite common, especially during the pilot workshop. In general it seemed hard for participants to express their thoughts freely. This issue is difficult to address, but essential for this research. In the second workshop it was addressed with a stronger moderation of the interpretations in a open discussion setting. This requires a lot of time but seems to be absolutely necessary. Last but not least, even after extensive introductions into sound design vocabularies and strategies to communicate about sound, a comprehensive, systematic yet accessible vocabulary for everyday sounds remains one of the bottlenecks for such investigations.

6.5 The Impact of Performativity

While the focus of the design and research method described here was on sonic narratives and their interpretation it was felt that performance and agency could be an important aspect in the process of meaning making and interpretation. According to Erika Fischer-Lichte [40], cultural actions and events are not only interpreted as meaningful signs, but the quality of their execution, the performance itself, comments, modifies, strengthens or weakens the function and meaning of their referential aspects. This understanding of the impact of performance on meaning has been first systematically explored by John L. Austin [41] and has been adopted in many contexts by several scholars in linguistics, ethnology, sociology and media theory in what has been called *performative turn* [40]. In particular, theatre studies were confronted with avant-gardistic art movements, such as Fluxus, where performers do not only act out and represent, but actually *do* things, with all consequences to mind and body. This led to a fundamental reconsideration of traditional hermeneutic aesthetics and text-driven semiotics, where an ultimately passive audience tries to make sense of what an actor, or an author, "wants to tell". These studies provide valuable conceptual frameworks for considering performativity in the analysis and design of sonic interactions.

Narrativity vs. Performativity. As mentioned earlier, there is an essential difference between the perception and interpretation of an interaction which is audio-visually represented and a "live" performance. This fact could clearly be observed in the presentations sessions of the projects. For instance, the interpretation of the Assembly Line and the Moody Hat demonstrations were strongly related to qualities of movement and expression on the part of the presenters. This hints at a limitation of a strictly hermeneutic approach to meaning and aesthetics. The specific qualities of performativity and its role in the interpretive process is of high relevance for the design of sounding interactive commodities and it is worthwhile to investigate the related theories from semiotics of theatre and performative arts.

The strategy of using theatrical representations of interaction scenarios in the workshop methods described here allowed to investigate this area. In the following sections, the topic and related findings will be discussed, in order to outline directions for future research.

Emergence of Meaning in Performance. A central characteristics of performative "live"-settings is, that we are dealing with autopoietics and feedback loops between performers, script, signs, actions and spectators. Even the difference between actors and spectators has been repeatedly resolved in postmodern performance art. The resulting ambivalence is caused by the liveness and the lack of a fixed, non-transitory artefact of reference. The situation encountered is judged in a dialogue with oneself and the (social) environment. Meaning thus emerges in the process and may or may not follow predetermined decoding processes. According to Petra Maria Meyer [42] the very act of displaying or showing (also acoustically), and also the way the act is executed, already change the sign and transform it. The performative action modifies and transforms the signs and expressive quality of an artefact, also the ones that are manifest sonically. This is in accordance with Meyer's statement, that a code required for semiosis can get lost through the "écriture", the "act of writing", of performing the sign [42]. That action is quite powerful in forcing a meaning on a sound, was also discovered in the experiment conducted during a workshop with actors at the University of York [43]. Performativity thus resists to some extend the claim of a "hermeneutic aesthetics", which aims at the decoding and understanding of an artwork [44]. This can be easily extended to interactive artefacts, which only can be fully understood in the light of the actual actions that they afford and when they actually are used in a performative act.

The link of such a performance-oriented semiotics and aesthetics to sound design is evident. Sound is inherently performative, bound to time, and can neither be easily decomposed nor fully represented (except through itself). It can be fixed using recordings, but - as in photography - this essentially transforms the content and meaning. Chion [15] emphasises that a fixed sound always constitutes a new "sonic object" and that "recording" as a conservation and representation of reality is not possible. The strong relationship between bodily performance and sound, and its impact on meaning and interpretation, has also been pointed out by van Leeuwen [9]: Some sound properties are strongly linked to the bodily action we need to perform in order to create them. For example, for a high-pitched, thin scream we need to put our body under tension and stretch ourselves. A warm, relaxed voice on the other hand, requires relaxed muscles and posture.

Three Levels of Emergent Meaning. According to Fischer-Lichte [44], the process of the emergence of meaning in performance consists of three elements:

- Semiosis as the process of decoding signs, based on conventions and a defined contextuality.
- Self-referentiality: The phenomenon draws attention to itself, to its qualities, e.g. when somebody performs a gesture very slowly.
- Association: Material, signifier and signified fall apart, and each can lead to new significations.

In the two latter cases conventions are de-stabilised or even become irrelevant. This also relates to the resolution of a stable contextuality, which is a common phenomenon designers have to deal with when creating products for everyday use which are not defined by a specific context of use (e.g. a dining or office table) but carried around or encountered everywhere, a feature that is quite common with ubiquitous information and communication technologies.

The relevant question thus is to realise, which order of perception is predominant (neither will be stable and exlusively effective at any stage): the representational or the presentational. To establish the former is the goal of classic theatre and immersive media like film. If the latter dominates, self-referentiality and association take places, and affective reactions take place, which again contribute to the autopoiesis of meaning in a social feedback loop. An example would be the circus, where the phenomenal experience of presence and action is the main constituent of meaning and aspects of representation are rather irrelevant.

The Principle of Perceptive Multistability. An important phenomenon in the emergence of meaning in performative situations is what Fischer-Lichte calls "perceptive multistability" [44]. The concept is familiar from Gestalt theory: The perceptual perspective can flip or shift in the very act of perception, just like figure-ground illusions. Likewise, an actor on stage one the one hand embodies a figure in a narrative, but at the same time still is himself in his phenomenal presence. This understanding abolishes the dichotomy between presence and representation, where the former is associated with authenticity and immediateness and the latter with mediation and subjection to a meaningful function. This principle can be easily related to sound. For instance, it is a common method in sound design to work with sound that oscillate between abstract, almost musical and concrete, indexical qualities.

In particular the schizophonic condition, as described in the introduction, shows that this principle has to be taken into consideration in conception, design and evaluation of sounds designed for interactive commodities: The sounds that emerge in a process of interaction can originate from the physical object itself, or from an electroacoustic source, while the latter, again, can simulate or suggest a physical artefact or process. Furthermore the causing process may be a manipulation, a mechanical process in the object or an action initiated by the objects "intelligence". These sounds also can be simultaneously a display of a function, or an information related to data changes in a process and the actual manifestation of the presence of a complex system at work. Levels of presentation and representation mix and intertwine, authenticity is resolved in favour of schizophonia, which requires an adequate theoretical framework [4].

A result of the oscillation between representation and presentation is that the state "in between" the two poles, the liminal state, emerges as meaningful in itself. If a transition is repeated (either forced by the performance or by a mental act of the perceiver) it becomes the focal point of attention and the defining source of meaning. The perceiver may not only focus on the quality of the transition itself but also on the fact that he or she actually perceives it. We start to observe and perceive our own perception. Also here we can see a strong parallel to sound, in particular to music, where it is the transitional moments, e.g. from tension to release, from dissonance to consonance, that play an important part in the experience. Taking an example from the projects presented here: In the Assembly Line prototype, the transitions between states were the focus of attention and the sound design experimented with the oscillation between abstract musicality and concrete materiality.

6.6 The Extraordinary and the "Everyday"

In all these considerations we must not forget that the artefacts in question are used in "trivial" everyday situations. They are tools, pragmatic means to an end. Therefore, we need to consider the properties of the "everyday": Noeth [45] summarises these as a) opposition to night and to Sundays or festivities, b) the present, the here and now, c) the familiar, ordinary and given, and also the contrary of the unique and the unexpected, with repetition, conventionalisation and automatisation of signs being further important aspects. At first view this seems to contradict the idea that a non-deterministic semiotics of performance is beneficial for the theory and design of interactive commodities. But while repetition and habituation may lead to automatised signs this does not mean that the everyday freezes to stereotypes [46]. In particular in the post-modern, individualised society, the everyday is incessantly remodelled by trends, fashions, and dynamic lifestyles. Also "trivial" everyday action, like washing dishes, brushing teeth, driving to work, etc., is never exactly the same. Every day is different from the day before, we are in different moods, in different spatio-temporal and social situations. We can thus follow, that exactly repetivity and conventionalisation need to be complemented by variation and change for a positive experience of the everyday. And even if we deal with exact repetitions, as they are caused for instance by industrialised proceses, the theory of performance clearly shows, that a repetition of the same performance is still never the same. As Meyer [42] has pointed out, the repetition itself produces a change and the introduction of a new quality that will in turn produce new meaning.

This is relevant for the discourse about sound design for interactive, schizophonic artefacts. On the sonic level, certainly, a door always sounds more or less similar, but it still offers a broad spectrum of variation in sonic expression, in relation to our actions and the environment. More complex systems offer even more potential for variation. And the dialectical relationship between repetition and variation is something sound designers and musicians can easily relate to. For instance, a rhythmical pattern, repeated over and over, starts to change in our perception, emphasis and the level of attention shift, back and forth, in and out. And even the exact repetition of a sound is not the same experience, even though we might decode it "correctly". Here lies the source of the potential annoyance through invariant sound. As already Murray Schafer [14] pointed out, the greatest problem for the acoustic ecology is posed by the repetivity and invariance of the sounds of mechanisation and electrification. Barry Truax [16] emphasises the importance of variety (between sounds but also between instances of the "same" sound), complexity and balance to achieve well-designed soundscapes. Both positions lead to an understanding that designing for absolute predictability and identical repetition is not suitable even for "everyday" sounds. For interactive commodities, the static, predictable auditory icon is an aesthetic dead end. Instead, we propose a design which establishes familiarities and spaces for variation, very much like it can be found in the structure of musical composition. To propose and evaluate actual ways of achieving this will have to be covered in future work.

6.7 "Hermeneutic Affordances" as Guiding Principle

The narrative design strategies outlined above form part of a hermeneutic approach to meaning, where the participants in an interaction interpret what they sense and experience in terms of a superordinate "message" or "story". As described, in interactions with the ephemeral quality of sound and the role of performativity in shaping this interpretation, the strictly hermeneutic approach is limited. A conceptual framework that is directed towards the interpretation of immediate and continuous sensorial experience-in-action is "affordance". In Gibson's [47] ecological understanding of perception, this is a quality of environments and objects which suggests certain interaction potentials[7]. This quality emerges in the process of interaction with the environment. Gibson proposed the concept also as an alternative to the understanding of meaning making and attribution of value being second to sensorial perception.

In terms of sound design for interactive commodities we can thus say, that we are designing "hermeneutic affordances": Artificial, designed sound as a composition in time potentially has a strong narrative, dramaturgic component, and at the same time the "materiality" of sound, and the possibility to shape the relationship between action and sound through real-time controls, enables the emergence of affordances, specifically through sound. In the interpretation process, as described above, the hermeneutic thread and the explorative, associative and self-referential thread of affordances intertwine. To recognise these two sides as part of a common system is an important step towards a theoretical and aesthetic framework for sound design for interactive commodities, and needs further elaboration.

7 Conclusions and Future Work

7.1 Process and Methodology

The design research process implemented in the workshop structure described in this paper helped not only to investigate how sounds of interactive commodities become meaningful and how a narrative approach performs for interactive commodities. It made possible to test and evaluate elements of a design process for the practice of Sonic Interaction Design. The concept of narrative metatopics emerged as useful help for formulating and evaluating sound design hypotheses. Several promising sound design strategies emerged which can now be subjected

[7] Note the interesting parallel to the notion of meaning potentials described by van Leeuwen [9], reported above.

to further refinement and evaluation by applying them in new projects. The process requires a few more iterations in order to be fine-tuned. In particular, the emergent topic of performativity and "soundmaking" needs to be addressed. Future workshops are planned for this purpose. Each one of these events provides a different setting that will have to be taken into account by adopting the overall process. The internal coherence of each step within the process allows the different formats of workshop to contribute results to the wider framework, providing multiple viewpoints on the same issue and rich data for further analysis. This process, grounded in structured participatory design experiences, will provide a solid basis for the development of design heuristics which then can be tested in experimental setups.

7.2 Narrativity Meets Performativity

The use of theatrical strategies for prototyping and evaluating schizophonic interactive commodities led to the identification of performativity as an important aspect related to meaning and aesthetics. We have outlined several key theories from performance and theatre studies that help to understand these phenomena. Future work will need to investigate the concrete implications and resulting heuristics for the sound design process of interactive commodities.

An important lesson to learn relates to semiotic epistemology: In non-immersive media of everyday use, the traditional idea of the sign as unambiguous and ultimate instance in meaning making is relativized. Also we can recognise the transitional rupture between hermeneutic and phenomenal "modes" of experience, as well as the dualistic nature of sound between presentation and representation as potential source of meaning.

It shall be noted, that, as discussed above, despite the impact of real-time performance on experience and interpretation, the transfer of sound design strategies derived from narrative metatopics worked in most of the cases. In some cases, performativity had a stronger, in some cases a lesser relevance. In practice, narrativity and performativity flow together and intertwine. For the sake of analysis and theorising it is necessary to separate the two but in the actual experience of sonic interactions they are inseparable. Sound design thus has to integrate narrative and performative aspects. The concept of "hermeneutic affordances" attempts to frame this complexity in a non-reductionist manner.

7.3 Implications for Sound Design Strategies

This research stresses the need for non-deterministic, procedural approaches to sound design and generation. A design approach which embraces variety, complexity and balance is more likely to overcome the "annoyance factor", which to a large extend relates to repetivity. There is a need for a new heuristics of everyday performance of sonic interactions. Sounds for interactive commodities have to offer a space of potential relationships with meaning and actions (of both humans and the autonomous artefact) which can be explored experimentally. An instrumental approach helps in this enterprise: Sonic manifestations of interactions are not anymore just static sounds that are repeated over and over, but a defined sonic system with familiarities that remain and interesting modulations, allowing sound to become a meaningful element in everyday interactions[8]. However, the hermeneutic thread must not be forgotten: The sound generator can not be arbitrarily designed, but has to build on narrative metatopics. The challenge of future research here will be to marry the narrative richness of time-bound, compositional sound design with the flexibility of procedural audio.

Acknowledgements. First and foremost, we would like to thank all the students involved in the workshops. Without their creativity, commitment and competence the richness and quality of prototypes required for this work could never have been achieved. Further, we would like to thank Inger Ekman and Cumhur Erkut for taking the risk and hosting the pilot workshop.

References

1. Franinovic, K., Hug, D., Visell, Y.: Sound embodied: Explorations of sonic interaction design for everyday objects in a workshop setting. In: Proceedings of the 13th international conference on Auditory Display (2007)
2. Barras, S.: Earbenders: Using stories about listening to design auditory interfaces. In: Proceedings of the First Asia-Pacific Conference on Human Computer Interaction APCHI 1996, Information Technology Institute, Singapore (1996)
3. Hug, D.: Towards a hermeneutics and typology of sound for interactive commodities. In: Proceedings of the CHI 2008 Workshop on Sonic Interaction Design, Firenze (2008)
4. Hug, D.: Genie in a bottle: Object-sound reconfigurations for interactive commodities. In: Proceedings of Audiomostly 2008, 3rd Conference on Interaction With Sound (2008)
5. Csikszentmihalyi, M., Rochberg-Halton, E.: The meaning of things - Domestic symbols and the self. Cambridge University Press, Cambridge (1981)
6. Vastokas, J.M.: Are artifacts texts? Lithuanian woven sashes as social and cosmic transactions. In: Riggins, S.H. (ed.) The Socialness of Things - Essays on the Socio-Semiotics of Objects, pp. 337–362. Mouton de Gruyter, Berlin (1994)
7. McCarthy, J., Wright, P., Wallace, J.: The experience of enchantment in human-computer interaction. Personal Ubiquitous Comput. 10, 369–378 (2006)
8. Chion, M.: Audio-Vision: sound on screen. Columbia University Press, New York (1994)
9. van Leeuwen, T.: Speech, Music, Sound. Palgrave Macmillan, Houndmills and London (1999)
10. Flückiger, B.: Sounddesign: Die virtuelle Klangwelt des Films. Schüren Verlag, Marburg (2001)
11. Sonnenschein, D.: Sound Design - The Expressive Power of Music, Voice, and Sound Effects in Cinema. Michael Wiese Productions, Studio City (2001)
12. LoBrutto, V.: Sound-on-film: Interviews with creators of film sound, Praeger, Westport, CT (1994)

[8] A comparable concept is the "Model Based Sonification" proposed by Hermann & Ritter [48]: Here a dataset is turned into a sound generating algorithm which then can be explored through interactions, for instance with physical interfaces.

13. Viers, R.: The Sound Effects Bible. Michael Wiese Productions, Studio City (2008)
14. Schafer, R.M.: The Soundscape: Our Sonic Environment and the Tuning of the World, 2nd edn., 1994 edn. Destiny Books, New York (1977)
15. Chion, M.: Le Son, Editions Nathan, Paris (1998)
16. Truax, B.: Acoustic Communication, 2nd edn. Greenwood Press, Westport (2001)
17. Goffman, E.: The Goffman Reader. Blackwell, Malden (1997)
18. Melles, G.: An enlarged pragmatist inquiry paradigm for methodological pluralism in academic design research. Artifact 2(1), 3–11 (2008)
19. Morse, J.M.: Principles of mixed methods and multimethod research design. In: Tashakkori, A., Teddle, C. (eds.) Handbook of Mixed Methods In Social and Behavioral Research, Sage, Thousand Oaks (2003)
20. Charmaz, K.: Constructing Grounded Theory: A Practical Guide Through Qualitative Analysis, Thousand Oaks (2006)
21. McCarthy, J., Wright, P.: Technology as Experience. MIT Press, Massachusetts (2004)
22. Gray, A.: Research Practice for Cultural Studies. Sage, London (2003)
23. Brix, A.: Solid knowledge: Notes on the nature of knowledge embedded in designed artefacts. Artifact 2(1), 36–40 (2008)
24. Zimmerman, J., Forlizzi, J.: The role of design artifacts in design theory construction. Artifact 2(1), 41–45 (2008)
25. Krippendorff, K.: The semantic turn - A new foundation for design. Taylor & Francis, Abington (2006)
26. Zimmerman, J., Forlizzi, J., Evenson, S.: Research through design as a method for interaction design research in hci. In: Proceedings of the Conference on Human Factors in Computing Systems, pp. 493–502. ACM Press, New York (2007)
27. Buchenau, M., Suri, J.F.: Experience prototyping. In: Proceedings of the Conference on Designing Interactive Systems, Brooklyn, New York, pp. 424–433 (2000)
28. Kelley, J.F.: An iterative design methodology for user-friendly natural language office information applications. ACM Transactions on Office Information Systems 2(1), 26–41 (1984)
29. Droumeva, M., Wakkary, R.: The role of participatory workshops in investigating narrative and sound ecologies in the design of an ambient intelligence auditory display. In: Proceedings of the 12th International Conference on Auditory Display, London, UK (2006)
30. Laurel, B.: Design improvisation - ethnography meets theatre. In: Laurel, B. (ed.) Design Research - Methods and Perspectives. MIT Press, Cambridge (2003)
31. Carroll, J.M.: Making Use - Scenario-Based Design of Human-Computer Interactions. MIT Press, Cambridge (2000)
32. Hornecker, E., Eden, H., Scharff, E.: In my situation, i would dislike thaaat!" - role play as assessment method for tools supporting participatory planning. In: Proceedings of PDC 2002: 7th Biennial Participatory Design Conference, Malmö (2002)
33. Oulasvirta, A., Kurvinen, E., Kankainen, T.: Understanding contexts by being there: case studies in bodystorming. Personal Ubiquitous Comput. 7(2), 125–134 (2003)
34. Schuler, D., Namioka, A. (eds.): Participatory Design: Principles and Practices. Lawrence Erlbaum Associates, Hillsdale (1993)
35. Franinovic, K., Gaye, L., Behrendt, F.: Exploring sonic interaction with artifacts in everyday contexts. In: Proceedings of the 14th International Conference on Auditory Display (2008)

36. Aarts, E., Marzano, S. (eds.): The New Everyday, Rotterdam, The Netherlands. 010 Publishers (2003)
37. Oleksik, G., Frohlich, D., Brown, L.M., Sellen, A.: Sonic interventions: Understanding and extending the domestic soundscape. In: Proceedings of the 26th Annual CHI Conference on Human Factors in Computing Systems, pp. 1419–1428 (2008)
38. Visell, Y., Franinovic, K., Hug, D.: Sound product design research: Case studies, participatory design, scenarios, and product concepts. nest-closed deliverable 3.1 (June 2007)
39. Böhme, G.: Acoustic atmospheres - a contribution to the study of ecological aesthetics. Soundscape - The Journal of Acoustic Ecology 1(1), 14–18 (2000)
40. Fischer-Lichte, E.: Ästhetische Erfahrung: Das Semiotische und das Performative. A. Francke Verlag, Tübingen und Basel (2001)
41. Austin, J.L.: How to do Things With Words, 2nd edn. Harvard University Press, Cambridge (1975)
42. Meyer, P.M.: Intermedialität des Theaters: Entwurf einer Semiotik der Überraschung, Parerga, Düsseldorf (2001)
43. Pauletto, S., Hug, D., Barras, S., Luckhurst, M.: Integrating theatrical strategies into sonic interaction design. In: Proceedings of Audio Mostly 2009 - 4th Conference on Interaction with Sound (2009)
44. Fischer-Lichte, E.: Ästhetik des Performativen, Suhrkamp (2004)
45. Nöth, W.: Handbuch der Semiotik. 2., vollständig neu bearb. und erw. aufl. edn. J. B. Metzler (2000)
46. Lefebvre, H.: Kritik des Alltagslebens: Grundrisse einer Soziologie der Alltäglichkei. Fischer, Frankfurt am Main (1987)
47. Gibson, J.J.: The ecological approach to visual perception. Houghton Mifflin, Boston (1979)
48. Hermann, T., Ritter, H.: Listen to your data: Model-based sonification for data analysis. In: Advances in intelligent computing and multimedia systems, Baden-Baden, pp. 189–194 (1999)

A Review of Methods and Frameworks for Sonic Interaction Design: Exploring Existing Approaches

Eoin Brazil

Department of Computer Science & Information Systems,
University of Limerick, Limerick, Ireland
eoin.brazil@ul.ie
http://www.csis.ul.ie

Abstract. This article presents a review of methods and frameworks focused on the early conceptual design of sonic interactions. The aim of the article is to provide novice and expert designers in human computer interaction an introduction to sonic interaction design, to auditory displays, and to the methods used to design the sounds and interactions. A range of the current best practices are analysed. These are discussed with regard to the key methods and concepts, by providing examples from existing work in the field. A complementary framework is presented to highlight how these methods can be used together by an auditory display designer at the early conceptual design stage. These methods are reflected upon and provides a closing discussion on the future directions of research that can be explored using these approaches.

Keywords: Sonic interaction design, auditory display, everyday sounds, earcons, auditory icons, conceptual design, design methods for sonic interaction.

1 Introduction

The focus taken by this article is on the existing methods. These methods are used to gather and analyse a range of issues including subjective experiences, sound identification, confusion of sounds, cognition of sound, pragmatic mental models[1], subjective experiences, perceived task difficulties, and a range of related aspects. Each method is focused on a single or small group of related facets. This article introduces these methods so that designers can use complementary methods to obtain a deeper and more detailed understanding of the sounds, requirements, and users with relation to a particular sonic interaction or auditory display design.

This type have knowledge has yet to be systematically included in the field of sonic interaction design and related fields such as auditory display, although

[1] In the case of this article, we refer to *pragmatic mental models* as those people use and form while interacting with an environment.

S. Ystad et al. (Eds.): CMMR/ICAD 2009, LNCS 5954, pp. 41–67, 2010.

work in the SMC[2] and the COST SID[3] research networks have begun to unify the existing knowledge. Within these fields and networks, there is an acknowledgement [1] that reductionistic approaches cannot reflect the rich variety of information within sound. We present the existing methods and frameworks including one of our formulation aimed at using multiple approaches to capture different aspects of the sounds under exploration as a means of providing a better reflection of their richness. The argument made in this paper is that the quantitative-analytical reductionist approach reduces a phenomenon into isolated individual parts which do not reflect the richness of the whole, as also noted by Widmer et al. [1]. The problem of methods and frameworks not satisfactorily addressing the complex socio-cultural context of sonic interaction design in a coherent and non-reductionist manner has been faced by most design focused research. Melles [2] has proposed a pragmatic stance with regard to methodology, where methods are chosen and blend with respect to their practicality for achieving set objectives. This stance has influenced methodological dialogues such as those related to multimethod research [3]. The framework presented in this article been structured to support the selection of sounds while allowing the exploration of specific aspects of the sounds and is based on earlier research that highlighted the potential for gathering design information using complementary techniques [4].

This paper focuses in particular on the methods related to sound description, to sound understanding, to sound synthesis/modelling and on those aimed for a wider interaction design audience. The main type of auditory resource explored by the methods presented are everyday sounds, however in many instances the methods would be equally applicable to music-like sounds such as earcons [5]. Both music and everyday sounds require multidimensional approaches and techniques from various domains to explore the complex interplay of their various facets. The focus on everyday sound arose from a general observation by many auditory display designers who found that designing auditory icons to be difficult [6,7]. The selection of methods and frameworks reviewed were chosen with an eye to acceptable cost, limited or no access to dedicated facilities such as anechoic chambers or listening booths, ease of use, ability to concisely present results, and timing requirements for both administration and analysis. This review explores a number of methods focused on the subjective experiences of participants with regard to the sounds and soundscapes presented.

Subjective experience methods gather and analyse the user's cognition of sound, their subjective experiences, and their pragmatic mental models with the goal of obtaining more complete awareness of the sounds and soundscapes. This type of tacit information and experience can be difficult to gather as objective measures are unlikely to return the same results for any two individuals. A design can be improved by including such tacit knowledge, inner needs, and desires of people. These results help bridge the understanding between the user and the designer who gains a better appreciation of the existing rich relationships

[2] http://www.smcnetwork.org/
[3] http://www.cost-sid.org/

between listeners' and their soundscapes. These relationships provide insights, which can help designers think creatively about alternative design possibilities. This type of design relevant information helps to answer elements of the question posed by Hug [8] about how to inform the sonic design for situations using ubiquitous technology, in particular how a specific meaning in a specific context is conveyed whilst forming an aesthetic unity between the operation and the object/s involved. Similar questions have been raised in the area of sound quality studies [9,10], where Blauert stated "*to be able to design other than intuitively, data should be made available to the engineer which tell him which meanings subjects tend to assign to specific sound and what the judgement on the desirability will possibly be*" [11].

2 Current Practise and Guidance

The existing guidance for those new to sonic interaction design or interested in designing auditory displays is limited and often scattered across a number of articles. The goal of this article is to summarise and collate the current state and practises. Many of the published frameworks are concentrated on a particular type of interaction, space, or user group. Aspects and methods can often be reused from these frameworks to deal with completely different objectives. The types of technique vary in outlook with some presenting the users view (user-centered), others focus on the experience of the interface or product (product-centered), and yet others concentrate on the interaction between user and interface (interaction-centered). Models reflecting user-centered approaches can be seen in the ideas of "*think, do, use*" [12] or "*say, do, make*" [13] and approaches that targeti motivations, actions, and contexts [14]. Interface or product-centric approaches explore experiences for product development [15] and for assessing the quality of experience of a product during its concept, planning, and use [16]. Interaction-centered models include Wright's [17] four threads of compositional, sensory, emotional and spatio-temporal, Pine's et al.'s model [18] and Overbeeke's [19] model of experiences. These models bring different aspects including the aesthetics of interaction with interface and action coupled into time, location, direction, modality, dynamics and expression relationships into more formal models. The starting point for all of these approaches requires that designers learn about basic interactions and experiences for the interface, scalability and unfolding of the experience. The next section presents a review of the existing methods and frameworks, which typically have a focus on prototyping or creating interfaces at the early evaluation stages and often do not place a high priority on empirical investigations of the created artefacts or follow up evaluations.

3 Existing Design Methods and Frameworks

This section is divided into an exploration of the existing methods and then the existing frameworks. Many of the methods are incorporated in a number of different frameworks, understanding the methodology, rationale, inputs, and outputs.

Understanding each method, its advantages and disadvantages allows designers to appreciate how best to use these methods. This understanding is helpful in pointing where such methods may be complementary within the frameworks and why. There is no one size fits all technique or approach that can address all the issues or insights raised. The only approach is to use a range of complementary techniques as a means of collating the information from different perspectives. This article will highlight aspects where methods are complementary and will further discuss structured approaches in the form of frameworks where designers can more easily use this range of techniques to achieve a particular goal.

The methods presented in the following paragraphs span a range of complementary viewpoints and provide the atomic building blocks upon which the wider frameworks are build. This article reviews a number of techniques including the repertory grid [20,21], similarity ratings/scaling [22,23], sonic maps & '*ear-witness accounts*' [24], and '*earbenders*' [25]. The repertory grid is one approach that can "*build up mental maps of the clients world in their own words*" [26], the similarity ratings/scaling method explores the attribute or perceptual space for a set of stimuli, and sonic mapping summaries the salient features of an acoustic environment and documents the key sounds within it. Earbenders are a technique used to classify narratives into task, information, and data categories as a means of providing inspiration for new designs and as an approach to help structure how and where sonic interactions may be used.

3.1 The Repertory Grid

This method [20,21] is based on Kelly's view that the world can be seen through the contrasts and similarities of events. The repertory grid uses direct elicitation and asks each individual participant to describe events using their own words. It does not require extensive prior training of subjects or lengthy group discussions to create a vocabulary of a given set of sounds. Direct elicitation is founded upon the opinion that a link exists between a sensation and the verbal argument used by a person to describe it. It infers the criteria used by the listeners' in making their judgements about the sounds presented. The criteria are derived from their descriptions and the meaningful construct labels that they attach to the sounds as part of the method. The method can be problematic when a listener's language is vague or varied, however, the relationship between sounds helps provide relevant information about the entire set of sounds. The written descriptions provided by this methods requires a classification or sorting stage as part of the analysis process to codify the responses from the participants [27,28]. These studies have shown that only within a small number of sounds does any large degree of dissimilarity exist for the classification or sorting categories. An example of the range of results that is produced as part of the analysis process for this method is shown in fig. 1. The method has been applied in a number of domains including food technology, acoustics, and medicine. In food technology [29,30], it has joined a range of sensory profiling techniques that explore subjective and hedonic aspects of experience. A typical example explored how consumers

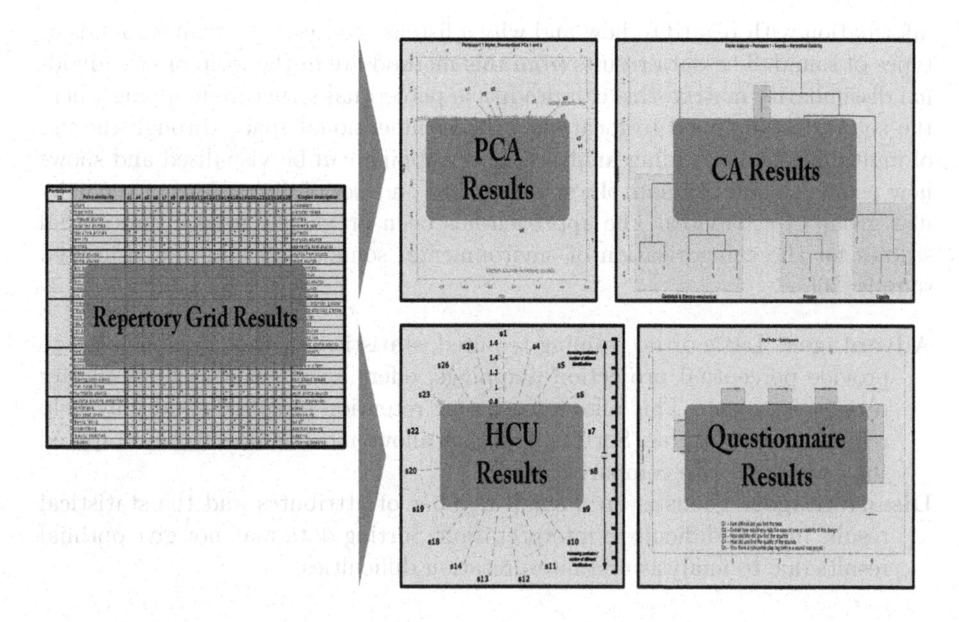

Fig. 1. The repertory grid [20,21] provides a range of results, which are useful for early design stages

perceive both novel and familiar fruits [30]. In acoustics, it have been used to study spatial and multichannel audio experiences of loudspeakers [31,32]. It has been used in medicine to provide the patient perspectives on the physician-patient relationship [33].

Advantages. Little or no training required, individual unbiased responses, statistical methods can be used to provide perceptual projection mappings, and the descriptors can be combined with other textual interpretative methods such causal uncertainty [34].

Disadvantages. Individual rather than group / consensus language and specifics, however the statistical visualisations may be difficult to interpretation. Linguistic or semantic knowledge needed to interpret responses.

3.2 Similarity Rating / Similarity Scaling

The similarity rating technique [35] asks participants to scale and sort stimuli according to acoustical or physical dimensions. The similarity scaling technique [36] is a derivative of this method which presents the sounds or stimuli and use multidimensional scaling or sorting rather than the single dimension at a time approach of similarity rating. These methods are suited to similar sounds as an implicit conjecture is made that the sounds only vary on a small set of continuous auditory dimensions or attributes. The technique allows for examining an individual's perceptual structure of a particular dimension with regard to a particular stimuli set. The results from this method provides useful information with regard to how and why a listener confuses particular sounds or types of sound. The data results from this method are in the form of an individual dissimilarity matrix, this can produce a perceptual structure mapping where the sounds are mapped to locations on a n-dimensional space through the use of multidimensional scaling analysis. The mapping can be visualised and shows how people cluster or relate the sounds based on a set of dimensions or a particular group of attributes. The approach has been previously used in perceptual studies for the categorisation of environmental sounds [22] and of synthesised sounds [36].

Advantages. Little or no training required, statistical methods can be used to provide perceptual projection mappings, offers an alternative to similarity ratings approach. This allows for better retention of sounds and may help reduce listener fatigue. Sorting approach allows more sounds to be compared than with pairwise comparison test.

Disadvantages. Focus is on a small number of attributes and the statistical results may be difficult to interpretation. Sorting data may not give optimal results due to analysis and interpretation difficulties.

3.3 Sonic Maps

Written accounts of soundscapes and of environments have been used to help identify key features and aspects. In this article we discuss two techniques in this category, sonic mapping [24] and ear-witness accounts [25]. These help provide detailed information on how the original auditory environment was experienced and summarise the salient sonic details. The sonic mapping technique consists of listening and analysing the sonic elements of an everyday environment by defining the major sounds and the meanings associated to them by listeners. The first stage creates a description classification grouping based on the vocabulary used to the describe the sounds. The schema produced was similar to work by Özcan and van Egmond [37] with 11 semantically distinct categories of: known source, unknown source, source properties, event, visibility/location, onomatopoeia, simile/metaphor, meaning, temporal, vocal, or emotional. A second stage divides sounds into three grouping of background, contextual, and foreground sounds with visible/hidden aspects. Background sounds provide information about the state of the world, examples include a computer's fans or extract fan in a kitchen. Contextual sounds are sounds that help orient you to the particular environment; examples include seagulls or trawlers indicating a sea or sea related context. Foreground sounds provide details about actions such as the sound effect indicating the emptying of your computer's trash can/folder. The visible/hidden aspect allows a sound to be a member of both aspects and still fit within one of the three groupings of background, contextual, and foreground. The sounds are further classified into one of three information categories:

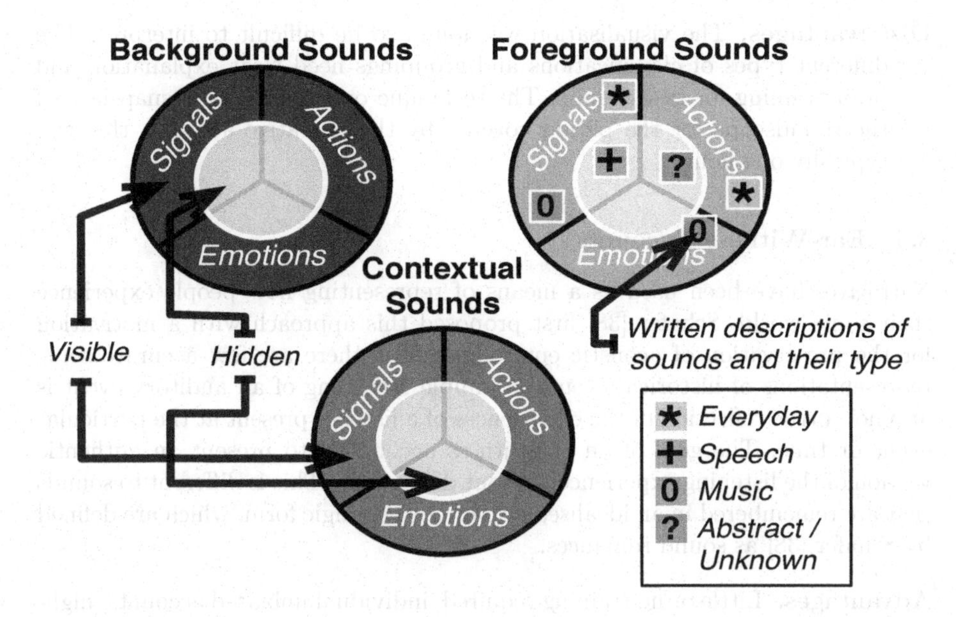

Fig. 2. The sonic mapping [24] is one approach for the analysis and graphical representation of written soundscape accounts

Emotional. These are sounds expressing an emotion or which arouse an emotion. Examples include laughter or crying but can also include sounds like bird song.

Action. These are sounds that do not provide immediate information to a listener. Examples include fan noise, traffic in the distance, or the rustling of wind through trees.

Signals. These are sounds provide specific information about a particular event. Examples include a car alarm, a mobile phone ringtone, or door bell.

The sounds themselves were classified into one of four types. The four types were speech, music, everyday/environmental, and abstract/unknown. This method is easy used by simply listening to the environment, analysing the elements, and with pen and paper or a computer application filling in the Sonic Map diagram. An outline of the diagram with categories and elements is shown in fig. 2, each small square represents a note or post-it verbal description of a particular sonic element categorised by colour.

Advantages. The description classification grouping can help novice listeners in describing the sounds, however this may impact on the richness of these descriptions. An easy to teach technique for introducing novice interaction designers to soundscapes and sonic concepts.

Disadvantages. The visualisation was found to be difficult to interpret. The different types of classifications and groupings need clear explanation and prior training for best results. The technique only produces a 'snapshot' of the soundscape for the period covered by the narrative account, this was typically one hour.

3.4 Ear-Witness Accounts

Narratives have been used as a means of representing how people experience their sonic world. Schafer [38] first proposed this approach with a motivation for the preservation of acoustic environments as there are few accurate sonic representations of historical events. A simple recording of an auditory event is also not a true reflection of the experiences of a listener present at the particular event or time. The goal of an ear-witness account is to present an authentic version of the listening experience without distortions. This is different to sounds that are remembered in an idealised account or nostalgic form, which are defined by Schafer [38] as sound romances.

Advantages. Little or no training required, individual unbiased accounts, highlights the salient auditory sounds in a soundscape.

Disadvantages. It provides a useful overview of the set of sounds but little design or guidance on how these link to one another or what are the key subjective aspects a listener attaches to the sounds.

3.5 Earbenders

Earbenders are an approach for mapping accounts of a soundscape into a narrative form using a case-based approach of design by example [25]. The case is typically a top-down example, which addresses the key aspects of the problem and provides an account of those aspects. Reusing previous successful designs can help in designing better solutions and allows for the features of several prior designs to be integrated to create a new design. Barrass [25] used everyday stories about listening experiences to create the accounts. Sounds and their were found in the stories by reformulating each story as a series of questions. The answers and their organisation given by people help identify the syntax of events in the story, an example is that of the Boiling Kettle story indicated by the *rumble, bubbling, click* sequence. If this sequence did not occur in the same order, then you know something is wrong. The stories contain many elements including semantic, syntactic, and pragmatic aspects. Three different analysis approaches (metonymic, metaphoric, and pattern) were used in this method to provide design insights. The metonymic approach is similar to Gaver's approach [39] for mapping events and actions in the real world to those in a computer interface. It indicates the whole by using part of that whole, an example would be the use of a car engine sound to function as a metonym for car. The metaphoric approach expresses an unfamiliar concept using a familiar concept or metaphor. The unfamiliar concept is typically an event or action with no existing sound such as as those that occur within a computer. This approach is supported by a TaDa analysis (see section 4.1), which describes the requirements for the particular design scenario. The pattern approach identifies aspects that are shared by multiple solutions to the same or similar problems. These aspects can be used to generate patterns of auditory characteristics and includes varies sub-characteristics such as sound type (natural, musical, speech, etc.), sound level (part-of or whole sound), streaming/perceptual grouping, frequency/occurrence (once-off/intermittent), pattern (cyclic, sequential, unpredictable), spatial (movement to listener), perceptual type, textual descriptor, or compound (consists of a number of existing characteristics forming unique pattern).

Advantages. Set of predefined relations immediately available as starting points. Concrete guidelines and aspects of auditory design and mappings are produced.
Disadvantages. Provides a starting point but does not consider the users or domain unless added to the database. Super-ceded in many aspects by auditory design patterns.

4 Existing Design Frameworks

The existing frameworks within the Sonic Interaction Design (SID) field are typically concentrated on a particular type of interaction, artefact, or scenario. The frameworks range in topics from emotionally interactive artefacts [40] to interactive public spaces [41]. The different goals and origins of the frameworks are discussed to highlight the reasoning behind each of the methodologies. This helps to explain where and why particular aspects are highlighted over other facets in the approaches.

4.1 TaDA

The task and data analysis of information requirements [25] or TaDa as it is more commonly known, is a design approach to creating sounds that convey useful information based on the informational requirements of a particular design problem. It uses scenario analysis to capture the different aspects of the design problem by decomposing the problem then providing an analysis of each part. TaDa can be divided into three analysis parts, the first being the task analysis of the question, the second is the information analysis of the answers, and the third part is the data characterisation of the subject. An example of a TaDa analysis is given in fig. 3 for the dirt and gold scenario suggested by Sarah Bly [42].

This approach recasts the scenario as a question with a range of potential solutions, each question is phrased to elicit the particular activity and relationships between activities help refine the information required. The object of interest in the activity is the subject of the question. The question, answers, and subject details help provide the core information for the task, information, and data aspects of the scenario being developed. The task aspect questions the purpose, mode, type, and style of interaction for the scenario. The information aspect

Scenario	TaDa Analysis
Title: Gold	*Task:*
Storyteller: Sarah Bly	*Generic:* Is this it ?
Story:	*Purpose:* Identification
Can you find the gold? - There are six different aspects in the search space that determine whether gold may or may not be found. The first 20 data variables, each is a 6-dimensional record, are from from gold bearing sites. The second 20 data variables are from non-gold bearing sites. You need to listen to each of the remaining 10 data variables and decide whether that site has gold or not.	*Mode:* Interactive *Type:* Discrete / Procedure *Style:* Exploration *Info:* *Reading:* Direct *Type:* Boolean *Level:* Local *Organisation:* None
Keys:	*Range:* 2
Question: Does this sample contain gold ?	*Data:*
Answers: Yes or No	*Type:* 6-dimensional ratio
Subject: Soil samples	*Organisation:* None
Sounds: ????	*Range:* 0.0 - 3.0

Fig. 3. A TaDa analysis for Sarah Bly's dirt and gold scenario [42]

questions the reading, information type, level, and organisation of the scenario. The data aspect of the scenario is concerned with the type (ordinal, nominal, ratio, interval, or none) and organisation (categorical, time, location, mnemoric, continuum, etc.) of the scenario. The approach combines aspects from both visualisation and from human computer interaction task analyses as well as data characterisation methods.

4.2 Emotional Interactive Artefacts

Emotion as part of affective design within the field of sonic interaction design have been explored by deWitt and Bresin [40]. Their idea of affective design suggests that emotion and cognition are not solely phenomena that are personal and limited to the boundaries of oneself, rather that they manifest in a particular social context that is experienced by a person or a group of people. The paradigm of *embodied interaction* [43] was linked to the idea of continuous sonic interaction [27] were core elements of their design of the Affective Diary [40]. This approach deals with the concept of engaging interactions with artefacts in real time. deWitt and Bresin [40] state that the mechanism for sound perception constrain the possible sonic interactions to tightly coupled connections between the sound, the listener, and the event otherwise the listener will perceive an unrealistic or unnatural event.

Digital pets or avatars such as Furby (a bird-like toy with speech and speech recognition), Pleo (animatronic dinosaur), or the AIBO [44] robotic dog use non-verbal sounds to communicate and interact to produce affection feelings

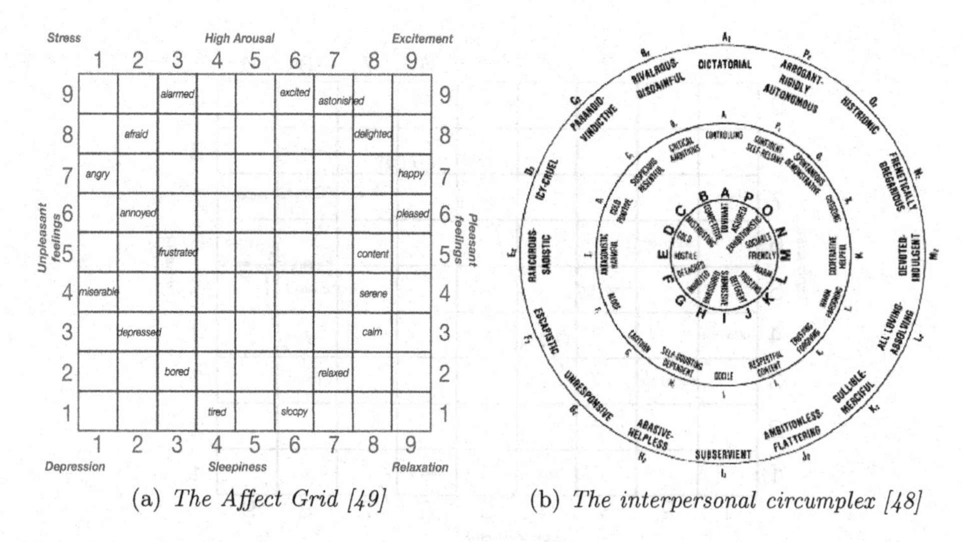

(a) *The Affect Grid [49]* (b) *The interpersonal circumplex [48]*

Fig. 4. (a) shows the Affect Grid [49] and (b) shows the interpersonal circumplex [48] both of which can be used for structuring the emotional dimensions of interactions

from children and even adults [45]. A five factor model has been suggested by Reeves and Nass [46] to analyse personality by breaking it into dimensions of dominance/submissiveness, friendliness, conscientiousness, emotional stability, and openness. An example of a digital avatar built using this model was SIG [47] which used variation represented by the interpersonal circumplex [48] and the dimensions of dominance/submissiveness and friendliness/hostility. These types of interactive and emotional artefacts highlight the need to analyse and consider emotions when design interaction that produce empathic or emotional responses. The Affect Grid [49] or the interpersonal circumplex [48], shown in fig. 4, can help structure emotional or empathic dimensions for those types of interactions.

This type of structured emotional and sonic interaction can be seen in ZiZi, the Affectionate Couch [50]. It was designed by Stephen Barrass, Linda Davy, and Kerry Richens. ZiZi was designed using an emotional design approach that treats emotions as the primary medium for designing interactions to produce empathy. The emotional design space for ZiZi is shown in fig. 5 and is based upon the Affect Grid [49]. Audio-Haptic interactions were mapped to emotions in the Grid as shown in fig. 5 and vary from sounds of boredom, of excitement, or of contentment with related action of sitting, stroking, or petting the couch. Sensors, rumblepacks, and large subwoofers embedded in the couch provide the haptic and sonic feedback as people move, sit, stroke, or otherwise interact with the couch. Purring and barking sounds from a variety of dogs were used to provide a sense of intimacy and pleasure. Stroking motions triggers sounds related to a particular level of excitement based on activity. The sounds express a spectrum of emotions including states such as '*I'm bored*', or '*I like that*', or '*This is heaven*'. This simple purring response produced a sense of character for ZiZi with

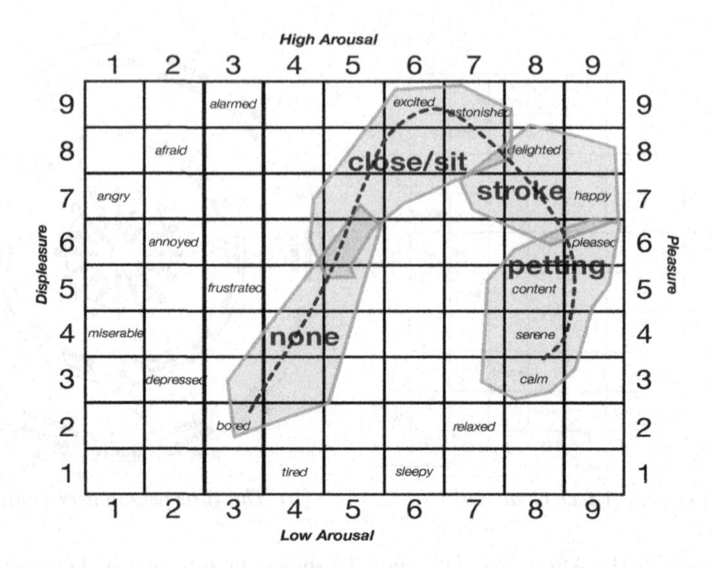

Fig. 5. The emotional design space used to design ZiZi, the Affectionate Couch [50]

Barrass [50] stating that often curators formed an emotional attachment with ZiZi, highlighting that a limited selection of behaviours with repeated exposure can successful create a affectionate interface.

4.3 Interactive Public Spaces

The sonic design of public spaces has been an area of growing interest with train stations [51], shopping centres, public squares [52], museums [53], and airports being explored. One such exploration, was the Shannon Portal [41] was an interactive public space installation designed for airport travellers waiting in the departures lounge in Shannon Airport, Co. Clare, Ireland. The aim was to create an interactive and engaging experience for travellers, which allowed them to send an electronic postcard incorporating either digital camera images from their trip to Ireland or stock postcard footage of Ireland and email it to friends, family, co-workers, etc. The sonic aspects of the installation helped travellers to navigate and browse the collection of stock postcard images using their own body movements coupled with sonic feedback. The scenario and further details are discussed by Fernström et al. [41].

The design framework used to develop this installation is shown in fig. 7 as sub-diagram 2. Ideas and concepts are firstly gathered through brainstorming and mood boards. The results of these stages are then sketched and video prototyped or role-played as a means of evaluating their potential. This provides a number of design candidates of further evaluation and testing using rapid audio prototyping tools such as PureData. In the case of the Shannon Portal [41], the approach allowed for the creation, evaluation, and analysis of four different iterations within a month.

4.4 Functional Artefacts

Functional artefacts and the sonic interaction design of such artefacts was explored in the EU FET Closed project [54], in particular it focused on the kitchen and the sounds of the kitchen. Example artefacts include an augmented coffee maker [55] and the *Spinotron* [56] device, which used a pumping interaction to explore the links between action and various sounds. A set of rolling and wheel/ratchet parameterised sound synthesis models[4] were linked to real-time sensor data in the *Spinotron*. The interaction concept of the *Spinotron* was based on the concept of a ratcheted wheel, the motion or pumping of the device controlled the rotation of the wheel. The methodology used interaction design techniques with basic design methods to structure ethnographic approaches. The goal was to facilitate the design of continuous auditory feedback in sonically augmented interfaces. The evaluation facets of the methodology use techniques such as material analysis and interaction gestalts as shown in fig. 7 as subdiagram 1. The methodology integrates this range of techniques and methods to provide an approach that supports a wider view of evaluation and helps to create products that better fit within their contexts of use. This wider views incorporates holistic measures of experience and of the functional performance of the device's end users. The dynamic interactivity presented by such interfaces is quite different to the passive listening situations where artefacts and their auditory feedback are not dynamically related to the actions of a user. The results from studies with the *Spinotron* [56] found that sonic interactions allowed users to better control an input device relative to the task posed even though the participants felt that the sounds did not contribute to the interaction.

The methodology includes a subjective accounting of the observed sounds where the actionhood of the sounds are analysed using a modified task analysis method [57], in conjunction with a number of specific action related aspects such as general parameters, action descriptors/examples, action label/category, the dynamics of the sound, and a number of acoustic descriptors. These aspects consisted of thirty primitives split into elementary or composite/complex action primitives. The results from these primitives can help highlight new ways of combining sonic feedback with physical action. An example for the kitchen context is shown in fig. 6.

4.5 Narrative Sound Artefacts

Film and game designers often use the concept of narrative. This has inspired a number of sonic methodologies including Back's micro narratives [58] and Hug's design oriented approach [59], which uses narrative approaches to create interactive sonic artefacts. The micro narrative approach [58] uses narration and well-chosen sounds to allow for cultural meanings to be evoked. Stories are small events with identifiable events, where each small event may be combined

[4] SDT impact and rolling models - http://closed.ircam.fr/uploads/media/SDT-0.4.3b.zip

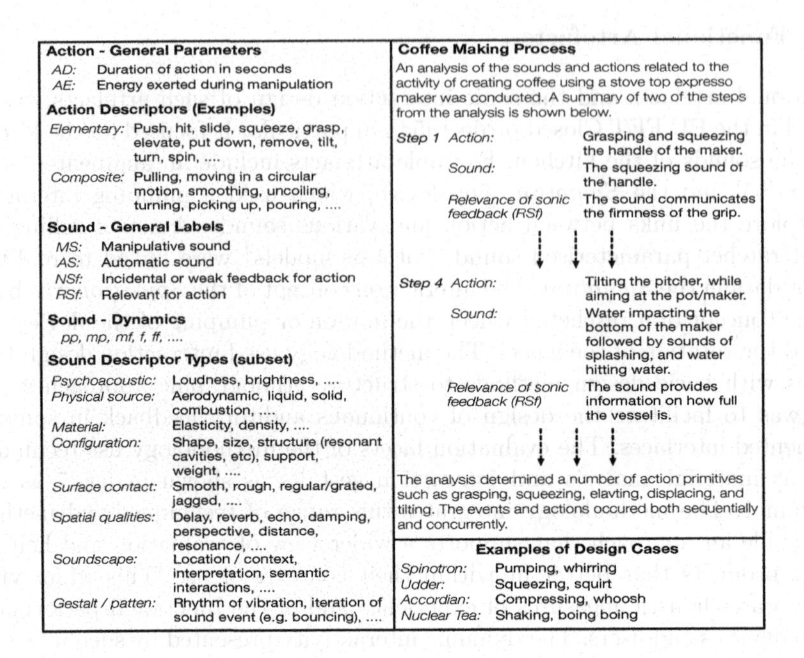

Action - General Parameters
AD: Duration of action in seconds
AE: Energy exerted during manipulation
Action Descriptors (Examples)
Elementary: Push, hit, slide, squeeze, grasp, elevate, put down, remove, tilt, turn, spin, ...
Composite: Pulling, moving in a circular motion, smoothing, uncoiling, turning, picking up, pouring, ...
Sound - General Labels
MS: Manipulative sound
AS: Automatic sound
NSf: Incidental or weak feedback for action
RSf: Relevant for action
Sound - Dynamics
pp, mp, mf, f, ff, ...
Sound Descriptor Types (subset)
Psychoacoustic: Loudness, brightness, ...
Physical source: Aerodynamic, liquid, solid, combustion, ...
Material: Elasticity, density, ...
Configuration: Shape, size, structure (resonant cavities, etc), support, weight, ...
Surface contact: Smooth, rough, regular/grated, jagged, ...
Spatial qualities: Delay, reverb, echo, damping, perspective, distance, resonance, ...
Soundscape: Location / context, interpretation, semantic interactions, ...
Gestalt / patten: Rhythm of vibration, iteration of sound event (e.g. bouncing), ...
Coffee Making Process
An analysis of the sounds and actions related to the activity of creating coffee using a stove top expresso maker was conducted. A summary of two of the steps from the analysis is shown below.
Step 1 Action: Grasping and squeezing the handle of the maker.
Sound: The squeezing sound of the handle.
Relevance of sonic feedback (RSf) The sound communicates the firmness of the grip.
Step 4 Action: Tilting the pitcher, while aiming at the pot/maker.
Sound: Water impacting the bottom of the maker followed by sounds of splashing, and water hitting water.
Relevance of sonic feedback (RSf) The sound provides information on how full the vessel is.
The analysis determined a number of action primitives such as grasping, squeezing, elavting, displacing, and tilting. The events and actions occured both sequentially and concurrently.
Examples of Design Cases
Spinotron: Pumping, whirring
Udder: Squeezing, squirt
Accordian: Compressing, whoosh
Nuclear Tea: Shaking, boing boing

Fig. 6. An example of the Sound Action Analysis used in the CLOSED project [54]

to create a larger ecology as a means for evoking a higher-order scheme. The systematic design process [59] uses a workshop setting to review existing narrative "metatopics" and design strategies as shown in fig. 7 as sub-diagram 3. This helps participants formulate new scenarios and artefacts using these metatopics. These concepts are then implemented as experience prototypes, where a sonic elaboration using wizard of oz or similar roleplaying is given priority over feasibility or technical implementation. The sonic elaboration is reviewed using a semi-structured protocol to provide material for discussion and design refinement. The view in this methodology is that artefacts are socio-cultural components within everyday life and are dynamic rather than static things. This approach generate a set of metatopics that can help create new scenarios and ideas.

4.6 Pattern Based

Pattern based design approaches to develop auditory displays have been suggested by a number of previous researchers[60,61,62]. Results of using *paco* framework [62] (pattern design in the context space) found that auditory display patterns helped expects to capture their design knowledge in a reusable form and helped novices to incorporate more advanced features and technique within their designs. It presented the use of a context space as a mechanism for organisation that allows for multi-dimensional classification according to context of use of the pattern. The aim of this space is to facilitate designers to match design problems with the existing design knowledge as represented by the patterns. The findings

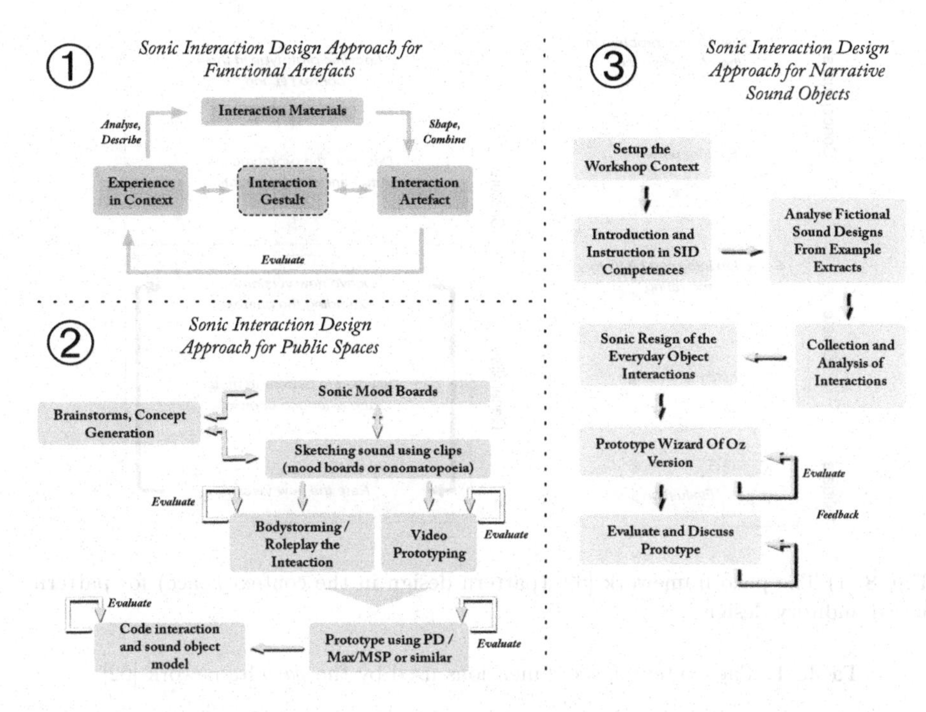

Fig. 7. 1) The design framework developed for interactive public spaces [41], 2) the design process developed by the CLOSED project [54] for creating functional artefacts, and 3) the design process for designing narrative sound objects [59]

pointed towards the need for more directed tool support with flexible formats to incorporate audio and sketching as a means of rapidly prototyping new designs based on existing patterns.

The *paco* framework [62] process is shown in fig. 8. The three major stages of this framework are conceptualise, create, and validate. The create stage is broken down into two further sub-stages of capture and generalise. The capture stage is where the rating, scoping and dimensions of the application are determine. Ratings are determined according to one of six values as shown in table 2. The patterns are scoped and placed within a number of dimensions as part of localising them to the context space with table 3 describing the scope of the application and with table 1 representing the dimensions within the context space. All of the dimensions are treated as tagging categories as with use of such tagging practices consensus can emerge given time and users.

Work by Halpin et al. [63] stated tagging gives "*90% of the value of a proper taxonomy but 10 times simpler*". Following the *paco* approach, a problem/pattern/prototype is assigned values and tags for each dimension creating a unique descriptor. This allows for structured pattern-mining and situates this descriptor close to similar descriptors in the context space helping to present

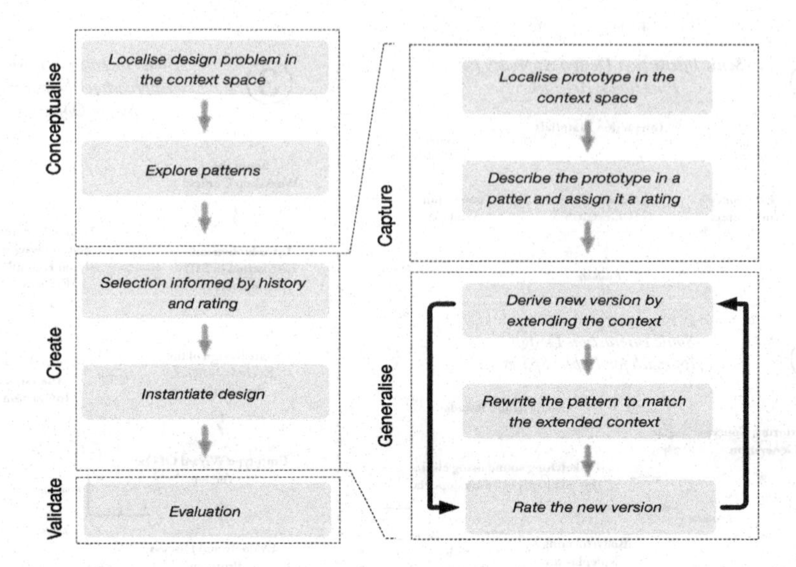

Fig. 8. 1) The paco framework [62] (pattern design in the context space) for pattern based auditory design

Table 1. The context space dimensions used by the *paco* framework [62]

Dimension	*Example of Tags*	*Scope*
User	Visually impaired, surgeon, teacher	1–5
Environment	Noisy, bright, classroom, office	1–5
Device	Mobile phone, web browser, headphones	1–5
Application domain	Mass media, neuroscience, sports	1–5
User experience	Fun, trust, home, cool, intuitive	none
Social context	Privacy, family, dating	none

Table 2. The ratings of patterns as used by the *paco* framework [62]

Values	*Rating Description*
1	Established, good practice
2	Several validated implementations
3	Valid solution for the specific problem
4	Strong indication that pattern is valid, however no evidence to support this
5	Informed guess or "gut-feeling"
6	Anti-pattern or bad practice

potentially relevant patterns to designers. This approach was found to helpful for expert designers in generalising their design knowledge of auditory displays and for novice designers who were attempting to design their first auditory display utilising existing patterns.

Table 3. The scope values and user scopes used by the *paco* framework [62]

Values	*User Scope*
1	Everybody, Mass-market product
2	Wide range of users, minor constraints
3	Users with specific needs, but loosely defined in other properties
4	Specific user groups, well defined properties
5	A single specific user group

5 Providing an Overview Framework for Novice Designers

In a further attempt to provide guidance for novice designers and newcomers to auditory display design, an overview design framework building upon earlier work [64]. This has collated many of the methods, techniques, and approaches discussed in this article. This framework, as shown in fig. 9, presents a structured approach to auditory display design highlighting where appropriate methods can complement each other. It focuses not only on the initial creation of sounds but also on their evaluation. The approach is divided into two stages of *sound creation* and of *sound analysis*, the first selects the sounds for a particular context or scenario and the second presents the methods for evaluating the selected sounds. The methods are designed to be complementary with each presenting details on a particular aspect or aspects. An example of this complementary approach can be seen in the use of causal uncertainty measures with textual descriptors from either sonic mapping or from the repertory grid as a mechanism for removing confusing sounds while classifying the sounds according to user defined taxonomies or clusters. Using this framework allows the selection of an appropriate mix of methods whether creating new sounds or using existing sounds, which can be followed by an examination and evaluation to determine various subjective properties and attributes of the sounds. The steps for the design framework are numbered in fig. 9 for convenience but this does not indicate a set sequence for method use. The *sound creation* stage will typically follow from steps 1 to 4 in sequence, however this does not have to be the case for the steps within the *sound analysis* stage. In the *sound analysis* stage, anything from a single path to all paths can be selected to meet the particular design needs, as shown in the bottom part of fig. 9.

5.1 Sound Creation

In this stage, sounds are defined, selected, created, and an ad-hoc evaluation of them is conducted. The workflow allows for their rapid creation and assessment by the designer. The approach is depend on the skill of the designer as inappropriate combinations of sounds or incorrect combinations of mappings may occur at this point. The ad-hoc nature of the evaluation at this stage does not fully prevent these possibilities. These issues are dealt with in the second stage,

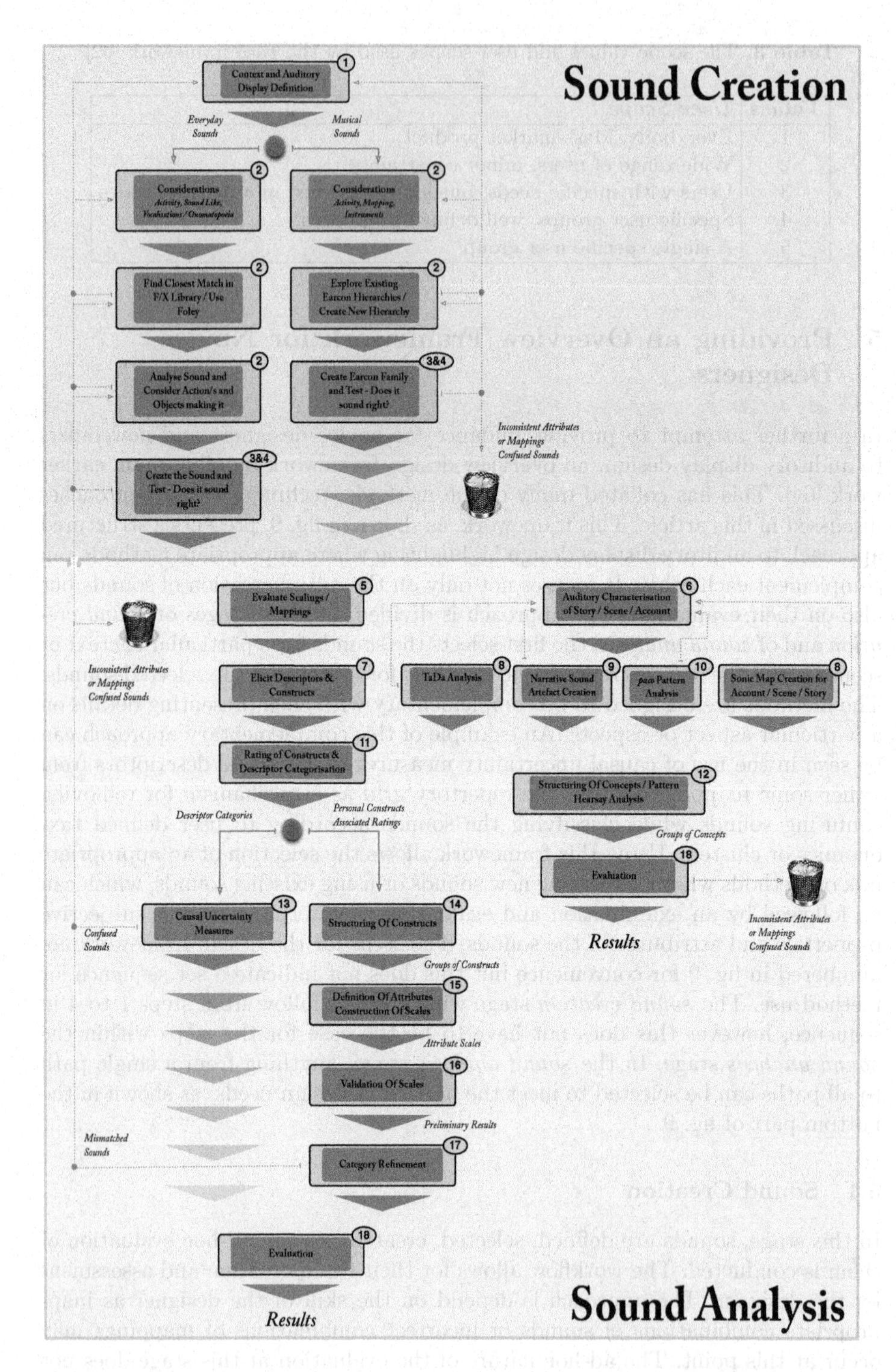

Fig. 9. An overview sonic interaction design process for novice designers [64]

which helps designers build upon the work of this first stage to ensure the best selection of sounds and mappings are made for the particular context.

1. *Context and Auditory Display Definition*: The purpose of the auditory display is defined, the context is determined, the initial conceptual design including possible sounds and mappings are created.
2. *Selection of Sounds*: A pool of sounds fitting the selected mappings are gathered and organised for evaluation. These sounds can be real, synthetic or a mix of both.
3. *Create the Sounds*: If necessary edit existing or create new sounds. These sounds can be real, synthetic or a mix of both.
4. *Listen to the Sounds*: If they do not sound right for the mapping or events, try again with other sounds.

5.2 Sound Analysis

The second stage in this framework presents methods that aid in the selection and understanding of sounds and their mappings. A number of perspectives are presented and allowing designers to focus on particular aspects of interest such as attributes / mappings, confusion metrics, or listeners' narratives. The framework is designed to be open for extension allowing new methods to be added or existing methods adapted depending on the particular design context. This approach allows many facets of the sounds to be explored and helps inform designers about the range of possibilities that may exist within the particular design space. This stage is shown as the bottom part of fig. 9 and is called the *sound analysis* stage.

5. *Evaluate Scaling / Mappings of the Sounds*: The subjects listen and compare the sounds and the mappings or attributes being used.
6. *Auditory Characterisation of Story/Scene/Account*: This is where a narrative for the sounds and environment are created.
7. *Elicit Descriptors & Constructs*: The participants created textual descriptors for the sounds presented.
8. *TaDA & Sonic Mapping*: The narrative / descriptors are analysed and broken down into the different types and aspects of sounds occurring.
9. *Narrative Sound Artefact Creation*: Use the narrative sound artefact approach to design the auditory display using narrative "metatopics".
10. *paco Pattern Analysis*: Creates new patterns and elicits design guidance from context space using *paco* approach.
11. *Rating of Constructs & Descriptor Categorisation*: Each subject rate the stimuli using these constructs created in the last stage.
12. *Hearsay Analysis / Structuring*: Take the auditory patterns and key sounds to create a short summary of salient points that could be reused in other auditory display contexts.
13. *Causal Uncertainty Measures*: The categorisation details are used to calculate the causal uncertainty of sounds.

14. *Structuring of Constructs*: Cluster analysis, multidimensional scaling and principal component analysis of the ratings data are used to clarify the attributes and reduce the dimensionality of the data as well as removing redundancy.
15. *Definition of Attributes, Construction of Scales*: The construct groups are analysed for their content. The appropriate descriptions for the participant identified attributes are then formulated. The rating scales are defined from these attributes.
16. *Validation of Scales*: The scales created are explored in terms of existing categorisations and taxonomies to test the appropriateness of the scales.
17. *Category Refinement*: The details from the earlier causal uncertainty measures and from the created scales can help suggest the removal of particular sounds as unsuitable for use in the particular sonic context.
18. *Evaluation*: The details and results are further analysed to produce the final evaluation results and summary of the evaluation.

5.3 Simplification of the Framework

Future methods and developments in auditory display and related disciplines will improve the design related knowledge of sounds and their subjective qualities, at which point some of the steps within this approach may be simplified, replaced, or even dropped. The framework is open-ended allowing new techniques and methods to be easily added to incorporate new approaches. The information and results derived from this framework are most helpful during the early design stages where the data can help designers better understand the sounds and subjective experiences of listeners with regard to the new auditory display design. Later design stages can use questionnaires or task analyses to further evaluate the auditory display and to provide information on the usability aspects of the auditory display. The framework can also be enhanced by using the existing guidelines [6,65,25,66] to help design better sonic interactions.

The use of these techniques, individually or collectively is still an acquired skill as there are no how-to or best practice guides for their use. Designers need to adopt these methods into their own design cycles, starting with one or two used in a design exploration and building iteratively upon each set of results. The aim of this article was to review these techniques, methods, and frameworks to help designers select the tools they need to address their design issues. The audience of these approaches is not limited to solely designers as researchers can combine these techniques and the approached suggested by Bonebright et al. [67] for evaluating the perceptual properties of acoustic signals. This combination provides researchers with a methodological toolkit that can help provide an understanding of the salient perceptual and cognitive aspects for a particular set of sounds within the specific design context.

The degree of '*richness*' of these methods varies but all help to provide additional insights on different aspects of the sounds or particular sound being explored. Several of the methods overlap in terms of experiment feedback from participants. This can allow a single experimental session to collect data which can be analysed using a number of the methods. The listening test approach [68], asks participants to write their own verbal descriptions of the sounds presented to them. These descriptions are somewhat similar to the personal constructs collected using the repertory grid method [21], the key sounds collected using sonic mapping [24], and if given in more detail are similar to the short stories collected by the earbenders method [25]. Previous research studies [4] have shown how the same set of collected responses can be analysed by both the repertory grid method [21] and the causal uncertainty method [34] to explore different aspects of the sound using a single experimental session's data. It should be noted that not all techniques are complementary, for instance the similarity rating / scaling technique [36] ask participants to sort and scale sounds rather than describe them, however this technique can easily be combined with a context-based rating [67] task, a sorting task or with a discrimination task [34].

The evaluation approach presented in this article is ambitious, it aims is to help designers but additionally it aims to raise the importance and awareness of these methods within the community. This will encourage further explorations using subjective experience methods in auditory display. This framework will continue to mature, presenting in this early development stage is an effort to elicit feedback and comments to assist future development of the framework and of these types of methods within auditory display design and sonic interaction design. New techniques are also being created with a focus on new areas or domain of research and future modifications of the framework will help to address these new areas.

6 New Areas of Research

The field of sonic interaction design is broadening to encompass wider scenarios, types of artefacts, and techniques. A number of new areas under exploration include early childhood artefacts and the inclusion of new techniques from domains such as theatre and film. This section presents two areas of research that are indicative of this trend, artefacts or interfaces aimed for children or young adults and the area at the intersection of film, theatre, and computer gaming.

6.1 Interactive Artefacts for Children or Young Adults

New areas for interactive artefacts for use in educational or in exercise contexts are emerging where sonic interaction design can help in creating interactions. Educational artefacts designed for children within natural settings are a growing research area [69]. Research in museum settings [70] has shown that encouraging children to explore and interpret artefacts can help them develop their conceptual understanding. Artefacts engaging their *somantic learning*, where all their sensory modalities are engaged and utilised. These types of artefacts can create a learning environment that is conductive to curiosity and exploration in an open-ended non-directive way.

Video games have become increasingly marketed as a games that help provide both fun and healthy exercise. An example of this type of interface is the

Nintendo Wii which uses motion sensors that players use to control avatars in a range of games including tennis, baseball, bowling, golf and boxing. Incorporating sonic elements can assist in making the game more attractive and in helping to ensure that the exercises are being effective undertaken. A review of this area [71] shows the potential for inclusion of informative sonifications or sonic interactions. Sonic feedback based on the events, actions or player responses can help players find their appropriate levels whilst making the game or exercise more enjoyable and more effective.

6.2 Theatre, Film and Computer Games

Existing work and theories from film, theatre, and games have developed processes for conveying meaning through the narrative qualities of sounds. These sound designs are often highly elaborate, tuned to the identities of characters, objects, and processes as well as being semantically rich. The possibilities for audio-only film has been shown by recent work by Lopez and Pauletto [72]. Aspects of diegesis from the domain of film sound have been used to create interactive sound objects that respond to haptic explorations, examples include a cocktail glass and teapot [73]. These areas offer rich insights from complementary domains that can help inform the design of sounds to help ensure they are both meaningful and informative. The approach of this areas can help in moving beyond "*magical*" [74] or anthropomorphised interfaces that used procedural interaction styles. Ideas such "*ergo audition*" when sound making is seen as an expressive act [75] entailing the experience of hearing oneself acting and the acoustic manifestation of this influence on the world. This helps moved beyond purely psychoacoustic or functional aspects of sound design to create new sonic interactions and experiences.

7 Discussion

This article has presented a spectrum of methods and techniques that can assist in sonic interaction design. Comparisons, textual descriptions, emotional responses, confusion ratings, or scaling information can highlight mapping, acceptance, or usage issues with a particular design or soundscape. These techniques have been used in automotive settings such as the work by Peugeot Citröen on interior car noises [76] and in Renault [77] to craft a 'sporty' engine sound. Food and drinks companies increasing use sensory profiling, projective mapping, and similar subjective explorations [29,30] to better understand consumers reactions to products and to develop new kinds of products. The purchasing decisions of customers have been show to be positively influenced by the perceived quality and meaning of products [78]. An example is where the drinks company, Tropicana® created the *Grovestand Orange Juice* product (now rebranded to *High Pulp*) based on research, which pinpointed that orange pulp was a key factor in the perception of the product [79,80]. These approaches have shown how products to better fit within a customer's perceptions or desires for that type of product.

Interaction design has begun to attribute more importance to these types of subjective aspects of products, interfaces, and interactions [81]. The use of the repertory grid in interaction design [82] has been found useful for exploring alternative design possibilities. It presented six different types of design relevant information broken down in design principles, quality of interaction, quality of presentation, hedonic quality, and adequacy concerns. These results helped point out directions for future development and highlighted factors of interest, individual beliefs, attitudes, and perceptions. The framework proposed in this article further builds upon the repertory grid and many other methods to incorporate complementary techniques to further supplement the results with findings on different aspects. Using this framework holds promise for investigating the design and selection of sounds that are intended to create new sonic interactions.

Acknowledgments

This work was part funded by the European COST IC0601 Action on Sonic Interaction Design (SID). A special thanks to Mikael Fernström, Stephen Barrass, Nicholas Miisdaris, Guilaume Lemaitre, Oliver Houix, Stephen Brewster and Daniel Hug for their comments and suggestions on this topic.

References

1. Widmer, G., Rocchesso, D., Välimäki, V., Erkut, C., Gouyon, F., Pressnitzer, D., Penttinen, H., Polotti, P., Volpe, G.: Sound and music computing: Research trends and some key issues. Journal of New Music Research 36(3), 169–184 (2007)
2. Melles, G.: An enlarged pragmatist inquiry paradigm for methodological pluralism in academic design research. Artifact 2(1), 3–11 (2008)
3. Morse, J.M.: Principles of mixed methods and multimethod research design. In: Handbook of Mixed Methods in Social and Behavioral Research. Sage, Thousand Oaks (2003)
4. Brazil, E., Fernström, J.M.: Investigating ambient auditory information systems. In: ICAD 2007, Montreal, Canada, June 26-29, 2007, pp. 326–333 (2007)
5. Brewster, S.A.: Non-speech auditory output. In: The Human Computer Interaction Handbook, pp. 220–239. Lawrence Erlbaum Associates, Mahwah (2002)
6. Mynatt, E.: Designing with auditory icons. In: Kramer, G., Smith, S. (eds.) Second International Conference on Auditory Display (ICAD 1994), Santa Fe, New Mexico, Santa Fe Institute, pp. 109–119 (1994)
7. Frauenberger, C., Stockman, T., Bourguet, M.L.: A survey on common practice in designing audio in the user interface. In: Proceedings of BCS HCI 2007, British HCI Group (2007)
8. Hug, D.: Towards a heremeneutics and typology of sound for interactive commodities. In: CHI'08 Workshop on Sonic Interaction Design: Sound, Information, and Experience, pp. 11–16 (2008)
9. Blauert, J., Jekosch, U.: Sound-quality evaluaiton: A multi-layered problem. Acta Acustica 83, 747–753 (1997)
10. Jekosch, U.: Assigning Meaning to Sounds - Semiotics in the Context of Product-Sound Design. In: Communication Acoustics, pp. 193–219. Springer, Berlin (2005)

11. Blauert, J.: Cognitive and aesthetic aspects of noise engineering. In: Proc. of Internoise 1996, Cambridge, USA (1996)
12. Cain, J.: Experience-based design: Towards a science of artful business innovation. Design Management Journal, 10–14 (Fall 1998)
13. Sanders, E.B.N., Dandavate, U.: Design for experience: New tools. In: Proc. of the First International Conference on Design and Emotion, pp. 87–92. Delft University of Technology, Delft (1999)
14. Mäkelä, A., Suri, J.F.: Supporting users' creativity: Design to induce pleasurable experiences. In: Proc. of the Conference on Affective Human Factors, pp. 387–391. Asean Academic Press, London (2001)
15. Jääskö, V., Mattelmäki, T.: Observing and probing. In: Proc. of DPPI 2003, ACM, New York (2003)
16. Alben, L.: Quality of experience: Defining the criteria for effective interaction design. Interactions 3(11) (May-June 1993)
17. Wright, P.C., McCarthy, J., Meekison, L.: Making Sense of Experience. In: Funology: From Usability to Enjoyment, pp. 43–53. Kluwer Academic Publishers, Dordrecht (2003)
18. Pine, B.J.I., Gilmore, J.H.: Welcome to the experience economy. Harvard Business Review 76(4), 97–105 (1998)
19. Overbeeke, C.J., Wensveen, S.A.G.: Reflection on pleasure: From perception to experience, from affordances to irresistibles. In: Proc. of DPPI 2003, pp. 92–97. ACM Press, New York (2003)
20. Kelly, G.A.: The Psychology of Personal Constructs, Norton (1955)
21. Fransella, F., Bell, R., Bannister, D.: A manual for repertory grid technique. John Wiley and Sons, Chichester (2004)
22. Bonebright, T.: Perceptual structure of everyday sounds: A multidimensional scaling approach. In: Hiipakka, J., Zacharov, N., Takala, T. (eds.) ICAD, Helsinki, Finland, Laboratory of Acoustics and Audio Signal Processing and the Telecommunications Software and Multimedia Laboratory, pp. 73–78. Univeristy of Helsinki, Espoo (2001)
23. Scavone, G.P., Lakatos, S., Cook, P.R.: Knowledge acquisition by listeners in a source learning task using physical models. J. Acoustic Society of America 107(5), 2817–2818 (2000)
24. Coleman, G.W.: The Sonic Mapping Tool. PhD thesis, University of Dundee (August 2008)
25. Barrass, S.: Auditory Information Design. PhD thesis, Australian National University (1997)
26. Tomico, O., Pifarré, M., Lloveras, J.: Needs, desires and fantasies: techniques for analyzing user interaction from a subjective experience point of view. In: Proceedings of NordiCHI 2006, pp. 4–9. ACM Press, New York (2006)
27. Fernström, M., Brazil, E., Bannon, L.: Hci design and interactive sonification for fingers and ears. IEEE Multimedia 12(2), 36–44 (2005)
28. Houix, O., Lemaitre, G., Misdariis, N., Susini, P.: Closing the loop of sound evaluation and design (closed) deliverable 4.1 everyday sound classification part 2 experimental classification of everyday sounds. FP6-NEST-PATH project no: 29085 Project Deliverable 4.1 Part 2, IRCAM (June 1, 2007)
29. Stone, H., Sidel, J.L.: Sensory Evaluation Practices, 3rd edn. Academic Press, London (2004)
30. Jaeger, S.R., Rossiter, K.L., Lau, K.: Consumer perceptions of novel fruit and familiar fruit: A repertory grid application. J. of the science of food and agriculture 85, 480–488 (2005)

31. Berg, J.: Opaque - a tool for the elicitation and grading of audio quality attributes. In: Proc. of the 118th Convention of the Audio Engineering Society, Barcelona, Spain (May 2005)
32. Choisel, C., Wickelmaier, F.: Extraction of auditory features and elicitation of attributes for theassessment of multichannel reproduced sound. In: Proc. of the 118th Convention of the Audio Engineering Society, Barcelona, Spain (May 2005)
33. Fleshman Murphy, S.M.: Transition from Diagnosis to Treatment: Changes in Cancer Patients' Constructions of Physicians and Self during the Early Phase of Treatment. PhD thesis, Texas Tech University, Texas, USA (December 2008)
34. Ballas, J.A.: Common factors in the identification of an assortment of brief everyday sounds. J. of Experimental Psychology 19(2), 250–267 (1993)
35. McAdams, S., Winsberg, S., Donnadieu, S., Soete, G.D., Krimphoff, J.: Perceptual scaling of synthesized musical timbres: common dimensions, specificities and latent subject classes. Psychological Research 58, 177–192 (1995)
36. Brazil, E., Fernström, M., Ottaviani, L.: A new experimental technique for gathering similarity ratings for sounds. In: ICAD 2003, pp. 238–242 (2003)
37. Özcan, E., Van Egmond, R.: Characterizing descriptions of product sounds. In: Proc. of ICAD 2005, Limerick, Ireland, pp. 55–60 (2005)
38. Schafer, R.M.: The soundscape- Our sonic environment and the tuning of the world. Destiny Books, Rochester (1977)
39. Gaver, W.W.: Using and creating auditory icons. In: Kramer, G. (ed.) Auditory Display: Sonification, Audification and Auditory interfaces, pp. 417–446. Addison-Wesley Publishing Company, Reading (1994)
40. DeWitt, A., Bresin, R.: Sound design for affective interactive. In: Paiva, A.C.R., Prada, R., Picard, R.W. (eds.) ACII 2007. LNCS, vol. 4738, pp. 523–533. Springer, Heidelberg (2007)
41. Fernström, M., Brazil, E.: The shannon portal: Designing an auditory display for casual users in a public environment. In: Proc. of ICAD 2009, Copenhagen, Denmark (2009)
42. Bly, S.: Multivariate data analysis. In: Auditory Display: Sonification, Audification and Auditory interfaces, pp. 405–416. Addison-Wesley Publishing Company, Reading (1994)
43. Dourish, P.: Where the Action Is: The Foundations of Embodied Interaction (Bradford Books). The MIT Press, Cambridge (2004)
44. Fujita, M.: On activating human communications with pet-type robot aibo. Proc. of the IEEE 92(11), 1804–1813 (2004)
45. Turkle, S., Taggart, W., Kidd, C.D., Dasté, O.: Relational artifacts with children and elders: the complexities of cybercompanionship. Connection Science 18(4), 347–361 (2006)
46. Reeves, B., Nass, C.: The Media Equation: How People Treat Computers, Television, and New Media Like Real People and Places. Cambridge University Press, Cambridge (1996)
47. Okunu, H.G., Nakadai, K., Kitano, H.: Realizing personality in audio-visually triggered non-verbal behaviors. In: Proc. of IEEE 2003 International Conference on Robotics and Automation, Taipei, Taiwan, September 14-19, 2003, pp. 392–397 (2003)
48. Kiesler, D.J.: The 1982 interpersonal circle: A taxonomy for complementarity in human transactions. Psychological Review 90(3), 185–214 (1983)
49. Russell, J.A., Weiss, A., Mendelsohn, G.A.: Affect grid: A single-item scale of pleasure and arousal. J. of Personality and Social Psychology 57(3), 493–502 (1989)

50. Barrass, S.: Clothing the homunculus. Visual Communication 7(3), 317–329 (2008)
51. Pellarin, L., Böttcher, N., Olsen, J.M., Gregersen, O., Serafin, S., Guglielmi, M.: Connecting strangers at a train station. In: Proc. of the 2005 conference on New Interfaces For Musical Expression (NIME), pp. 152–155 (2005)
52. Franinovic, K., Visell, Y.: Recycled soundscapes. In: Proc. of Designing interactive systems: processes, practices, methods, and techniques (DIS), pp. 317–317 (2004)
53. Rocchesso, D., Bresin, R.: Emerging sounds for disappearing computers. In: Streitz, N.A., Kameas, A.D., Mavrommati, I. (eds.) The Disappearing Computer. LNCS, vol. 4500, pp. 233–254. Springer, Heidelberg (2007)
54. Visell, Y., Franinovic, K., Scott, J.: Closing the loop of sound evaluation and design (closed) deliverable 3.2 experimental sonic objects: Concepts, development, and prototypes. FP6-NEST-PATH project no: 29085 Project Deliverable 3.2, HGKZ, Zurich (2008)
55. Rocchesso, D., Polotti, P.: Designing continuous multisensory interaction. In: Sonic Interaction Design: a Conf. Human Factors in Computing Systems (CHI) Workshop. COST Action IC061, April 2008, pp. 3–9 (2008)
56. Lemaitre, G., Houix, O., Visell, Y., Franinovic, K., Misdariis, N., Susini, P.: Toward the design and evaluation of continuous sound in tangible interfaces: The spinotron. International Journal of Human-Computer Studies 67, 976–993 (2009)
57. Diaper, D.: Understanding task analysis for human-computer interaction. In: Diaper, D., Stanton, N.A. (eds.) The Handbook of Task Analysis for Human-Computer Interaction, pp. 5–47. Lawrence Erlbaum Associates, Mahwah (2003)
58. Back, M., Des, D.: Micro-narratives in sound design: Context, character, and caricature in waveform manipulation. In: International Conference on Auditory Display ICAD 1996 (1996)
59. Hug, D.: Using a systematic design process to investigate narrative sound design strategies for interactive commodities. In: ICAD 2009, Copenhagen, Denmark, pp. 19–26 (2009)
60. Barrass, S.: A comprehensive framework for auditory display: Comments on barrass icad 1994. ACM Trans. On Applied Perception 2(4), 403–406 (2005)
61. Adcock, M., Barrass, S.: Cultivating design patterns for auditory displays. In: Proc. of ICAD 2004, Sydney, Australia, July 6-9 (2004)
62. Frauenberger, C., Stockman, T.: Auditory display design—an investigation of a design pattern approach. International Journal of Human-Computer Studies 67(11), 907–922 (2009)
63. Halpin, H., Robu, V., Shepherd, H.: The complex dynamics of collaborative tagging. In: Proc. of the 16th International Conference on World Wide Web (WWW 2007), pp. 211–220 (2007)
64. Brazil, E., Fernström, M.: Empirically based auditory display design. In: Sound and Music Computing Conference (SMC 2009), Porto, Portugal, pp. 7–12 (2009)
65. Papp, A.L.: Presentation of Dynamically Overlapping Auditory Messages in User Interfaces. Phd thesis, University of California (1997)
66. McGookin, D.: Understanding and Improving the Identification of Concurrently Presented Earcons. PhD thesis, University of Glasgow (2004)
67. Bonebright, T.L., Miner, N.E., Goldsmith, T.E., Caudell, T.P.: Data collection and analysis techniques for evaluating the perceptual qualities of auditory stimuli. ACM Transactions on Applied Perceptions 2(4), 505–516 (2005)
68. Vanderveer, N.J.: Ecological Acoustics- Human Perception of Environmental Sounds. Phd, University of Cornell (1979)

69. Jensen, J.J., Skov, M.B.: A review of research methods in children's technology design. In: Proc. of IDC 2005, Boulder, Colorado, USA, June 8-10 2005, pp. 80–87 (2005)
70. Hall, T.: Disappearing technology, emerging interactivity: a study of the design of novel computing to enhance children's learning experience in museums. PhD thesis, Department of Computer Science and Information Systems, University of Limerick (November 2004)
71. Sinclair, J., Hingston, P., Masek, M.: Considerations for the design of exergames. In: GRAPHITE 2007: Proceedings of the 5th international conference on Computer graphics and interactive techniques in Australia and Southeast Asia, Perth, Australia, pp. 289–295. ACM Press, New York (2007)
72. Lopez, M.J., Pauletto, S.: The design of an audio film for the visually impaired. In: Proc. of ICAD 2009, Copenhagen, Denmark (2009)
73. Barrass, S.: Haptic and Audio Interaction Design (chapter Haptic-Audio Narrative: from Physical Simulation to Imaginative Stimulation). In: McGookin, D., Brewster, S. (eds.) HAID 2006. LNCS, vol. 4129, pp. 157–165. Springer, Heidelberg (2006)
74. Hug, D.: Genie in a bottle: Object-sound reconfigurations for interactive commodities. In: Proc. of Audio Mostly Conference, Pitea, Sweden, pp. 56–63 (2008)
75. Chion, M.: Audio-Vision: Sound on Screen, 1st edn. Columbia University Press (1994)
76. Richard, F., Roussaire, V.: Sound design of the passenger compartment - process ad tool for the control of engine sound character. In: Proc. of "Les journées du design sonore (IRCAM, SFA)", Centre Georges Pompidou, Paris, October 13–15 (2004)
77. Le Nindre, B.: Brand sound identity: the case of sporty vehicles. In: Proc. of Les journées du design sonore (IRCAM, SFA)", Centre Georges Pompidou, Paris, October 13–15 (2004)
78. Tourila, H., Monteleone, E.: Sensory food science in the changing society: opportunities, needs, and challengers. In: Trends in Food Science and Technology (2008) (article in Press, Corrected Proof)
79. Moskowitz, H.R., Gofman, A.: Selling Blue Elephants: How to make great products that people want before they even know they want them. Wharton School Publishing, Upper Saddle River (2007)
80. Moskowitz, H.R., Hartmann, J.: Consumer research: creating a solid base for innovative strategies. Trends in Food Science and Technology 19, 581–589 (2008)
81. Wensveen, S.A.G., Overbeeke, K., Djajadiningrat, T.: Push me, shove me and i show you how you fell: recognising mood from emotionally rich interaction. In: Proc. of DIS 2002, pp. 335–340. ACM Press, London (2002)
82. Hassenzahl, M., Wessler, R.: Capturing design space from a user perspective: the repertory grid revisited. International Journal of Human-Computer Interaction 12(3,4), 441–459 (2000)

Designing a Web-Based Tool That Informs the Audio Design Process

Johan Fagerlönn and Mats Liljedahl

Interactive Institute, Sonic Studio, Acusticum 4, 94128 Piteå, Sweden
johan.fagerlonn@tii.se, mats.liljedahl@tii.se

Abstract. Research on auditory displays has shown how various properties of sounds can influence perception and performance. However, a challenge for system developers is how to find signals that correspond to specific user situations and make sense within a user context. This paper presents a web-based tool called AWESOME Sound design tool. The fundamental idea with the tool is to give end users control over some aspects of the auditory stimuli and encourage them to manipulate the sound with a specific user scenario in mind. A first evaluation of the tool has been conducted in which aspects of both usefulness and usability were addressed.

Keywords: audio design, auditory display, interface, usability.

1 Introduction

Sound can be an attractive means for communication, not least in safety critical and visually demanding contexts. However, designing sounds for user environments is not necessarily an easy and straightforward task. We know that sound can impact the listener in diverse and complex ways. Scientific studies have for instance shown how fundamental properties of a signal can influence perceived urgency [1,2,3]. In recent years there has been a growing interest in research on how sounds can impact users' emotional state, and how that in turn can impact performance [4]. A body of research have compared the usefulness of different "sound types" to convey information [5,6,7,8]. Auditory signals that are distracting may be an issue in demanding user situations [9,10]. Also, a general challenge in applied audio design is how to find sounds that will not be annoying.

In a typical controlled scientific study some aspect of a treatment is manipulated and given to the participant who responds. When evaluating signals for user environments responses can for instance be measured in terms of response time, response accuracy or task completion time. Subjective experiences and internal responses can be collected using questionnaires, interviews or by monitoring the participants physiological state. Results from these kinds of experiments can give system developers some ideas of which aspects of the sounds that are important to consider when designing auditory displays. However, a challenge for developers is how to find successful sounds that correspond to specific situations and make sense within the user context.

S. Ystad et al. (Eds.): CMMR/ICAD 2009, LNCS 5954, pp. 68–79, 2010.

One way to find appropriate sounds could be to invite users to give their opinions on a number of sound design suggestions and suggest design modifications. Experienced users can hopefully both identify and judge the situation appropriately. However, users who have little or no experience in sound design do not necessarily feel comfortable judging sounds and suggest modifications on a detailed level. Lack of a common taxonomy for sonic experiences can also be an issue. For instance, in a study by Coleman et al [11], one of the problems identified was that some participants were frustrated by the process of trying to relate and map their listening experience to the relatively abstract terminology used in the test.

An alternative strategy is to give users control over the auditory stimuli and encourage them to modify the sound with some particular user situation in mind. These design suggestions could potentially be used to inform the design process and help developers to better understand the correlations between properties of the sound and characteristics of a context and situation.

In the project AWESOME (Audio WorkplacE SimulatiOn MachinE) the Interactive Institute Sonic Studio is developing a number of tools with aim to help and facilitate the work of designers of sounding environments. The tool presented in this paper is referred to as the AWESOME Sound design tool. This utility shares some fundamental ideas with REMUPP (RElations between Musical Parameters and perceived Properties), an application which was developed by Interactive Institute and which has been evaluated within the area of music psychology [12,13]. Like REMUPP, the idea with the present tool is to make participants respond to an experience using the medium of sound itself, without having to translate it into other modes of expression. But in contrast to REMUPP, the present tool has been specifically developed for system developers working in a design process.

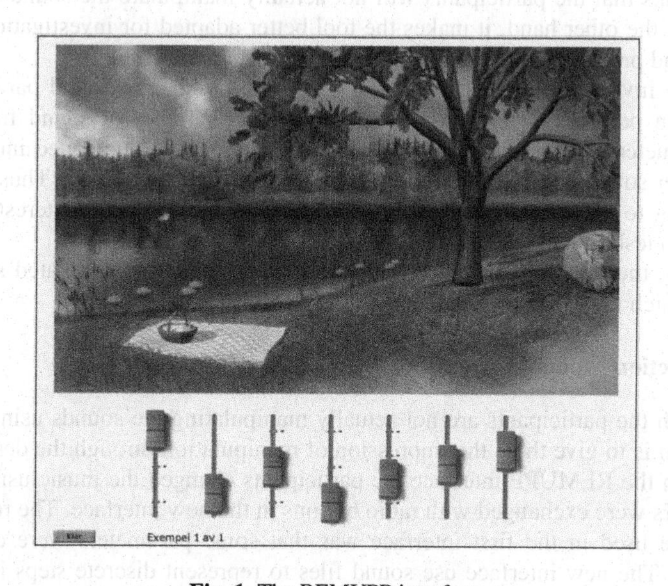

Fig. 1. The REMUPP interface

The REMUPP interface, presented in figure 1, allows participants to change the expression of a musical piece in real time. By using sliders the participant can manipulate a variety of musical parameters such as tempo, rhythm and articulation. The experimenter prepares the application with a number of contexts represented by an image or animation. The combination of visual representation of a context and variable musical stimuli in one interface allows scientists to investigate correlations between musical properties and characteristics of a context.

REMUPP was not designed to meet the requirements of system developers intending to design sounds for interfaces and user environments. This group of users may typically have other needs and requirements than most music psychologists. These potential demands resulted in a number of design suggestions and decisions for the new tool.

2 Design Decisions

2.1 Other Sound Formats

REMUPP was primarily designed to handle musical sounds. The sound player used MIDI data that was manipulated in real time by the participant. A system developer might not just want to investigate musical sounds but any kind of sounds and parameters such as acoustic features, fundamental sound properties, different "sound types" or voice timbres. To make this possible a decision was made not to use MIDI input for the music player but instead prepare the system for pre-recorded sound files.

This means that the participants will not actually manipulate the sounds using the tool. But on the other hand, it makes the tool better adapted for investigations of any kind of sound properties and combinations of sound parameters.

This also involves potential issues. First, a large number of sound parameters or levels within parameters will require an even larger number of sound files. Three sound parameters in three levels will require 27 sounds to be uploaded into the system, but four sound parameters in four levels will require 256 sounds. Thus, developers intending to use the tool will need to carefully select the most interesting sound properties to test.

Secondly, the system developer needs to provide already manipulated sounds for the tests, which may require some audio design skills.

2.2 Interaction Model

Even though the participants are not actually manipulating the sounds using the tool, the intention is to give them the impression of manipulation through the design of the interface. In the REMUPP interface the participants changed the music using sliders. These sliders were exchanged with radio buttons in the new interface. The reason why sliders were used in the first interface was that some parameters were continuous parameters. The new interface use sound files to represent discrete steps in parameters. Thus, no continuous parameters will be used.

The radio buttons are ordered in columns which are separated with different background colors. The idea is to give the impression that each column of buttons is used to control one unique parameter of the sound.

Another consideration was whether to allow the developer to name the columns/parameters. A text string explaining each parameter would probably make it easier for participants to understand how the interface relates to the design space. It was decided however not to implement this possibility. The reason was to encourage participants to "use their ears" when selecting the sounds instead of making them rely on text describing the sound parameters and the design space.

2.3 Web-Based Interface

Users of systems and interfaces might be found in different cultures and parts of the world. Making the interface web based allow developers to more easily gather data from users of different backgrounds. The tool can be used to investigate global similarities as well as cultural differences. REMUPP requires various software synthesizers to render the MIDI data. This is not an issue in the new tool since it uses pre-recorded sound files.

3 Implementation

Awesome sound design tool is a client-server application. The server side of the system is a database accessed via PHP that stores sound files and configured user situations. The database has four tables: Session, Example, Trial and Sound. Prior to a test one sets up a number of "Examples", or situations, using an administrator interface. An "Example" contains a picture or movie, a text string explaining the situation and an optional sound file for ambient sound. An "Example" serves as a situation and context for the participant to relate the proposed sound design. The table "Sound" is used to store a matrix of audio files that constitutes the sound design space. The number of dimensions in the matrix corresponds to the number of sound parameters that forms the design space. In the presented configuration of the tool three dimensions are used which can be manipulated in the levels.

The client side has been developed using Macromedia Flash and has two interfaces. One interface is used to prepare test sessions and to extract data from completed tests. The other interface is used to present user situations and select sounds. A screenshot of this interface is presented in figure 2. Context and situation are represented on the screen by an image or video and text string. The radio buttons used to select sounds are organized in the lower part of the screen. When a user clicks on a radio button the sound corresponding to that button configuration is played automatically. The participant can also play the current configuration by using the "Play button". Pushing the "OK button" advances the subject to the next situation and design task. The order of the situations and the default configuration of the radio buttons are randomized automatically by the application.

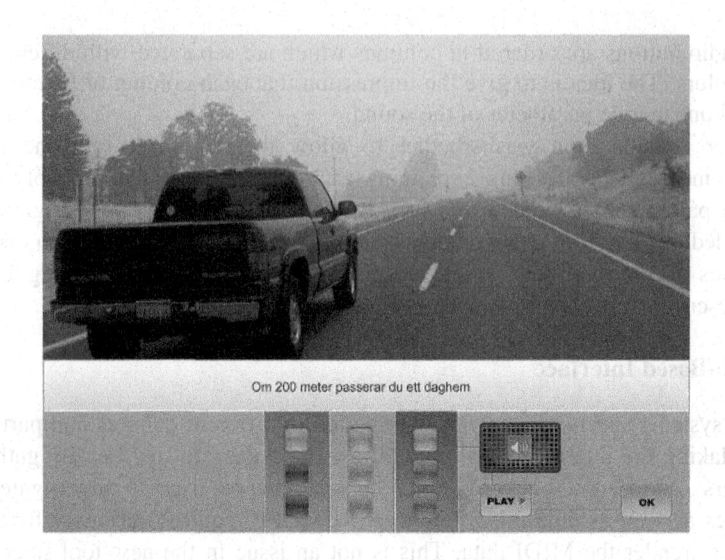

Fig. 2. Interface for participants

3.1 Data Analysis

Data generated by the participants are stored in the database for analysis. By using the administrators interface one can export the data to a tab separated text file (.txt). This format is recognized by common spreadsheet applications such as Microsoft Excel. Excel spreadsheet documents can in turn easily be imported in a variety of statistical software programs for more extensive analysis.

4 Pilot Study

Evaluation of the AWESOME Design Tool is conducted in an ongoing collaboration project between Interactive Institute, Scania CV AB and Luleå university of technology. The present work addresses how audio design can meet the requirements of safe and effective Intelligent Transport System (ITS) communication in heavy vehicles. Using AWESOME Sound design tool can be helpful when designing auditory warning signals for new and existing systems. One aspect here would be to find signals that express an appropriate level of urgency in specific driving situations. Inviting drivers to design sounds for traffic situations could be a way to find appropriate levels in a number of sound parameters and investigate differences between drivers.

4.1 Aims

One aim of the pilot study was to get a first indication of whether using the tool could be an appropriate way to get input from experienced users. By presenting traffic situations that differ significantly in urgency and a design space with parameters correlated with perceived urgency, we should expect to find patterns for which drivers select sounds for driving situations. These patterns will give us some insight in whether the drivers were able to express themselves using the tool.

Another aim was to investigate potential usability issues that can be can be avoided in future versions of the interface. The following questions were addressed in the usability evaluation.

- Do the drivers intuitively understand the interaction model?
- Do the drivers feel comfortable expressing themselves using the tool and the medium of sound?
- Do the drivers feel satisfied or restricted by the design space?

4.2 Participants

41 drivers, 23 males and 18 females, participated in the pilot study. Ages ranged between 25 and 65 years (mean 36,5). None of the drivers had any previous experience of using the tool. All drivers had self-reported normal hearing.

4.3 Apparatus

The interface was presented on a Macintosh MacBook 2.0 GHz laptop (Apple Inc., CA, USA). The sounds were played to participants at a comfortable listening level through a pair of KOSS Porta Pro headphones (Koss Corporation, WI, USA).

4.4 Design Space

Drivers were able to change three aspects of the warning signal: number of tones / speed (A), musical dissonance (B) and register (C). Each parameter had three levels as presented in figure 3. The first bar in each parameter (A-C) illustrates the low level in that parameter, the second bar illustrates the medium level and the third bar illustrates the high level. In total the subjects could change the parameters in 27 combinations. Recordings from a string quartet were used for all sounds.

Fig. 3. Design space used in the pilot study. A. Number of tones /speed, B. Musical dissonance. C. Register. The first bar in each parameter (A-C) shows the low level, the second bar shows the medium level, and the third bar shows the high level.

4.5 Driving Situations

Five driving situations were selected for the study. Two of the situations assumed to be perceived as very urgent by drivers. In one situation (*child*), a young girl was running in front of the vehicle. In the other urgent situation (*car*) another car in oncoming traffic was driving towards the own car. Two of the driving situations were assumed to be less urgent. In one situation (*school bus*), a school bus was parked behind a road crest. In the other situation (*cyclists*) a number of cyclists were standing in fog on top of a road crest. The final situation (*speed camera*) differed from the other driving situation, as it was not directly related to traffic safety. In this situation the driver was approaching a speed camera. This situation was assumed to be perceived as the least urgent situation by drivers. The images used to represent the traffic situations were all taken from the driver position.

4.6 Procedure

Instructions presented on the screen introduced the drivers to the design task. Neither the sound parameters nor the characteristics of the traffic situations were revealed in the instructions. By clicking a button the drivers continued to the first traffic situation and design task. The five situations were presented to the participants in random order. The experimenter was present during the test but participants were not allowed to ask any questions unless they got stuck. After the last situation the drivers were given a questionnaire containing 15 statements related to the usability questions addressed in the study. The statements were presented in mixed order. For each statement the drivers marked their answer on a 5 point scale ranging from "do not at all agree" to "completely agree". The drivers were also allowed to write freely about any issues experienced and suggest improvements of the interface.

4.7 Results

Complete data was collected from 41 participants. Figure 4-6 shows the results from the design tasks. For the most urgent situations (*car* and *child*) a majority of the

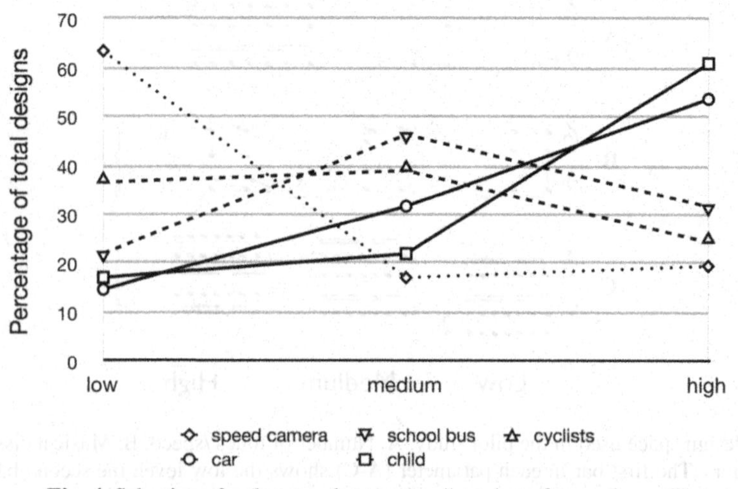

Fig. 4. Selections for the sound parameter "number of tones / speed"

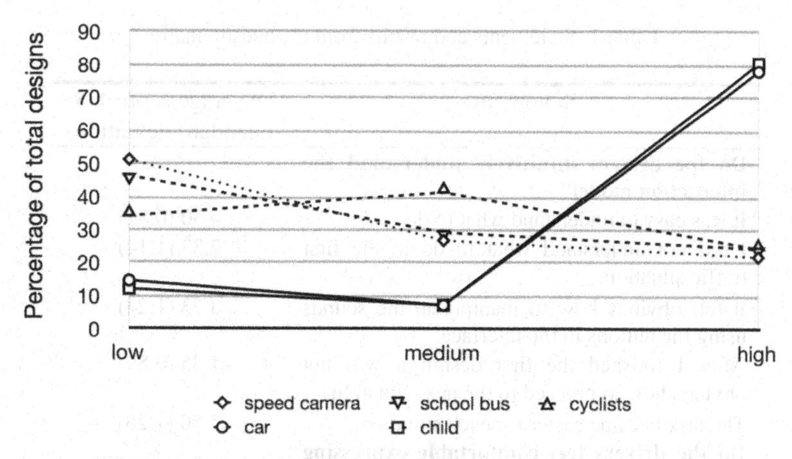

Fig. 5. Selections for the sound parameter "musical dissonance"

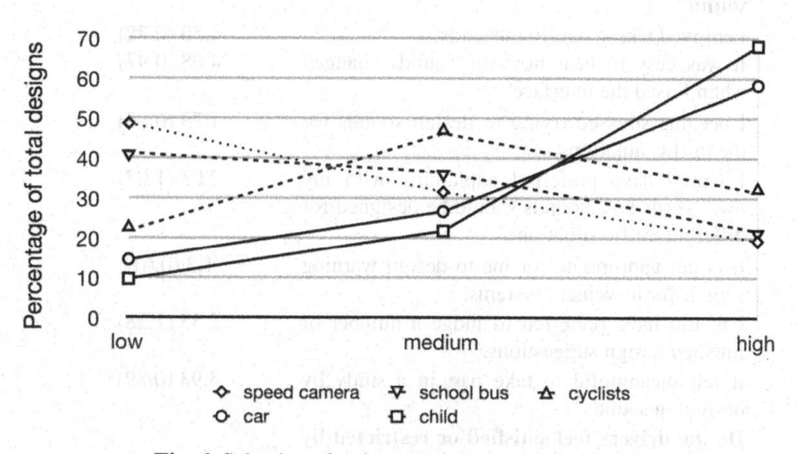

Fig. 6. Selections for the sound parameter "register"

drivers preferred the high level in all three parameters. There was a clear tendency among drivers to choose the highest level in all parameters in the most urgent situations. About 80% of the drivers selected the high level of dissonance in the urgent situations. Similar patterns were not seen for the less urgent situations *cyclists* and *school bus*. These less urgent situations tended to have relatively even distribution with a small peak in either the low or medium levels of the parameters. It was assumed that drivers would perceive the *speed camera* situation as the least urgent situation. This situation showed opposite patterns compared to the most urgent situations in all three sound parameters.

The distribution patters seen for the traffic situations are not surprising. Scientific experiments have reported positive correlations between perceived urgency and the parameters frequency, musical dissonance and number of tones/speed [1,3]. The clear tendencies seen in the designs indicate that the drivers actively manipulated the sounds to make them correspond to the traffic situations.

Table 1. Statements and results from the questionnaire

Statement	Average score 1-5 (standard deviation)
Do the drivers intuitively understand the interaction model?	
It was easy to understand what to do.	4.30 (0.82)
I did not understand what to do in the first traffic situation.	2.33 (1.14)
It felt obvious how to manipulate the sounds using the buttons in the interface.	3.73 (1.24)
After I finished the first design it was not obvious how to proceed to the next situation.	1.35 (0.89)
The task became easier for each situation.	3.70 (1.26)
Do the drivers feel comfortable expressing themselves using the tool and the medium of sound?	
I enjoyed taking part in the study.	4.50 (0.72)
It was easy to hear how the sounds changed when I used the interface.	4.08 (0.97)
I became stressed trying to design sounds for the traffic situations.	1.73 (0.93)
I would have preferred explaining with my own words how sounds should be designed for different traffic situations.	2.03 (1.07)
It is not appropriate for me to design warning signals for in-vehicle systems.	1.3 (0.61)
I would have preferred to judge a number of finished design suggestions.	2.55 (1.28)
It felt meaningful to take part in a study by designing sounds.	3.93 (0.89)
Do the drivers feel satisfied or restricted by the design space?	
I successfully designed sounds for the situations.	3.33 (0.76)
It would have preferred the possibility to change the sound even more.	2.88 (1.26)
The design space was well adapted for the task.	3.23 (0.83)

Results from the user evaluation are based on results of the questionnaire, written statements by drivers, and notes made by the experimenter during the trials. Table 1 shows the results from the questionnaire.

4.7.1 Did the Drivers Understand the Interaction Model?

According to the results of the questionnaire most of the participants found the interface easy to use. However, some drivers experienced difficulties in the very first traffic situation. 5 drivers wrote that they did not understand right away how to change the sound using the interface. 4 subjects wrote that they could not understand from the beginning how the buttons in the interface were related to each other. 3 drivers got stuck in the first example and had to ask the experimenter for assistance. One of the drivers suggested that a practice example at the beginning of the test would have been helpful.

In summary, the results indicate that the drivers found the interface easy to use. However, the lack of visual clues about how buttons were related to each other and the design space caused problems for some participants in the first design task.

4.7.2 Did the Drivers Feel Comfortable Expressing Themselves Using the Tool and the Medium of Sound?

The results of the questionnaire indicate that the drivers were comfortable expressing themselves using the tool. They seemed to like the interactive and nonverbal concept. Most of them liked taking part in the study (average score 4,5/5). 5 drivers wrote explicitly that it was fun, inspiring and felt creative to take part in the study. In the questionnaire only 6 of 40 participants reported that they would have preferred expressing themselves verbally instead of using the tool. 10 of 40 participants would have preferred judging finished design suggestions instead of using the interactive tool. A majority of the subjects also reported that it is appropriate for drivers to be part of the design process by designing sounds.

4 drivers wrote that they would have preferred a possibility to go back and change previous designs. The current version of the tool does not have that feature. In the present study the idea was to collect spontaneous opinions from drivers. In other studies it may be better if subjects are allowed to compare both situations and designs during trials.

In summary, the results suggest that the concept of using the interactive tool, and an "audio design space", can be an alternative strategy to get user opinions in a design process.

4.7.3 Did the Drivers Feel Comfortable with the Design Space?

Most issues reported by drivers were regarding the design space. 11 drivers wrote that they felt restricted by the use of string instruments. Timbre might be an important and dominant sound parameter in this particular kind of design tasks. Timbre was considered as a parameter for the pilot study. It was however not implemented, mainly because of the unclear relation between musical timbre and perceived urgency. Some drivers found it impossible to design sounds for the most urgent situations. They simply could not find (design) a sound that sounded urgent enough. For the parameter "musical dissonance" about 80% of the drivers selected the highest level. One can almost assume that drivers would have designed even more dissonant sounds for the urgent situations if that had been possible.

In summary, even though drivers successfully design more urgent sounds for the urgent traffic situations, there were issues reported regarding the design space. Finding appropriate parameters and levels within parameters can be a challenging task for developers intending to use the tool.

5 Discussion

In this paper we presented AWESOME Sound Design tool, a web-based utility that allows participants to respond to an experience using the medium of sound itself, without having to translate it to other modes of expression. The interface allows experienced users to take part in a design process but without requiring any prior experience in sound design. The software utilises pre-rendered sound files that can be easily exchanged by the experimenter depending on the type of study or parameters of a sound that are to be investigated.

At the ICAD 08 conference Mustonen [14] pointed out that the same sound could be listened to with different outcomes in different situations and orientations. "*When listening to interface elements we intuitively recognise familiar parts from the sound and construct the meaning from their relation to the situation*". AWESOME Sound design tool brings together experienced users' knowledge about context and situations and their perception and interpretation of sound. A better understanding of these correlations might allow developers to find more appropriate auditory solutions for user interfaces and audio environments.

A pilot study was conducted which addressed both the usefulness and usability aspects of the tool. In the study 40 drivers designed warning signals for traffic situations with different levels of urgency. In the results we saw clear patterns in how drivers designed sounds for the situations. These results indicate that the drivers actively and successfully used tool to design sounds that correspond to the traffic situations.

The results of the usability evaluation indicated that the tool was easy to use, and that the participants felt comfortable expressing themselves using the tool and the medium of sound.

The evaluation also exposed some usability issues. First, it was challenging for some drivers to understand how the radio buttons in the interface were related to each other and the design space. In the design process we made an active choice not to reveal the names of the sound parameters in the interface. The reason was to encourage participants to actually listen to the sounds. One potential solution can be (like one participant stated) to set up a practise example before the actual tests begins. Another solution could be to design more visual clues explaining the relationships between the structure of the button set and the intended design procedure. Organising buttons in columns with different colour is perhaps not representative and redundant enough.

Secondly, some participants had problems with design space. Selecting parameters and appropriate levels within parameters may be a challenge for developers intending to use the tool. In the pilot study we realized that many participants would have liked the possibility to change the sounds more than was possible. It is likely that developers would like to offer users a more extensive design space with more parameters and/or levels within parameters. One should remember that such tests may require a large number of sound files that needs to be uploaded into the system. Instead, we emphasize the use of pre-tests that can highlight important sound parameters and reasonable levels within parameters before using the tool in a design process.

AWESOME Sound design tool has been designed both for use in controlled tests and remote studies using Internet. Online data collection can be an effective way to gather opinions from a large number of end users, or users with different cultural backgrounds. The web-based interface offers developers a chance to maintain some control in remote studies. Future evaluations should address the use of AWESOME Sound design tool for online data collection.

Acknowledgements

Thanks to Stefan Lindberg at the Interactive Institute Sonic Studio for helping us with the sound design. Thanks to Håkan Alm at Luleå University of Technology, and Robert Friberg and Josefin Nilsson at Scania CV AB for support and inspiration.

References

1. Edworthy, J., Loxely, S., Dennis, I.: Improving auditory warning design: relationship between warning sound parameters and perceived urgency. Human Factors 33(2), 205–231 (1991)
2. Hellier, E.J., Edworthy, J., Dennis, I.: Improving auditory warning design: Quantifying and predicting the effects of different warning parameters on perceived urgency. Human Factors 35(4), 693–706 (1993)
3. Russo, F.A., Lants, M.E., English, G.W., Cuddy, L.L.: Increasing effectiveness of train horns without increasing intensity. In: Proceedings of ICAD 2003, Boston (2003)
4. Tajadura-Jiménez, A., Väljamäe, A., Kitagawa, N., Västfjäll, D.: Affective multimodal displays: Acoustic spectra modulates perception of auditory-tactile signals. In: Proceeding of ICAD 2008, Paris (2008)
5. McKeown, D.: Candidates for within-vehicle auditory displays. In: Proceedings of ICAD 2005, Limerick (2005)
6. Walker, B.N., Nance, A., Lindsay, J.: Spearcons: speech based earcons improve navigation performance in auditory menus. In: Proceedings of ICAD 2006, London (2006)
7. Leung, Y.K., Smith, S., Parker, S., Martin, R.: Learning and retention of auditory warnings. In: Proceedings of ICAD 1997, Paolo Alto (1997)
8. Bussemakers, M.P., Haan, A.: When it Sounds like a duch and it looks like a dog...auditory icons vs. earcons in Multimedia environments. In: Proceedings of ICAD 2000, Atlanta (2000)
9. Baldwin, C.L.: Acoustic and semantic warning parameters impact vehicle crash rates. In: Proceedings of ICAD 2007, Montreal (2007)
10. Fagerlönn, J., Alm, H.: Auditory signs to support traffic awareness. In: Proceedings of ITS World Congress 2009, Stockholm (2009)
11. Coleman, G.W., Macaulay, C., Newell, A.F.: Sonic mapping: Towards engaging the user in the design of sound for computerized artifacts. In: Proceedings of NordiCHI 2008, Lund (2008)
12. Wingstedt, J., Liljedahl, M., Lindberg, S., And Berg, J.: REMUPP – An Interactive Tool for Investigating Musical Relations. In: Proceedings of NIME 2005, Vancouver (2005)
13. Wingstedth, J., Brändström, S., Berg, J.: Young adolescents' usage of narrative functions of media music by manipulation of musical expression. Psychology of Music 36(2), 193–214
14. Mustonen, M.: A review-based conceptual analysis of auditory signs and their design. In: Proceeding of ICAD 2008, Paris (2008)

Auditory Representations of a Graphical User Interface for a Better Human-Computer Interaction

György Wersényi

Széchenyi István University, Department of Telecommunications, Egyetem t. 1,
H-9026 Győr, Hungary
wersenyi@sze.hu

Abstract. As part of a project to improve human computer interaction mostly for blind users, a survey with 50 blind and 100 sighted users included a questionnaire about their user habits during everyday use of personal computers. Based on their answers, the most important functions and applications were selected and results of the two groups were compared. Special user habits and needs of blind users are described. The second part of the investigation included collecting of auditory representations (auditory icons, spearcons etc.), mapping with visual information and evaluation with the target groups. Furthermore, a new design method for auditory events and class was introduced, called "auditory emoticons". These use non-verbal human voice samples to represent additional emotional content. Blind and sighted users evaluated different auditory representations for the selected events, including spearcons for different languages. Auditory icons using environmental, familiar sounds as well emoticons are received very well, whilst spearcons seem to be redundant except menu navigation for blind users.

Keywords: auditory icon, earcons, blind users, spearcons, GUIB.

1 Introduction

Creating Graphical User Interfaces (GUIs) is the most efficient way to establish human-computer interaction. Sighted people benefit from easy access, iconic representation, 2D spatial distribution of information and other properties of graphical objects such as colors, sizes etc. The first user interfaces were text-based, command line operation systems with limited capabilities. Later, hierarchical tree-structures were utilized mostly in menu navigation, since they enable a clear overview of parent-child relations, and causality. Such interfaces are still in use in simple mobile devices, cell phones etc. For the most efficient work the GUIs proved to be the best solution. Nowadays almost all operation systems offer a graphical surface and even command line programs can be accessed by such an interface. Some GUIs also include sounds but in a limited way as an extension to the visual content or for feedback only.

However, the blind community and the visually disabled do not benefit from a GUI. Access to personal computers became more and more difficult for them as the GUIs took over the former command line and hierarchical structures [1]. Although

S. Ystad et al. (Eds.): CMMR/ICAD 2009, LNCS 5954, pp. 80–102, 2010.

there is a need for transforming graphical information to auditory information for blind users, most so-called "auditory displays" are audio-only interfaces creating a virtual sound-scape, where users have to orientate, navigate and act. These virtual audio displays (VADs) have limited quality, spatial resolution and allow reduced accessibility.

As a result, blind users often use only textual representation of a screen. These text-to-speech (TTS) applications or screen-readers nowadays offer good synthesised speech quality, but they are are language-dependent and only optimal for reading textual information. Some programs, such as the most frequently used Job Access With Speech (JAWS) [2] or the Window-Eyes [3] also read icon names and buttons. The user moves the cursor with the mouse or navigates with the TAB-button over the icons and across the screen and information will be read about objects that he crosses. Unfortunately sometimes confusion is created when the objects are read phonetically. A TTS system can not follow a GUI, it is more disadvantageous than helpful in translating graphical information into textual. Tactile translations have encountered many of the same difficulties in representing graphical information in a tactile way [4, 5].

The overriding goal is to create an audio environment where blind users have the same or almost the same accessibility as sighted colleagues do. To achieve this, the most important considerations are the following:

- accessibility and recognition: blind users have to be able to use the interface, recognize items, programs, identify and access them. Some issues to be resolved are: what are the objects, what is the name/type, where are they, what attributes do they have?
- iconic representation: short, easily identifiable sounds, that can be filtered, spatially distributed etc. They have to be interruptable even if they are short.
- safe manipulation: safe orientation and direct manipulation with auditory feedback.

Screen readers and command line interfaces do not currently offer these possibilities. Some stumbling blocks have been

- In contrast to graphics, auditory signals cannot be presented constantly.
- It is hard with auditory displays to get an overview of the full screen and users have to use their short-time memory to remember the content of the screen. Concurrent sound sources are hard to discriminate and/or long term listening to synthesised speech can be demanding (synthesised speech overload).
- Blank spaces of the background (without sound) can lead to disorientation.
- Other graphical information can also be relevant: relatively bigger buttons, fontsizes, different colors or blinking may indicate relative importance that is hard to translate into auditory events.
- Grouping of information: the spatial allocation of similar functions and buttons is also hard to map to an auditory interface.
- The static spatial representation of a GUI seems to be the most difficult to transfer and the cognitive requirements for a blind user are quite demanding. Hierarchical structures are easily abstracted but they represent discrete values (menu items). Sonification of continuous data, such as auditory graphs is also in interest [6, 7].

The most critical issue is here navigation: good overall performance by using an auditory display is strongly related to good and fast navigation skills. Navigation without the mouse is preferred by blind users. Keyboard short-cuts and extended presentation of auditory events (spatial distribution, filtering etc.) are useful to expert users. Spatial models are maybe preferable as opposed to hierarchical structures, but both seem to be a good approach to increase accessibility. Learning rates are also an important consideration, because everybody needs time to learn to use an auditory interface.

It is impossible to transfer all the information in a GUI to an auditory interface, so we have to deal with some kind of an "optimal" procedure: the most important information should be transferred, and details and further information (for trained and expert users) can extend the basic auditory events. The goal is that blind users can work with computers, create and handle basic text-oriented applications, documents, e-mails, and also browse the internet for information. They have to be able to save, open, copy, delete, print files. After basic formatting and file managing then sighted users may do any remaining formatting.

1.1 Some Previous Results

Earlier investigations tried to establish different auditory interfaces and environments for the visually impaired as early as the 1990s. The *SonicFinder* [8] was an Apple program which tried to integrate auditory icons into the operating system for file handling, but it was not made commercially available primarily because of memory usage considerations. Mynatt and colleagues presented a transformed hierarchical graphical interface, utilizing auditory icons, tactile extension and a simplified structure for navigation in the *Mercator* project [4]. The hierarchical structure was thought to best to capture the underlying structure of a GUI. The project focused on text-oriented applications such as word-processors, mailing programs but neglected graphical applications, drawing programs etc. The TTS module was also included. A basic set of sounds were presented the users as seen in Table 1.

Furthermore, they used filtering and frequency manipulations to portray screen events, e.g. appearing of pop-up windows, selecting items or the number of objects. These were mostly chosen intuitively and were sometimes not very helpful at all, because some sounds are ambiguous (closing a pop-up window can have the same sound as "close" or even some speech feedback) or the related events are not really important (pop-up blocking reduces pop-ups to a minimum). A more general problem is that there are no standards or defined ways to use the simplest modifications in volume, pitch, timbre or spectral content of an auditory event. For instance, the sound of paper shuffling in Mercator represented "switching between applications" but this sound is clearly not good in Windows, where a similar sound is mapped with the recycle bin. Different operating systems may require different sound-sets, but the overriding concern is to find the most important applications, functions and events of the screen that have to be represented by auditory events.

Involving sighted people in this quest is desirable both for comparison with blind users, and because it can be advantageous for the sighted user as well: they can examine the efficiency of transition from GUI to auditory interface and finally, they could also benefit from auditory feedback during work.

Table 1. Auditory icons introduced by Mynatt for the Mercator [4, 9]

Interface Object	Sound
Editable text area	Typewriter, multiple keystrokes
Read-only text area	Printer printing out a line
Push button	Keypress (ca-chunk)
Toggle button	Pull chain light switch
Radio button	Pa pop sound
Check box	One pop sound
Window	Tapping on glass (two taps)
Container	Opening a door
Popup dialog	Spring compressed then extended
Application	Musical sound

Later, the *GUIB project* (Graphical User Interface for Blind persons) tried a multi-modal interface, using tactile keyboards (Braille) and spatial distributed sound, first with loudspeaker playback on the so-called sound-screen, then using headphone play-back and virtual simulation [5, 10, 11, 12]. In this project the Beachtron soundcard was used with real-time filtering of the Head-Related Transfer Functions (HRTFs) to create a spatial virtual audio display. A special 2D surface was simulated in front of the listener instead of the usual "around the head" concept. This should create a better mapping of a rectangle computer screen and increase in navigation accuracy with the mouse as well. Listening tests were carried out first with sighted and later with blind users using HRTF filtering, broadband noise stimuli and headphone playback. The results showed an increased rate of headphone errors such as in-the-head localization and front-back confusions, and the vertical localization was almost a complete failure. A follow-up study used additional high-pass and low-pass filtering to bias correct judgements in vertical localization (Fig. 1.) and achieved about 90% of correct rates [13, 14].

Simulation of small head-movements without any additional hardware also seemed very useful in reducing of errors [15, 16]. Spatial distributed auditory events can be used in a special window-arrangement in different resolutions according to the users' experience and routine. In addition, distance information can be used for overlapping windows or other parameters.

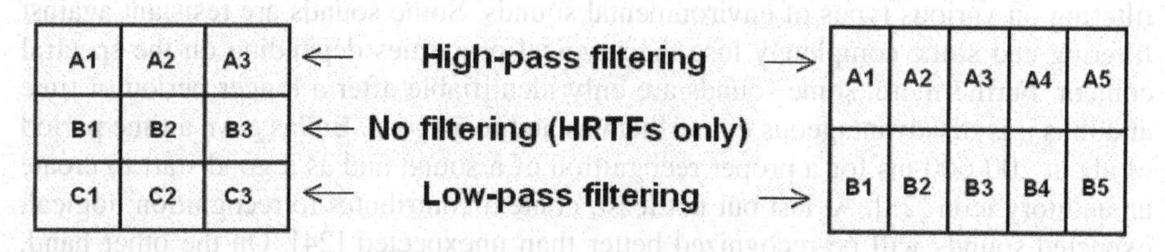

Fig. 1. A possible scheme for increasing vertical localization judgments. Input signals can be filtered by HPF and LPF filters before or after the HRTF filtering.

In [4] it was reported that blind users have positive response to the project, but they were skeptical about hierarchical navigation schemes. A spatial one seems to be better, primarily for blind people who lost their vision later in life. Users who were born blind have more difficulties in understanding some spatial aspects of the display, but tactile extensions can be helpful to understand spatial distribution and forms.

2 Auditory Representations

What kind of auditory events can be used in an auditory display? Different basic sound types have different considerations:

- Speech is sometimes too slow, language-dependent, and syntheised speech overload can happen. A TTS is neccessery for textual information but not optimal for orientation, navigation and manipulation.
- Pure tones are easily confused with each other, are not very pleasant to listen to them and mapping is intuitive that needs more learning time.
- Musical instrumentation is easier to listen to, but also needs learning and absraction because of the intuitive mapping.
- Auditory icons, earcons, spearcons and auditory emoticons, or structured combination of environmental sounds, music, non-speech audio or even speech can create good iconic representations. Iconic everday sounds can be more intuitve than musical ones [4].

Auditory icons and earcons were the first introduced by William Gaver, followed by others [17-19]. These sounds are short „icon-like" sound events having a semantic connection to the physical event they represent. Auditory icons are easy to interpret and easy to learn. Users may connect and map the visual event with the sound events from the initial listening. A typical example is the sound of a matrix dot-printer that is intuitively connected with the action of printing. Gaver provided many examples of easily learned auditory icons. Unfortunately, there are other events on a screen that are very hard to represent by auditory icons.

Environmental sounds are very good for auditory icons, because they are easily identifiable, learnable, they have a semantic-nomic connection to (visual) events. There are numerous factors that affect the useability of environmental sounds as auditory icons: a brief overview was provided in [20-22]. Among these are the effects of filtering on various types of environmental sounds. Some sounds are resistant against filtering and some completely lose their typical properties depending on the spectral content. Furthermore, some sounds are only identifiable after a longer period of time and thus it is disadvantageous to use them as auditory icons. Ballas gave a time period of about 200-600 ms for a proper recognition of a sound and as a good start to create an auditory icon [23]. At last but not least, context contributes to recognition: logical, expected sounds will be recognized better than unexpected [24]. On the other hand, unexpected sounds do not have to be too loud to get attention to. Realistic sounds sometimes are inferior to other but more familiar versions of them. Cartoonification may help, or e.g. a gunshot is much different in the real life as it is in movies [25, 26].

On the other hand, earcons are „meaningless" sounds. The mapping is not obvious, so they are harder to interpret and to learn, and have to be learned together with the event they are linked to. An example: the sounds that we hear during start-up and shut down the computer or during warnings of the operation system are well-known after we hear them several times.

Spearcons have already proved to be useful in menu navigations and in mobile phones because they can be learned and used easier and faster than earcons [27-30]. Spearcons are time-compressed speech samples which are often names, words or simple phrases. The optimal compression ratio, required quality and spectral analysis was made for Hungarian and English language spearcons [31]. For the study described later, the Hungarian and German spearcon databases for our study were created with native speakers.

Furthermore, some sound samples can not be classified into the main three groups mentioned above. Based on the results of a user survey, we will introduce a new group of auditory events called auditory emoticons. Emoticons are widely used in e-mails, chat and messenger programs, forum posts etc. These different smileys and abbreviations (such as brb, rotfl, imho) are used so often that users suggested that they be represented with auditory events as well.

Auditory emoticons are non-speech human voice(s), sometimes extended and combined with other sounds in the background. They are related to the auditory icons the most, using human non-verbal voice samples with emotional load. Auditory emoticons – just like the visual emoticons - are language independent and they can be interpreted easily, such as the sound of laughter or crying can be used as an auditory emoticon.

All the above auditory events are intended for use in auditory displays both for sighted and blind users as feedback of a process or activation, to find a button, icon, menu item etc.

3 Evaluation and Comparison of User Habits

After many years of research, the Hungarian Institution of Blind Persons is involved in our survey and we have access to blind communities in Germany as well. The first part of the investigation was to find out how blind persons use personal computers nowadays, what their likes and dislikes are, or their needs for a better accessibility. In order to do this we created a questionnaire both for blind people and for people with normal vision. Based on the answers we selected the 30-40 most important and frequently accessed programs and functions. The second part of the project included the selection and evaluation of sound events (auditory icons, earcons or spearcons) representing these functions. Furthermore, user habits of different age groups and user routines were also evaluated. Details of the survey and some preliminary results of sighted users were presented and described in [31, 32].

The survey included 100 persons with normal vision and 50 visually impaired (from Hungary and Germany). Subjects were categorized based on their user routines on their ages. Eighty-three percent of the sighted subjects were "average" or "above average" users but only forty percent of the blind users were. It is clear that a large number of blind users often restrict themselves to basic computer use.

Table 2. Ranking points for applications and services

Points	
1	Unknown by the user
2	Known, but no use
3	Not important, infrequent use
4	Important, frequent use
5	Very important, everyday use

The average age of sighted users was 27,35 years and 25,67 for blind participants. Subjects had to be at least 18 years of age and they had to have at least basic knowledge of computer use. Usage was ranked on a scale from 1 to 5, detailed in Table 2. Mean rankings above 3,75 correspond to frequent use. On the other hand, mean rates below 3 points are regarded not to be very important. Because some functions appear several times on the questionnaire, these rates were averaged again (e.g. if „print" has an mean value of 3,95 in Word; but only 3,41 in the browser then a mean value of 3,68 will be used).

Mean results are listed in Table 3. Light grey marked fields indicate important and frequently used programs and applications (mean ranking 3,00 – 3,74). Dark grey fields indicate everyday use and higher importance (mean ranking above 3,75 points). At the end of the table some additional ideas and suggestions are listed without rating. Additional applications suggested by sighted users were: Wave editor, remove USB stick, date and time. Blind users mentioned DAISY (playback program for reading audio books), JAWS and other screen-readers. As mentioned above, the frequent use of emoticons (smileys) in e-mails and messenger applications brought up the need to find auditory representations for these as well.

3.1 Blind Users

Blind users have different needs sometimes when using personal computers. We observed that:

- Blind users like the icons, as well as programs that are on the desktop by default, such as My Computer and the My Documents folder. They use these more frequently than sighted users, because sighted can easily access other folders and files deeper in the folder structure as well.
- Programs that use graphical interfaces (e.g. Windows Commander) for ease of access are only helpful for sighted users.
- Image handling, graphical programs, movie applications are only important for sighted users. However, the Windows Media Player is also used by the blind persons, primarily for music playback.
- Select and highlighting of text is very important for the blind, because TTS applications read highlighted areas.
- Blind users do not print often.
- Acrobat is not popular for blind persons, because screen-readers do not handle PDF files properly. Furthermore, lots of web pages are designed with graphical contents (JAVA applications) that are very hard to interpret by screen readers.

Table 3. Averaged points given by the subjects

Programs/ applications/ functions	**sighted**	**blind**
Number of subjects	**100**	**50**
	Total avg.	Total avg.
Internet Browser (icon/starting of the program)	4,62	4,67
E-mail client	4,11	4,67
Windows Explorer	3,98	3,25
My Computer	3,58	3,94
Windows/Total Commander	2,75	1,88
Acrobat (Reader)	4,26	3,33
Recycle Bin	4,41	3,67
Word (Word processor)	4,53	4,56
Excel	3,81	2,94
Power Point	3,14	2,47
Notepad/ WordPad	2,6	2,61
FrontPage (HTML Editor)	2,09	2,24
CD/DVD Burning	3,84	3,94
Music/Movie Player	4,09	4,17
Compressors (RAR, ZIP etc.)	3,41	3,22
Command Line, Command Prompt	2,84	1,83
Printer handling and preferences	3,87	2,64
Image Viewer	3,98	1,62
Downloads (Torrent clients, DC++, GetRight)	2,7	2,06
Virus/Spam Filters	4,29	4,39
MSN/Windows Messenger	3,29	3,17
Skype	2,91	3,39
ICQ	2,33	1,78
Chat	2,06	1,5
Paint	3,05	1,67
Calculator	2,82	2,61
System Preferen., Control Panel	3,6	3,17
Help(s)	2,55	2,98
Search for files or folders (under Windows)	3,32	3,11
My Documents folder on the Desktop	3,52	4,11
JAWS, Screen-readers	1	4,6
FUNCTIONS		
Home button (Browser)	3,53	3,61
Arrow back (Browser, My Computer)	4,22	4,33
Arrow forward (Browser, My Computer)	4,22	4,33
Arrow up („one folder up", My Computer)	3,53	3,53
Re-read actual site (Browser)	3,31	3,47
Stop loading (Browser)	2,79	2,98
Enter URL address through the key-board (Browser)	4,02	4,05
Favorites, Bookmarks	3,55	3,88
New register tab (Browser)	3,95	2,94
New window (Browser)	3,78	3,53
Search/find for text on the screen (Browser, Docs)	3,35	3,88
Save/open image and/or location (Browser)	3,59	3,19
Print	3,41	2,9

Table 3. (*Continued*)

Cut	4,11	3,88
Paste	4,26	4,41
Copy	4,49	4,56
Move	4,26	4,38
Delete	4,14	4,56
New folder (My Computer)	4,24	4,38
Download mails/ open E-Mail client (Mail)	4,1	4,59
Compose, create new mail (Mail)	4,2	4,67
Reply (Mail)	4,14	4,67
Reply all… (Mail)	2,91	2,98
Forward (Mail)	3,26	4,19
Save mail/drafts (Mail)	3,18	3,47
Send (Mail)	4,22	4,67
Address book (Mail)	3,51	3,72
Attachment (Mail)	3,7	4,06
Open	4,43	4,71
Save	4,28	4,71
Save as…	4,29	4,71
Close	4,27	4,65
Rename	4,14	3,94
Restore from Recycle Bin	3,04	3,44

Empty Recycle Bin	3,74	3,83
New Document	4,29	4,53
Spelling (Docs)	3,47	3,41
Font size (Docs)	3,79	3,41
Format: **B**/*I*/U (Docs)	3,98	3,47
Select, mark, highlight text (Docs)	3,91	4,29
Repeat	3,78	2,94
Undo	3,78	2,94
OTHERS		
Waiting… (hour-glass)		
Start, shut-down, restart computer…		
Resize windows (grow, shrink)		
Frame/border of the screen		
Scrolling		
Menu navigation		
Actual time, system clock		
EMOTICONS		

- Word is important for both groups, but Excel, Power Point use mainly visual presentation methods, so these latter programs are useful for sighted users.
- For browsing the Internet, sighted users are more likely to use the "new tab" function, while blind persons prefer the "new window" option. It is hard to orientate for them under multiple tabs.
- The need for gaming was mentioned by the blind as a possibility for entertainment (specially- created audio games).

The idea of extensions or replacements of these applications by auditory displays was welcomed by the blind users, however, they suggested not to use too much of them, because this could lead to confusion. Furthermore, they stated spearcons to be unnecessary on a virtual audio display because screen-readers offer speeded up speech anyway.

Blind users mentioned that JAWS and other screen readers do not offer changing the language "on the fly"; so if it is reading in Hungarian, all the English words are pronunciated phonetically. This is very disturbing and makes understanding difficult. However, JAWS offers the possibility to set such words and phrases for a correct pronunciation one by one. An interesting note is that JAWS 9.0 does not offer yet Hungarian language, so Hungarian blind users use the Finnish module although, the reputed relationship between these languages has been questioned lately. Another complaint was that JAWS is expensive while the free version of a Linux-based screen reader has a low quality speech synthesizer.

The best method for a blind person to access applications would be a maximum of a three-layer structure (in menu navigation), alt tags in pictures, and the use of the international W3C standards (World Wide Web Consortium) [33]. Only about 4% of the internet web pages follow these recommendations.

As mentioned before, there is a strong need among blind users for audio-only gaming and entertainment. There are currently some popular text-based adventure games using the command line for navigation and for actions. But there is more need for access to on-line gaming, especially for on-line table and card games, such as Poker, Hearts, Spades or Bridge. This could be realized by speech modules, if the on-line website would tell the player the cards he holds and are on the table.

One of the most popular is the game Shades of Doom, a trial version of which can be downloaded from the internet [34]. In a three dimensional environment, the user guides a character through a research base and shuts down the ill-fated experiment. It features realistic stereo sounds, challenging puzzles and action sequences, original music, on-line help, one-key commands, five difficulty levels, eight completely navigable and explorable levels, the ability to create Braille-ready maps and much more. This game is designed to be completely accessible to blind and visually impaired users, but is compatible with JAWS and Window-Eyes if desired.

On the topic of using environmental sounds in auditory displays for the blind, it should be noted that in one comparative study blind people did not perform better in recognizing environmental sounds than sighted people do: the two groups both performed at a relatively low level of about 76-78% of correct answers. However, blind subjects can be more critical about how auditory icons should sound [10, 35]. Our current investigation (in preparation) about virtual localization of blind persons also showed that in a virtual environment they may not hear and localize better than sighted people.

4 Evaluation of Auditory Events

After determining the most important functions and applications, a collection of sound samples was developed and evaluated based on the comments and suggestions of blind and sighted users. Below is listed the collection of sounds that was previously selected by the users as the "winning" versions of different sound samples. The rating procedure for Hungarian, German and English spearcons and sound samples is based on an on-line questionnaire with sound playback [36]. Figure 2 shows a screenshot of the website, where users rated a sound sample to be *bad* (3 points), *acceptable* (2 points) or *very good* (1 point). According to the German system, the less points are given, the better the results are. Detailed results and evaluation rates are shown here for the auditory icons only (right column in Table 6). All the sound samples can be downloaded from the Internet in wave or mp3 format [32].

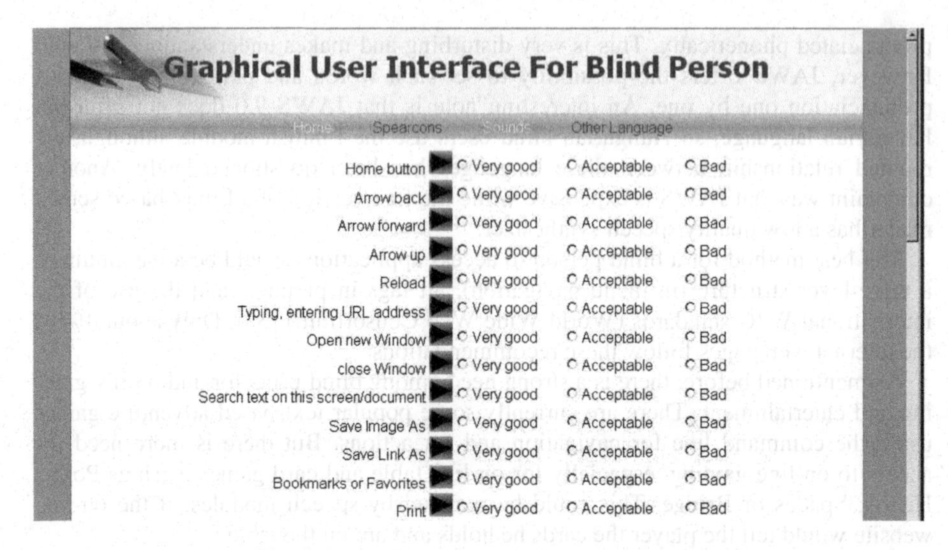

Fig. 2. Screenshot of the website for evaluation

The approach has been to create sound samples the majority of which have a length of about 1,0-1,5 sec. The actual durations are between 0,6 and 1,7 sec with an mean of 1,11 sec. There are two types of sounds: normal sounds that have to be played back once; and sounds to be repeated (in a loop). The first represents icons, menu items or short events. Looped signals are supposed to be played back during a longer action or event (e.g. copying, printing).

Sound files were recorded or downloaded from the Internet and were edited by Adobe Audition software in 16 bit, 44100 Hz mono wave format [37, 38]. Editing included simple operations of amplifying, cutting, mixing, fade in/out effects. At the final stage, all samples were normalized to the same loudness level (±1 dB).

A collection of wave data of about 300 files was categorized, selected and evaluated. Subjects were asked to identify the sound (what is it) and judge them by "comfortability" (how pleasing it is to listen to it). Subjects evaluated different sound samples (types) and variations to a given application or event. For example, a sound had to be applied to the action of opening a file. Thus, it had to be determined what "open" sounds like? Possible sound samples include a slide fastener (opening the zipper on a trouser), opening a drawer, opening a beer can or pulling the curtains. We presented different versions of each to insure inclusion of an appropriate representation. In addition, subjects were asked to think about the sound of "close" – a representation in connection with „open". Therefore, we tried to present reversed versions of opening sounds (simply played back reversed) or using the squeezing sound of a beer can. The reverse playback method can not be applied every time; some samples could sound completely different reversed [39]. Subjects could make suggestions for new sounds as well. If there was no definite winner or no suggested idea at all, a spearcon version was used (e.g. for Acrobat).

The sound files listed in Tables 4-7 (right columns) are included in a ZIP file that can be directly downloaded from http://vip.tilb.sze.hu/~wersenyi/Sounds.zip.

4.1 Applications

Table 4 shows the most important applications and programs that have to be represented by an auditory event. These were selected if both blind and sighted users ranked them as "important" or "everyday use" (having a mean ranking of at least 3,00 on the questionnaire), except for the My Documents folder and JAWS because these were only important for blind users.

The sound for internet browsing includes two different versions, both were accepted by the users. It is interesting that the sound sample "search/find" contains a human non-speech part that is very similar in different languages, and is easy to relate to the idea of an "impatient human". Subjects could relate the intonation to a sentence of "Where is it?" or 'Wo ist es?" (in German) or even "Hol van már?" (in Hungarian). It appears a similar intonation is used in different languages to express the feeling during impatient searching. As a result, the same sound will be used in other applications where searching, finding is relevant (Browser, Word, Acrobat etc.). Another idea was a sound of a sniffing dog.

The table does not contain some other noteworthy samples, such as a modified sound for the E-mail client, where the applied sound is extended with a frustrated "oh" in case there is no new mail and a happy "oh" if there is a new mail. Since mail clients do have some kind of sound in case of existing new mails this sample was not used.

Table 4. Collection of the most important programs and applications (MS Windows based). Sound samples can be found under the given names.

Application	Description	Filename
Internet browser (1)	Door opening with keys	Browser1
Internet browser (2)	Knocking and opening a door	Browser2
E-mail client	Bicycle and P.O. box	Mail1
Windows Explorer	Spearcon	S_Explorer
My Computer	Computer start-up beep and fan noise	My Computer
Acrobat	Spearcon	S_Acrobat
Recycle Bin	Pedal of a thin can with the recycle bin sound	Pedal
MS Word	Spearcon	S_Word
CD/DVD burning	Burning flame	Burn
Movie/music player (MS MediaPlayer)	Classic movie projector	Projector
Compressors (ZIP, RAR)	Pressing, extruding machine	Press
Virus/Spam killer	Coughing and "aaahhh"	Cough
MSN Messenger	Spearcon	S_Messenger
Control Panel	Spearcon	S_ControlP
My Documents folder on the desktop	Spearcon	S_MyDocs
Search for files etc.	Seeking and searching with human voice (loop) or dog sniffing	Search_(loop)
JAWS/TTS/Screen Reader appl.	Spearcon, speech	

The events related to the recycle bin also have sound events related to the well-known sound effect of the MS Windows "recycle bin.wav". This is used if users empty the recycle bin. We used the same sample in a modified way to identify the icon, opening the recycle bin or restore a file from it. The application identification uses the "paper noise" and a thin can pedal together. Restoring a file utilizes the paper noise with human caw. The caw imparts the feeling of a false delete earlier. This thematic grouping was very helpful to identify connected events.

For compressor applications, we used samples of human struggling while squeezing something, e.g. a beer can, but similar sounds appear later in open, close or delete. Similarly, a ringing telephone was suggested for MSN/Windows Messenger, but this sound is used by Skype already. Finally, two different samples for "Help" were selected: a whispering human noise and a desperate "help" speech-sample. Because Help was not selected as a very important function, and furthermore, the first sample was only popular in Hungary (Hungarian PC environments use the term "whispering" instead of "help", an analog to theatrical prompt-boxes) and the second contains a real English word, these samples were culled from the final listing.

4.2 Navigation and Orientation

The sounds in Table 5 were judged to be the most important for navigation and orientation on the screen, primarily for blind persons. Although, blind users do not use the mouse frequently, sometimes it is helpful to know where the cursor is. The movement of the cursor is a looped sound sample indicating that it is actually moving. The actual position or direction of moving could be determined by increasing/decreasing sounds (such as by scrolling) or using HRTF synthesis and directional filtering through headphones [12-14]. This is not implemented yet. Using this sound together with a "ding" by reaching the border of the screen allows a very quick access to the system tray, the start menu, or the system clock which are placed bottom left and right of the screen.

In case of menu navigation spearcons have been already shown to have great potential. Modifications to spearcons to represent menu structures and levels can be used, such as different speakers (male, female) or different loudness levels etc. In case

Table 5. Collection of important navigation and orientation tasks (MS Windows based). Sound samples can be found under the given names.

Other sounds	Description	Filename
Moving the mouse (cursor)	Some kind of "ding" (loop)	Mouse_(loop)
Waiting for... (sand-glass turning)	Ticking (loop)	Ticking_(loop)
User intervention, pop-up window	Notification sound	Notify
Border of the screen	Some kind of "ding"	Ding (Border)
Scrolling	Increasing and decreasing freq.	
Menu navigation	Spearcons with modifications	
System clock	Speech	S_SystemClock
Start menu	Spearcon, speech	S_StartMenu

of short words, such as Word, Excel, or Cut the use of a spearcon is questionable, since these words are short enough without time-compression in. Users preferred the original recordings instead of the spearcons in such cases. We did not investigate thoroughly what the limit is, but it seems that speech samples with only one syllable and with a length shorter than 0,5 sec. are likely too short to be useful as a spearcon. On the other hand, long words with more vowels become harder to understand after having compressed them into spearcons.

4.3 Functions and Events

Table 6 contains the most important and frequently used sub-functions in several applications. The second column indicates where the given function can be found and some common visual representations (icons) can also be seen. Finally, the last column shows mean values given by blind and sighted users on the homepage by ranking them from 1 to 3 points.

The sounds related to internet browsing have something to do with "home". Users liked the home button being represented by a doorbell and a barking dog together – something that stereotypically happens when one arrives home. Arrows forward, back and up also have something to do car actions: start-up, reverse or engine RPM boost. Similarly, mailing events have stamping and/or bicycle bell sounds representing a postman's activity. This kind of thematic grouping is very important in creating auditory events and sound sets. It results in increased learnability and less abstraction is needed. Some of the auditory icons and thematic grouping methods have to be explained but after the users get the idea behind them they use it comfortably. It is recommended to include a short FAQ or user's manual in a help menu for such sound sets.

Bookmarks/favorites in a browser and the address book/contacts in the e-mail client share the same sound of a book, turning pages and a humming human sound. This is another good example for using a non-speech human voice sample interacting with a common sound and thus creating a better understanding and mapping.

The sound for printing can be used in a long version or looped in case of ongoing printing (in the background this can be more quiet) or as a short sound event to represent the printing icon or command in a menu. The same is true for "copy": a longer version can be used indicating the progress of the copying action (in the background), and a shorter to indicate the icon or a menu item.

The sound for "paste" is one of the most complex samples. It uses the sound of painting with a brush on a wall, a short sound of a moving paint-bucket and the whistling of the painter creating the image of a painter "pasting" something. This works best for English because in Hungarian and in German a different expression is used and the idea behind this sound has to be explained.

In case of "move" there are two versions: the struggling of a man with a wooden box, and a mixed sound of "cut and paste": scissors and painting with a brush.

Based on the comments, the action of "saving" has something common with locking or securing, so the sound is a locking sound of a door. As an extension, the same sound is used for "save as" with an additional human "hm?" sound indicating that the securing process needs user interaction: a different file name to enter.

Table 6. Collection of the most important actions and functions (MS Windows based). Sound samples can be found under the given names.

Events, Functions	Where?	Visual Representations	Description	Filename	Mean Values
Home button	Internet Browser	,	Doorbell and dog barking	Homebutton	1,34
Arrow back	Internet Browser, My Computer	,	Reverse a car with signaling	Backarrow	1,53
Arrow forward	Internet Browser, My Computer		Starting a car	Forwardarrow	2,15
Arrow up	My Computer, Explorer	[..]	Car engine RPM increasing	Uparrow	2,68
Re-read, Re-load actual page	Internet Browser	,	Breaking a car and start-up	Reread	2,31
Typing, entering URL address	Internet Browser	http:// http://	The sound of typing on a keyboard	Keyboard	1,46
Open new /close Browser Window	Internet Browser		Opening and closing sound of a wooden window	Window_open Window_close	1,9
Search/ find text on this screen	Internet Browser, E-mail, Documents	,	Seeking and searching with human voice (loop)	Search_(loop)	1,87
Save link or image	Internet Browser		Spearcon	S_SaveImageAs S_SaveLinkAs	2,46
Bookmark Favorites	Internet Browser	,	Turning the pages of a book with human sound	Book	1,99
Printing (action in progress)	Everywhere		Sound of a dot-matrix printer	Print	1,2
Cut	Documents, My Computer, Browser		Cutting with scissors	Cut	1,11
Paste	Documents, My Computer, Browser		Painting with a brush, whistle and can chatter	Paste	2,46

Table 6. (*Continued*)

Copy	Documents, My Computer, Browser		Sound of a copy machine	Copy_(loop)	1,57
Move	Documents, My Computer, Browser		Wooden box pushed with human struggling sound or	Move1	2,0
			cutting with a scissor and pasting with a brush	Move2	2,3
Delete	Documents, My Computer, Browser		Flushing the toilet	Delete	1,32
New Folder…	My Computer		Spearcon	S_New	
New mail, create/ compose new message	E-mail		Breathing and stamping	Composemail	2,25
Reply to a mail	E-mail		Breath and stamp (once)	Replymail	2,49
Forward mail	E-mail		Movement of paper on a desk	Forwardmail	2,74
Save mail	E-mail		Sound of save and bicycle bells	SaveMail	2,18
Send mail	E-mail		Bicycle bell and bye-bye sounds	Sendmail	1,99
Address-book	E-mail		Turning the pages with human sound	Book	1,97
Attachment to a mail	E-mail		Stapler	Attach	1,32
Open	Documents, Files		Zip fly up or opening beer can	Zip_up	1,22
				Beer_up	1,43
Save	Documents, Files		Locking a door with keys	Save	1,72
Save as…	Documents, Files		Locking a door with keys with human "hm?"	Save_as	1,88

Table 6. (*Continued*)

Close	Documents, Files		Zip fly down or squeezing beer can	Zip_down Beer_down	1,56 1,82
Rename	Documents, Files		Spearcon	S_Rename	
Restore from the recycle bin	Recycle bin		Original "paper sound" of MS Windows and human caw	Recycleback	2,0
Empty recycle bin	Recycle bin		Original sound of MS Windows (paper sound)	Recycle	1,53
New Document (create)	Documents		Spearcon	S_New	
Text formatting tools	Documents	F D A	Spearcons	S_Fontsize S_Formatting S_Bold S_Italic S_Underline S_Spelling	
Mark /select (text)	Documents, Browser, E-mail		Sound of magic marker pen	Mark	1,82

Opening and closing is very important for almost every application. As mentioned earlier, the sounds have to be somehow related to opening and closing something and they have to be in pairs. The most popular version was a zip fly of a trouser to open up and close. The same sound that was recorded for opening was used for closing as well: it is simply reversed playback. The increasing and decreasing frequency should deliver the information. The other sample is opening a beer can and squeezing it in case of closing.

Based on the mean values a total mean value of 1,86 can be calculated (the lower the point the better the sound is). The best values are as low as 1,1-1,5. Only two sounds have worse results than 2,5. This indicates a successfully designed sound set for these functions. A comparison between languages showed only little differences. An explanation phase regarding thematically grouped sounds helped the users to associate the events with the sounds, so this resulted in better ranking points.

4.4 Auditory Emoticons

Table 7 contains the auditory emoticons together with the visual representations. Smileys have the goal of representing emotional content using a few keystrokes and as a result some of them appear to be similar. As smileys try to encapsule emotions in an easy but limited (graphical) way, the auditory emoticons also try the same using a brief sound. As in real life, some of them express similar feelings. In summary, it can be said that auditory emoticons:

- reflect emotional status of the speaker
- are represented always with human sounds, non-verbal and language independent
- can also contain other sounds, noises etc. for a deeper understanding.

Table 7. Collection of the most important emoticons. Sound samples for female and male versions can be found under the given names.

Auditory Emoticon	Visual Representation	Description	Filename (Female)	Filename (Male)
Smile	☺, :-), :),	chuckle	Smile_f	Smile_m
Laughter	:-D	laughing	Laugh_f	Laugh_m
Wink	;-)	Short "sparkling" sound and chuckle	Wink_f	Wink_m
Mock (tongue out)	:-P	Typical sound of tongue out	Tongue_f	Tongue_m
Surprise	:-o	"oh"	Surprise_f	Surprise_m
Anger	,	"grrrrrrr, uuuhhh"	Anger_f	Anger_m
Perplexed, distracted	:-S,	"hm, aaahhh"	Puzzled_f	Puzzled_m
Shame, "red face"		"iyuu, eh-eh"	Redface_f	Redface_m
Sadness, sorry	☹, :-(, :(,	A sad "oh"	Sad_f	Sad_m
Crying, whimper		Crying	Cry_f	Cry_m
Kiss	:-*,	Sound of kiss on the cheek	Kiss_f	Kiss_m
Disappointment	:-I,	"oh-hm"	Dis_f	Dis_m

Although there is no scientific evidence that some emotions can be represented better by a female voice than by a male voice, we observed that subjects prefer the female version for smiling, winking, mocking, crying and kissing. Table 7 contains both female and male versions. Users especially welcomed these emoticons.

4.5 Presentation Methods

All the auditory representation presented above can be played back in the following ways:

In a direct mapping between a visual icon or button: the sound can be heard as the cursor/mouse is on the icon/button or it is highlighted, and the auditory event helps first of all the blind users to orientate (to know where they are on the screen).

During an action in progress, e.g. during copying, deleting, printing etc. in loop.

After an action is finished and completed as a confirmation sound.

The sounds have to be tested further to find which presentation method is the best for a given action and sound. It is possible that the same sound can be used for both:

e.g. first, the sound is played back once as the cursor is on the button "back arrow", and after clicking, the same sound can be played back as a confirmation that the previous page is displayed.

4.6 Spearcons

Spearcons, as a version of speeded up speech, were introduced to the Hungarian and German blind and sighted users as well. A MATLAB routine was used to compress original recordings of Hungarian and German words and expressions related to computer usage. Table 8 shows some of the spearcons (here translated in English), duration of original and compressed samples and the compress ratio. Different resolutions of original recordings were tried, from 8 bits to 16 bits and from 8000 Hz to 48000 Hz sampling frequency. Furthermore, the final evaluation regarding the quality of spearcons includes native English speakers and TTS versions as well.

Table 8. List of services and features for Hungarian spearcons introduced to blind users. The length and compress ratio is also shown. Original recording was made by a male speaker in 16 bit, 44100 Hz resolution using a Sennheiser ME62 microphone.

Spearcon	Duration (original) [sec]	Duration (compressed) [sec]	Compress ratio [%]
Close	0,87	0,302	65,3
Open	0,812	0,288	64,5
Save	0,687	0,257	62,6
Save as	1,125	0,362	67,8
Search	0,694	0,258	62,8
Copy	0,818	0,289	64,7
Move	0,748	0,272	63,6
Delete	0,661	0,25	62,2
Print	0,752	0,273	63,7
Download	0,853	0,298	65
Stop	0,908	0,311	65,8
Word	0,576	0,228	60,4
Excel	0,599	0,234	60,9
My Computer	0,805	0,286	64,5
Start Menu	0,734	0,268	63,5
Browser	0,845	0,296	65
E-Mail	0,545	0,22	59,6

Spectral evaluation of the spearcons showed that 16-bit resolution and at least 22050 Hz sampling frequency is required. Using 44100 Hz is actually recommended to avoid noisy spearcons [31]: compression has effect on the frequency regions at 4-5 kHz and 16 kHz, so decreasing of the sample frequency or resolution (bit depth) results in a noisy spectrum. A text-to-speech application (SpeakBoard) was also used to save wave files, but listeners preferred original recordings of a human speaker.

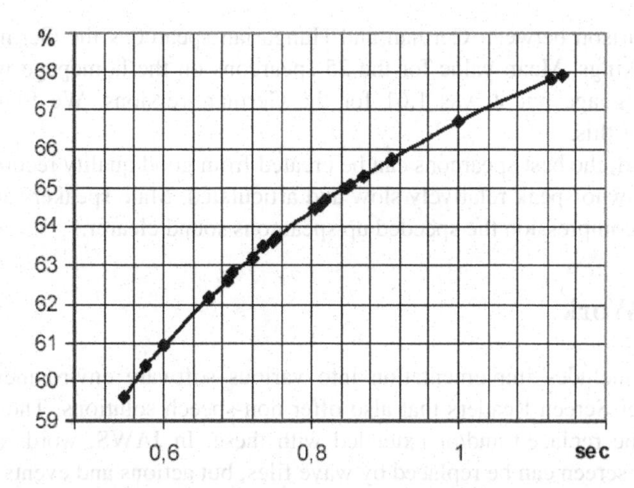

Fig. 3. Compression rates (%) as function of duration of the original sample (sec)

The compression ratio is almost linear from 59% to 68% of the duration of the original sample: the longer the sample the higher the compression (Figure 3.). It is always recommended to truncate the samples before compression to remove unnecessary silence at start.

For German spearcons we recorded four male native speakers. One set was accent-free, while the other speakers had typical German accents (Saxonian, Bavarian, Frankonian). A current investigation is examining the effects of different accents for German spearcons. All spearcons are made from original recordings in an anechoic chamber using Adobe Audition software and Sennheiser microphones. The Hungarian database was recorded by a native male speaker of 33 years of age. The databases contain 35 words (spearcons) respectively but on the homepage there are 25 for evaluation.

We observed that longer words (having more vowels) are harder to understand after creating the spearcons. Longer sentences (more than 3-4 words) become unintelligible after compression, so this method is not suited for creating spearcons longer than 1-2 words. Although it is not required to understand the spearcon, subjects preferred those they have actually understood. Independent of the fact, whether a spearcon was used or not, all were tested and judged by the subjects. All spearcons were played back in a random order. A spearcon could be identified and classified as follows:

- the subject has understood it the first time,
- the subject could not understand it, and he had a second try,
- if the subject failed twice, the spearcon was revealed (the original recording was shown) and a final try was made.

The evaluation showed that only 12% of the spearcons were recognized on the first try. It was interesting that there was no clear evidence and benefit for using accent-free spearcons: e.g. recognition of the spearcon sometimes was better for the Saxonian version (across all German speakers). Blind persons tend to be better in this task than sighted persons.

In a comparison between German and Hungarian spearcons the German versions got better rankings. Mean value for the 25 spearcons on the homepage was 2,07 for Hungarian language but it was 1,61 for the German versions. We found no clear explanation for this.

Summarized, the best spearcons can be created from good quality recordings of native speakers, who speak relatively slow and articulated. Male speakers are preferred because after compression the speeded up spearcons sound clearer.

5 Future Work

Future work includes implementation into various software environments such as JAWS or other Screen Readers that also offer non-speech solutions. The pre-defined samples can be replaced and/or extended with these. In JAWS, words and phrases written on the screen can be replaced by wave files, but actions and events usually can not be mapped with external sound files.

Furthermore, a MS Windows patch or plug-in is planned (in Kernel-level or using the Microsoft Automation or another event logger). This executable file can be downloaded, extracted and installed. It will include a simple graphical user interface with check-boxes for activate and deactivate the sounds and simple environmental settings (e.g. auto start on start-up, default values etc.) and all of the default sound samples, probably in mp3 format.

6 Summary

Fifty blind and hundred users with normal vision participated in a survey in order to determine the most important and frequently used applications, and furthermore, to create and evaluate different auditory representations for them. These auditory events included auditory icons, earcons and spearcons of German and Hungarian language. The German spearcon database contains original recordings of a native speaker and samples with different accents. As a result, a new class of auditory events was introduced: the auditory emoticons. These represent icons or events with emotional content, using non-speech human voices and other sounds (laughter, crying etc). The previously selected applications, programs, function, icons etc. were mapped, grouped thematically and some sound samples were evaluated based on subjective parameters. In this paper the "winning" sound samples were collected and presented. Based on the mean ranking points and informal communications, both target groups liked and welcomed the idea and representation method to extend and/or replace the most important visual elements of a computer screen. This is mostly true for environmental sounds; spearcons are only interesting for blind users in menu navigation tasks, because the screen-reader software offers speeded up speech already. However, becoming an expert user and benefit from all these sounds requires some accommodation and learning time and a guiding explanation or FAQ can ease this process.

References

1. Boyd, L.H., Boyd, W.L., Vanderheiden, G.C.: The Graphical User Interface: Crisis, Danger and Opportunity. Journal of Visual Impairment and Blindness, 496–502 (December 1990)
2. http://www.freedomscientific.com/fs_products/software_jaws.asp last viewed (December 2009)
3. http://www.gwmicro.com/Window-Eyes/ last viewed (December 2009)
4. Mynatt, E.D.: Transforming Graphical Interfaces into Auditory Interfaces for Blind Users. Human-Computer Interaction 12, 7–45 (1997)
5. Crispien, K., Petrie, H.: Providing Access to GUI's Using Multimedia System – Based on Spatial Audio Representation. J. Audio Eng. Soc. 95th Convention Preprint, New York (1993)
6. Nees, M.A., Walker, B.N.: Encoding and Representation of Information in Auditory Graphs: descriptive reports of listener strategies for understanding data. In: Proc. of the 14th International Conference on Auditory Display (ICAD 2008), Paris, p. 6 (2008)
7. Nees, M.A., Walker, B.N.: Listener, Task, and Auditory Graph: Toward a Conceptual Model of Auditory Graph Comprehension. In: Proc. of the 13th International Conference on Auditory Display (ICAD 2007), Montreal, pp. 266–273 (2007)
8. Gaver, W.W.: The SonicFinder, a prototype interface that uses auditory icons. Human Computer Interaction 4, 67–94 (1989)
9. Mynatt, E.D.: Designing Auditory Icons. In: Proc. of the International Conference on Auditory Display (ICAD 1994), Santa Fe, pp. 109–120 (1994)
10. Petrie, H., Morley, S.: The use of non-speech sounds in non-visual interfaces to the MS Windows GUI for blind computer users. In: Proc. of the International Conference on Auditory Display (ICAD 1998), Glasgow, p. 5 (1998)
11. Wersényi, G.: Localization in a HRTF-based Minimum Audible Angle Listening Test on a 2D Sound Screen for GUIB Applications. J. Audio Eng. Soc. 115th Convention Preprint, New York (2003)
12. Wersényi, G.: Localization in a HRTF-based Minimum-Audible-Angle Listening Test for GUIB Applications. Electronic Journal of Technical Acoustics 1 (EJTA), 16 (2007), http://www.ejta.org
13. Wersényi, G.: What Virtual Audio Synthesis Could Do for Visually Disabled Humans in the New Era. AES Convention Paper, presented at the AES Tokyo Regional Convention, Tokyo, Japan, pp. 180–183 (2005)
14. Wersényi, G.: Localization in a HRTF-based Virtual Audio Synthesis using additional High-pass and Low-pass Filtering of Sound Sources. Journal of the Acoust. Science and Technology 28(4), 244–250 (2007)
15. Wersényi, G.: Effect of Emulated Head-Tracking for Reducing Localization Errors in Virtual Audio Simulation. IEEE Transactions on Audio, Speech and Language Processing (ASLP) 17(2), 247–252 (2009)
16. Wersényi, G.: Simulation of small head-movements on a virtual audio display using headphone playback and HRTF synthesis. In: Proc. of the 13th International Conference on Auditory Display (ICAD 2007), Montreal, pp. 73–78 (2007)
17. Gaver, W.W.: Auditory Icons: using sound in computer interfaces. Human-Computer Interactions 2(2), 167–177 (1986)
18. Blattner, M.M., Sumikawa, D.A., Greenberg, R.M.: Earcons and Icons: Their structure and common design principles. Human-Computer Interaction 4, 11–44 (1989)

19. Gaver, W.W.: Everyday listening and auditory icons. Doctoral thesis, Univ. of California, San Diego (1988)
20. Gygi, B., Shafiro, V.: From signal to substance and back: insights from environmental sound research to auditory display design. In: Proc. of the 15th International Conference on Auditory Display (ICAD 2009), Copenhagen, pp. 240–251 (2009)
21. Gygi, B.: Studying environmental sounds the watson way. The Journal of the Acoustical Society of America 115(5), 2574 (2004)
22. Gygi, B., Kidd, G.R., Watson, C.S.: Spectral-temporal factors in the identification of environmental sounds. The Journal of the Acoustical Society of America 115(3), 1252–1265 (2004)
23. Ballas, J.A.: Common factors in the identification of an assortment of brief everyday sounds. Journal of Exp. Psychol. Human 19(2), 250–267 (1993)
24. Gygi, B., Shafiro, V.: The incongruency advantage in elderly versus young normal-hearing listeners. The Journal of the Acoustical Society of America 125(4), 2725 (2009)
25. Fernström, M., Brazil, E.: Human-Computer Interaction design based on Interactive Sonification – hearing actions or instruments/agents. In: Proc. of 2004 Int. Workshop on Interactive Sonification, Bielefeld Univ. (2004)
26. Heller, L.M., Wolf, L.: When Sound Effects Are Better Than The Real Thing. The Journal of the Acoustical Society of America 111(5/2), 2339 (2002)
27. Vargas, M.L.M., Anderson, S.: Combining speech and earcons to assist menu navigation. In: Proc. of the International Conference on Auditory Display (ICAD 2003), Boston, pp. 38–41 (2003)
28. Walker, B.N., Nance, A., Lindsay, J.: Spearcons: Speech-based earcons improve navigation performance in auditory menus. In: Proc. of the International Conference on Auditory Display (ICAD 2006), London, pp. 63–68 (2006)
29. Palladino, D.K., Walker, B.N.: Learning rates for auditory menus enhanced with spearcons versus earcons. In: Proc. of the 13th International Conference on Auditory Display (ICAD 2007), Montreal, pp. 274–279 (2007)
30. Dingler, T., Lindsay, J., Walker, B.N.: Learnabiltiy of Sound Cues for Environmental Features: Auditory Icons, Earcons, Spearcons, and Speech. In: Proc. of the 14th International Conference on Auditory Display (ICAD 2008), Paris, p. 6 (2008)
31. Wersényi, G.: Evaluation of user habits for creating auditory representations of different software applications for blind persons. In: Proc. of the 14th International Conference on Auditory Display (ICAD 2008), Paris, p. 5 (2008)
32. Wersényi, G.: Evaluation of auditory representations for selected applications of a Graphical User Interface. In: Proc. of the 15th International Conference on Auditory Display (ICAD 2009), Copenhagen, pp. 41–48 (2009)
33. `http://www.w3.org/` (Last viewed December 2009)
34. `http://www.independentliving.com/prodinfo.asp?number=CSH1W` (Last viewed December 2009)
35. Cobb, N.J., Lawrence, D.M., Nelson, N.D.: Report on blind subjects' tactile and auditory recognition for environmental stimuli. Journal of Percept. Mot. Skills 48(2), 363–366 (1979)
36. `http://guib.tilb.sze.hu/` (last viewed December 2009)
37. `http://www.freesound.org` (last viewed December 2009)
38. `http://www.soundsnap.com` (last viewed December 2009)
39. Gygi, B., Divenyi, P.L.: Identifiability of time-reversed environmental sounds. Abstracts of the Twenty-seventh Midwinter Research Meeting, Association for Research in Otolaryngology 27 (2004)

PhysioSonic - Evaluated Movement Sonification as Auditory Feedback in Physiotherapy

Katharina Vogt[1], David Pirrò[1], Ingo Kobenz[2], Robert Höldrich[1], and Gerhard Eckel[1]

[1] Institute for Electronic Music and Acoustics, University of Music and Performing Arts Graz, Austria
[2] Orthopaedic Hospital Theresienhof, Frohnleiten, Austria
vogt@iem.at, pirro@iem.at, hoeldrich@iem.at, kobenz@theresienhof.at, eckel@iem.at

Abstract. We detect human body movement interactively via a tracking system. This data is used to synthesize sound and transform sound files (music or text). A subject triggers and controls sound parameters with his or her movement within a pre-set range of motion. The resulting acoustic feedback enhances new modalities of perception and the awareness of the body movements. It is ideal for application in physiotherapy and other training contexts.

The sounds we use depend on the context and aesthetic preferences of the subject. On the one hand, metaphorical sounds are used to indicate the leaving of the range of motion or to make unintended movements aware. On the other hand, sound material like music or speech is played as intuitive means and motivating feedback to address humans. The sound material is transformed in order to indicate deviations from the target movement.

PhysioSonic has been evaluated with a small study on 12 patients with limited shoulder mobility. The results show a clear benefit for most patients, who also report on PhysioSonic being an enrichment of their therapeutic offer.

1 Introduction

Body movements do not cause other feedback than given by the proprioception, the usually unconscious perception of movements and spatial orientation. Additional feedback is helpful in different disciplines of motor learning and physical exercise. In training, sports men and women or patients of orthopaedic rehabilitation are usually limited to visual feedback. Acoustic feedback has been used, but was limited to partial data or indirect measurements until recently, when real-time full-body tracking became possible.

An optimal learning and training process depends on the permanent comparison of an actual value to the desired value, in the sense of a closed-loop feedback. In learning of motor skills or re-learning them in the course of rehabilitation, the learner or patient needs information on his or her distance to a (therapy) goal

S. Ystad et al. (Eds.): CMMR/ICAD 2009, LNCS 5954, pp. 103–120, 2010.

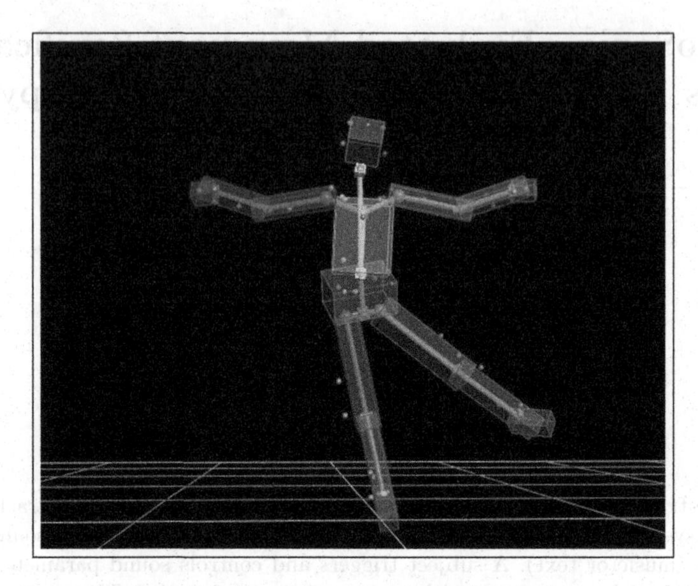

Fig. 1. Example of a full-body tracking, used in EGM [22] project at the IEM

and on the quality of the movement. This feedback can be inherent information, intrinsically coming from the human proprioception. But athletes or rehabilitation patients often require augmented, extrinsic feedback, which is usually given by the patient or the therapist. The learner then needs to know about quantitative and qualitative movement features the knowledge of results vs. the knowledge of performance. With this background it is possible to compare the actual to the desired value, and achieve a continuous performance increase. The therapy tool as presented in this paper allows the patient to access both information channels, the knowledge of results and the knowledge of performance. The feedback is in real-time, exact and without the subjective filter of the therapist or trainer.

In this paper we describe a sonification system for auditory feedback of different movements of the human body. We will shortly summarize the state of the art and outline the novelties in our approach. A technical and medical background will describe the system in more detail for an application in physiotherapy. Shoulder patients have difficulties in lifting their arms laterally in the coronal plane. For this context, several training scenarios have been developed. A pilot test was performed, where the training gain of our sonification system was tested.

2 State of the Art

We found different applications of sonification of movement data. Some focus on sports and physiotherapy, some serve as a basis for artistic research. Next to pure sonification approaches, the International Society for Virtual Rehabilitation [2] does research on virtual reality assisting in rehabilitation and for disabled people. See also the end of this section.

In sports, approaches to sonification vary largely. There is, e.g., a training system to learn the perfect golf swing supported by auditory feedback [3], [20]. Sonification is also used to analyse the dynamics of individuals interacting in a sport's team [18].

Several sonifications based on human motion data in sports were conducted by Effenberg et al., see [12].

In physiotherapy, one approach was the sonification of EMG (Electromyography), data on the electric potential of muscle cells [21].

A direct motion-sonification system is AcouMotion [14], thought for application in various disciplines from sports to physiotherapy. An accelerometer is used, for instance representing a virtual racket, allowing for training of *blindminton*, a game for people with visual impairment [16]. In a different application, a pilot study was conducted on the learning effects of people throwing a virtual object, when the thrower was supported with visual or auditory feedback or both. The latter lead to best results [17]. The change of proprioception of the test persons has been refered to augmented reality in this context [25].

Within the SonEnvir project [4], the movement of juggling clubs in relation to a juggling person was tracked with a motion tracking system and sonified in real-time [9]. The goal was to provide an additional feedback for the juggler and elaborate a multimedia performance. The same tracking system is used in the research project EGM - Embodied Generative Music [22]-, where the dancers are tracked directly. An interactive sonification of these movements allows to study the change of the proprioception of the dancers exhibited with new auditory feedback. Also, aesthetical interrelations between sound and movement is object of research in the project.

In MotionLab [13], kinematic data of a walking person is gathered with a motion tracking system and deduced quantities, as the expenditure of energy calculated. These data are sonified in order to analyse the movement. At Stanford university, motion tracking is used in different contexts ranging from dance and instrument playing to sports in a multidisciplinary course [5].

In general, the linking of movement to sound or other modalities is captured by the concept of aesthetic resonance in virtual rehabilitation. It is applied in a variety of cases, and can be classified as musculo-skeletal, post-stroke and cognitive Virtual Rehabilitation [11]. Classical rehabilitation, on the one hand, has disadvantages: it is repetitive, thus sometimes boring, provides few and often subjective data, is costly as a one-to-one treatment and it cannot be monitored at home. Virtual rehabilitation, on the other hand, has as a major advantage economy of scale. Furthermore, it is interactive and motivating and provides objective and transparent data. The main drawbacks, as cost-intensive equipment and disturbance of interfaces, that have not been developed for a therapeutic application, become less and less important with technological development.

The proceedings of the International Conference on Disability, Virtual Reality and Associated Technologies gives testimony of a growing number of applications in this field. As one example focussing on sound [10], the EU project CARESS [1] studied the interaction of children of various ability and disablity with sound.

3 PhysioSonic

Many of the pure sonification approaches refered to in section 2 use abstract sounds. Their meaning has to be learned by the subjects or the sonification is only meant for movement analysis by trained staff. Usually, the exact repetition of a movement leads to the exactly same sound. In a training context, this may cause fatigue or annoyance of the listener.

In PhysioSonic, we focus on the following principles:

- The tracking is done without a disturbance of the subject, as markers are fixed to an overall suit or the skin directly (and the subject does not, e.g., have to hold a device).
- In collaboration with a sport scientist, positive (target) and negative (usually evasive) movement patterns are defined. Their range of motion is adapted individually to the subject. Also a whole training sequence can be pre-defined.
- The additional auditory feedback changes the perception including the proprioception of the subject, and the movements are performed more consciously. Eyes-free condition for both the subject and an eventual observer (trainer) free the sight as an additional modality. In addition to the target movement, pre-defined evasive movements are made aware.
- We sonify simple and intuitive attributes of the body movement (e.g., absolute height, velocity). Thus the subject easily understands the connection between them and the sound. This understanding is further enhanced by using simple metaphors: e.g., a spinning-wheel-metaphor keeps the subject

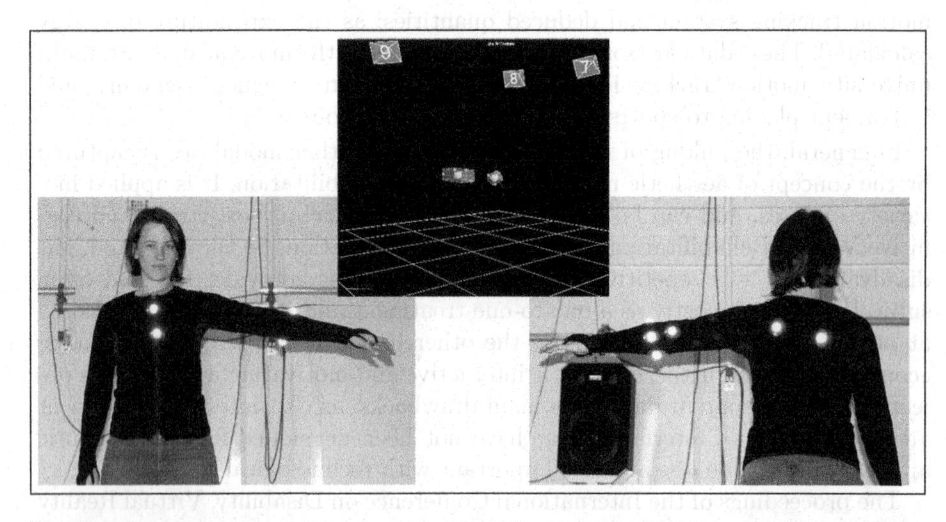

Fig. 2. The placement of 8 markers: 4 define the thorax and 4 the arm. The tracking system calculates location and rotation of each object. The screenshot shows the tracked markers grouped as these objects by the software. The numbers on top refer to some of the cameras.

moving and thus *'pushing'* the sound; if the subject does not accumulate enough *'energy'*, the looping sound of a disk-getting-stuck metaphor is used.
– The sounds are adapted to each subject. Music and spoken text that are used as input sounds can be chosen and thus enhance the listening pleasure. Also, the input sounds change over time and have a narrative structure, where the repetition of a movement leads to a different sound, thus avoiding fatigue or annoyance. (Still, the sonification design is well defined, and the quality of the sound parameters changes according to the mapping.)

3.1 Technical Background

Our sonification system was developed at the Institute of Electronic Music and Acoustics in Graz, Austria, in the CUBE [27]. Next to full audio equipment, this space is wired up with the VICON motion tracking system [7]. It allows spatial resolution in the millimeter range and a temporal resolution and reaction time below 10 ms: therefore the high quality of the tracking ensures that all relevant aspects of the body movement will be captured. This is the key feature that allows us to use this data to drive the sound synthesis and sound manipulation.

The 15 infrared cameras (see Fig.3) which are part of the system are sampled at 120 frames per second and are equipped with a ring of infrared emitting leds. They are installed at ca. 3 meters height in order to provide a homogeneous tracking quality within a volume of approximately 6 meters diameter and 3 meters height. In this space, every object or person with infrared reflecting markers is clearly visible to the cameras.

The video data coming from the cameras is collected by a data station and then processed in the VICON software which reconstructs the coordinates and orientation of the objects, representing different parts of the body.

The system has now also been installed at the Orthopaedic Hospital Theresienhof in Frohnleiten, Austria. The hospital has an own tracking system with 6 cameras.

Fig. 3. Infrared camera of the VICON motion tracking system

In the example that will be described in Sec. 4, we worked with a specifically designed marker set which defines single objects tracked in the space (see Fig. 2). But the system allows also a full-body tracking mode in which the software reconstructs the whole skeleton of the test person providing 19 reliably tracked body segments from head to toe using inverse kinematics fitting in real-time (see Fig.1). The tracking data, including position and orientation data either in absolute or relative form, is then streamed via the *Vicon2OSC* [8] program developed at the IEM to a SuperCollider[6] client. Developing the mapping of the tracking data to the audio synthesis control parameters, we make use of SuperCollider classes and operators that handle tracking data and were implemented in the context of the EGM [22] project.

The system allows a very detailed continuous temporal and spatial control of the sound produced in real time. We establish thereby a closed-loop feedback between the subject's movements and the sound produced.

3.2 Generic Model for Movement Sonification

A block diagram of the sonification is shown in Fig. 4. A specific example of the sonification design for shoulder rehabilitation is described in Section 4.4. A movement model is defined, assessing movement patterns and an optional training sequence. Sound material and synthesis are chosen. The movement model is adjusted individually and used to evaluate the data of the tracking system. This data is used for parameter mapping, the samples are transformed, sound is synthesized and/or spatialized.

Development of movement and sonification models. For a certain training goal, *positive* (target) and *negative* (evasive) movement patterns are defined. Optionally, also an evolution over time can be specified, supporting a training sequence. Sounds are classified as motivating or inhibitory. The sound synthesis is defined and sound samples, music or speech files, are loaded.

Individual Adjustment. The model parameters are adjusted to the subject (stationary position, range of motion, training goal, etc.). The aesthetic taste of the subject is taken into account in the sample selection.

Movement evaluation and auditory feedback. The motion tracking system captures the movement data, and reconstructs the body parts as objects, using a body model, in real-time. This data is evaluated on the basis of the movement model, defined above. Inside the individually adjusted tolerance range, any deviation is ignored. Outside the range, the samples are sonified or sounds synthesized according to the parameter mapping (see below). Auditory feedback in real time enables the subject to correct the movement.

Parameter mapping. The mapping uses the sound material and transforms them or generates sounds according to the evaluated data. The sound samples are transformed by filtering or playing at a different rate. Sounds are synthesized depending on the data. Finally, a spatialization of both, sound samples and abstract sounds, gives additional cues.

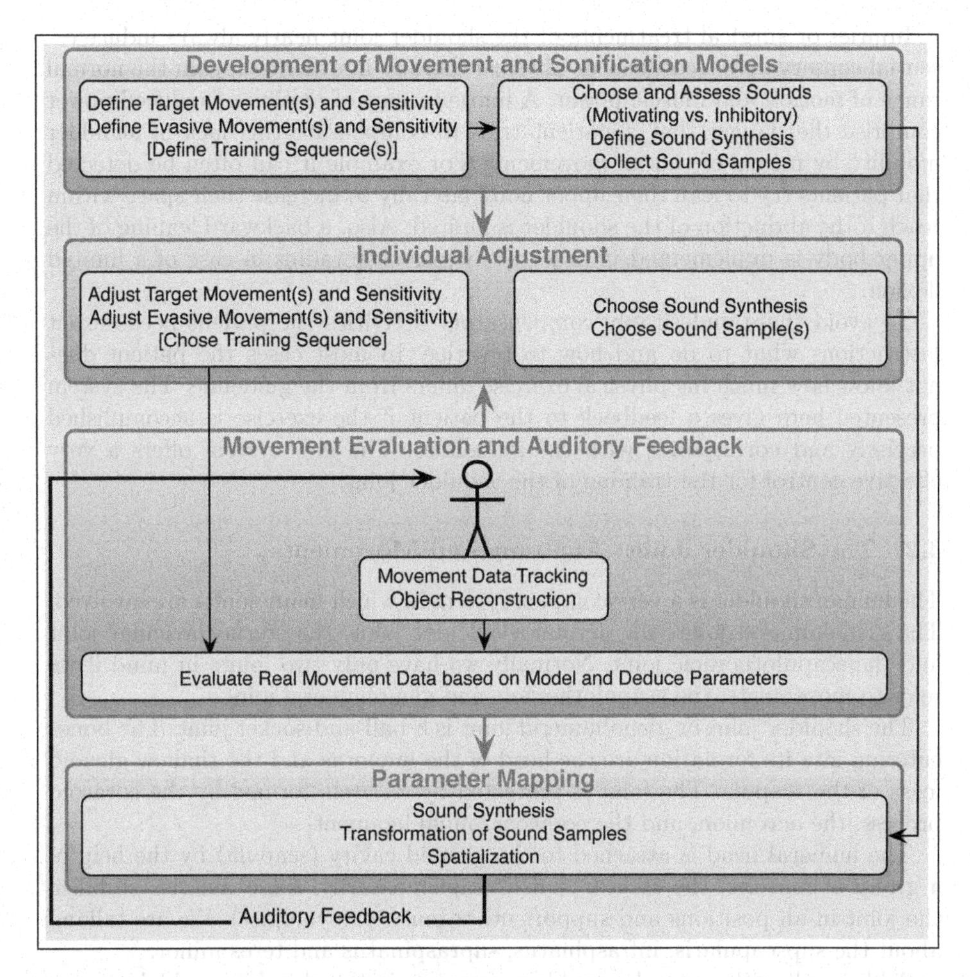

Fig. 4. Block diagram of the project design

4 Example: Shoulder

4.1 Medical Background

From a series of tests we know that movement analysis of the upper extremity by means of 3 dimensional kinematics is a very exact and objective method to quantify the joint angles of wrist, elbow or shoulder [15], [24]. Correspondingly, monitoring the functional ranges of upper limb activity could become a very important method to evaluate the improvement of a physical therapy.

The study presented in this paper describes a new way to transform the results of 3D movement analysis of the shoulder joint kinematics into a special feedback which can be used to correct false postures or coordinate a therapeutic exercise for the shoulder joint.

Injuries or surgical treatments of the shoulder joint nearly always induce essential conservative treatment or physical therapy in order to regain the normal range of motion and muscle power. A limited range of motion or reduced power comprise the danger that a patient tries to compensate this lack of shoulder mobility by means of thorax movements. For example it can often be detected that patients try to lean their upper body laterally to increase their space within reach if the abduction of the shoulder is limited. Also, a backward leaning of the upper body is implemented in order to increase the radius in case of a limited flexion.

To avoid these undesirably compensatory activities the patient needs exact instructions what to do and how to practice. In most cases the patient does not know how much his physical exercise differs from the guideline. The system presented here gives a feedback to the patient if the exercise is accomplished precisely and corresponds with the guidelines. The new system offers a very effective control for the training of the shoulder joint.

4.2 The Shoulder Joint: Anatomy and Movements

The human shoulder is a very complex system in which many joints are involved: The glenohumeral joint, the acromioclavicular joint, the sternoclavicular joint and the scapulothoracic joint. Normally we have only two joints in mind if we refer to movements: the scapulothoracic and glenohumeral joint.

The shoulder joint or glenohumeral joint is a ball-and-socket joint. The bones entering into its formation are the head of the humerus and the shallow glenoid fossa of the scapula. The joint is protected by an arch, formed by the coracoid process, the acromion, and the coracoacromial ligament.

The humeral head is attached to the glenoid cavity (scapula) by the help of a group of muscles, the rotator cuff. Complementarily, these muscles stabilize the joint in all positions and support other muscles (synergist). We are talking about the supscapularis, infraspinatus, supraspinatus and teres minor.

Additionally other muscles enable movements in the glenohumeral joint: teres major, subcapularis, deltoid, biceps femoris, triceps femoris, corachobrachialis. Muscles moving the scapula are trapezius and rhomboids e.g. Pectoralis major and latissimus dorsi both move the scapula as well as the humerus.

Scapular and the glenohumeral muscles enable the patient to fulfill a range of different movements:

- to lift his arm laterally (movements in the coronal plane, abduction-adduction)
- to move the arm forward (movements in the sagittal plane, flexion-extension)
- to rotate the arm around the long axis of the humerus (movements in the transverse plane, medial-lateral rotation).

There is no doubt that the quantification of these complex joint movements by means of a 3D analysis system is very difficult.

According to Rab [23] in our simplified biomechanical model which serves as the basis for this study the shoulder complex is reduced to the movement of the humerus in relation to the thorax.

4.3 A Special Movement: The Abduction

What is the therapeutical profit of this new system?

According to Kaspandji [19] the abduction can be split into 3 phases:

- Within the first phase the arm is abducted from the neutral position to 90 degree abduction. Deltoideus and supraspinatus are mainly responsible for the work. The end of this phase is the contact between tuberculum major and the scapula which limits the movement. If this abduction is combined with a small anteversion of about 30 degree the arm follows a physiological track [26]. This guideline could be transmitted to the patient by acoustic feedback. If the patient sticks exactly to the rules his effort can be controlled by the system. As a result it is possible to train certain muscles in special ranges. Resistance can be added according to demand.
- Within the second phase (from 90° to 150°) the scapula must move additionally to guarantee the aspired abduction. Trapezius and serratus anterior can be trained in this phase because they are responsible for the scapula movement.
- In the last phase of abduction (150° to 180°) the thorax has to move to the contralateral side. If this movement is detected within phase I or II it could be signalled with the help of the feedback system. The signal should encourage the patient to increase the space within reach by an enhanced movement of the shoulder joint.

A series of informal analyses proved that the new system is an excellent means to guide patients during their training program. By the help of acoustic feedback the patient practises within the allowed extent of abduction and does not exceed beyond the desired range of motion. A pilot study will be set up to judge on the efficiency of this new training method vs. traditional ones.

4.4 Sonification Design and Metaphors

As discussed in the previous sections, in orthopaedic rehabilitation patients train shoulder abduction and adduction. In the training sequences it is important that the subjects do the required movements without inclining the thorax too far (see Fig.5a). Another requirement is that the movement is done laterally in

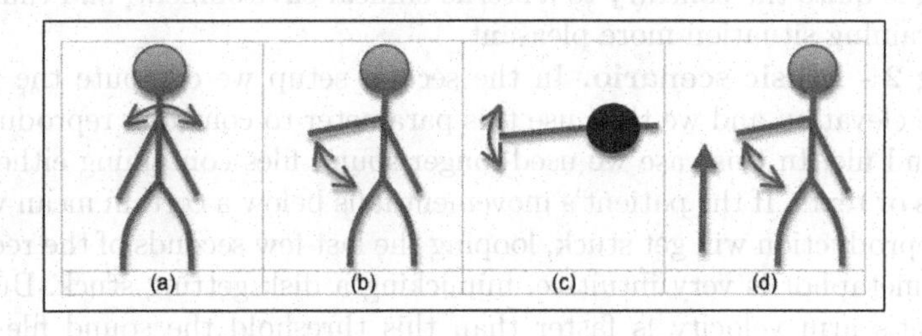

Fig. 5. Relevant movement patterns

the coronal plane, also inside a certain range of tolerance (see Fig.5c, seen from above). Given these constrains we concentrated in designing two specific training scenarios.

1. In the first one the elevation range of the arm is divided into a number of previously defined subject-specific ranges. The aim for the patient is to reach each of these ranges in sequence (see Fig.5b). The therapist may adjust the posture with the patient and give him/her a feeling of the right position.
2. In the second training sequence the patient is required to repeat the abduction and adduction of the arm continuously within a pre-defined velocity range and possibly also reaching a specific minimal height (see Fig.5d). Thus s/he can perform a training independently.

In order to capture the movements' features that are relevant in these two scenarios we used 8 markers grouped in 2 sets defining 2 different objects (see Fig.2). Thus from the motion tracking system we receive position and rotation information of thorax and arm. The first allows us to detect whenever the patient tilts the thorax and in which direction. The arm position and orientation give us the possibility to compute the elevation of the arm relative to the thorax. Furthermore, we use the azimuth value of the arm to show whenever the movement of the patient leaves the required plane for the exercise. For all different movements, ranges of motion are defined with a certain tolerance region in which deviations are ignored. All data is relative, thus independent of the absolute position in the room.

Training 1 - Wood scenario. As described above, in the first training sequence, we divided the elevation range of the arm into slices. To each of these slices corresponds a particular sound sample which is then looped. When the patient moves the arm higher or lower into another elevation range another sound is triggered. We experimented with this setup using natural sound textures. The lowest range corresponds to the sound of a rustling leaves, in the next slice we find animal sounds, croaking frogs, grunting pigs and chirring crickets, then birds' songs and, in the highest range, the sound of the wind through the trees (see Fig.6). This sound was chosen, because it allows a very intuitive estimation of the height of the arm, and motivates the patient to *reach*, e.g., the birds. Also it reminds to a wooden surrounding, which is quite the contrary to a sterile clinical environment, and thus makes the training situation more pleasant.

Training 2 - Music scenario. In the second setup we compute the velocity of the elevation and we then use this parameter to control a reproduction of a sound file. In this case we used longer sound files containing either music pieces or texts. If the patient's movement falls below a certain mean velocity, the reproduction will get stuck, looping the last few seconds of the recording. This metaphor is very intuitive, mimicking a disk getting stuck. But if the patient's arm velocity is faster than this threshold the sound file will be reproduced normally allowing to hear it to the end.

Fig. 6. Training scenario 1: Forestal wildlife as a pleasant and simple metaphor to absolute heights of the arm. The different animal sounds overlap and give a continuous impression.

Another metaphor that we want to implement is the spinning wheel: if the sound is not pushed enough, e.g. by reaching a pre-defined height, the playback slows down, i.e., the pitch is lowered, and the sound eventually stops.

Evasive movements - thorax inclination. Looking at the evasive movement of the thorax (see Fig.5a), first we decided to use synthesized white noise to sonify the inclination. The volume of this noise is coupled to the computed displacement and thus becomes louder when bending away from the correct position. The direction of the displacement (front, back, left, right) is taken into account by the spatialisation of the sound i.e. having the produced sound come from a loudspeaker placed in the same direction as the evasive movement. That means when bending forward the noise will come from a loudspeaker placed in front of the patient. Next to simple masking white noise, we used a creaking sound as a metaphor to underline the negative quality of the evasive movement. Noise or creaking can be chosen (or muted) and are then played together with the training scenarios.

Evasive movements - arm evasion. Finally we take into account the case in which the patient's arm leaves the optimal movement plane we defined above (see Fig.5c). When the arm moves more to the front or to the back of the patient, the sounds produced according to the elevation or velocity of the arm will be strongly filtered thus becoming more unrecognizable the farther the arm is from the desired movement plane.

All metaphoric sounds and sonification designs described above allow to hear the correct sound only with the correct movement. Else, the music or text files are masked (by noise or creaking), filtered, get stuck and loop, or slow down and loose their pitch. All this destroys the original *gestalt* and is very easy to perceive also for non-trained people.

Both the music and text files and the ambient sounds can be chosen by the patient. Until now, we provide a set of different music styles ranging from J.S. Bach to Paolo Conte, as well as texts from Harry Potter to Elfriede Jelinek. In a later stage, the patient shall have the opportunity to bring his or her own favorites along to the therapy.

5 First Evaluation

A first study on the practicability of PhysioSonic was conducted in the orthopedic hospital Theresienhof. It investigated the influence on augmenting the shoulder abduction with in the same time stabilizing the thorax in the frontal plane. The compensatory movement of the upper limb should be avoided. The focus of the study was set on the practicability of the software package from the point of view of the therapist and the acceptance by the patients.

5.1 Test Persons

12 in-patients of the hospital took part in the study, out of which were 5 women and 7 men. These had indicated in their admission the enhancement of mobility of the shoulder joint as primary therapy goal. Criterion for exclusion of the study were situations of structurally limited mobility of the shoulder joint because of muscle or tendinous contractures or a restrained mobility caused by neuronal damages. Also patients with reduced hearing ability and lack of compliance were not taken into account.

The average age was 55,2 years (standard deviation 8,9 years). The diagnoses in particular: 4 total ruptures of the rotator cuff, 3 ruptures of the supraspinatus tendon, 2 fractures of the humerus and 1 inverse shoulder prosthesis delta. The average of the postoperative interval of these patients was about 89,4 days (std.dev = 49,9 days). 2 patients with a shoulder impingement syndrome were treated conservatively.

All patients had a higher passive mobility in the shoulder joint than the active range during therapy.

5.2 Methodology

Before and after the PhysioSonic training the active mobility of the shoulder joint was recorded with the VICON motion tracking system. For each patient, the maximal abduction, maximal flexion and the maximal outward rotation was measured before and after PhysioSonic (pre- and post-measurement). A comparison of these results was done to measure the outcome of Physiosonic. This allowed to compare the mobility in the shoulder joint. For calculating the dynamics, the upper limb model following Oxford metrics was used (see Fig. 7).

After the pre-measure the marker setup was reduced: two rigid segments were marked: thorax and upper arm (see Fig. 8). The relative movement between these segments and their absolute location and orientation in space are the

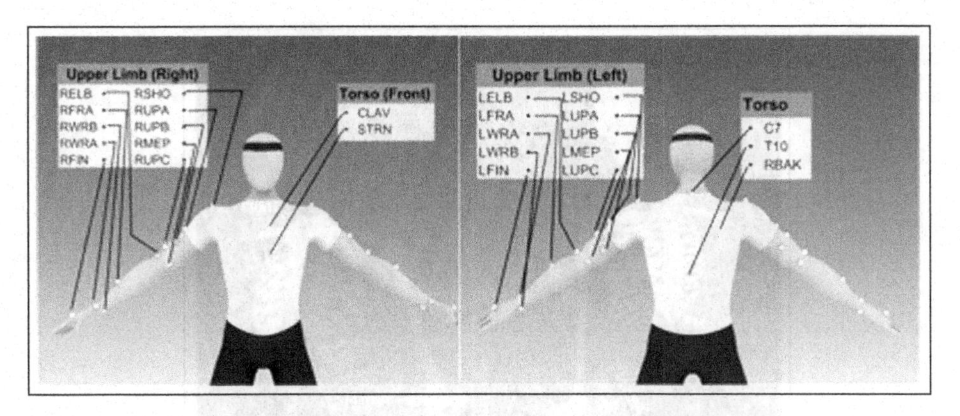

Fig. 7. Marker placement of the upper limb model (Oxford metrics)

basics for PhysioSonic (see section 4.4). The placement of the markers is described in detail in the following table:

Definition of the markers for PhysioSonic.

THORAX Segment:

Marker Label	Definition	Marker Placement
THO	Thorax Origin	Processus Xiphoideus
TH1	Thorax M1	Jugular notch manubrium
TH2	Thorax M1	Processus Spinosus C7
TH3	Thorax M1	Processus Spinosus TH10

ARM Segment:

Marker Label	Definition	Marker Placement
ARO	Arm Origin	Epicondylus lateralis
AR1	Arm M1	On the lateral upper arm
AR2	Arm M2	Epicondylus medialis
AR3	Arm M3	Most prominent part of the olecranon process

The patient then was explained the different acoustic feedbacks of PhysioSonic and how to control them. The sensitivity of the thorax parameters was adjusted to each patient with a short training. Then, the music scenario was played for continuous training, followed by the wood scenario played for 3-5 times. In the wood scenario, the parameter for the arm's elevation range was re-adjusted several times in order to find a goal range that leads to a performance increase. Then the training in the music scenario was repeated.

Finally, the measuring marker setup was re-set for the post-measurement. The placement of the markers had been color-marked in order to achieve an identic measuring situation.

The time for the pretest, the PhysioSonic therapy and the posttest was about 30 minutes, the evaluation of the data not included. Using only PhysioSonic for therapy without data analysis took about 15 minutes. Therefore, as mentioned, a very simple marker setting was used (see Fig. 8).

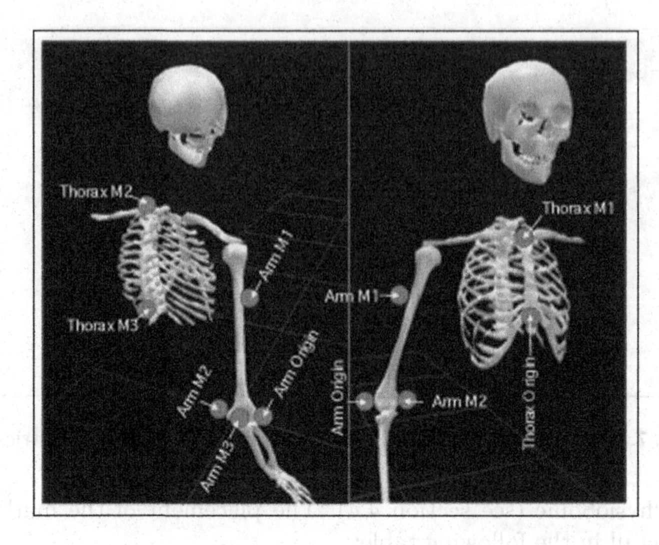

Fig. 8. Marker placement for PhysioSonic

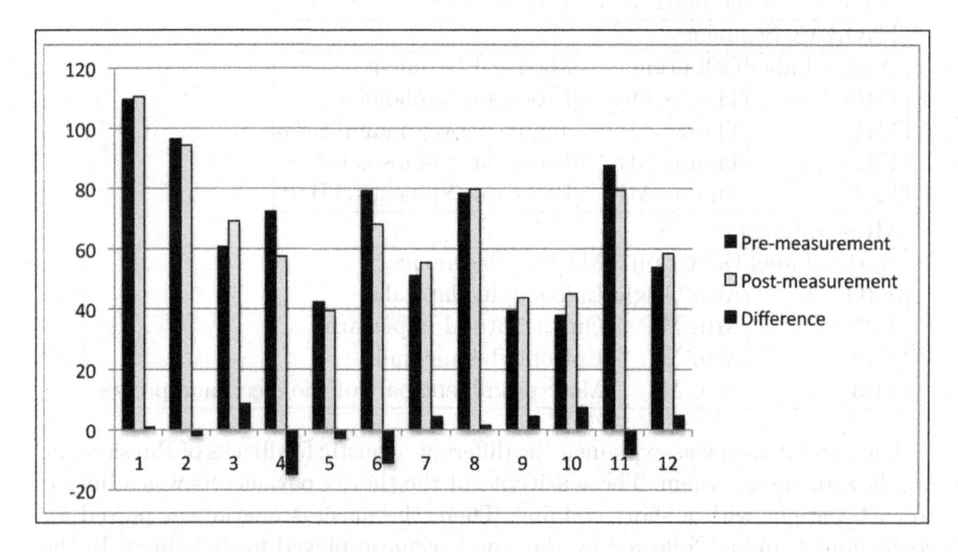

Fig. 9. Shoulder abduction before and after the therapy with PhysioSonic

5.3 Results

Study indicators were the stability of the upper limb and the change of abduction range in the shoulder joint. The use of acoustic feedback for movements of the arm in the sagittal plane was neglected, as all patients showed no significant deviations and conducted the movement in the right way.

Figure 9 shows the shoulder abduction range before and after the PhysioSonic therapy, and the difference between the two. 7 out of the 12 test persons showed

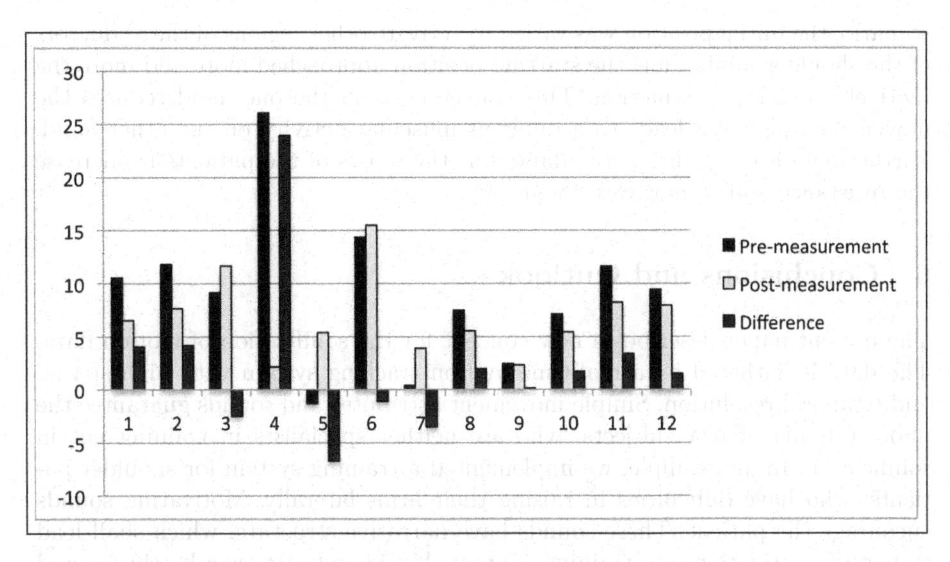

Fig. 10. Verticalisation of the thorax using Physiosonic

an increased abduction after PhysioSonic (58,3%). The mean improvement was 4,5°. Figure 10 shows the verticalization of the thorax before and after the PhysioSonic training and their difference. In the post-measurement, 8 out of 12 patients (66,6%) showed an optimized, thus reduced, evasive movement of the upper limb to the contra-lateral side.

5.4 Discussion and Additional Results

The application of PhysioSonic in the rehabilitation as described above leads to the discussed increased performances and optimizations. The observed worsening of the abduction for 3 patients can be explained by an overload of the relevant musculature. This shows, that PhysioSonic can lead to a muscular demand that excesses the subjective optimal load limit, and thus reduces the mobility. The reduced verticalization of the upper limb with 3 patients can also be explained by this effect.

Additionally to this study, two patients have used PhysioSonic at their own request several times a week during their stay at the hospital. A clear amelioration of mobility and strength in the shoulder joint could be observed with both of them. Still, a mono-causal assignment can neither be referred to PhysioSonic nor to other therapeutic measures alone.

With single patients, further implementation options of PhysioSonic were tested. During the music scenario, the arm movement was influenced transversally leading to a force distracting from the desired movement plane. The additional muscular activity necessary to stay within the reference plane constituted a therapeutic measure. The specific invigoration can be optimally detected with PhysioSonic and controlled with different resistances. Furthermore, in the music scenario, the initial position was varied to activate other regions of the abductors of the shoulder joint. Thus the starting position approached more and more the (actively) maximal abduction. This convergence on the one hand reduces the movement range and leads to a different muscular activity on the other hand. Furthermore free weights were adapted at the wrists of the patients to increase the resistance and to motivate them.

6 Conclusions and Outlook

The present paper describes a new concept for the sonification of motion data. The data is gathered by a real-time motion tracking system with high spatial and temporal resolution. Simple movement attributes and sounds guarantee the understanding of test subjects, who are neither specialists in training nor in sonification. In an example, we implemented a training system for shoulder patients, who have difficulties in raising their arms laterally. Motivating sounds encourage the patient. These sounds have narrative structure, which shall lead to further motivation in a training context. Music and texts can be chosen and thus follow the aesthetic preferences of the patient. Negatively defined sounds are inhibitory or mask the other sounds. They are linked to evasive and often unconscious movements and will be avoided by the patient.

We implemented several training scenarios of the lateral abduction and adduction of the arm for orthopaedic rehabilitation. The system has been set up at the hospital Theresienhof in Frohnleiten, Austria. A pilot study is in preparation that will evaluate the sonification designs and the efficiency gain in the training.

PhysioSonic was throughout accepted positively with the patients and enriched the therapeutic offer according to their statements. The use of auditory feedback was completely new for all of them, which lead to high motivation in the conduction of the movement. For the therapist, PhysioSonic is an easy to handle tool. It has various application options that allow using it in practice for different goals. Still, modifications in the software are necessary to extend PhysioSonic to a variety of possible other movements and human joints. These modifications are under development.

7 Documentation and Acknowledgements

Videos of the described system can be found at
https://iem.at/Members/vogt/physiosonic.
A short television report *in german* can be found at
http://steiermark.orf.at/magazin/immergutdrauf/wissen/stories/335626/

We would like to thank the Austrian Science Fund for their support in the QCD-audio project and the EGM project, where basic tools used for PhysioSonic have been developed.

References

1. Creating aesthetically resonant environments in sound, http://www.bris.ac.uk/caress
2. International society for virtual rehabilitation, http://www.isvr.org
3. Sonic golf, http://www.sonicgolf.com
4. Sonification environment research project, http://sonenvir.at
5. Stanford move, http://move.stanford.edu
6. Supercollider3, http://www.audiosynth.com
7. Vicon motion tracking, http://www.vicon.com
8. Vicon2osc, http://sonenvir.at/downloads/qvicon2osc
9. Bovermann, T., Groten, J., de Campo, A., Eckel, G.: Juggling sounds. In: International Workshop on Interactive Sonification, York (February 2007)
10. Brooks, T., Camurri, A., Canagarajah, N., Hasselbad, S.: Interaction with shapes and sounds as a therapy for special needs and rehabilitation. In: Proc. of the Int. Conf. on Disability, Virtual Reality and Associated Technologies (2002)
11. Burdea, G.: Keynote address: Virtual rehabilitation-benefits and challenges. In: International Medical Informatics Association Yearbook of Medical Informatics. Journal of Methods of Information in Medicine, pp. 170–176 (2003)
12. Effenberg, A.: Bewegungs-Sonification und Musteranalyse im Sport - Sportwissenschaft trifft Informatik. E. Cuvillier (2006)
13. Effenberg, A., Melzer, J., Weber, A., Zinke, A.: Motionlab sonify: A framework for the sonification of human motion data. In: Ninth International Conference on Information Visualisation, IV 2005 (2005)
14. Hermann, T., Höner, O., Ritter, H.: Acoumotion – an interactive sonification system for acoustic motion control. In: Gibet, S., Courty, N., Kamp, J.-F. (eds.) GW 2005. LNCS (LNAI), vol. 3881, pp. 312–323. Springer, Heidelberg (2006)
15. Hill, A., Bull, A., Wallace, A., Johnson, G.: Qualitative and quantitative description of glenohumeral motion. Gait and Posture 27, 177–188 (2008)
16. Höner, O., Hermann, T.: Listen to the ball!' - sonification-based sport games for people with visual impairment. a.p.a.: a discipline, a profession, an attitude. In: Proc. of the 15th International Symposium Adapted Physical Activity, Verona (2005)
17. Höner, O., Hermann, T.: Der goalballspezifische leistungstest tamp: Entwicklung und evaluation einer virtuellen ballwurfmaschine. In: für Sportwissenschaft, G.I. (ed.) Jahrestagung der Sektion Sportmotorik der Deutschen Vereinigung für Sportwissenschaft. Abstract-Band, Motorik, vol. 10, pp. 15–16 (2007)
18. Höner, O., Hermann, T., Grunow, C.: Sonification of group behavior for analysis and training of sports tactics. In: Hermann, T., Hunt, A. (eds.) Proc. of the International Workshop on Interactive Sonification (2004)
19. Kapandji, A.: Funktionelle Anatomie der Gelenke. Thieme (2003)
20. Kleiman-Weiner, M., Berger, J.: The sound of one arm swinging: a model for multidimensional auditory display of physical motion. In: Proceedings of the 12th International Conference on Auditory Display, London (2006)
21. Pauletto, S., Hunt, A.: The sonification of emg data. In: Proceedings of the 12th International Conference on Auditory Display, London (2006)
22. Pirrò, D., Eckel, G.: On artistic research in the context of the project embodied generative music. In: Proceedings of the International Computer Music Conference (2009)

23. Rab, G.: Shoulder motion description: The isb and globe methods are identical. Gait and Posture 27, 702 (2008)
24. Rab, G., Petuskey, K., Bagly, A.: A method for determination of upper extremity kinematics. Gait and Posture 15, 113–119 (2002)
25. Schack, T., Heinen, T., Hermann, T.: Augmented reality im techniktraining. In: für Sportwissenschaft, G.I. (ed.) Motorik, 10. Jahrestagung der Sektion Sportmotorik der Deutschen Vereinigung für Sportwissenschaft (2007)
26. Steindler, A.: Kinesiology of the human body. Charles C. Thomas (1955)
27. Zmölnig, J., Ritsch, W., Sontacchi, A.: The iem cube. In: Proc. of the 9th International Conference on Auditory Display (2003)

Sonification and Information Theory

Chandrasekhar Ramakrishnan

Native Systems Group, ETH Zürich
Zürich, Switzerland
Institut für Design- und Kunstforschung, Hochschule für Gestaltung und Kunst
Basel, Switzerland
cramakrishnan@acm.org

Abstract. We apply tools from the mathematical theory of information to the problem of sonification design. This produces *entropy sonification*, a technique for sonification, as well as a framework for analyzing and understanding sonifications in general.

Keywords: Sonification Design, Interaction Design, Information Theory, Granular Synthesis.

1 Introduction

The mathematical theory of information was developed by Claude E. Shannon at Bell Labs and presented in his seminal paper "The Mathematical Theory of Communication," published in the Bell System Technical Journal in 1948. It extended work done by Nyquist and Hardley to analyze the problems of data compression and data transmission over a noisy channel, such as messages sent by telegraph ([15] p. 31).

Information theory has since proven to be a powerful framework and has been successfully applied to problems in a variety of disciplines. It is not only the natural language for discussing highly technical domains like compression, network transmission, and error correction, but it is has also been used to develop models for understanding "softer" problem domains like human-computer interaction and graphical interface design. Jef Raskin, for example, presents such a model in his book *The Humane Interface* [13].

Although information theory was a central to early work on auditory displays (see [10]), it today occupies only a peripheral place in the auditory display toolkit. We wish to remedy this situation. When creating a process to convert data to sound, one is confronted with many design decisions, and, as we argue in this paper, information theory can help us quantify and compare the consequences of these decisions, and thus inform the choices made to arrive at a more effective sonification.

In this paper, we present two applications of information theory to sonification design. The first application, *entropy sonification*, is presented in the first part of the paper, along with the necessary mathematical prerequisites and illustrative examples (Sections 2 – 8). In the second part of the paper, we build on the

S. Ystad et al. (Eds.): CMMR/ICAD 2009, LNCS 5954, pp. 121–142, 2010.

definitions from the first half to develop a framework for quantitatively analyzing and comparing sonifications using information theoretical machinery (Sections 9 – 13). This framework can be used to perform back-of-the-envelope calculations to determine the viability and effectiveness of a sonification, as well as inform the search for an appropriate sonification model, given the data to be displayed. We present expository examples contrived for didactic purposes, as well as analyses of sonifications found in the literature.

2 Entropy Sonification

In his three volume classic on visualization, Edward Tufte distinguishes between visualization strategies based on the kinds of information they are designed to convey. *The Visual Display of Quantative Information* [20] focuses on visualizations that display numbers; *Envisioning Information* [19] focuses on visualizations that depict nouns; and *Visual Explanations* [21] focuses on visualizations that illustrate verbs.

Making a similar distinction in the domain of sonification lets us distinguish between different purposes and uses of sonification techniques. Hermann and Ritter provide a taxonomy of sonification techniques [6], breaking them down into the following categories:

1. Auditory Icons
2. Earcons
3. Audification
4. Parameter Mapping
5. Model-Based Sonification

For each of these techniques, we can ask, "What kind of information is it designed to present?" Alberto de Campo's design space map provides a guide for answering this question [2]. The technique of auditory icons [5], for example, is a sonification technique well-suited for presenting a discrete collection of nouns and verbs. Audification, on the other hand, is a technique for presenting numbers, as is model-based sonification.

Sound, being an inherently time-based medium, contains another posibility for presentation that has no analogue in domain of static visualization: *narrative*. A narrative functions by emphasising particular events and aspects that are key to understanding a story and glossing over other ones that are less central. Thus, to construct a narrative, we need a measure of importance in order to distinguish between events that are interesting, and thus deserving of increased emphasis, and those that are uninteresting, and thus good candidates for elision. This is exactly the kind if measure information theory provides.

Entropy sonification is a technique for imbuing a sonification with a narrative structure derived from the underlying data itself. The goal of this narrative structure is to focus attention on the interesting and salient aspects of the data.

Huffman coding vs. arithmetic coding.

3 Entropy Sonification Construction

Information theory treats data as the product of an information source, which is simply modeled as a stochastic process ([15], section on discrete noiseless systems). This can be justifiably criticized as a gross oversimplification. Nonetheless, information theory has been successfully applied to a variety of problems in data transmission and compression and has proven to be a powerful framework for creating and analyzing algorithms for communicating information.

The information source model forms the basis of entropy sonification. This model is used to "tell" the sonification which data points to highlight and which ones to push into the background. The highlighting/backgrounding of data points is accomplished though rhythmic variation: important points are accented and have a longer duration than others. Rhythmic variation was chosen since it is known that rhythm is a more salient aspect of music cognition than melody or timbre. Dowling and Harwood summarize research on the importance of rhythm vs. melody as follows: "It is noteworthy that the rhythmic aspects of these stimuli overrode their pitch-pattern aspects in determining listeners' responses. This clearly demonstrates the importance of rhythm in music cognition." ([3] p. 193)

To control rhythmic variation, we need to vary the loudness, duration, and attack of the sound. Granular synthesis is a natural technique for realizing this. The urparameters of a grain are duration and amplitude envelope or window (Roads' *Microsound* [14], discussion in Chapter 3), which map perfectly to the variables we need to control. Figure 1 presents a graphical depiction of this scheme.

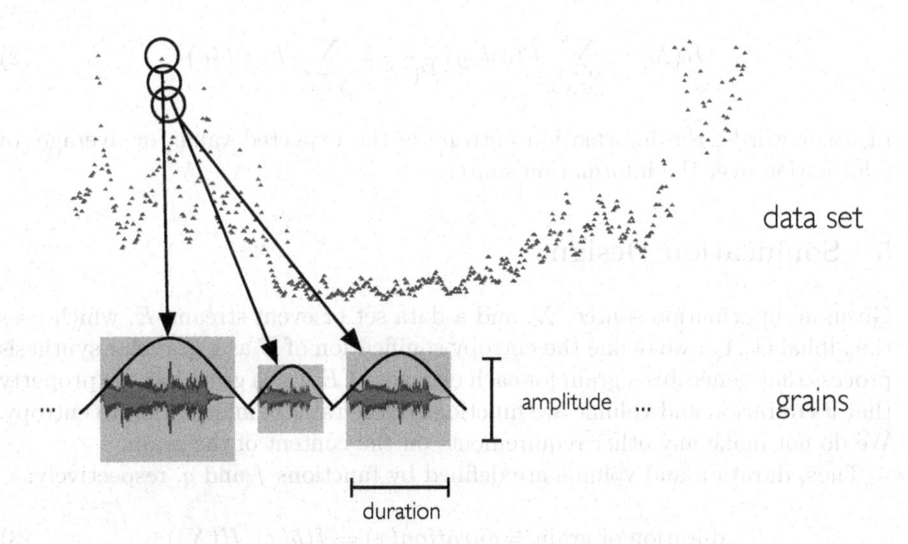

Fig. 1. Schematic Depiction of Entropy Sonification. Amplitude and duration are functions of the ratio of information content to information entropy.

One final issue must be settled – the order in which the grains are played. In some cases, as in the examples we present, there is a natural order to the data which can be used by the sonification as well. In situations where no natural order exists, an ordering must be artificially produced, for example, using the distance from a user selected point.

4 Information Theory Background

We provide the minimum fundamentals of information theory necessary for the presentation of the sonification technique. For a more extensive treatment, we refer the reader to the source, Shannon and Weaver's *The Mathematical Theory of Communication* [15], or David MacKay's *Information Theory, Inference, and Learning Algorithms* [7] for a broader and more modern presentation.

An **information source** X is determined by two sets of N elements: an alphabet $A_X = \{a_1, a_2, ..., a_N\}$, which defines the set of possible values, and a probability distribution $P_X = \{p_1, p_2, ..., p_N\}$, which defines a probability for each element of A_X. We use the notation $P(a)$ to denote the probability of an element of the alphabet. In particular, $P(a_i) = p_i$.

The **Shannon information content** of an element a of the alphabet A_x is defined as

$$h(a) = log_2 \frac{1}{P(a)}. \tag{1}$$

We will just use "information content" or "information" to mean Shannon information content.

The **information entropy** of an information source X is defined as

$$H(X) = \sum_{a \in A_X} P(a) log_2 \frac{1}{P(a)} = \sum_{a \in A_X} P(a)h(a) \tag{2}$$

In other words, the information entropy is the expected value, or average, of information over the information source.

5 Sonification Design

Given an information source, X, and a data set or event stream, E, which uses the alphabet A_X, we define the entropy sonification of E as a granular synthesis process that generates a grain for each element of E. Each grain has the property that its duration and volume are functions of the ratio of information to entropy. We do not make any other requirements on the content of the grain.

Thus, duration and volume are defined by functions f and g, respectively:

$$\text{duration of grain} = duration(x) = f(h(x)/H(X)) \tag{3}$$

$$\text{volume of grain} = volume(x) = g(h(x)/H(X)) \tag{4}$$

5.1 Sonification Duration

Let us consider the special case where the duration function is multiplication by a constant, d:

$$duration(x) = d\frac{h(x)}{H(X)} \tag{5}$$

In this case, the expected value of the duration of a grain is:

$$\mathcal{E}[duration] = \frac{d}{H(X)} \sum_{a \in A_X} P(a)h(a) = d \tag{6}$$

Thus, given N points of data, $e_1 ... e_N$, the expected length of the sonification is:

$$\mathcal{E}[length] = \sum_{i=1}^{N} \mathcal{E}[duration(e_i)] = \sum_{i=1}^{N} d = Nd \tag{7}$$

In other words, if grains are played one after another without overlap, the total duration of the sonification should be the duration per grain times the number of data points, assuming the information source accurately represents the data. This is a nice property which can be used to gauge the accuracy of the information source description.

5.2 Uniform Distribution

Another special case to consider is an information source based on the uniform distribution; thus all elements of the alphabet have equal probability. Using the notation $|A_X|$ to represent the number of elements in A_X, this gives the following expressions for information content of an element of the alphabet and entropy of the information source:

$$h(a) = log_2 \frac{1}{P(a)} = log_2 |A_X|, a \in A_X \tag{8}$$

$$H(X) = \sum_{a \in A_X} P(a)h(a) = log_2 |A_X| \tag{9}$$

Since every symbol has the same information, which is equal to the entropy of the information source, this results in a sonification in which every grain has the same duration and envelope. This information source model, the uniform information source, is useful as a basis for comparison.

5.3 Examples and Limitations

In the following two sections, we present examples of entropy sonification. For the purposes of exposition, we restrict ourselves to the simple case of discrete information sources and two dimensional data sets. This is not a limitation of the theory, but a desire to keep the exposition simple. Entropy sonification can be applied to any dimensionality of data; only a description of the data as an information source is necessary.

Our examples use simple histograms as the basis of information-source models. In general, any stochastic process can serve as an information source. And though we only present and discuss discrete information sources, the theory extends to continuous information sources [15], and entropy sonification extends naturally to this situation as well.

6 Example 1: Text

Taking our lead from Claude Shannon, we first turn our attention to a sonification of English-language text. Text has two attributes that make it a good candidate for entropy sonification: it is a discrete information source, as opposed to a continuous information source such as one defined by a Gaussian distribution; and letter frequency tables to define the model are widely available.

The source code and audio for this example may be downloaded from our web site [12].

6.1 Data

The "data," in this case text, is taken from the beginning of Thomas Pynchon's *Gravity's Rainbow* [11]:

> A screaming comes across the sky. It has happened before, but there is nothing to compare it to now.
> It is too late. The Evacuation still proceeds, but it's all theatre. There are no lights inside the cars. No light anywhere. Above him lift girders old as an iron queen, and glass somewhere far above that would let the light of day through. But it's night. He's afraid of the way the glass will fall – soon – it will be a spectacle: the fall of a crystal palace. But coming down in total blackout, without one glint of light, only great invisible crashing.

6.2 Grain Waveform Design

The text sonification architecture is implemented in version 3 of James McCartney's programming language, SuperCollider [8]. Though this is a sonification, not a text-to-speech system, it does take a few cues from speech synthesis. Letters are converted to sound differently based on whether the letter is a vowel or consonant. Vowels are synthesized as a combination of a fundamental frequency and two formants with frequencies specified by Peterson and Barney's table [9]. Consonants are synthesized as filtered noise.

The uniform information source applied to this grain design yields the sonification displayed in Figure 2.

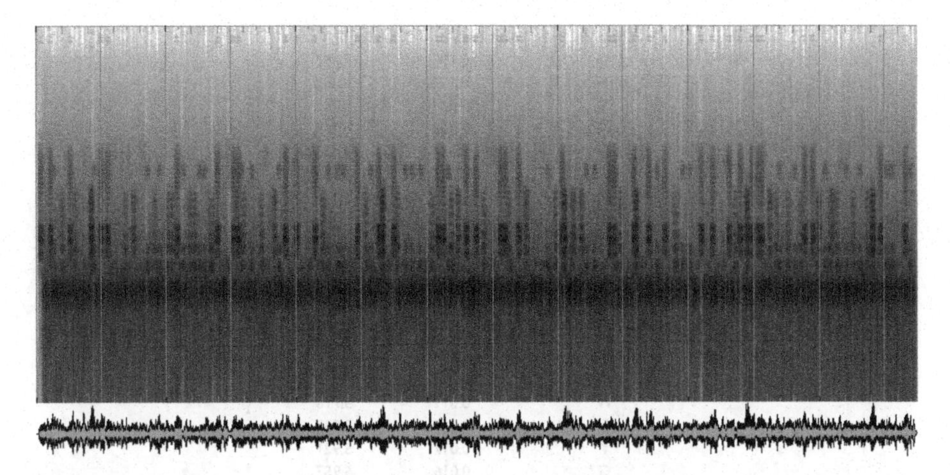

Fig. 2. Text Sonification Using the Uniform Distribution in Frequency and Time Domains; Duration: 0:34

6.3 Model

Using a letter frequency table [17] that provides frequencies for 27 characters, the 26 letters of the English alphabet plus space, we can compute the information content for each letter, plus the information entropy for the alphabet as a whole, as displayed on the next page in Table 4. In our model, we ignore punctuation marks such as commas, periods, etc.

This model applied to the text produces the sonification shown in Figure 3.

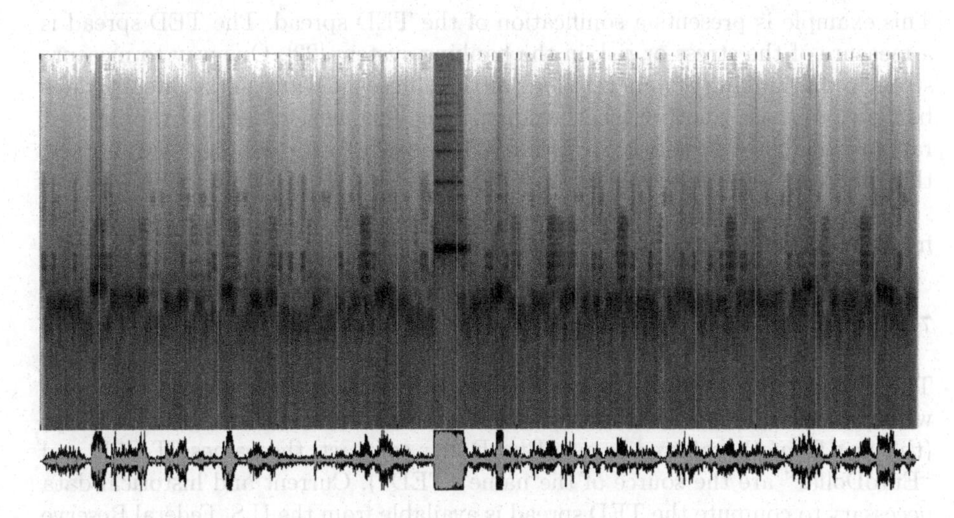

Fig. 3. Text Sonification in Frequency and Time Domains; Duration: 0:37. Letters with lower probability stand out, in particular the "q" near the middle of the sonification.

Character	Probability	Information
[ ]	0.190	2.393
E	0.102	3.298
T	0.074	3.762
A	0.066	3.919
O	0.061	4.035
I	0.056	4.153
N	0.056	4.166
H	0.054	4.206
S	0.051	4.299
R	0.046	4.449
D	0.037	4.760
L	0.033	4.943
U	0.023	5.455
M	0.021	5.608
C	0.019	5.703
W	0.019	5.718
F	0.018	5.837
Y	0.017	5.921
G	0.016	5.957
P	0.013	6.254
B	0.012	6.442
V	0.009	6.828
K	0.007	7.243
X	0.001	9.480
J	0.001	10.288
Q	0.001	10.288
Z	0.001	10.966
Entropy		4.078

Fig. 4. English Letter Probabilities and Information

7 Example 2: TED Spread

This example is presents a sonification of the TED spread. The TED spread is a measure of the stress or risk in the banking system [22]. One way to measure stress in the banking system is to look at the difference between the interest rates banks charge one another when lending money to each other and the interest rate provided by a "low-risk" investment. If banks are safe and trust one another, this difference should be small.

As with the previous example, the source code and audio may be downloaded from our web site [12].

7.1 Data

The TED compares the returns provided by U.S. T-Bills, the low-risk investment, with the LIBOR (London Interbank Offered Rate) rate banks charge one another (this was formerly referred to as a EuroDollar contract; the names "T-Bill" and "EuroDollar" are the source of the name "TED"). Current and historical data necessary to compute the TED spread is available from the U.S. Federal Reserve [1]. This data is graphed in Figure 5.

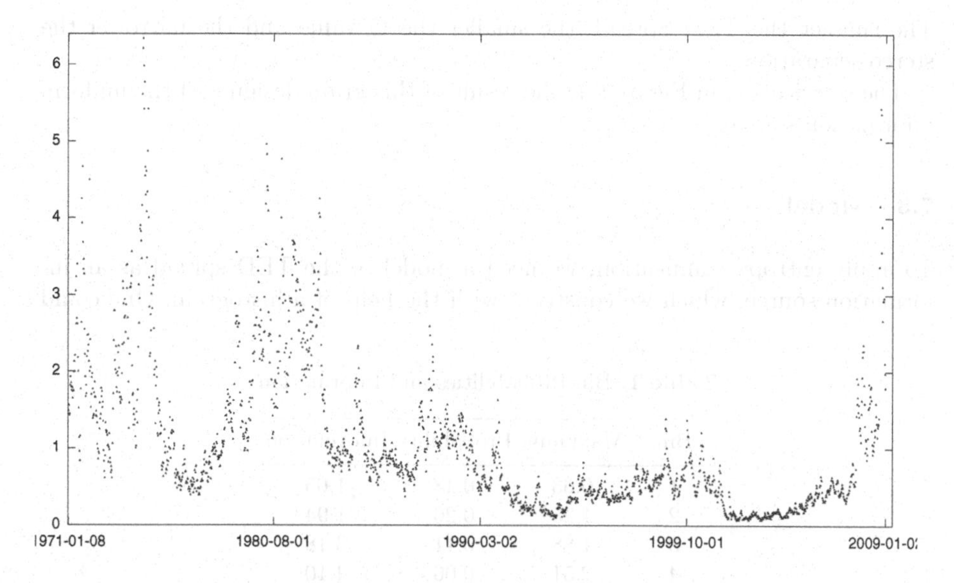

Fig. 5. The Weekly TED Spread, 1971 – 2009

7.2 Grain Waveform Design

Grains are noise filtered by two bandpass filters, one for each component of the TED Spread. The center frequencies of the filters are determined by the interest rate on T-Bills and LIBOR contracts, respectively. The quality, Q, parameter of the filter and stereo pan are determined by the value of the TED Spread itself.

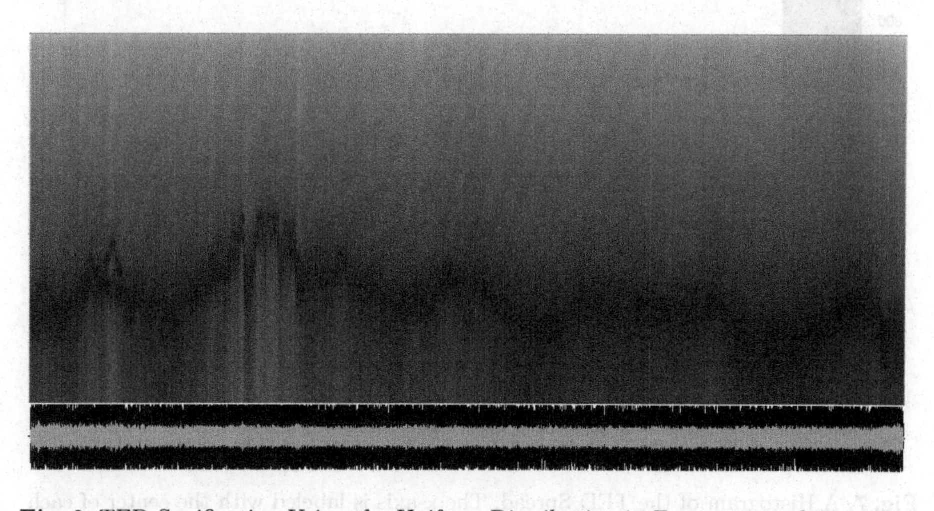

Fig. 6. TED Sonification Using the Uniform Distribution in Frequency and Time Domains; Duration: 2:04

The smaller the TED Spread, the smaller the Q value and the narrower the stereo separation.

The sonification in Figure 6 is the result of this grain design and the uniform information source.

7.3 Model

To apply entropy sonification, we need a model of the TED spread as an information source, which we construct with the help of a histogram. One could

Table 1. Bin Probabilities and Information

Bin	Maximum	Probability	Information
1	0.63	0.48	1.05
2	1.25	0.26	1.94
3	1.88	0.11	3.18
4	2.51	0.06	4.10
5	3.14	0.05	4.41
6	3.76	0.02	5.47
7	4.39	0.01	6.72
8	5.02	0.005	7.64
9	5.64	0.002	8.64
10	6.27	0.002	8.97
Entropy			2.08

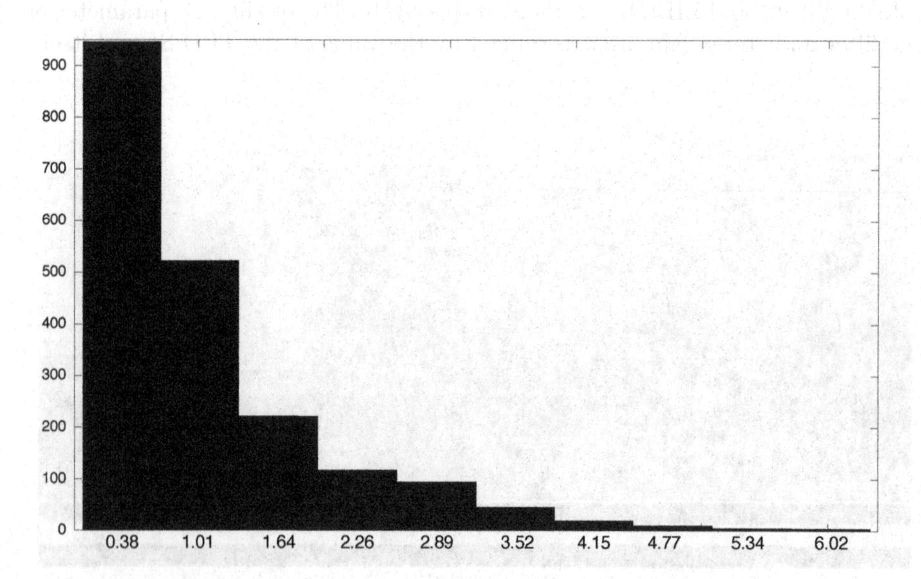

Fig. 7. A Histogram of the TED Spread. The x-axis is labeled with the center of each bin; the y-axis is the number of values that fall within the bin.

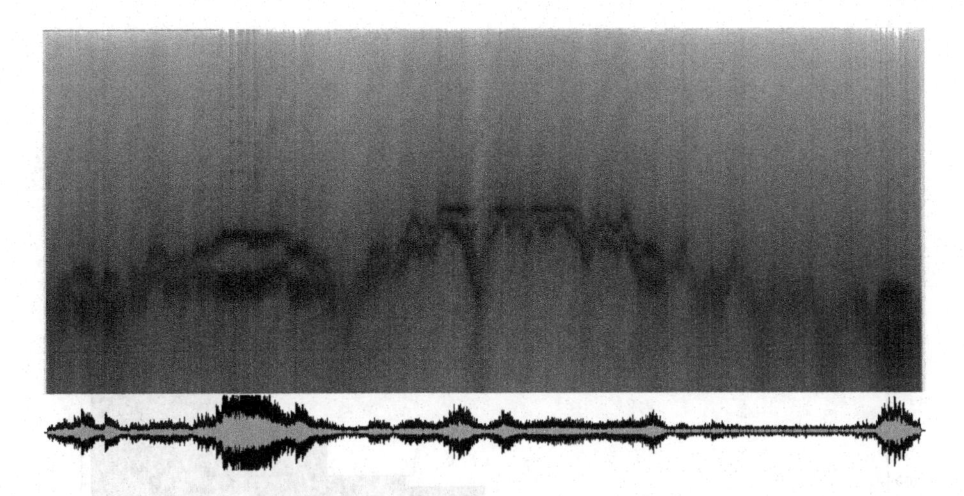

Fig. 8. TED Sonification in Frequency and Time Domains; Duration: 1:57

construct a more precise model, but even this simple one yields a sonification that brings out many salient features of the data set.

Figure 7 shows the distribution of values in the TED spread series. For the purposes of computing information content, we concern ourselves solely with which band in the histogram each value falls. This model is summarized in Figure 1.

The resulting sonification is shown in Figure 8.

8 Contrived Models

In the examples above, we have constructed information sources based on empirical analysis of the subject of the sonification. There are, however, also other ways of producing an information source. In some situations, for example, it might be appropriate to use an *a priori* model derived from other principals and not *a posteriori* analysis of data. One such application of this idea is entropy zooming, a technique for modifying an entropy sonification to focusing on a particular aspect of the data set chosen by the user.

8.1 Entropy Zooming

The goal of entropy sonification is to make interesting elements of a data set stand out. An artifact of using Shannon information as the basis for entropy sonification is the implicit assumption that interesting elements are rare ones. This may not actually be the case. The interesting elements could actually be the common ones, and the rare ones might be uninteresting.

This problem is solved with entropy zooming. Entropy zooming involves constructing a model of a data set as an information source with the goal of bringing out particular characteristics of the data. A model used for entropy zooming is

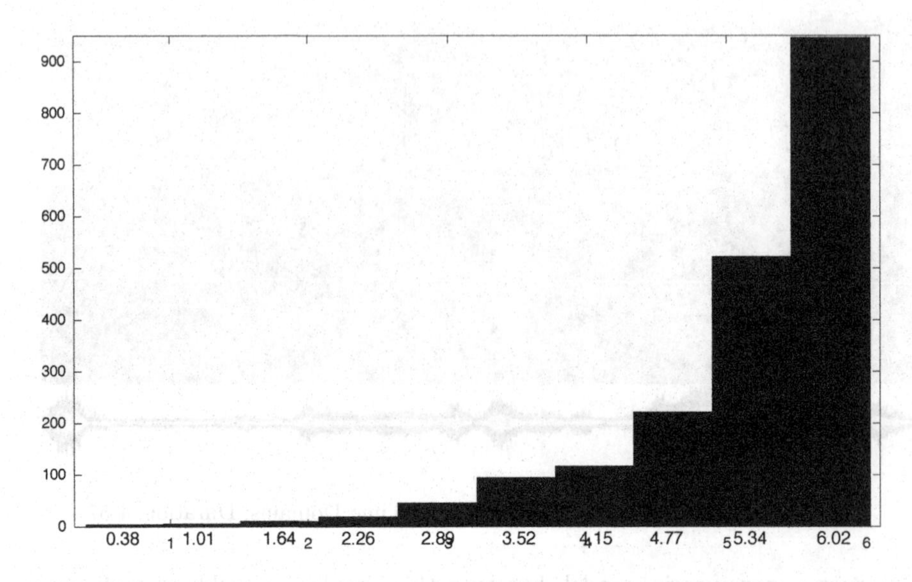

Fig. 9. Alternative Histogram of the TED Spread for Zooming

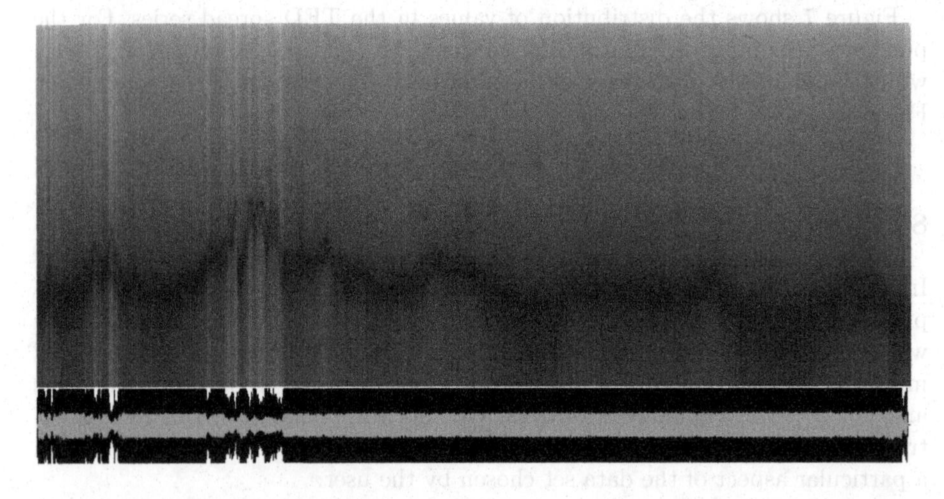

Fig. 10. Zoomed TED Sonification in Frequency and Time Domains; Duration: 8:34

not based on the actual probability distribution of elements in the data set, rather it is intentionally constructed by assigning high information to the elements of the data set we want to concentrate on and low information to other elements.

One way to create such a model is to start with one derived from probabilistic analysis and then *reduce* the probability of the data with the characteristics we want to focus on and increase the probability of the data we want to ignore. Decreasing the probability of an element increases its Shannon information and, in turn, gives it more emphasis in an entropy sonification. To be clear, entropy zooming does not involve altering the data; entropy zooming creates an alternative model of the data that is not based on the actual probabilistic distributions, but rather on the user's notion of what is interesting.

To illustrate, we take the TED spread example and define a new information source model to focus on the smaller values of the TED spread by making them less likely according to our new, "zoomed" model. This gives us a new model, summarized in Figure 9, which results in the sonification in Figure 10.

9 Information Theoretic Analysis of Sonifications

In the preceding sections, we modeled data as an information source and used this model to inform a sonification process. Now, we add some additional mathematical tools and model not just the data, but the reception of sound by the listener as well. This allows us to analyze entire sonification processes, from data to listener interpretation, to quantitatively compare effectiveness and efficiency.

Following Shannon, we segment the communication process into several sections, each with their own statistical characteristics. These sections, shown in Figure 11, break down to the following: an information source, a transmitter that encodes the data for transmission over a communications channel, a noise source that injects noise into the communications channel, a receiver that decodes the (noisy) data from the communications channel, and a destination that accepts the decoded data. In the case of a sonification, the information source is the original data, the transmitter is the sonification process, the communications channel is sound, the noise source is determined by processes that govern psychoacoustics and music cognition, and the receiver and destination are in the brain of the listener.

To model the listeners interpretation of the sound produced by a sonification, we treat the sound – the communications channel – as an information source which is dependent on the data source. In the ideal situation, the communications channel conveys the data source perfectly, without any errors. In the real

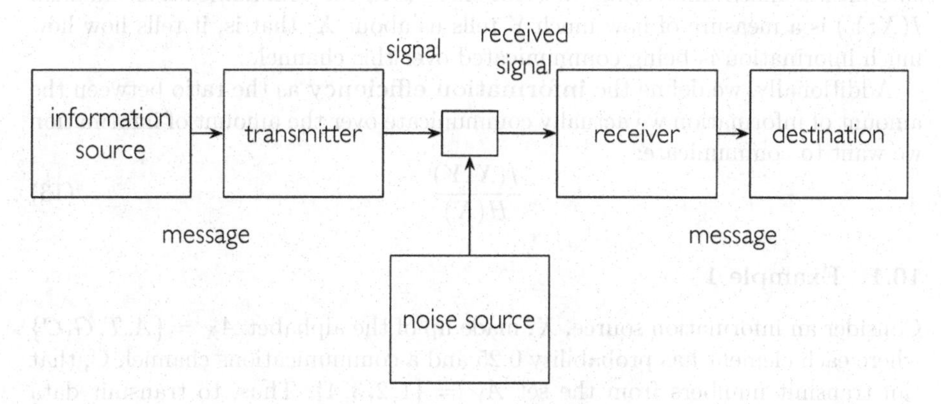

Fig. 11. Schematic Depiction of Communications Process, after Shannon

world, though, most communications channels have some level of intrinsic noise: they introduce errors as they transmit data. They also have a maximum capacity – there is a limit to how much data they can transmit. By quantifying the characteristics of the communications chanel, we can refine the sonification process and build in some necessary redundancy to improve accuracy in conveying the data.

We begin by introducing several mathematical definitions.

10 Conditional Entropy

In Section 4, we defined the information content of an element in an information source. Now, we consider two information sources X and Y, and investigate how they are related. Again, we follow MacKay [7], and refer the reader to his book for a more thorough presentation.

The **conditional information content** of an element x of the alphabet A_x and an element y of the alphabet A_y is defined as

$$h(x|y) = log_2 \frac{1}{P(x|y)}. \tag{10}$$

The **conditional entropy** of information sources X and Y is defined as

$$H(X|Y) = \sum_{y \in A_Y} P(y) \sum_{x \in A_X} P(x|y) log_2 \frac{1}{P(x|y)} = \sum_{y \in A_Y} P(y) \sum_{x \in A_X} P(x|y) h(x|y) \tag{11}$$

The **mutual information** between information sources X and Y is defined as

$$I(X;Y) = H(X) - H(X|Y) = H(Y) - H(Y|X) \tag{12}$$

These quantities can be used to measure the reliability of a communications channel. If X is an information source representing the data source and Y is an information source representing the communications channel, $H(X|Y)$ is a measure of how much uncertainty or error there is in the communications channel. $I(X;Y)$ is a measure of how much Y tells us about X; that is, it tells how how much information is being communicated over this channel.

Additionally, we define the **information efficiency** as the ratio between the amount of information we actually communicate over the amount of information we want to communicate:

$$\frac{I(X;Y)}{H(X)} \tag{13}$$

10.1 Example 1

Consider an information source, X, made up of the alphabet $A_X = \{A, T, G, C\}$ where each element has probability 0.25 and a communications channel, C, that can transmit numbers from the set $A_Y = \{1, 2, 3, 4\}$. Thus, to transmit data generated by X over the channel C, we convert the letter to a number following

Table 2. $P(X|Y)$

A_Y	A	T	G	C
1	1.0	0	0	0
2	0	1.0	0	0
3	0	0	1.0	0
4	0	0	0	1.0

the rule $A \to 1, T \to 2, G \to 3, C \to 4$ and send it over the channel C. If C is a noisy communications channel, it might alter the transmitted data.

The data received from transmission determines a new information source, Y, governed by the probabilities of receiving $y \in A_Y$ after transmission. These probabilities are in turn determined by the probabilities of the elements of the data source along with the noise characteristics of the communications channel, which can be represented by $P(x|y)$, the marginal probability that a received y was an attempt to transmit an x.

If the communications channel is noise free, then the relationship between Y and X is specified in Table 2.

In this case, $H(X|Y) = 0$ bits and $I(X;Y) = H(X)$ – the information is being perfectly communicated. This is a very good communications channel.

10.2 Example 2

Now consider the same information source as in Example 1, but with the data being transmitted over a noisy channel. In the worst case, this noise is so bad that a received number could map to any letter of the data source with equal probability. This is summarized in Table 3.

Table 3. $P(X|Y)$

A_Y	A	T	G	C
1	0.25	0.25	0.25	0.25
2	0.25	0.25	0.25	0.25
3	0.25	0.25	0.25	0.25
4	0.25	0.25	0.25	0.25

In this case, $H(X|Y) = 2$ bits and $I(X;Y) = 0$. Since X and Y are independent random variables, the information received after being transmitted over the channel bears no connection at all to the original information. So, even though 25% of the 1's will, in general, indeed map back to an A as its source, no information is actually being communicated. This is not a very good communications channel.

In practice, our communications channels will be somewhere between these two extremes.

11 Sonification Analysis

The preceding definitions give us the tools necessary to model the communication process of a sonification all the way from data to listener. Instead of considering communications systems that map data sources to numbers, we will focus on systems that map data sources to sounds: sonifications. The key strategy we employ is treating both the source data and the *listeners perception* of a sonification as information sources. By taking these two information sources into account, we can compute the amount of information, in bits, a sonification communicates.

Creating an information source model of data is straightforward; creating an information source model of the listener's perception requires more finesse. The difficulty is that modeling the receiver requires making assumptions about the accuracy of the listener's audio perception capabilities. For a concrete example, consider a sonification that converts a symbol, say T into a sound, say a quarter note C_4 on a piano. To model the reception of the sonification, we need to know how often a listener, when presented with the sound of a C_4 on a piano will recognize the sound as representing an T.

For some particularly simple sonifications, it might be possible to estimate an answer to this question by referring to literature in sound perception, music cognition, or music theory. More complex sonifications will require experimental data to determine how consistently listeners follow the mapping from data to sound.

Let us now return to the task of defining a model and rephrase this task in the language of probability theory. For each sound y in the domain A_Y of sounds produced by a sonification, we want the conditional probabilities $P(x|y)$ representing the likelihood that the listener will associate sound y with source symbol x. Once we have these conditional probabilities, we can compute the amount of information a sonification communicates, and using this knowledge, we can determine what proportion of the amount in information in the data source the sonification conveys and generate ideas for improving the sonification.

We explore two examples in further detail.

12 Example: DNA Sonification

Let us develop a sonification of the nucleobases of DNA. Our data source is X, made up of the alphabet $A_X = \{A, T, G, C\}$ where each element has probability 0.25. We consider several mappings summarized in Figure 12. These mappings are inspired from examples of actual sonifications [for example, Ferguson's sonification process which won the scientific category of the ICAD 09 genome sequence sonification competition [4]] and have been constructed to let us explore the interaction of sonification design and information theory. We begin with Mapping 1, $A_Y = \{C_4, F_4, G_4, C_5\}$ – 1, 4, 5, 8.

A listener with perfect pitch or good relative pitch would have no trouble decoding the sonification. For her, the sonification would have the characteristics of the communications channel presented in Example 1 above – a noiseless channel.

Fig. 12. Mappings of Nucleobases to Notes

There are, however, many potential listeners that do not have such a good sense of relative pitch. Let us model the sonification for a subject who can correctly distinguish the two Cs that differ by an octave, but has difficulty with the F and G (fourth and fifth). Let us assume he mixes them up with equal probability. Table 4 summarizes this subject's association of sound to source symbol.

The conditional entropy $H(X|Y) = 0.5$ bits. This means that our effective rate of communication is $I(X;Y) = H(X) - H(X|Y) = 2 - 0.5 = 1.5$ bits per quarter note. The amount of data we want to send (2 bits / quarter note) is limited by the the information transfer capacity of the sonification (1.5 bits / quarter note). For this subject, this encoding has an efficiency of .75.

To improve the rate of communication, we need to either use the capabilities of the channel more effectively (by trying to send less information, for example), or more cleverly (by coding the data better). The first possibility is explored in Mapping 2, which drops one of the nucleobases. Table 5 presents the subject's association of sound to source symbol.

This gives us an effective rate of communication of $I(X;Y) = H(X) - H(X|Y) = log_2 3 - 0 = 1.6$ bits per quarter note. We try to communicate less

Table 4. $P(X|Y)$ for Mapping 1 (Fourth / Fifth Confusion)

A_Y	A	T	G	C
C_4	1.0	0	0	0
F_4	0	0.5	0.5	0
G_4	0	0.5	0.5	0
C_5	0	0	0	1.0

Table 5. $P(X|Y)$for Mapping 2 (Three Bases)

A_Y	A	T	C
C_4	1.0	0	0
F_4	0	1.0	0
C_5	0	0	1.0

information and, thus, can better use the characteristics of the channel. We are now able to communicate more information per symbol and do so more efficiently (efficiency = .80). Sometimes less is more.

Mapping 3 takes an alternative approach. It tries to communicate the full range of information of the data source, but does so by coding the data differently. Here, we use rhythm to help distinguish between cases where pitch alone causes confusion. Assuming this is sufficient for our listener to disambiguate the these two possibilities, we have a sonification where the listener can consistently recognize the mapping between sound and source symbol, as shown in Table 6.

Table 6. $P(X|Y)$ for Mapping 3

A_Y	A	T	G	C
C_4	1.0	0	0	0
F_4	0	1.0	0	0
G_4, G_4	0	0	1.0	0
C_5	0	0	0	1.0

In this case, we have an effective rate of communication of $I(X;Y) = H(X) - H(X|Y) = log_2 4 - 0 = 2$ bits per quarter note and an efficiency of 1.

13 Example: Auditory Graphs

The previous example highlights the power of information theory to model subtle aspects of how individual listeners may perceive a sonification. This knowledge can be used to understand the limits of the amount of information that can be communicated using a given mapping, taking the capabilities of individual users into consideration, and it can also provide insight into how to alter the mapping to communicate more information.

Now we turn to an analysis of a sonification described in the literature. Smith and Walker, in their article "Tick-marks, Axes, and Labels: the Effects of Adding Context to Auditory Graphs," [16] describe approaches to enhancing auditory graphs with contextual sounds, analogues of tick marks, axes, or labels. In the beginning of the article, they depict their struggle to find an appropriate terminology for describing "the ability of a user to accurately extract information from a display" (ibid., Section 1.2). This section forms our response to this conundrum.

The authors perform an experiment, presenting an auditory graph of the fluctuations of a stock during the course of a 10-hour trading day to 52 subjects. The subjects were given a randomly-selected time during the day and asked to estimate the value of the stock at that time. Using information theory, we can quantify the value of the additional contextualization and, by doing this, generate new research questions.

13.1 A Model of the Data Source

To begin, we need to model the communications process. The stock in question starts the day with a value of $50, has a maximum value of $84, a minimum of $10. For the purposes of our model, we restrict ourselves to integer values and assume that each integer between 10 and 84 is an equally probable stock price. We thus have a data source, X with $H(X) = log_2 74 = 6.21$ bits of information.

13.2 An Assessment of the Communication Process

In the first part of the experiment, subjects were presented the stimulus without any contextualizing information. In the second part of the experiment, the subjects were presented the same stimulus, but with different kinds of contextualizing data (axes and tick marks). Of these, two kinds of contextualization, providing a dynamic Y-axis and providing an X-axis with a dynamic Y-axis, significantly improved the accuracy of the estimates made by subjects.

Raw data about the subjects' answers is not provided, but Smith and Walker do provide mean absolute errors (MAEs) compared to the correct value and standard deviations of the answers from the mean absolute errors. Within the context of information theory, the questions to ask are, how much information does the sonification communicate ($I(X;Y)$) and what percentage of the total information is communicated ($\frac{I(X;Y)}{H(X)}$). We can use the Smith and Walker's data, in particular the standard deviations, to answer these questions.

Let us assume the guesses conform to a normal distribution about the mean absolute error μ. The entropy of the normal distribution is related to the standard deviation, σ by the following formula (e is the base of the natural logarithm):

$$H(X) = \frac{1}{2} log_2 2\pi\sigma^2 e \tag{14}$$

Using this formula, we can compute the conditional entropy of the different sonification strategies $H(X|Y)$, as well as $I(X;Y)$, the amount of information communicated by the communications channel, and $\frac{I(X;Y)}{H(X)}$, the amount of information communicated relative to the total amount of information as well. These values are summarized in Table 7.

Table 7. Channel Characteristics for Different Graphs

Graph	MAE	Std. Dev($)	$H(X\|Y)$	$I(X;Y)$	$\frac{I(X;Y)}{H(X)}$
No Axes	12.56	3.189	3.72	2.49	0.40
Dynamic Y-Axis	6.42	1.28	2.40	3.81	0.61
X-Axis + Dynamic Y-Axis	7.18	2.113	3.13	3.08	0.49

From the table, we see that the unadorned sonification communicates less than half the information of the data source, but by adding contextualization, we can convey 60% of the information.

There are several natural questions to ask at this point. One question relates to the bias in the guesses. The guesses made by the subjects are, in some cases, quite a bit off from the actual value of the stock. Nonetheless, the guesses themselves are relatively close to each other. This shows that, even though the listeners were systematically decoding the information incorrectly, quite a bit of information can be communicated using these sonifications. If the subjects were made aware of the bias in their interpretation of the sounds (i.e., if the receiver's decoding algorithm was improved), they could perform much better on the tests.

Another question is, how accurately does one need to know the price of a stock or the movement trend of a stock in order to make decisions? In other words, how much information does this interface need to convey in order to be useful? For some situations, having a ballpark estimate of the order of magnitude is enough; in other cases, precise information is necessary. Knowing which case is applicable influences the requirements for the efficiency of the sonification.

Also worth investigating is the possibility of designing a higher-capacity channel by using other musical parameters beyond pitch, rhythm or timbre. How much of an improvement is possible?

14 Conclusions and Future Work

In this article, we present two applications of information theory to sonification design. The first, entropy sonification, is a technique for focusing the user's attention on particularly interesting portions of data with enough built-in flexibility to let the sonification designer, or even user herself, decide what she finds interesting and wants to focus on.

In the second, we apply Claude Shannon's model of communication to study sonification processes from data to reception by a listener. This lets us quantitively capture the amount of information communicated by a sonification and thereby compare sonifications to one another as well as determine how much of the information in a data source is conveyed by a sonification.

This methodology can be used to derive limits on the amount of information a sonification can communicate. It can also hint at ideas for improving sonifications by highlighting sources of confusion.

Though the research is presented here in an abstract form, it was developed to support *eMotion*, a multi-disciplinary research project at the Institut für Design- und Kunstforschung in the Hochschule für Gestaltung und Kunst Basel FHNW [18]. In this project, we have been investigating how museumgoers interact with artworks by analyzing realtime data collected in June and July 2009 from visitors to the Kunstmuseum in St. Gallen, Switzerland as they wandered through an exhibition conceived specifically for this purpose.

The application of information theory to sonification design shows much potential. We have only begun to scratch the surface. Thus far, our analyses have been restricted to rather simple models. The technique can however be easily extended to more complex models of data, such as those based on Markov chains, more complex models of sound and music perception and cognition, and more complex sonification techniques, like model-based sonification [6]. Our initial experience suggests that further work in this direction has to potential to be quite illuminating.

Acknowledgements

The authors would like to thank Prof. Curtis Roads, Dr. Oliver D. King, and Prof. Alberto de Campo as well as the eMotion team for their helpful criticisms and comments. This work was generously funded by the Schweizerischer Nationalfonds and Ubisense.

References

1. Board of Governors of the Federal Reserve System: Federal Reserve Statistical Release, `http://www.federalreserve.gov/releases/H15/data.htm` (accessed January 9, 2009)
2. de Campo, A.: Toward a Data Sonification Design Space Map. In: Proc. of the 2007 International Conference on Auditory Display, Montréal, Canada (June 2007)
3. Dowling, W.J., Harwood, D.L.: Music Cognition. Academic Press, San Diego (1986)
4. Ferguson, S.: Genetic Sonification Production, `http://web.arch.usyd.edu.au/~sfer9710/SonProcess.txt` (accessed October 29, 2009)
5. Gaver, W.W.: The SonicFinder, a prototype interface that uses auditory icons. Human Computer Interaction 4(1), 67–94 (1989)
6. Hermann, T., Ritter, H.: Listen to your Data: Model-Based Sonification for Data Analysis. In: Int. Inst. for Advanced Studies in System Research and Cybernetics, pp. 189–194 (1999)
7. MacKay, D.J.C.: Information Theory, Inference, and Learning Algorithms. Cambridge University Press, Cambridge (2003)
8. McCartney, J.: SuperCollider: a New Real Time Synthesis Language. In: Proc. of the 1996 International Computer Music Conference, Hong Kong (1996)
9. Peterson, G.E., Barney, H.L.: Control Methods Used in a Study of the Vowels. Journal of the Acoustical Society of America 24(2), 175–184 (1952)
10. Pollack, I.: The Information of Elementary Auditory Displays. Journal of the Acoustical Society of America 25(4), 765–769 (1953)

11. Pynchon, T.R.: Gravity's Rainbow, 1987th edn. Penguin Putnam Inc, New York (1973)
12. Ramakrishnan, C.: Entropy Sonification Implementation in SuperCollider, http://www.illposed.com/research/entropy_sonification (accessed October 29, 2009)
13. Raskin, J.: The Humane Interface. Addison-Wesley, Boston (2007)
14. Roads, C.: Microsound. MIT Press, Cambridge (2004)
15. Shannon, C., Weaver, W.: The Mathematical Theory of Communication. University of Illinois Press, Illinois (1949)
16. Smith, D.R., Walker, B.N.: Tick-marks, Axes, and Labels: the Effects of Adding Context to Auditory Graphs. In: Proc. of the 2002 International Conference on Auditory Display, Kyoto, Japan (July 2002)
17. von Sydow, B.: Mono-, Bi and Trigram Frequency for English. (October, 17 2008), http://www.cs.chalmers.se/Cs/Grundutb/Kurser/krypto/en_stat.html (Accessed December 22, 2008)
18. Tröndle, M., Greenwood, S., et al.: eMotion, http://www.mapping-museum-experience.com/ (accessed March 20, 2009)
19. Tufte, E.: Envisioning Information. Graphics Press, Cheshire (2006)
20. Tufte, E.: The Visual Display of Quantitative Information. Graphics Press, Cheshire (2007)
21. Tufte, E.: Visual Explanations. Graphics Press, Cheshire (2005)
22. Wikipedia contributors: "TED Spread," Wikipedia, The Free Encyclopedia, (December 13, 2008), http://en.wikipedia.org/w/index.php?title=TED_spread\&oldid=257712580 (accessed January 8, 2009)

A Sound Design for Acoustic Feedback in Elite Sports

Nina Schaffert[1], Klaus Mattes[1], and Alfred O. Effenberg[2]

[1] Department of Human Movement Science, University of Hamburg, Mollerstr. 2, 20148 Hamburg, Germany
nina.schaffert@uni-hamburg.de
[2] Institute of Sport Science, Leibniz University Hannover, Am Moritzwinkel 6, 30167 Hannover, Germany

Abstract. Sound (acoustic information) is the naturally evocative, audible result of kinetic events. Humans interact with the world by the everyday experience of listening to perceive and interpret the environment. Elite athletes, especially, rely on sport specific sounds for feedback about successful (or unsuccessful) movements. Visualization plays the dominant role in technique analysis, but the limitations of visual observation (of time related events) compared with auditory perception, which represents information with a clearer time-resolution, mean that acoustic displays offer a promising alternative to visual displays. Sonification, as acoustic representation of information, offers an abundance of applications in elite sports for monitoring, observing movement and detecting changes therein. Appropriate sound is needed to represent specific movement patterns. This article presents conceptual considerations for a sound design to fulfill the specific purpose of movement optimization that would be acceptable to elite athletes, with first practical experience with elite athletes in rowing.

Keywords: Sonification, acoustic feedback, sound perception, sound design, auditory display, elite sport, rowing, movement optimization, music, rhythm, movement-synchronization, motion-sounds.

1 Introduction

Sounds and different acoustic formats as well as music have influenced human movements in all known cultures and at all times during history from tribal cultures to modern civilisation, e.g. for social functions (music and dance), therapeutic monitoring (using external rhythm to enhance the timing of movements) or in sport gymnastics for the rhythmic synchronisation of movement patterns [1]. The theory of music since the ancient world has incorporated the term 'musical movement' as an expression of the alliance between the perception of music and physical movement [2]. "Music involves the manipulation of sound" [3] and thus, both seem to be ideal synchronisation devices. It was assumed that music evolved as a cooperative method for the coordination of actions and promotion of group cohesion [4]. The reason is that sound, as well as the human ear, deals with physical objects. Movements as well as sounds have a physical 'nature' that is reflected in their inherent time structure;

S. Ystad et al. (Eds.): CMMR/ICAD 2009, LNCS 5954, pp. 143–165, 2010.

both appear in and depend on time, which makes them inseparable [5] [6] [7]. Accordingly, both are results of a temporal sequence of events. The existence of this relationship and several advantages out of it were described more in detail elsewhere [6] [8] [9] [1] [10].

Movements are always accompanied by natural sounds, e.g. in sport situations, where the impact-sound of the ball reflects the velocity and impact force (e.g. tennis). However, not every movement-produced sound is perceivable by human ears [5]. Those sound waves within the human hearing range (16 Hz – 20 kHz [11]) are generated during the contact phases of the sport equipment with a surface (e.g. the ground). Any source of sound involves an interaction of materials [6] [12] or, in other words, sound is the acoustic consequence of kinetic events.

In everyday situations the relationship between sound[1] (with its rhythmic structure) and movements become evident, when humans tap or clap to an audible rhythm which often happens unconsciously. Movement 'follows' sound because of the strong connection between 'sensory input and motor output' [10][2]. Rhythm particularly evokes motor responses which are caused by the time-based analysis of nerve impulses. Rhythm and tone pitch, therefore, allow a very high grade of systematisation with regards to those structural elements which were temporally analysable [13]. When designing acoustic feedback that utilises an auditory display it is necessary to know about the function and mechanisms of the auditory perception in humans which includes abilities to encode and retain acoustic information.[3]

Humans interact with the world by means of their everyday listening [14] [15] using the auditory system to perceive the surrounding environment which is characterised as the oldest and elementary function of listening [7]. This highly automated ability was (and still is) essential for surviving from the earliest time of human development to locate sound sources in the environment as well as to interpret them [16]. The environment is full of sounds that provides humans with awareness and addresses attentional focus to potentially dangerous events.[4]

The human sensory system has an enormous task to process sensory stimuli efficiently. Therefore, most of the information humans are faced with every day, were filtered out subconsciously by the brain. That is important in view of the fact that otherwise the sensory system would have an information overload. Because of this, the human brain is characterised as the "most brilliant data mining system" [7] that filters the relevant information out of numerous sensory stimuli. A sudden noise (or sound) with high intensity may indicate strong activity close-by and can be amplified when it is accompanied by high frequencies which corresponds to tone pitch in the resulting sound. Tone pitch is the perceptual correlate to frequency in acoustics [18]

---

[1] For ease of exposition, the terms 'sound' and 'auditory feedback' will be used interchangeably in this article henceforward.

[2] The meaning was translated by the authors.

[3] This article focuses on the perception of acoustic information in terms of the physiological properties of auditory perception besides aspects of culture-specific influences.

[4] For example, Indians put their ear on the ground to get sound information about potentially approaching enemies or herds of buffalos. (The reason therefore is that sound wave propagation is faster in solid material (e.g. ground) than in fluid environs (e.g. air). (For details: [17].)

[3].[5] A thunderstorm for example can evoke imminent danger when it is in the immediate vicinity (and perceptibly louder). Knowledge of the characteristics in sound enables humans to interpret events regarding its endangerments. So, many object-related actions can be recognised because of the sound they make [20]. The physical properties of different sound sources become evident in the sound result; besides the sound location, as well the direction of the sound source, its size, shape and material can be identified with a synaesthetic interpretation. "Hearing is an intimate sense similar to touch: the acoustic waves are a mechanical phenomenon and somehow they 'touch' our hearing apparatus" [21] (cf. [22]). The human ear is characterised as a 'high-tech measurement system' with its 'combined temporal-frequency-resolution' whose accuracy is hardly replicated by other measurement systems [23].

Studies in ecological acoustics showed that humans often rely on surrounding sounds for information about what happens around them. A major advantage of acoustic feedback is the possibility to improve effectiveness in interaction without distracting the focus of attention of the human. Furthermore, acoustic feedback can effectively convey information about a number of attributes [24] [14].

The structural equivalence between sport movements and sound appears in rhythm and tempo, which is used to clarify the time-structure of movement patterns. One of the advantages of auditory perception is its sensitiveness particularly "to temporal characteristics, or changes in sounds over time" [25]. It furthermore enables the listener to interpret multiple information-streams simultaneously [26]. That affords the possibility of guiding athletes' attention by the rhythm to specific sections in the sound while perceiving the natural and accustomed soundscape[6] (e.g. of rowing) at the same time. According to Bregman [27], listeners are able to switch between heard sound sequences as one or two streams by changing the focus of attention. Taking the example of listening to a classical concert, a listener is able to concentrate on one instrument and follow its melody through a symphony, but he is able to switch attention to another instrument at the same time. Thus, the human ear is able to filter out and concentrate on several sound sources as a 'selective hearing process' which is known as the 'cocktail-party-effect' [28] [11]. Early investigations in sport science used the strong audio-motor relationship by providing acoustic feedback to the athletes in order to sensitize them to the time-dynamic structure of movements (e.g. in tennis, javelin throwing, hurdle run, swimming and ergometer rowing, [5] [6] for an overview).

Another promising feature of sound and its inherent structures is that they are not always explicitly processed. Time-critical structures are perceived subliminally which is of crucial importance for precision when modifying movement execution (which was shown by Thaut et al. [29] (cf. [30]), with "subliminal entrainment" on finger tapping). As soon as someone starts to think how to execute a movement in detail, the movement stumbles [31] (for an overview: [32]).

[5] Frequency is just one part of the complex function of tone perception. (For a more detailed description [18] [19].)

[6] The term 'soundscape' is meant as a counterpart of landscape (according to Schafer [21]) and denotes sound in its environmental context (both naturalistic and within urban scenarios). A natural sound in rowing is the splash sound the blade makes when it enters the water.

Moreover, audio-motor imitation of movement execution is more precise than would be possible when consciously intended (ibid.). With sound information, motor behaviour can be modified unconsciously [33], making it a promising tool for implicit learning.[7] Furthermore, there is empirical evidence, that acoustic sound can be an alternative to graphs in the analysis of complex data (e.g. [35]).

1.1 Acoustic Feedback

As already mentioned, sound plays an important role in sport [36], especially for elite athletes who rely on sport specific sounds, which are caused by their actions, for feedback about the success (or failure) of the movement and, if necessary, to adjust their movement sequence. In rowing, the sound of the oarblade[8] and, even more, the sound of the boat during the boat run play an important role for elite rowers, who listen to the sound of the boat to control their actions [37].

"In crews we listen to the tone and texture of catches and releases and also for their synchrony, a key element in speed. (…) At the beginning of the stroke (called the catch (…)), we listen for a clean *plash* – identical and synchronised on port and starboard – as the oars, properly squared up, drop into the water. This sonority pleases a rower's ears just as the throaty *thwock* of a tennis ball on a racquet's sweet spot delights a tennis player. At the end of the stroke (the release (…)), we listen for a quiet, crisp sort of suction, like the tearing of a fresh lettuce leaf." [38]

Subjective descriptions by elite rowers (i.e. their verbalised and conscious cognition of the structure of motor behaviour and the changes in their thoughts and feelings during the training process), bring an additional insight to biomechanical analysis in terms of a meditative dimension [39] [40], and hence, for technique training; knowing what the athletes perceive while rowing and how they react implicitly to the sound, generated by their actions, raises the possibility to furnishing the athletes with the feedback they need in order to adjust their movements without concentrating too much on the information received. Furthermore, it expands biomechanical aspects and criteria with a "perceptual instrument" (or alternatively a "smart perceptual mechanism" [41]) or a "smart sensor" [42] and thus integrates athletes' mind and movements.

Providing acoustic feedback to athletes in order to sensitize them to the time-dynamic structure of the movement, doubtlessly improves its execution [6] [37] [43]. Empirical motor learning studies have already proved effective for enhancing the process of learning new movement techniques [44].

Acoustic feedback is especially helpful, if the visual perception-channel is already blocked [45]. Existing feedback systems in elite rowing have been used successfully for more than ten years [46]. However, they have mainly utilised visual techniques. In doing so the visual focus is occupied and the eyes have to be directed continuously towards the display. Besides the restrictive condition of focusing on the display visually, effectiveness of visual observation decreases with increasing movement intensities (stroke frequency) and corresponding boat velocity. In principle, visual observation is limited by the properties of visual perception which is extremely selectiveness in temporal resolution. The eyes scan a picture at its own speed, sound in contrast, is heard

[7] According to Seger [34] implicit learning processes enable the acquisition of highly complex information without complete verbalisable knowledge of what has been learned.

[8] The term 'blade' refers to an 'oar-blade' or a 'scull-blade'.

as it is revealed. Acoustic information, however, can be received by the athletes as a side effect. And moreover, it enhances the performance of the human perceptual system by increasing the awareness of every athlete in the boat, rather than only that of a single athlete.

With the positive attributes of the sound it is possible to implement acoustic feedback into training processes of elite athletes in order to optimise their movements. This offers new possibilities, especially for motor control and motor learning [43] as well as for rhythmic education in racing boats and, furthermore, coordination (as the most challenging factor that influences the rowing movement -besides force and endurance- [47]) is enhanced for both, coordination between athletes (interpersonal) as well as within the single athlete (intrapersonal). Moreover, reproduction of several movement patterns is facilitated and monitoring is eased [48].

To model the source (boat motion) in terms of its physical behaviour (acceleration trace) with sonification,[9] several design features are taken into account (e.g. [25] [51] [52] [53]). Barrass [54] proposed different approaches for an auditory display design. One of them was a pragmatic approach which was mainly task-oriented. The task analysis and data characterisation make it possible to define the requirements for the representation of specific information and address them to a specific group of users (e.g. athletes). To reconcile the demands of the physics-basis of biomechanical technique with athletes' perception of it, the sound design has to fulfil several basicrequirements. First, structures were needed which were directly extracted from the physical energy changes, together with knowledge about how this would be perceived by the athletes [55]. As a consequence of this, the sonification has to convey the intended information, or, in other words, the sound has to represent the movement appropriately; thus, it is important to represent characteristic movement parameters as well as different movement intensities in the sound result. Furthermore, qualitative changes (such as increases or decreases in the boat motion) must be perceivable and differences have to be identifiable. To quantify and analyze the motion sequence in detail (in order to find its characteristic rhythm), the movement has to be separated into its subparts and afterwards recomposed to the movement in full. Finally, the sound result has to be basically aesthetic; it should be pleasant to listen to the 'melody' of the movement. Only a purposeful design would turn a series of tones into a meaningful expression and establish connections between different parts [56].

According to its definition, rhythm organises time intervals by events (according to Auhagen [2]). The rhythm of a movement is an ordered structure of actions in a temporal process [57] which becomes evident during the movement execution. Rowing, as a cyclic motion, has repetitive characteristics which appear reliably from cycle to cycle which consist of the drive phase, recovery phase, front and back reversal (the catch and finish turning points). In rowing, the rhythm of the movement is represented chronologically by the relation between the individual phases of the rowing stroke. In other words, the rhythm represents the continuous recurrence of phases of exertion and relaxation.

---

[9] According to Kramer et al. [25] sonification "... is the use of nonspeech audio to convey information. More specifically, sonification is the transformation of data relations into perceived relations in an acoustic signal for the purposes of facilitating communication or interpretation." (For a deepening discussion: [49] [50].)

Thus, the boat motion is obviously well suited for representing the cyclic motion of rowing with its characteristic time-dynamic and repetitive structure. To monitor qualitative changes in the boat motion, the propulsive acceleration trace was chosen. Acceleration by definition is the change in velocity over time [58], which is determined as a complex parameter by the combined effect of all internal and external active forces acting on the system (boat and athletes) making it ideal for an acoustic mapping. Sonified to a sound sequence, the boat motion implies the characteristic rhythm of the rowing cycle and different stroke rate steps [59]. With it, the sound conveys information inherently and can help to improve the feeling for the rhythm and time duration of the movement for both the single rowing stroke and the cycle series. The rhythm experienced and the duration of the movement creates a feeling for the desired outcome. Furthermore, an 'idea' (according to Mulder [60]) or perception arises about the sounds associated with different sections of the rowing cycle. The different intensities could be thought of as a characteristic 'sound print' [21] or an acoustic footprint of the respective movement pattern or intensity (for different stroke rate steps), assuming that a feeling for the motion-sequence develops together with the synthesised motion-sounds experienced [61]. The sound result therefore remains in relation to kinaesthesia and movement performance [62] [5]. Moreover, as a result of the synchronisation of the components of the movement (the measured boat-related rhythm as well as the subjectively perceived athlete-rhythm), a common team rhythm evolves with its communicative, compulsive and intoxicating effects. Therewith, everybody is drawn into and moves with the same rhythm. Individual athlete-rhythms become subordinated to the rhythm of the group [57]. From a psychological-physiological point of view, rhythm is transferred because of the principles of the 'ideomotor effect' (also known as the 'Carpenter effect') that describes the phenomenon wherein a subject performs motions unconsciously as a result of a movement mentally simulated or watching another person moving. Thus, the motor reactions are in principle not different from real movements (ibid.). Athletes get into the group flow that is characterised as a state of completely focused motivation in which attention, motivation and action merges [63] while receiving direct and immediate feedback.

Since a real-time model for presenting acoustic feedback with a first sound version was designed, the next step included the examination of how athletes and coaches perceived the sound result and whether it represented the boat motion appropriately. Furthermore, and most important for the purposes of the study, would the sound result be comprehensible to elite athletes and their coaches as a means of optimising the boat motion.

2 Methods

2.1 Subjects

The subjects who participated in the pilot study were male, elite junior athletes (N=4), between 16 and 17 years (body height: 193.8cm ± 3.2; body weight: 86.3kg ± 3.9).

In the follow-up study, male and female elite athletes (N=8) participated, whose age was below 23 (between 18 and 21) years. Male (n=4): 178.8cm ± 4.6, 68.9kg ± 3.4; Female (n=4): 169.5cm ± 3.4; 59.5kg ± 0.7).

DRV[10] licensed coaches (N=42) (A-C licence) answered the questionnaire for coaches.

[10] DRV = Deutscher Ruderverband (German Rowing Association).

2.2 Test Design

In the pilot study, a training run of elite junior athletes was measured in different stroke rate (sr) steps (18, 20, 22, 24, 30, 36 strokes per minute) for the overall duration of thirty rowing strokes. The lower frequencies equate to the average stroke rate of regular extensive training runs, where the higher stroke rates rather occur during spurt phases in a rowing race (or training session). Stroke rate 36 equates to the average pace of a rowing race.

The follow-up-study included additional, on-line[11] auditory information as direct acoustic feedback and was transmitted to the athletes via loudspeakers which were mounted inside the boat. The sound information could be (selectively) changed and switched on or off by remote-control, and was presented during selected periods of three to five minutes. Simultaneously, the coaches received the same acoustic feedback on-line transmitted into their motorboat but in contrast to the athletes the coaches could listen to the sound during the whole training session, and during periods in which the athletes did not receive the sound information.

Additionally, standardised questionnaires were used for athletes and coaches to examine the acceptance of the auditory information and its assessment in regards of effectiveness in on-water training sessions. The athletes answered the questionnaire directly after the training session. For the coaches, the sonified training run was synchronised with the video and presented in an investigation room. To give the coaches enough time to understand the function of the sound information in relation to the video, they could listen to the sound as often as they needed.

2.3 Measurement System

The training and measurement system Sofirow was developed in cooperation with BeSB GmbH Berlin to measure kinematic parameters of the propulsive boat motion: boat acceleration (a_{boat}) with an acceleration sensor (piezo-electric) (100 Hz) and horizontal boat velocity (v_{boat}) with the global positioning system (GPS) with a 4-Hz sampling rate. Sofirow was developed to represent the boat motion appropriately and to create movement-defined sound sequences of the boat motion as on-line auditory information. Standardised questionnaires for athletes and coaches were used to examine acceptance and effectiveness of the acoustic information.

2.4 Data Sonification

The sound result was created by using a traditional approach, normally used for rendering sonification from data: Parameter Mapping [49] [25]. Parameter attributes (frequency, amplitude, etc) were mapped by numeric values of the given data vector. Thus, the parameter of the propulsive boat motion was conveyed into auditory information by describing the relationship mathematically; each piece of acceleration data was related to a specific tone on the musical tone scale according to tone pitch. Every integer (whole number) equated to a specific semitone. The sound data represented

[11] The term "on-line" used in this article means the amount of time needed for activities in real-time (without any time lag).

changes in the boat motion by changes in tone pitch. The point of zero acceleration was at middle C on the tone scale (261.63 Hz). In other words, the resulting sound (melody) changed as a function of the boat motion.

The sound data created on-line was transmitted to the athletes as direct acoustic feedback via loudspeakers while they were rowing. For the analysis, selected rowing strokes were sonified and synchronised with the video of the training run to make audible the changes and variations in measured data which were not discernable on the video.

3 Results

3.1 Data

First of all, it was important to demonstrate that the basic requirements for the acoustic sonification of the boat motion, as an appropriate representation of the measured data, had been fulfilled. Therefore, different movement intensities needed to be identified which should display a distinct difference from one stroke rate step to the next. Furthermore, characteristic movement patterns as well as qualitative changes (such as increases or decreases of the boat motion) must be perceivable and differences have to be identifiable. The results showed significant differences between the various stroke rate steps (p=0.00), although the prescribed stroke rates were not achieved exactly by the athletes (Tab.1). Boat velocity increased (3.9 m/s – 5.4 m/s) with an increase in stroke rate (18 - 36 strokes per minute). This defined the required differences in movement intensities.

Table 1. Nominal stroke rate (sr) steps, recorded stroke rate steps (mean and standard deviation) and boat velocity (v_{boat})

	sr [strokes/minute]		v_{boat} [m/s]	
sr-steps	**mean**	**sd**	**mean**	**sd**
18	20.6	0.3	3.9	0.5
20	21.1	0.9	4.0	0.5
22	21.8	0.9	4.2	0.6
24	23.8	1.6	4.4	0.6
30	29.3	2.1	5.0	0.8
36	35.1	1	5.4	1

To detect characteristic movement patterns and any qualitative changes, it was important to display the information that is included in the acceleration trace. Therefore, the rowing cycle was divided into two characteristic main phases, drive (d) and recovery (r), that appear in every stroke cycle of the periodic boat acceleration trace and that are characterised as the two fundamental reference points in the rowing stroke (Fig.1).

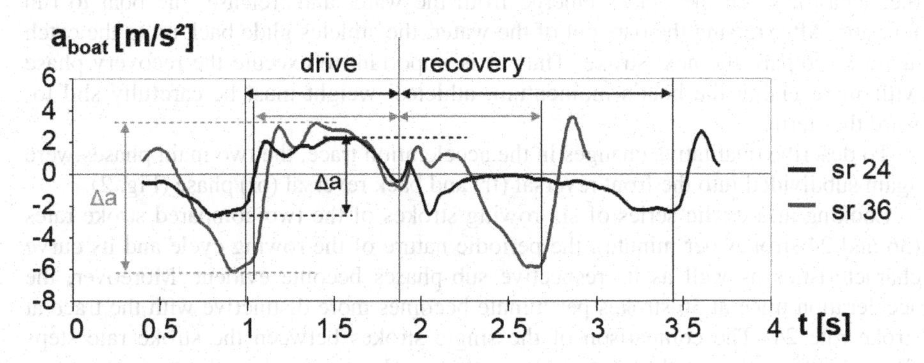

Fig. 1. Acceleration traces of one complete rowing cycle at stroke rates 24 and 36 strokes per minute with drive and recovery phase and characteristic curve(s)

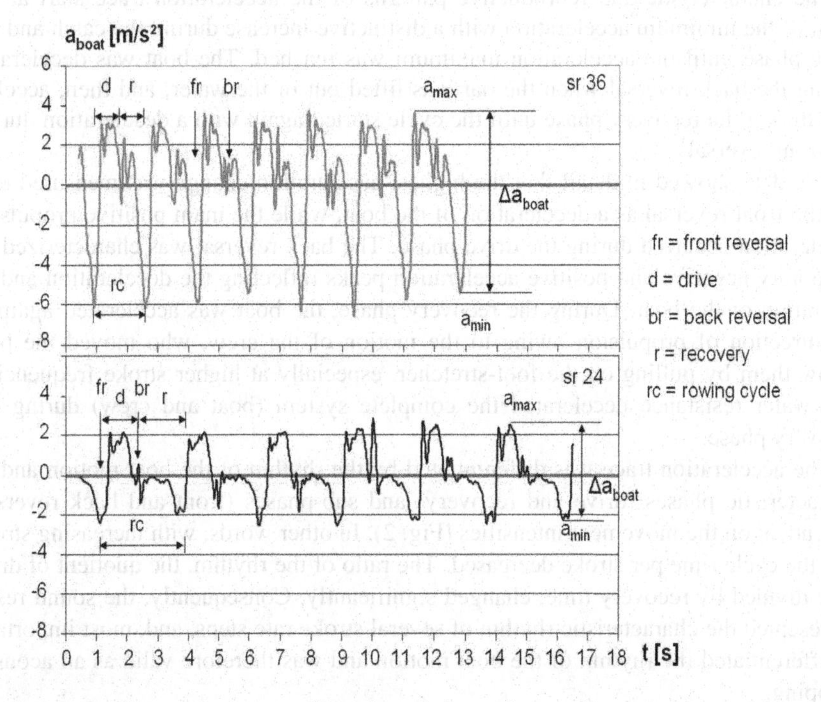

Fig. 2. Acceleration traces in a cyclic series of six rowing strokes at stroke rate 36 and 24 strokes per minute

Figure 1 showed a complete single rowing stroke for the stroke frequency of 24 strokes per minute compared to 36 strokes per minute. The characteristic curves with the two main phases are evident. The drive phase starts with the catch that is when the oarblades first 'catch' the water and accelerates the boat instantly, giving it the driving force. The end of the stroke is characterised by the recovery phase (also known as the release), when the blades emerge from the water and 'release' the boat to run forward. After raising the oars out of the water, the athletes glide back up to the catch again to prepare the next stroke. Thus, it is important to execute the recovery phase without reversing the boat's momentum; athletes' weight must be carefully slid toward the stern.

To describe qualitative changes in the acceleration trace, the two main phases were again subdivided into the front reversal (fr) and back reversal (br) phase (Fig. 2).

Looking at a cyclic series of six rowing strokes of the two compared stroke rates (36 and 24 strokes per minute), the periodic nature of the rowing cycle and its curve characteristics as well as its respective sub-phases become evident. Moreover, the acceleration trace at 36 strokes per minute becomes more distinctive with the trace at stroke rate 24. The comparison of the single strokes between the stroke rate steps showed differences in the magnitude of the acceleration and in the time structure; with increasing stroke rate the magnitude of the acceleration increased whereas the cycle time per stroke decreased.

The characteristic and reproductive patterns of the acceleration trace start at the point of the minimum acceleration with a distinctive increase during the catch and the drive phase until the acceleration maximum was reached. The boat was decelerated during the back reversal when the oar was lifted out of the water, and then, accelerated during the recovery phase until the cycle started again with a deceleration during the front reversal.

The data showed in detail that the highest acceleration change was measured during the front reversal as a deceleration of the boat, while the main positive propulsive acceleration occurred during the drive phase. The back reversal was characterized by temporary negative and positive acceleration peaks reflecting the deceleration andacceleration of the boat. During the recovery phase, the boat was accelerated again in the direction of propulsion owing to the motion of the crew, who moved the boat below them by pulling on the foot-stretcher, especially at higher stroke frequencies. The water resistance decelerated the complete system (boat and crew) during the recovery phase.

The acceleration trace was differentiated by the rhythm of the boat motion and its characteristic phases (drive and recovery) and sub-phases (front and back reversal) dependent on the movement intensities (Fig. 2). In other words, with increasing stroke rate the cycle time per stroke decreased. The ratio of the rhythm, the quotient of drive time divided by recovery time, changed significantly. Consequently, the sound result represented the characteristic rhythm of several stroke rate steps, and, most important, it differentiated the rhythm of the boat motion and was therefore valid as an acoustic mapping.

3.2 Coaches' Questionnaire

All coaches (100%[12]) answered the questionnaire regarding the acceptance of the concept and the acoustic feedback. They rated the sonified boat motion in the direction of propulsion and identified the characteristic phases of the rowing cycle (100%) (Fig. 3).

---

[12] Percentage descriptions must be interpreted in relation to the small number of tested subjects.

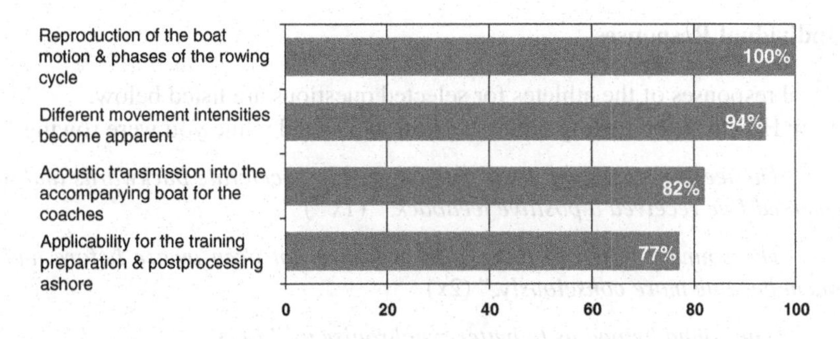

Fig. 3. Percentage of answers (coaches' questionnaire) taken in the follow-up-study

Different movement intensities (represented by the stroke rate steps) became apparent for 94% of the coaches and assistance for the coaches by transmitting the acoustic information into the accompanying boat was appreciated (82%). Using the sound result to support the preparation and post-processing training phases as a conceivable implementation, 77% of the coaches would take the benefit out of it.

3.3 Athletes' Questionnaire

All athletes (100%) answered the questionnaire positively regarding the acceptance of the concept and the acoustic feedback. They recognized the reproduction of the boat motion to 100% and every responding athlete (100%) could focus his attention on the various movement sections as a result of the guidance from the sound. Furthermore, 87.5% detected characteristic sub-phases of the boat motion reproduced in the sound result. For 75% of the athletes, provision of acoustic information during on-water training sessions is a promising training aid, and 50% of the athletes saw the sound result as not distracting their attention from their actual perception (Fig. 4).

Moreover, 50% of the athletes felt assisted and supported with regard to their individual perception of the movement in the way the sound provided the time structure. As an outcome of this, athletes derived a feeling for the time duration of the rowing cycle.

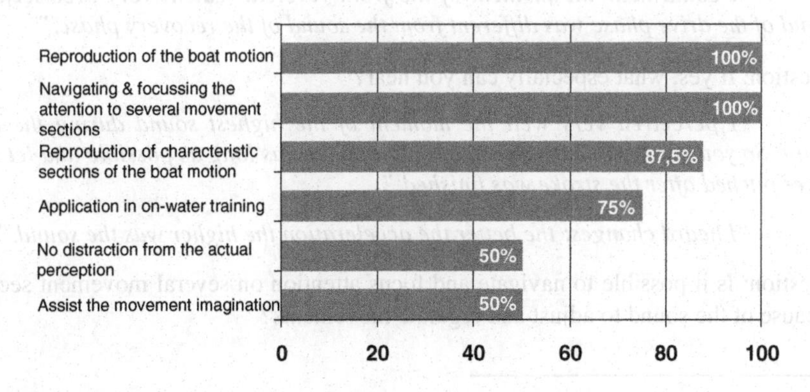

Fig. 4. Percentage of answers (athletes' questionnaire) taken in the follow-up-study

3.4 Individual Responses

Individual responses of the athletes for selected questions are listed below.
Question: How was the feeling with the additional sound while you were rowing?

"The feeling was good. I was encouraged to accelerate backwards and when implemented I've received a positive feedback." (1x[13])

"The sound pointed to things that we have not been aware before and the movement became more consciously." (2x)

"The sound helped us to better synchronise us." (4x)

"Everybody was more concentrated." (8+)

"The boat run is enhanced." (8+)

"It felt like playing with a Gameboy." (1x)

Question: Are the different movement intensities mapped clearly (by the several stroke rate steps)?

"During the drive phase there was a high pitched sound whereas during the recovery phase the sound was low pitched and, just before the blades touched the water it sounded very low pitched."

"You recognized the sound getting higher and shorter at higher frequencies."

"You recognized the shorter recovery phase at higher stroke frequencies in the sound."

Question: Is it possible to hear variations of successful and less successful strokes (e.g. if a stroke didn't feel successful, but another one did)? We mean variations between single strokes of the same stroke rate step.

"Especially a break at the front reversal (catch) was very good perceptible."

"I could hear the moment of the front reversal (catch) very precisely. The sound of the drive phase was different from the sound of the recovery phase."

Question: If yes, what especially can you hear?

"I perceived very well the moment of the highest sound during the drive phase. So you can try to keep the sound high pitched as long as possible and let it get lower pitched after the stroke was finished."

"I heard changes: the better the acceleration the higher was the sound."

Question: Is it possible to navigate and focus attention on several movement sections because of the sound to adjust and regulate movements?

---

[13] Terms used for the classification of the different types of boats: 1x: single scull; 2x: double scull; 4x: quad (or quadruple) scull; 8+: eight (always coxed).

"You can concentrate to the sound especially for the moment before the oar blades enter the water: trying not to decrease the sound too much to get the best boat acceleration."

"Especially at the finish position it was possible to 'play with the sound': you can try to keep the decrease of the sound as minimal as possible."

"You try to accelerate backwards to minimize the decrease of the sound before the catch."

"You can try to extend the rising sound at the end and keep it high pitched before the oar blades entered the water."

"Mistakes were easier detectable and become more distinctive. Furthermore it is possible to concentrate on specific movement sections."

Question: In which modality is the sound applicable for on-water training?

"Not during the whole session, sort of phased getting the chance to implement the 'heard strokes', then, turning the sound on again to compare the strokes. It can be helpful to listen to the sound recorded after the training session to compare the different strokes."

4 Discussion

This article presented conceptual considerations concerning a sound design for the optimization of sport movements with first practical experiences in elite rowing. The boat motion was verified and described acoustically (sonified). The propulsive acceleration trace was sonified using the training and measurement system Sofirow in order to make the measured differences in intensity between several stroke rate steps audible. Sofirow represented the acceleration trace acoustically in real-time by audibly differentiated intensity steps which were chronologically synchronised. Sonification in general offers an abundance of applications (e.g. [25]) and as acoustic feedback in particular, it seems to be a promising training aid for supporting elite sports, however there is still not enough practical experience. To get the desired benefit, it is important to utilise the sound information in a meaningful way in the training process. Therefore, conceptual knowledge is necessary of the design as well as of the way in which athletes perceive the sound data. That implies knowledge about the way acoustic information is perceived by the human brain and this above all involves knowledge about the relation of sound to its meaning [7].

Humans interact with the world by perceiving information all the time and learn to interpret as well as to filter it. As already mentioned, human senses fulfil an enormous task by managing sensory overload. In interaction with the world, humans learn to associate daily objects and events with various sounds that evoke as an inherent cause and imply at the same time a specific meaning. The sound perception while filling an empty bottle with water illustrates the direct physical relationship and the associated meaning: the timbre variations convey the information about the level of water in the bottle.

Another example shows that we can learn to differentiate between objects because of their sounds: a car being driven, stopped and parked sounds different owing to the sounds (aside from its being visibly motionlessness in the last case). It sounds louder with a more highly pitched tone the faster the car is driven and it does not make any sound if it is stationary – as simple as that. As soon the car starts to move, acceleration increases which is perceived by a higher tone pitch. Another example is the experience when hitting an object harder; this causes a louder sound which is learned as an association between the energy that was put into a system and its response. Several acoustic attributes were associated with objects and events with an acoustic result which might further be associated with danger. (A moving car can cause an accident.) There is evidence that humans are particularly sensitive to an increasing or decreasing tone pitch from one note to another and were able to remember the relationships between the pitches [3]. Moreover, humans are able to store and interpret meaning that is conveyed by sound determined by the physics of acoustic systems.[14] The human brain is tuned to exploit such relations and becomes the actual medium between sound and sense [64]. Accordingly, we learn unconsciously (not only from the car example), that if there is no sound, there is no movement and that with increasing movement, the sound increases in tone pitch which is perceived at the same time as an increase in loudness according to the Weber-Fechner law[15] [11]. The principles of pitch perception in tones are well documented (e.g. [66] [67] [68]) and the use of tone pitch for displaying data is facilitated because most of the listeners are familiar with changing pitches in everyday sounds, and furthermore can detect minimal changes in tone pitch. Moreover, pitch represents more evenly a broader range of values than other perceived dimensions of sound (e.g. loudness)[16] [69].

4.1 Data

The sonified boat motion (thought of as the 'melody' of rowing) changed as a function of the acceleration trace which means that athletes perceived an increasing tone pitch the more the boat was accelerated. Tone pitch by definition "is the perceptual correlate of periodicity in sounds" [3] whose waveforms repeat in a time period comparable to the periodic rowing cycle. One period is characterised as the smallest repeating unit of a signal and describes the periodic signal completely. To detect the periodic characteristics, at least two concatenated periods have to form a repeating pattern in the signal. 'Repetition' (in music) is described as the most elementary and at the same time emphatic and formative force [56]. The periodic recurrence of defined sections which are known as a refrain (referring to a piece of music) awakens a sensitivity for defined details in the sequence which is linked to thinking processes resembling perception (in general) which is selective and constructive and based on earlier individual experiences (for details see [70]). To overcome information overload, individuals tend to focus attention on those stimuli or events that are of crucial

[14] As Lambert described regarding rowing: "Naturally, more oars make more sounds (…)." [38]

[15] The psychophysical law of perception from Weber and Fechner describes the relationship between stimuli intensity and stimuli sensation. (For details see [11] [65].)

[16] Disadvantages e.g. of loudness (at its extremes): it does not provide an effective display dimension, because soft sounds are masked by e.g. ambient noise, and loud sounds can be potentially disturbing.

relevance for the current situation.[17] It is possible to focus the attention consciously to specific stimuli or events in the environment as soon someone gets sensitized (apart from stimuli that are unconsciously controlled).

The sonified boat motion, as a short tonal sequence, represents the rowing cycle which follows one cycle to another like a mosaic that emerges during perception and that does not need to contain any oral or textual explanation. Awareness of the structure emerges solely from the knowledge of the movement and the visual and auditory interaction. The circular representation corresponds to the form of the rowing movement whose elements return in a recurrent and circular way. Each sub-part of the movement as a mosaic of the total movement corresponds to a short sound sequence as a piece of the total sound sequence. Every change in tone pitch represents an attribute of the movement (e.g. higher tone pitch for an increasing acceleration). Nuances of pitch change differentiate the variations inside the cyclic movement sequence (intra-cyclic variations) and make it is possible to detect changes in the cycle that also repeat in every rowing stroke with minimal variability. The sound sequence takes into account the cognitive processes of attention and memorisation. It also makes it possible to establish bonds of similarity and consequently gives direction to the formal structure which proceeds under the ears of the athlete (and coach). That opens the possibility to guide athletes to focus their attention to specific sequential events in the movement.

Since the functional attribution of tone pitch is a function of changes in acceleration, the sound result appeared equivalent to the way the auditory information is encoded in the human brain. The physical model of the boat motion as it runs through the water makes the sound particularly informative; detailed information about velocity and periods of acceleration and deceleration were conveyed as a result of the parameters and the instantaneous states of the objects in contact (boat and water resistance) and interacting. Thus, the sound result represents more than the natural sounds of the boat motion (which were already used from the athletes [37]). The dynamic behaviour of the macroscopic characteristic profile of the complex model was represented in the periodic patterns (of tone pitch) and intensity which are caused by the boat acceleration. With it, the sound data remains in correspondence to the boat motion. This follows the principle of simplification of the physical model (according to De Götzen et al. [22]), and besides a higher intelligibility, it creates the possibility of using the complexity of human nature in order to control the parameters of the model. Listening to the movement as an acoustic totality, the sound data becomes intuitively comprehensible to and applicable by the athletes.

The dynamic relationship of tone pitch to acceleration facilitates an intuitive understanding of the sound sequence that corresponds to experiences in everyday situations (e.g. as previously mentioned, a car's engine noise). Humans develop sensitivities to the regularities of a tonal system just by listening to noises in everyday life without being necessarily able to verbalise them [19].

[17] When someone gets sensitized to runners or to a special running style for example (may be because he started running recently), he tends to focus his attention in the future to other runners or to the special running style and suddenly it seems that he meets runners everywhere or people who have this particular running style, although there are not more than before – he just was not aware of them.

In this way, the physics-based algorithm which was used to create the movement defined sound sequences, simplifies and abstracts the complex parameters and facilitates efficient implementation for the athletes. The sound result maps the data and several events in the rowing cycle effectively in a way that is perceptually and cognitively meaningful.

4.2 Questionnaires

The results of the questionnaires showed basic agreement amongst coaches and athletes in their responses regarding the acceptance and effectiveness of the concept in general and the sound result in particular. The individual perception of the sound result was different, which became evident in athletes' responses, and confirmed the concept of 'subjective theories' [71] that contrast the outside view (as the visible movement e.g. from the coach) with the inside view (as the conscious cognition of the athlete) [72]. With it, human perception and its interpretation depends on the subject and this "is shared with other perceptualization techniques that bridge the gap between data and the human sensory system" [49]. It must be noticed, that an objective description of a situation (e.g. the description of a musical structure) does not always correspond with the actual perception of a subject [73]. Two people can watch (or listen to) the same thing, but see (or hear) it differently. Subjects' views can differ, because humans interact and communicate with the world by using the knowledge of earlier experiences. In doing so, in a metaphoric sense, every situation (including sensory stimuli) passes through an internal filter that compares and combines the situation with previous individual experiences of situations that were remembered because of their significance for the individual [70]. Thus, perception is always an active and selective process. Every individual creates his[18] own perception of reality as a result of his own experiences. Furthermore, research in human physiology showed that perception in principle is limited by the properties of the human sensory system: in this way, the brain has to fill a gap left open by the perception sensors, the visual nerves for example. Humans are only able to see a picture as a whole because the brain puts together the incomplete information it receives from the eyes that consists of two displaced pictures (for details [11]). Humans are in principle not able to see a thing or a situation in an objective way (aside from its individual background). It is in every case a reconstruction of the so called 'reality' which is always the reality of every single individual [74]. "Perception is a complex interplay between the internal knowledge (i.e. what is already stored in memory) of each individual and the external sensory stimuli" [75]. According to Piaget, perception depends on the individual and his world view which is described as the result of arranged perceptions into existing images by imagination. In summary, perception is a part of imagination.

According to this, every athlete has his own way to experience the feeling of rowing and realizes rowing technique differently [37] and with it, the optimisation of his movements. Learning approaches differ from subject to subject and, consequently, every athlete realises the rowing movement and technique differently and in a personal way and time. There are many ways to reach the same goal. That consequently

[18] For ease of exposition, only the masculine form is used in this article. The female form is always implied.

means that every athlete needs a different way of instruction and feedback. One may prefer visual input, such as figures or graphs or just watching the blades, while another prefers the acoustic stimuli by hearing the boat. An additional (externally) presented sound can not affect every athlete in the same way, except aesthetically. However, it can help getting an enhanced feeling for the rhythm and duration of the movement and can bridge the gap in the psychological interaction between coach and athlete; the presented information (sound) is consistent with its cause (the physical movement) and it is conveyed by the same modality of senses [76]. It is, in contrast to verbal instructions, intelligible to all. Out of this it is possible to form an impression of the movement on the basis of earlier experiences and to create so called 'memory traces' [77] which become more pronounced the more they are used. It becomes easier to recover a specific trace with the number of repetitions of use.[19] Mulder [60] names such traces 'ideas' that are a result out of the relation between movements and feedback information. With it, it is possible to create expectations that can be satisfied by the received sensory information while executing the movement.

A further important aspect for the purposes of the study is the regularity of the feedback; it should neither be used in every training session nor during the whole time, but rather selectively, to sensitise the athletes' focus of attention towards specific sections of the rowing movement, or, more generally, towards the rhythm of the movement; e.g., in specific situations, such as when athletes change boat classes (from a pair to a four). The sound data may help the athletes to find the characteristically rhythm faster. Despite this, the sound data should be presented in addition to the natural soundscape.

A longer-term use of the sonified boat motion needs athletes' acceptance and further experiences in training sessions. The positive responses of the athletes in regards to the sound sequence indicated that functionality may be a primary aspect in the aesthetic appreciation of sonification in elite sports [78]. The presented sound sequence fulfilled according to Henkelmann [79] 'at least bearable if not enjoyable' aspects. But this again, depends on the individual perception and its interpretation, and therewith individuals' aesthetic sensations.

5 Conclusions and Prospects

The many ways that sound (in all its variations) influences human perception offers promising possibilities and opens up new vistas in different scientific fields (e.g. [25] [80] [22]). With sonification, that was intended as information representation by means of sound, interdisciplinary work is possible precisely because it is still an open research field (including communication theory, sound design, cognitive psychology, psychoacoustics and other disciplines). Recent work in sonification across different disciplines had and still has trouble to convey the information in an optimal way and to bring it appropriately to the target users. Probably, the main difficulty with sonification practice is still the artificial, or fatiguing and annoying, or even irritating and distracting, and in the worst case misleading effect of the sound result. Therefore, De Götzen et al. [22] ask for the development of clear strategies and examples of sound

[19] Metaphorically spoken, like a trail that becomes to a beaten path.

designs that can convey information at least comprehensibly and in the best cases in a pleasurable way.

This article has tried to contribute to research in sport as well as in sonification for the purposes of movement optimisation in elite sports using the example of rowing. As already mentioned sonification offers an abundance of applications in elite sports. The sound result as acoustic information, presented here, offers a promising application for technique training and provides an alternative to existing feedback systems in elite rowing [46]. Additionally, external acoustic information, produced from the movement, introduces a new approach to analysis and optimisation of technique in elite rowing. Therefore, it is possible to describe the ideal boat movement by the rhythm of the boat acceleration and that, in turn, implies an option for rhythm education in racing boats. It opens new opportunities for technique training and tactics in racing.

Further studies will pursue and evaluate the effectiveness of the sonified boat motion and its longer-term usage. Furthermore, the sound of the 'optimal movement' will be created and used in feedback training as a guideline or 'reference sound', bringing the athletes closer to their main goal: to cover the race distance in the shortest time possible which is mainly influenced by minimising periods of negative boat acceleration.

To create a comprehensible and maybe even pleasurable sound result, movement (in general) is well suited to provide its data because movement is characterised as an 'ancestral impulse' of music. It is a propulsive force which expresses itself in melodic structures and is borne by the force of the rhythm [56].

The positive responses to the sound result in this study were encouraging but also, at the same time, raised questions about different designs and hypotheses. Just what are the aesthetic dimensions of sonification in general and in particular for the purposes of elite sports?

Does the sound affect long-term usage? And do the aesthetics of sonification include the psychological and emotional effects that music has on athletic performance? Should different sports or different teams sound different?

Do the nature of the sonification sounds in principle have any effect on performance? And last but not least, could the sonification presented here, become a standard for all sports sonifications?

In a further approach, aspects of mental training will be involved as preparation for training and competition since it is known that many elite athletes take advantage of mental training strategies (e.g. [70]). Therefore, the creation of 'reference sound sequences' is considered [81] that could be heard prior to the race while imagining the race situation mentally without being physically in motion at the same time (according to Eberspächer [82]). There is evidence, that imagination also plays a key role in learning processes (e.g. [83]).[20] Thus, athletes create a 'picture' of the movement as an 'imaginary anticipation' on the basis of previous experienced movements and involved feelings. While trying to re-enter a specific situation mentally, people remember details that thought it was forgotten [84].

[20] An example (which is at the same time an ideal training for) imagination is to listen to a narrative storytelling, where the exactness of the chosen words is the basic factor to 'evoke worlds' [84].

A completely different approach in further studies could include the creation of different pieces of music for each respective elite sport; e.g. for rowing, that could motivate the athletes in a different way for long lasting (extensive endurance) training sessions, since it is known that listening to a piece of music inhibits sensations of fatigue and make the exercise experience more pleasurable [85]. Intermingling each day's rowing training with sections in which pieces of music are presented to the athletes, helps them to synchronise their strokes to the tempo of the music and according to this, the interaction and synchronisation among the crew can be enhanced, whilst expecting the athletes to move together with the 'flow' which put them into an optimal state of absorption [39] that include a distorted sense of time. 'Being in the zone' lets the athletes experience a feeling of optimal boat run, apparently effortlessly. "When strokes synchronise perfectly, the crew pulls in phase, like light waves in a laser beam, and, as with a laser, the energies reinforce each other and multiply. To the crew, an eight-oared boat in peak form feels rowed by a single oar, and in a sense it is." [38]

Thus, the flow state implies a kind of focused attention in a sense of control and concentration that causes a transcendence of normal awareness (e.g. the loss of self-consciousness) [86]. This intrinsically rewarding effect let the athletes perceive a personal control over the activity and with it they fully internalise mastery of the sport.

Acknowledgements. First, we would like to thank the German Federal Institute of Sport Science (BISp) for financially supporting this research project (IIA1-070802/09). Second, we thank the engineers of BeSB GmbH Berlin (http://www.besb.de/), namely Edelbert Schaffert, Reiner Gehret and Sebastian Schroeter, for developing the measurement system and sound device Sofirow and their technical support at every time. We want to thank as well Bruce Grainger for his help with the revision of some English expressions. And, finally, we thank the coaches and athletes of the German junior national rowing team and the German Rowing Association (DRV) for the great cooperation.

References

1. Bruhn, H., Kopiez, R., Lehmann, A.C. (eds.): Musikpsychologie. Das neue Handbuch. Reinbek bei Hamburg (2008)
2. Auhagen, W.: Rhythmus und Timing. In. Bruhn, H., Kopiez, R., Lehmann, A.C (eds.). Musikpsychologie. Das neue Handbuch. Rowohlt Verlag. Reinbek bei Hamburg, pp. 437–457 (2008)
3. McDermott, J.H., Oxenham, A.J.: Music perception, pitch, and the auditory system. Curr. Opin. Neurobiol. 18, 1–12 (2008)
4. Brown, S.: Biomusicology, and three biological paradoxes about music. Bulletin of Psychology and the Arts 4, 15–17 (2003)
5. Effenberg, A.O.: Sonification – ein akustisches Informationskonzept zur menschlichen Bewegung. Verlag Schorndorf, Hofmann (1996)
6. Mechling, H., Effenberg, A.O.: Perspektiven der Audiomotorik. In. Praxisorientierte Bewegungslehre als angewandte Sportmotorik. Leipziger Sportwissenschaftliche Beiträge. Sankt Augustin, pp. 51–76 (1998)
7. Hermann, T., Ritter, H.: Sound and meaning in auditory data display. Proceedings of the IEEE 92(4), 730–741 (2004)

8. Effenberg, A.O., Mechling, H.: Akustisch-rhythmische Informationen und Bewegungskontrolle. Von der rhythmischen Begleitung zur Sonification. Motorik, Schorndorf 22, 4 (1999)
9. Effenberg, A.O.: Der bewegungsdefinierte Sound: Ein akustisches Medium für die Darstellung, Vermittlung und Exploration motorischer Prozesse. In: Altenberger, A., Hotz, U., Hanke, K. (eds.) Medien im Sport – zwischen Phänomen und Virtualität, pp. 67–76. Hofmann, Schorndorf (2000)
10. Fischinger, T., Kopiez, R.: Wirkungsphänomene des Rhythmus. In: Bruhn, H., Kopiez, R., Lehmann, A.C. (eds.) Musikpsychologie. Das neue Handbuch, pp. 458–475. Rowohlt Verlag, Reinbek bei Hamburg (2008)
11. Birbaumer, N., Schmidt, R.F.: Biologische Psychologie, Auflage, vol. 6. Springer Medizin Verlag, Heidelberg (2006)
12. Avanzini, F.: Interactive Sound. In: Polotti, P., Rocchesso, D. (eds.) Sound to Sense, Sense to Sound. A State of the Art in Sound and Music Computing, pp. 345–396. Logos Verlag, Berlin (2008)
13. Fricke, J.P., Louven, C.: Psychoakustische Grundlagen des Musikhörens. In: Bruhn, H., Kopiez, R., Lehmann, A.C. (eds.) Musikpsychologie. Das neue Handbuch, pp. 413–436. Rowohlt Verlag, Reinbek bei Hamburg (2008)
14. Gaver, W.W.: What in the world do we hear? An ecological approach to auditory event perception. Ecological Psychology 5(1), 1–29 (1993a)
15. Gaver, W.W.: Synthesizing auditory icons. In: Proc. of the INTERACT and CHI Conf. on Human Factors in Computing Systems, Amsterdam, The Netherlands, pp. 228–235 (1993b)
16. Hellbrück, J.: Das Hören in der Umwelt des Menschen. In: Bruhn, H., Kopiez, R., Lehmann, A.C. (eds.) Musikpsychologie. Das neue Handbuch, pp. 17–36. Rowohlt Verlag, Reinbek bei Hamburg (2008)
17. Cremer, L., Heckl, M., Petersson, B.A.T.: Structure-Borne Sound: Structural Vibrations and Sound Radiation at Audio Frequencies, 3rd edn. Springer, Berlin (2005)
18. Cremer, L., Hubert, M.: Vorlesungen über Technische Akustik, 4th edn. Springer, Heidelberg (1990)
19. Krumhansl, C.L.: Cognitive foundation of musical pitch. Oxford University Press, New York (1990)
20. Kohler, E., Keysers, C., Umilta, M.A., Fogassi, L., Gallese, V., Rizzolatti, G.: Hearing sounds, understanding actions: action representation in mirror neurons. Science 297, 846–848 (2002)
21. Schafer, M.: Soundscape - Our Sonic Environment and the Tuning of the World. Destiny Books, Rochester (1994)
22. De Götzen, A., Polotti, P., Rocchesso, D.: Sound Design and Auditory Displays. In: Polotti, P., Rocchesso, D. (eds.) Sound to Sense, Sense to Sound. A State of the Art in Sound and Music Computing, pp. 397–445. Logos Verlag, Berlin (2008)
23. Eckel, G., De Campo, A., Frauenberger, C., Höldrich, R.: SonEnvir – Eine Sonifikationsumgebung für wissenschaftliche Daten. Uni Graz (2005), `http://www.interactive-sonification.org/ISon2007/`
24. Gaver, W.W.: How do we hear in the world? Explorations of ecological acoustics. Ecological Psychology 5(4), 285–313 (1993c)
25. Kramer, G., Walker, B., Bonebright, T., Cook, P., Flowers, J., Miner, N., Neuhoff, J.: Sonification Report: Status of the Field and Research Agenda. NSF Sonification White Paper– Master 12/13/98, 5 (1999) `http://sonify.psych.gatech.edu/publications/pdfs/1999-NSF-Report.pdf`

26. Höner, O., Hermann, T., Grunow, C.: Sonification of Group Behavior for Analysis and Training of Sports Tactics. In: Proc. Int. Workshop of Interactive Sonification (ISon), Bielefeld, Germany (2004)
27. Bregman, A.S.: Auditory scene analysis: Hearing in complex environments. In: McAdams, S., Brigand, E. (eds.) Thinking in sound: The cognitive psychology of human audition, pp. 10–36. Clarendon Press/Oxford University Press, New York (1990)
28. Cherry, E.C.: Some experiments on the recognition of speech, with one and with two ears. Journal of the Acoustical Society of America 25, 975–979 (1953)
29. Thaut, M.H., Tian, B., Azimi-Sadjadi, M.R.: Rhythmic finger tapping to cosine-wave modulated metronome sequences: Evidence of subliminal entrainment. Human Movement Science 17(6), 839–863 (1998)
30. Effenberg, A.O.: Bewegungs-Sonification und multisensorische Integration: Empirirsche Befunde und zukünftige Perspektiven. In: Effenberg, A.O. (ed.) Bewegungs-Sonification und Musteranalyse im Sport – Sportwissenschaft trifft Informatik, pp. 42–47. Cuvillier Verlag, Göttingen (2006)
31. Wulf, G., Prinz, W.: Directing attention to movement effects enhances learning: A review. Psychonomic Bulletin & Review 8, 648–660 (2001)
32. Wulf, G.: Attentional focus and motor learning: A review of 10 years of research. In: Hossner, E.-J., Wenderoth, N. (eds.) Gabriele Wulf on attentional focus and motor learning [Target article]. E-Journal Bewegung und Training, vol. 1, pp. 4–14 (2007)
33. Bigand, E., Lalitte, P., Tillmann, B.: Learning Music: Prospects about Implicit Knowledge in Music, New Technologies and Music Education. In: Polotti, P., Rocchesso, D. (eds.) Sound to Sense, Sense to Sound. A State of the Art in Sound and Music Computing, pp. 47–81. Logos Verlag, Berlin (2008)
34. Seger, C.A.: Implicit learning. Psychological Bulletin 115, 163–169 (1994)
35. Pauletto, S., Hunt, A.: A comparison of audio & visual analysis of complex timeseries data sets. In: Proc. of ICAD 11th Meeting of the International Conference on Auditory Display, Limerick, Ireland, July 6-9 (2005)
36. Loosch, E.: Allgemeine Bewegungslehre. Wiebelsheim (1999)
37. Lippens, V.: Inside the rower's mind. In: Nolte, V. (ed.) Rowing faster, pp. 185–194. Human Kinetics, Inc (2005)
38. Lambert, C.: Mind over water. Lessons from life from the art of rowing. Mariner book. Houghton Mifflin Company, Houston (1999)
39. Csikszentmihalyi, M. (ed.): Optimal experience: psychological studies of flow in consciousness. Cambridge Univ. Press, Cambridge (1992)
40. Jackson, S.A., Csikszentmihalyi, M. (eds.): Flow in sports. Human Kinetics Pub Inc. (1999)
41. Runeson, S.: On the possibility of "smart" perceptual mechanisms. Scandinavian Journal of Psychology 18, 172–179 (1977)
42. Burt, P.J.: Smart sensing. In: Freeman, H. (ed.) Machine vision, pp. 1–30. Academic Press, San Diego (1988)
43. Effenberg, A.O.: Movement Sonification: Effects on Perception and Action. IEEE Multimedia, Special Issue on Interactive Sonification 12(2), 53–59 (2005)
44. Effenberg, A.O., Weber, A., Mattes, K., Fehse, U., Kleinöder, H., Mechling, H.: Multimodale Informationen beim Bewegungslernen. In: SportStadtKultur, Hamburg, Germany, September 26-28. Sportwissenschaftlicher Hochschultag der Deutschen Vereinigung für Sportwissenschaft, vol. 18, pp. 210–211. Verlag, Czwalina (2007)

45. Walker, B.N., Kramer, G.: Sonification. In: Karwowski, W. (ed.) International Encyclopedia of Ergonomics and Human Factors, 2nd edn., pp. 1254–1256. CRC Press, New York (2006)
46. Mattes, K., Böhmert, W.: Biomechanisch gestütztes Feedbacktraining im Rennboot mitdem Processor Coach System-3 (PCS-3). In: Krug, J., Minow, H.-J. (eds.) Sportliche Leistung und Techniktraining. Schriften der deutschen Vereinigung für Sportwissenschaft, vol. 70, pp. 283–286. Academia, Sankt Augustin (1995)
47. Anderson, R., Harrison, A., Lyons, G.M.: Rowing. Sports Biomechanics 4(2), 179–195 (2005)
48. Effenberg, A.O.: Multimodal Convergent Information Enhances Perception Accuracy of Human Movement Patterns. In: Proc. 6th Ann. Congress of the ECSS, p. 122. Sport und Buch, Strauss (2001)
49. Hermann, T.: Taxonomy and Definitions for Sonification and Auditory Display. In: Proc. 14th Int. Conference on Auditory Display (ICAD), Paris, France, June 24-27 (2008)
50. Sonification.de, `http://www.sonification.de/main-def.shtml`
51. Brown, L.M., Brewster, S.A., Ramloll, R., Burton, M., Riedel, B.: Design guidelines for audio presentation of graphs and tables. In: International Conference on Auditory Display (ICAD), Boston, MA, pp. 284–287 (2003)
52. Nees, M.A., Walker, B.N.: Listener, task, and auditory graph: Toward a conceptual model of auditory graph comprehension. In: 13th International Conference on Auditory Display (ICAD), Montreal, Canada, June 26-29, pp. 266–273 (2007)
53. Erkut, C., Välimäki, V., Karjalainen, M., Penttinen, H.: Physics-Based Sound Synthesis. In: Polotti, P., Rocchesso, D. (eds.) Sound to Sense, Sense to Sound. A State of the Art in Sound and Music Computing, pp. 303–343. Logos Verlag, Berlin (2008)
54. Barrass, S.: Auditory Information Design. Unpublished Ph.D. Thesis, Australian National University (1997) `http://thesis.anu.edu.au/public/adt-ANU20010702`
55. Leman, M., Vermeulen, V., De Voogdt, L., Taelman, J., Moelants, D., Lesaffre, M.: Correlation of gestural musical audio cues and perceived expressive qualities. In: Camurri, A., Volpe, G. (eds.) GW 2003. LNCS (LNAI), vol. 2915, pp. 40–54. Springer, Heidelberg (2004)
56. Kühn, C.: Formenlehre, vol. 8. Aufl. Bärenreiter-Verlag Kassel, Basel (2007)
57. Meinel, K., Schnabel, G.: Bewegungslehre – Sportmotorik. Abriss einer Theorie der sportlichen Motorik unter pädagogischem Aspekt, Aufl. Aachen, vol. 1. Verlag, Meyer & Meyer Sport (2007)
58. Crew, H.: The Principles of Mechanics, p. 43. BiblioBazaar, LLC (2008)
59. Schaffert, N., Gehret, R., Effenberg, A.O., Mattes, K.: The sonified boat motion as the characteristic rhythm of several stroke rate steps. In: O'Donoghue, P., Hökelmann, A. (eds.) Book of Abstracts. VII. World Congress of Performance Analysis of Sport, Magdeburg, Germany, September. 3rd-6th, p. 210 (2008)
60. Mulder, T.: Das adaptive Gehirn. Thieme, Stuttgart (2007)
61. Böger, C.: Der Bewegungsrhythmus – grundlegendes Prinzip beim Lernen und Lehren von Bewegungen? In: Moegling, K. (ed.) Integrative Bewegungslehre Teil II. Wahrnehmung, Ausdruck und Bewegungsqualität. Bewegungslehre und Bewegungsforschung, Kassel, Germany, vol. 14, p. 161 (2001)
62. Meinel, K.: Bewegungslehre. Volk und Wissen Berlin (1977)
63. Csikszentmihalyi, M.: Beyond Boredom and Anxiety. Jossey-Bass Inc., San Francisco (1975)
64. Leman, M., Styns, F., Bernardini, N.: Sound, Sense and Music Mediation: a Historical-Philosophical Perspective. In: Polotti, P., Rocchesso, D. (eds.) Sound to Sense, Sense to Sound. A State of the Art in Sound and Music Computing, pp. 15–46. Logos Verlag, Berlin (2008)

65. Goldstein, B.: Wahrnehmungspsychologie: eine Einführung, vol. 2. Spektrum, Heidelberg (2001)
66. Stevens, S.S.: On the psychophysical law. Psychological Review 64, 153–181 (1957)
67. Deutsch, D. (ed.): The psychology of music. Academic Press, New York (1982)
68. Moore, B.C.J.: An introduction to the psychology of hearing, 3rd edn. Academic Press, London (1989)
69. Walker, B.N., Ehrenstein, A.: Pitch and pitch change interact in auditory displays. Journal of Experimental Psychology 6, 15–30 (2000)
70. Mayer, J., Hermann, H.-D.: Mentales Training. Springer, Heidelberg (2009)
71. Groeben, N.: Explikation des Konstrukts "Subjektive Theorie". In: Groeben, N., Wahl, D., Schlee, J., Scheele, B. (eds.) Forschungsprogramm Subjektive Theorien. Eine Einführung in die Psychologie des reflexiven Subjekts, pp. 17–24. Francke, Tübingen (1988)
72. Adam, K.: Die Entstehung der modernen Trainingsformen. Rudersport, Minden, 80(31), Lehrbeilage 2 (1962)
73. Schaeffer, P.: Traité des Objets Musicaux, Paris, France. Éditions du Seuil (1966)
74. Roth, G., Menzel, R.: Neuronale Grundlagen kognitiver Leistungen. In: Dudel, J., Menzel, R., Schmidt, R.F. (eds.) Neurowissenschaft – Vom Molekül zur Kognition, pp. 543–562. Springer, Heidelberg (2001)
75. Spitzer, M.: Selbstbestimmen. Gehirnforschung und die Frage. Was sollen wir tun?. Spektrum, Heidelberg (2003)
76. Mechling, H.: Lerntheoretische Grundlagen von Feedback-Prozeduren bei sportmotorischem Techniktraining. In: Daugs, R. (ed.) Medien im Sport. Die Steuerung des Technik-Trainings durch Feedback-Medien, Berlin, pp. 9–33 (1986)
77. Spitzer, M.: Lernen. Gehirnforschung und die Schule des Lebens. Spektrum, Heidelberg (2002)
78. Schaffert, N., Mattes, K., Barrass, S., Effenberg, A.O. (accepted): Exploring function and aesthetics in sonifications for elite sports. In: 2nd International Conference on Music Communication Science (ICoMCS2), Sydney, Australia, (December 3rd-4th, 2009)
79. Henkelmann, C.: Improving the Aesthetic Quality of Realtime Motion Data Sonification. Computer Graphics Technical Report CG-2007-4, University of Bonn (2007)
80. De Campo, A., Dayé, C., Frauenberger, C., Vogt, K., Wallisch, A., Eckel, G.: Sonification as an interdisciplinary working process. In: Proceedings of the 12th International Conference on Auditory Display (ICAD), London, UK, June 20–23 (2006)
81. Schaffert, N., Mattes, K.: Sonification of the boat motion to improve the boat run during technique training and racing in rowing. In: Book of Abstracts. 13th Annual Congress of the ECSS, Estoril, Portugal, July 9-12, p. 469 (2008)
82. Eberspächer, H.: Mentales Training. Das Handbuch für Trainer und Sportler. München Copress (2001)
83. Kieran, E.: Imagination in Teaching and Learning. University of Chicago Press, Chicago (1992)
84. Sallis, J.: Force of Imagination: The Sense of the Elemental. Indiana University Press (2000)
85. Karageorghis, C.I., Mouzourides, D.A., Priest, D.L., Sasso, T.A., Morrish, D.J., Walley, C.J.: Journal of Sport & Exercise Psychology 31(1), 18–36 (2009)
86. Hunter, J., Csikszentmihalyi, M.: The Phenomenology of Body-Mind: The Contrasting Cases of Flow in Sports and Contemplation. Anthropology of Consciousness 11(3-4), 15 (2000)

Surface Interactions for Interactive Sonification

René Tünnermann, Lukas Kolbe, Till Bovermann, and Thomas Hermann

Ambient Intelligence Group
Cognitive Interaction Technology - Center of Excellence (CITEC)
Bielefeld, Germany
{rtuenner,lkolbe,tboverma,thermann}@techfak.uni-bielefeld.de
http://www.cit-ec.org/

Abstract. This paper presents novel interaction modes for Model-Based Sonification (MBS) via interactive surfaces. We first discuss possible interactions for MBS on a multi-touch surface. This is followed by a description of the *Data Sonogram Sonification* and the *Growing Neural Gas Sonification Model* and their implementation for the multi-touch interface. Modifications from the original sonification models such as the *limited space scans* are described and discussed with sonification examples. Videos showing interaction examples are provided. Furthermore, the presented system provides a basis for the implementation of known and novel sonification models. We discuss the available interaction modes with multi-touch surfaces and how these interactions can be profitably used to control spatial and non-spatial sonification models.

Keywords: Sonification, Model-Based Sonification, Data Mining, Interactive Surfaces.

1 Introduction

Exploratory Data Analysis aims to develop techniques for users to better *grasp* the hidden structure in complex data. If we take this statement literally, we might not only ask how we could implement techniques to *manually* interact and get our hands on data, but also how it sounds – or should sound – if we interact physically with data. Real-world acoustic responses that we experience when touching (hitting, scratching, tapping, etc.) an object or surface are often very useful and reveal a whole range of information about the object's properties (material, stiffness, surface properties, etc.). We often underestimate the utility of such direct feedback since it is omnipresent and at the same time effortlessly integrated into our multi-modal perceptions.

The arising questions are how can we inherit the benefits of action-perception loops for a better understanding of complex data and how can we structure surface-based interfaces in a way that users obtain an informative acoustic reaction on arbitrary interactions? Model-Based Sonification takes these aspects of interaction particularly into account [11]. *Sonification models* according to MBS can be excited by the user. For this excitatory process many different interaction interfaces beyond the mouse, such as the audio-haptic ball interface,

S. Ystad et al. (Eds.): CMMR/ICAD 2009, LNCS 5954, pp. 166–183, 2010.

or the malleable user interface, have been presented [12,13]. These are primarily input interfaces and the sonification in many implementations has been the only output modality in the interaction loop.

In this paper, we investigate the above research questions by using *interactive surfaces*. We start by presenting the tDesk system, a device developed within the Ambient Intelligence Group that combines the possibilities of Multi-Touch Interactions and Tangible Interactions in a desk-based system for simultaneous multi-user use. Our multi-touch system allows to create tightly coupled audiovisual interaction loops to represent the temporal evolution of sonification models while at the same time allowing real-time complex manual interaction with a sonification model. The system has been developed to serve as a sound basis to fuse and explore the potential of multi-touch interactions together with tangible interactions, while using truly multi-modal output media. In Sec. 4 we provide categories for flat surface-based interaction and then use these to discuss how interactions can be connected to the excitation for Model-Based Sonifications. We demonstrate the system together with two specific sonification models. The *Data Sonogram Sonification Model* allows the user to use multi-point interactions to set centers of excitation waves that spherically pass through data space. The *Growing Neural Gas Sonification Model* allows the user to listen to the growing neural gas during its adaption process and to visually and auditorily explore the state of the network with the help of multi-touch interaction.

In comparison to the mouse-based interaction used in previous implementations of these two, the multi-touch interaction provokes new interaction styles such as rapid A/B-comparison and simultaneous excitations in different regions. Furthermore the real-time visualization supports a better cross-modal binding.

Beyond the demonstration of new interaction modes for holistic data experiences as exemplified with the use of our system for the interaction with Model-Based Sonifications, we see diverse application fields where sonification can be plugged-in to enhance the experience. For instance, in didactic multi-touch applications such as an interactive visualization of electromagnetic fields, sonification may represent the electric field as sound while the user moves electric charges or touches the surface. In the area of interactive games, sonification could enable games between sighted and visually impaired users where each receives the modalities she could use best.

2 Model-Based Sonification

Model-Based Sonification (MBS) is a framework for the development of sonification techniques [10]. *MBS* starts from the observation that humans are well trained to interpret the complex acoustic signals in the world with respect to sound source characteristics. To give an example in everyday interaction, imagine to fill a thermos flask with water. By the pitch rise, due to the changing resonance of the bottle, we are aware of the flask's fill level. There is a large variety of situations when we use sound to gain insight into complex systems

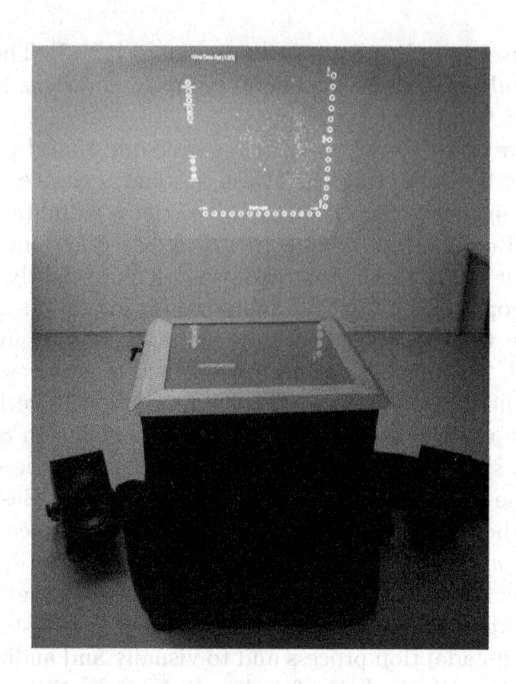

Fig. 1. The multi-touch enhanced tDesk platform with an additional projection on the opposing wall

(e.g. engineers listening to machine sounds or physicians using the stethoscope to support diagnosis) [17,18].

Most important, sound is connected to the underlying physical system by means of a dynamic (physical) *model*. The model mediates between a system's state and its acoustic response. The human brain is trained to infer source properties from sound that results from physical models. This principle provides the basis for *Model-Based Sonification* which defines in analogy dynamic processes between elements that are typically parameterized by the data. As in physical systems, a *sonification model* is silent without any excitation. Via interaction, the user is able to excite the model which connects MBS to the field of *interactive sonification* [14]. Guidelines and examples for creating interactive sonification models are provided in [9,10,11].

For the definition of a sonification model according to MBS, six aspects need to be addressed: At first, the *setup* of dynamical elements and the *initial state* of the model have to be determined. This is followed by the *dynamics*, which define how dynamic elements behave in time. The *excitation* defines how the user is able to interact with the model. The *Link Variables* used as transducers between the model and the audible domain have to be chosen and designed before the listener can be positioned in the setting. Finally, depending on the model, the listener needs to be positioned with respect to the data.

2.1 Excitation Modes for Sonification Models

The above definition for MBS has already shown that excitation plays a crucial role in the use of sonification models. Designers can take inspiration from all real-world interactions where the world responds acoustically, from 'foot steps' to 'hitting of objects'. If we focus on manual interaction we see that in most situations we either hit, pluck, shake, deform/squeeze or scrub objects. Most of these interactions have several degrees of freedom, e.g. the location of the interaction, the strength, the detailed direction relative to the surface. Depending on the details, interaction with real-world objects provides sonic feedback which includes information about the object. Obviously this richness is far beyond what can be obtained by simple mouse or keyboard interfaces. For those reasons, new interaction devices have been developed to better explore our manual interaction abilities [12,13].

If we consider the interaction with surfaces in general (e.g. consider to search a wall for hidden holes) we often use tapping, scratching (to examine the surface) and (think of drumheads) bimanual interactions where one hand hits while the other changes an aspect of the surface. Similarly, interactions are natural for interacting with surfaces, and with interactive multi-touch systems we now have the chance to define audiovisual surface reactions, so that an almost as natural utilization of manual interaction procedures may occur by users that explore complex data.

Our long range aim is to implement examples for all available surface-based interactions to explore the potential of MBS to connect manual interactions with exploratory excitations to support the understanding of data under analysis. In this paper we start this investigation with *tapping interactions* as excitations for Sonification Models.

3 Multi-touch Technology for the Tangible Desk (tDesk)

As a basis for the development, we started off using the tangible desk (tDesk) [1], a tabletop environment for tangible interaction (see Fig. 2). The tDesk is assembled using aluminium strut profiles. It consists of a 70 cm × 70 cm glass surface resting on four aluminum poles in about 80 cm height (see Fig. 2). The chosen extent of this table allows to conveniently work either alone or collaboratively within a group on and with touchable and tangible applications. Any spot on the surface can be reached with ease regardless of the user's deskside position. Since modularity was a major design issue of the tDesk, the current glass surface is easily exchangeable. We designed a drop-in replacement surface, enabling the tDesk to recognize fingers touching the surface. The used setup consists of the tDesk platform, the constructed acrylic surface with attached aluminium frame, lighting modules covering the pane edges, projector, camera, speakers and a computer system for image processing and multi-modal feedback (see Fig. 1). Basically, the constructed surface is a spatially resolved 2D-sensor recognizing multiple touch inputs. The physical sensor pane is made out of acrylic glass. The display is provided using a screen foil and an inside mounted projector.

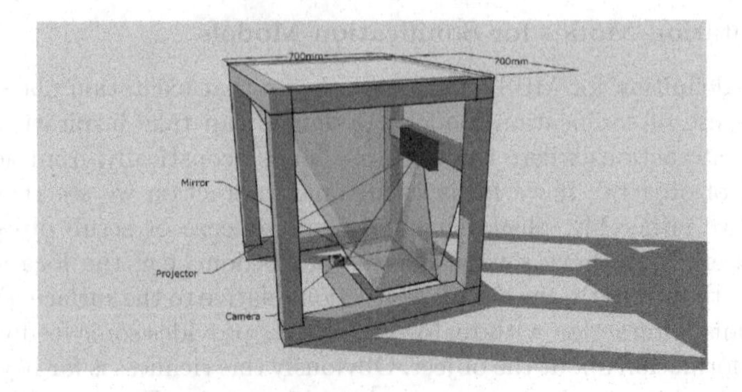

Fig. 2. The tangible desk (tDesk) platform provides the basis for the multi-touch system. It already contains a projector and a firewire camera.

The designed surface allows simultaneous interactions by at least four people in a closed-loop and direct manner. When designing the surface the following aspects where considered:

- Low Latency
  Closed-loop interactions require a low latency and therefore a high sampling rate. Reducing latency is of utmost importance, as rising latency can disturb closed-loop human-computer interaction. There has to be immediate feedback while the user interacts with the surface, otherwise the feedback can not be associated with the previous action.
- Input Points
  The sensor should support at least 40 input points so that up to four users can use all fingers simultaneously.
- High Resolution
  The sensor's resolution should be close to the display's resolution to support detection of the characteristics of touch, such as the shape or the orientation of the touched spot on the surface.
- Backprojection
  To intensify the degree of immersion, the projection should have its source inside the table. Top projection would lead to disturbing occlusions caused by the user's hands and fingers.

We used the *FTIR* technique (as proposed by Han [5]) to sense contacts on the surface. By using this technique we where able to sense almost as many input contacts as fit on the surface, achieving a high resolution and sufficient latency. To apply *FTIR* sensing, an acrylic pane is *flooded* with *near-infra-red (NIR)* light. When a finger touches the surface a bright *NIR* blob is caused on the surface. A camera, mounted underneath the surface, is used to capture the resulting blobs. To reduce the latency in the loop, a firewire camera capturing images at a frequency of 60 Hz is used. To improve the image quality, an optical bandpass filter was mounted in front of the camera. Finally, the display is provided by an underneath mounted projector.

To calibrate the camera and projector coordinate system we have chosen a mapping to resolve:

- camera trapezoid and pincushion distortion
- translation, rotation and scaling
- projector trapezoid and pincushion distortion

Optimal parameter values of the transformation map are determined by minimizing a quadratic error function using a least-squares minimization. Matching point-pairs from both coordinate systems are needed to compute the residuals for each iteration of the optimization process. The Levenberg-Marquardt Algorithm (LMA) [15,16] was used to train the mapping's parameters.

4 Multi-touch Interactions for MBS Excitation

Since the appearance of publicly available *multi-point* displays, more advanced displays allowing *absolute-spatial*, *multi-point* and *multi-user* interaction by the use of one's *fingers* are announced frequently, but the characteristics of touch (applied force, angle of approach) are mostly ignored. Whereas orientation and force are commonly exploited in graphics tablets such as Wacom's devices, these characteristics are sparsely used in other surfaces. An introduction and some framing of that subject by Buxton is available on his website [2]. In general, interactive surfaces can be considered as transducers between the digital and the analog interaction space. The following categories were a useful scope for us to better differentiate and discuss characteristics of surface-based interaction, particularly to excite sonification models. These are similar and partly based on previous work by Buxton [2].

Point vs. Touch: Existing '*multi-touch*' displays often offer *multi-point* instead of *multi-touch* input. The touch of a finger's tip is only used for a mere pointing, neglecting the details of touch. In addition, touching the surface with the hand or the arm will often lead to undefined behavior.

Single- vs. Multi-Spot: Old fashioned *touchpads*, which are still quite common, support only single-point input. Whereas *single* connotes just one spot input, *multi-spot* refers to devices capable of sensing more than one spot, for example all of the user's fingers. With *single-* and *multi-spot* as two sides of the continuum, in between there are n-spot devices capable of sensing a fixed number n of spots.

Collaborative Use: Even though newer notebook computers offer multiple input surface devices, these can hardly be used by more than one person at a time. Even if those pads theoretically could be used by more than one person at a time, in most cases this will lead to odd experiences.

Degrees of Freedom: When using spots on the surface only as pointing input, the surface provides input with two degrees of freedom. The transducer gains degrees of freedom by adding information about the pressure-vector of touch and direction of approach or other information.

Feedback: Traditional *touchpads* give no direct active feedback at all. *Touchscreens* and multi-spot displays feature visual feedback and thereby create the illusion of being able to manipulate digital items directly. Visual feedback can be enhanced by sound to intensify the degree of immersion. Digital objects with an auditory behavior can create sounds when triggered or when several objects interact with each other. Additionally, vibration motors could be used to create haptic feedback.

Relative vs. Absolute: *Touchpads* are, like mice, relative input devices: If touched, they take the cursors position. The position in the real world is set equal to the position of the cursor in the screen space. Touchscreens on the other hand feature *absolute* input. The user does not have to move the cursor from its current position to the target, but approaches the object *directly* by using a finger or a tool.

Direct vs. Abstract: When an object is moved with a finger or a pen-like tool, the interaction with it can be in a direct manner. If there is a relative transducer in the input chain, for example a mouse, the interaction becomes more abstract. There are a lot of discussions about when interactions are to be seen as abstract or not. We do not doubt that for someone who is familiar with mouse interaction, the relative transducer is ubiquitous and therefore virtually *ready-to-hand* (see dimension *Tools* and [8]). In this work, the term *direct* is used if the input chain of an interaction is free of relative transducers and the application allows the user to touch or move digital items.

Point vs. Gesture: Irrespective of the above-mentioned properties, an application can depend on the actual position, on the trajectory of the input spot, or both. Most common relative *point*-input devices are just using the actual position of the cursor. However, gesture-like input can be used to scale and rotate objects on the surface. Further it can be used to trigger certain actions such as to open a navigation menu.

Discrete vs. Continuous: Discrete interactions can be seen as single actions or events. A continuous interaction can be described by a trajectory of actions or events. Imagine typing on an on-screen-keyboard, or *pressing* displayed buttons, the interaction would be discrete. By moving an object from one position to another, the interaction becomes continuous.

Tools: Surface input devices can be designed to be used with different parts of the body such as a finger or with external tools. A tool can be *ready-to-hand* or *present-at-hand* to the user. A pen for example, when used for sketching or drawing tasks, is *ready-to-hand.* The user does not have to think about how to handle the pen, he just spends time on the drawing task itself [8], [3].

According to the above definition we have implemented direct and *absolute-spatial*, multi-point *Sonification Models*, which allow discrete and continuous use in a collaborative manner. We plan to exploit attributes such as force sensitive input, the use of tools and touch characteristics since the constructed surface already provides these informations.

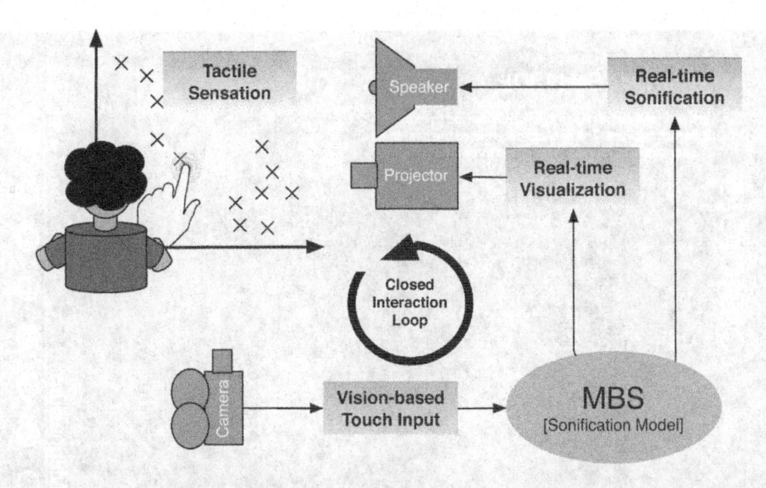

Fig. 3. Multi-Touch Interactions as melting pot for multi-modal coherent binding. The interface causes visual, tactile and auditory perceptions which are bound together via their synchronization to multi-modal perceptual entities.

5 The Data Sonogram Sonification Model

5.1 Overview

The *Data Sonogram Sonification Model* can be described by the following five categories:

Setup: Data points are used as point masses in a model space of the same dimension as the data space. The Data points' coordinates are used as fix point location for a virtual spring connected to the point mass.

Dynamics: (a) oscillation of spring-mass systems modeled by classical mechanics, and (b) the propagation of shock waves in the high-dimensional model space.

Excitation: The user can excite multiple 'shock waves' to emanate at certain locations in the model space.

Link Variables: The kinetic energy of all point masses is used to generate the sound signal, which represents the sonification. Alternatively the elongation from equilibrium can be used as link variable.

Listener: A two-dimensional view of the data and visual controls to navigate the data exploration are provided to the listener on the interactive surface. In the original model, the virtual listener is positioned in model space at the point where the shock wave is initiated. In this implementation, however the listener is centered at the x-axis of the 2D plot in front of the table, since only a stereo panning is used for sound spatialization.

A two-dimensional scatter plot of the data serves as the interaction area for the user to excite data sonograms by selecting a position on the plot. The speed of the shock wave can be adjusted by the user interactively. The resulting shock wave then passes through the data set within some seconds, so that the user can examine the data step-by-step.

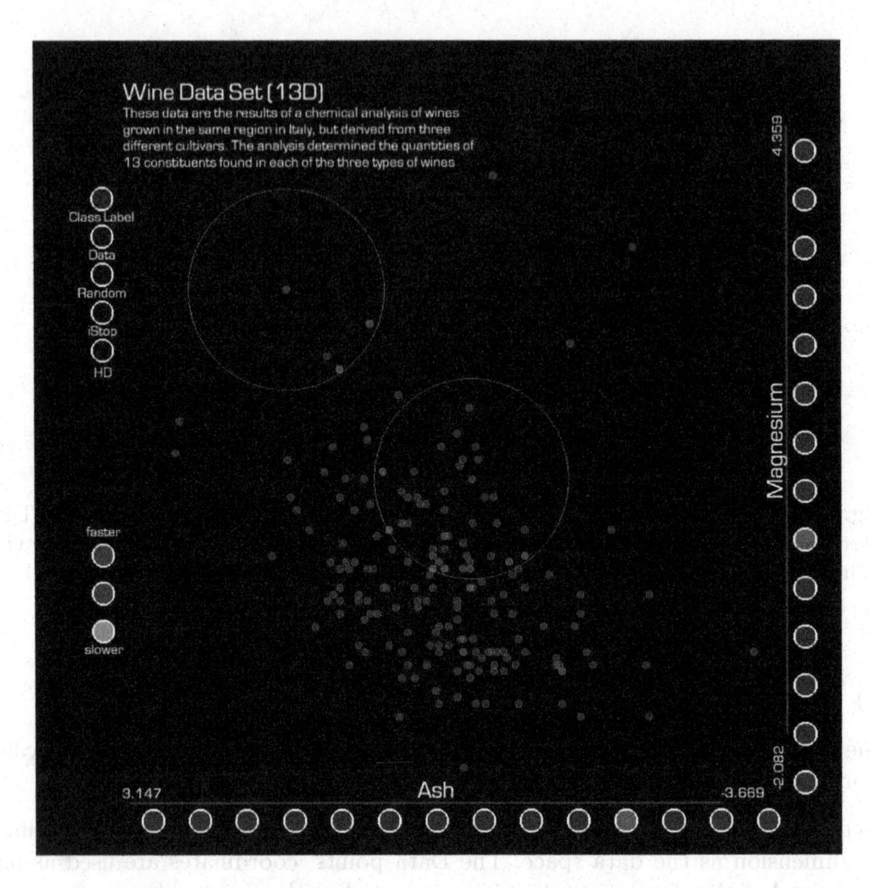

Fig. 4. The graphical user interface of the Data Sonogram application. The 13D *wine data set* is selected. The dimensions *Ash* and *Magnesium* are chosen for the visual feedback. Dimensions are chosen by a tap on a button in the button group for the corresponding axis. In a 13-dimensional data set, 13 buttons are displayed along each axis. Two shock waves are active, propagating through the two-dimensional data space.

5.2 Multi-touch Adaptations for Data Sonograms

We added several features for the multi-touch Data Sonogram implementation. At the point of excitation in a sonogram, a virtual shock wave is initiated. For the sake of usability and in contrast to physical constraints, the speed of the propagating wave can be adapted while the wave is traveling. A *shock wave* has a center and a source point. The distinction has to be made because the point of excitation is not always the center of the shock wave. In case the user initiates a high-dimensional shock wave, the shock wave center will be located at the coordinates of the data point that is nearest to the excitation point in the two-dimensional display.

In our original implementation [10], no visual feedback was given during the excitation of the data sonogram. Here we have added an interactive visualization

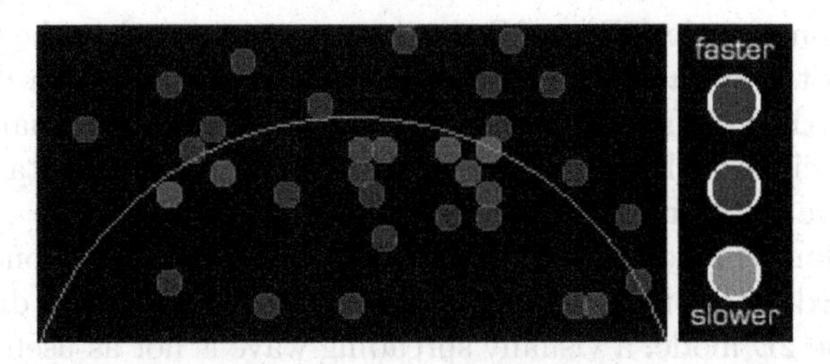

Fig. 5. Left: Point instances in data space with different energy, indicated by the spot's brightness. Energy is induced by the augmented shock wave. Right: An example of a button group used to control the velocity of the traveling wave.

of the shock wave front while the sonification is computed in real-time. Thereby the user knows where the sound comes from at the time they are perceived. A visual shock wave front is only meaningful in the case of a two-dimensional shock wave expanding on the table. If a high-dimensional shock wave is propagating data points might be reached by the shock wave in a less predictable pattern. In that case we update the data points' visual blobs by increasing their intensity at the time the high-dimensional shock wave passes through them. Again, these synchronized visual and auditory events help the user to better interpret the sound with respect to the data.

With the increased directness that multi-touch interactions serve, we discovered that users were interested to 'probe' the data sets frequently at different locations. The continuation of the shock wave after it has been triggered is then irritating and thus not helpful for this interaction pattern. We therefore implemented a new interaction style, where a shock wave immediately stops at the time when the user lifts the finger. In result, users can tap on the scatter plot freely and always get and compare their impression about local neighborhood distribution (from the temporal organization of the first sound events at the excitation point). This behavior turned out to be a useful optional feature for a step by step exploration of the data space.

The interface offers several options for the adjustment of parameters and the navigation of data in a coherent surface (see Fig. 4). Parameters such as the wave velocity, the sonic dimensionality and other features of the sonification model can be adjusted by the user interactively. The user is able to navigate in the data and can choose the to be displayed dimensions.

The chosen data dimensions (see Fig. 4) are presented on the interactive surface. Each dimension can be selected by a single touch on a corresponding button. Each axis is bound to only one dimension at a time. The two-dimensional scatterplot is used as the visual interface by which users are able to excite the sonification model. The system currently supports data sets with up to 16 dimensions, since the size of the interactive surface is limited. The users are able to trigger shock waves in two modes:

Two-dimensional mode (2D): a 2D shock wave is initiated at the touched coordinates on the surface. The mode has its main purpose as a didactic introduction to the system. The traveling shock wave front is visually augmented by a green circle with an increasing radius on the surface. Data points passed by the wave front are excited.

High-dimensional mode (HD): the user triggers a high-dimensional shock wave at the coordinates of the nearest 2D data point in the visual display. In contrast to the 2D mode, a visually spreading wave is not as useful in the high-dimensional mode. Instead of augmenting the propagating wave, passed data points to which energy is transferred are illuminated. The wave can be observed in the visual domain as a sequence of flashing data points (see Fig. 5).

The user can switch between these two modes through a button labeled *'HD'*, located at the left border where all control buttons are placed. At the lower left, a button group is placed consisting of three buttons to control the velocity (slow, normal, fast) of the propagating wave. The graphical user interface was written in *Processing* [7].

5.3 Auditory Components for Data Sonograms

The sonification sound signal is the superposition of all instantaneous elongations of masses from their equilibrium position. Since the spring forces the masses to a damped oscillation their representation becomes audible as a decaying sine tone. For the implementation of the spring-mass systems, unit generators for spring-mass systems in *SuperCollider* [19] have been used. This alleviates the problem of numerically integrating the dynamics of all mass-spring-systems, since they are well decoupled.

A stereo speaker setup is aligned towards the listener (as shown in Fig. 1). The virtual listener is centered in front of the table. When the shock wave front passes a data point, a sound event is spawned via the *OSC* protocol. Since the class label is used as spring stiffness, it can be perceived as pitch. The spatial location of the sound source can be estimated via the stereo panning.

5.4 Example Data Sets

There are three tutorial data sets available to the user by default:

wine data set: These data are the results of a chemical analysis of wines. They are derived from three different cultivars in the same region in Italy. The analysis measured the quantities of 13 constituents found in each of the three types of wines [4].

iris data set: This is perhaps the best known data set to be found in pattern recognition literature. It contains three classes of iris flowers, with four features each. 50 instances are included for each of the three classes. One class is linearly separable from the other two. The latter are not linearly separable from each other [6].

random data: The random data set contains uniformly distributed random data in four dimensions. It serves here as a benchmark distribution to train audiovisual exploration, i.e. to better learn to associate auditory and visually perceived elements in the multi-modal system.

Every time the data set is changed by the user's demand, a short description of the data set is displayed. This message shows the origin and history of the data set, its dimensionality and cardinality.

5.5 Interaction Examples

To discuss the approach, we provide a video showing a user interacting with the application on our website[1] In the first scene the user demonstrates different functions and explores the data space. The user chooses displayed dimensions and triggers shock waves in the high- and two-dimensional space. Then you can see and hear how the data sonogram evolves over time, starting with one of two pitch levels, depending on where the shock wave is initiated. Thereby the regions of different classes in the data set can be well discerned. Furthermore overlapping classes and class boundaries can be perceived.

6 The Growing Neural Gas Sonification Model

Growing Neural Gas (GNG), introduced by Fritzke in [21], is an undirected learning algorithm that incrementally 'grows' a network graph into the data distribution. The GNG is a network of neurons and connections between them. During the learning process, the neurons are moved to minimize the error with respect to the original data. New neurons are inserted and connections between them age or are reinforced.

The Growing Neural Gas Sonification Model introduced in [20] is described in brief below. It is categorized in the same way as the Data Sonogram Sonification Model before:

Setup: For the GNG Sonification Model, the connections in the GNG graph are used as transducers that transport energy between the neurons. The frequency of a neurons' tone is determined by the number of connections emanating from it: for each connection, a quint is added to the base frequency.

Dynamics: Using the energy flow equation (1), the energy for each neuron is calculated. It decays over time, depending on parameters g and q (which the user is able to adapt) and the current state of the GNG graph. The energy of each neuron determines the amplitude of the respective tone.

$$\frac{dE_i}{dt} = -gE_i(t) - \sum_{j \in I_N(i)} q(E_i(t) - E_j(t)) \qquad (1)$$

[1] `http://sonification.de/publications/TuennermannKolbeBovermannHermann2009-SIFSM/`

The parameter g steers the exponential energy decay, q determines the amount of energy that flows to every neighboring neuron each step. $E_i(t)$ is the energy of the neuron i, $I_N(i)$ is the set of neurons that are connected to neuron i.

Excitation: The user can induce energy into a neuron by tapping near it. This can be done at multiple points simultaneously or subsequently. The energy then propagates through the GNG until equilibrium is reached again.

Link Variables: The sonification is the superimposed sound signal from all existing neurons. This consists of one tone per neuron, with the frequency determined by the number of connections to other neurons and the amplitude determined by the current energy level of the neuron.

Listener: The resulting sonification for all neurons is presented to the user as well as the coupled visual feedback.

6.1 Overview

To benefit from the interaction capabilities of the tDesk, the GNG Sonification was reimplemented and simplified with multi-touch capabilities in mind. The goal was to be able to explore the GNG while it was growing, using only the fingers to deliberately excite the sonification. The user should not have to worry about setting up the program and initializing the GNG parameters, but be able to intuitively grasp the structure of the adapting GNG.

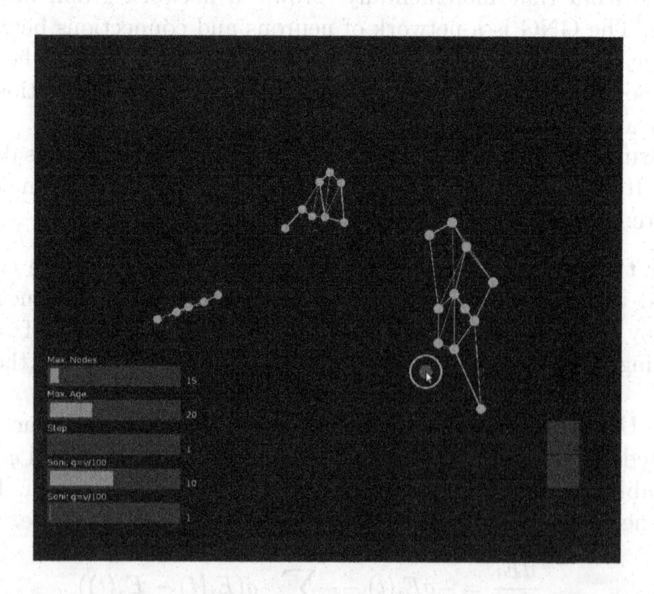

Fig. 6. The user interface of the GNG Sonification application, showing a two-dimensional scatterplot of the *three cluster* dataset overlayed with a GNG during its adaption process. Beneath the right cluster of data, the user induces energy into the network.

To start with, the user is presented with a two-dimensional scatterplot of the data. Five controls are available on the lower right corner, from top to bottom:

- the maximum number of neurons for the GNG
- the maximum age of connections between neurons
- the learning rate parameter
- the energy flow rate (parameter q in eq. (1))
- the energy dissipation rate (parameter g in eq. (1))

The first two parameters control the GNG algorithm itself, the third determines the speed of the learning process. The GNG has more tunables (see [21] for details), but they have well-working default values and most often do not need to be adapted. A future improvement will make it easier for the user to adjust them, without being overwhelmed by too many configuration options to choose from. The last two parameters define the energy decay and transport for the sonification. Finally, in the lower right corner are two buttons: The bottom one starts or pauses the adaptation process of the GNG, the upper one resets the GNG to its initial state, taking into account the changes the user made to the three initializations parameters.

Fig. 6 depicts an example of a GNG during the adaption process, using the *three cluster* dataset, which contains three cluster of an intrinsic dimensionality of two, four and eight each. It is a synthetic dataset used to evaluate the GNG Sonification Model, with 449 data points in total.

The bright circles represent the neurons, initially with an energy level of zero. The lines show the connections between them, with their thickness representing their age. In a future implementation of this sonification, we will take the age into account as well to further direct the energy flow to the neighboring neurons.

6.2 Implementation Details

The GNG sonification is implemented in python, utilizing the *Python Modular toolkit for Data Processing* [24] for the calculations. For the user interface and multi-touch interaction, *PyMT - A Multi-touch UI Toolkit for Pyglet* [23] is used. The sonification is synthesized in SuperCollider [19], utilizing Stinson's *OSC interface for Python* [25].

6.3 Interaction

When tapping near a neuron, the user induces energy into it. This leads to an energy flow within the network of neurons that is immediately sonified. At the same time, it is visualized – the sizes of the neurons indicate their current energy level (see Fig. 7). The user can influence the speed of energy decay within the network through independently adjusting the g and q parameters of the energy flow equation. For example, setting both to higher values leads to a faster decay and simultaneously reduces the distance the energy travels through the network, resulting in a more localized excitation of the sonification. In choosing lower

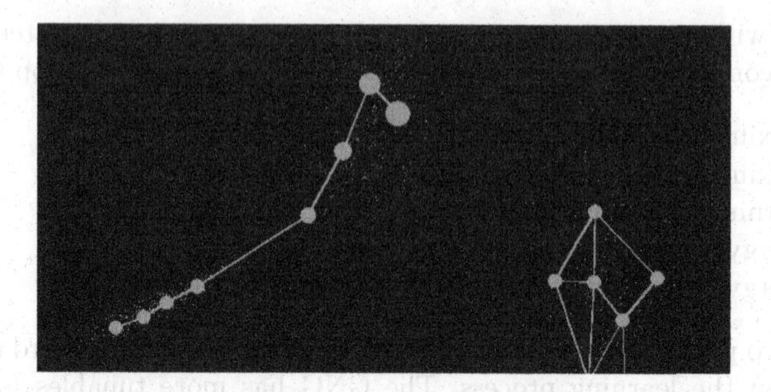

Fig. 7. An earlier state of the adaption process for the *three cluster* dataset. After the user induced energy into the rightmost neuron of the left cluster, the sonification begins to sound as the energy finds its way through the network. We visualize the current energy of a neuron through its size. Note that the right cluster does not emanate any sound, as the energy induced into the left cluster cannot possibly reach it at this point in the adaptation process.

values, the distance broadens and the decay slows down. In experimenting with different settings during different stages of the adaption process, or repeatedly inducing energy at similar points during the process, the user directly interacts with the sonification model and thereby is able to gain insights in the analyzed data.

For example, Fig. 6 shows an excitation of the GNG sonification on the lower right part. The resulting sound is very bright and slowly fading, as every neuron has between two to six connections to other neurons, resulting in the highest tone being the sixth quint over the base frequency. This indicates a high intrinsic dimensionality and, in fact, the underlying data distribution of this cluster is 8-dimensional. Would the user tap towards the left cluster of neurons, the highest frequency would be the second quint over the base frequency, indicating a less complex network and lower intrinsic dimensionality. The left cluster contains data that have only two independent variance dimensions.

In Fig. 7, the GNG is shown in an earlier state of the adaption process. The user induced energy into the rightmost neuron of the left cluster, so that it flows through the left part of the network only. The pitch of the sound is low, as each neuron has at most two neighbors. When excited while the GNG is adapting, new connections made to other neurons or newly added neurons are clearly audible through a rise in pitch. When connections are deleted, the sound suddenly becomes lower in pitch.

Growing Neural Gas is an undirected learning algorithm, but there exists no established decision criteria as to when it has fully grown into its dataset. After a while, overfitting occurs and ultimately the learned structure becomes diffused again. The user has to make an informed decision as to when to end the learning process, and the GNG Sonification Model provides a multi-modal and highly interactive tool to do just that.

7 Discussion and Conclusion

In this paper we have presented a multi-touch interface for the excitation of sonification models. We have reimplemented the Data Sonogram and the Growing Neural Gas sonification model and demonstrated multi-point multi-user exploration of scientific data via surface-based interaction.

The main advantage of our approach is that a very natural contact between the user and the surface (as the physical representation of the data) can be established. Interaction modes with typical real-world surfaces such as tapping, hitting, pushing and scratching provide examples of how interactions can be profitably used in the context of sonification models. With the two sonification models we have given first examples that show how spatially-resolved tapping on the surface can be utilized as a tapping into data spaces, using quasi-physical dynamic processes in the space of the sonification model to associate meaningful acoustic responses which then represent the data to the user. In result a qualitative experience is created from the ongoing continuous interaction.

An important aspect is that the interface connects the auditory and visual representation and binds them via the surface to multi-modal data perceptualization units. Synchronization is a key component for the user to be able to connect visual and auditory elements. Since the interaction occurs in the same frame of reference, and tactile sensations complement the experience, a tightly closed interaction loop is created.

In our future work we will particularly focus on sonification models that allow to explore yet *untouched* aspects of continuous interaction with data distributions. Instead of providing a trigger only, we want to enable users to continuously deform data representations in order to perceive the resulting tension by these deformations as informative sound. Sonification can be used in various multi-touch applications. For instance, for didactic applications, the real-time sonification of variables (e.g. stress, magnetic field strength, etc) while interacting with a simulation of a system can deliver complementary information to what is visible on the surface. Also, auditory games where the goal is to competitively or jointly shape sounds via physical interaction with the surface offer a great potential to explore tactile computing in a yet unseen way.

In summary, the presented multi-touch sensitive surface enriches the available modes to interact with complex data and to perceive structure-related features as sound via Model-Based Sonification. The tight coupling of visualization, sonification, tangible interfaces and continuous interaction in one interface contributes to a truly multi-modal experience and shows the potential of an increased level of understanding of structures in the data. The scope of our ongoing research is to explore and quantify the possibilities in this direction.

References

1. Bovermann, T., Hermann, T., Ritter, H.: A tangible environment for ambient data representation. In: First International Workshop on Haptic and Audio Interaction Design, August 2006, vol. 2, pp. 26–30 (2006)

2. Buxton, B.: Multi-touch systems that i have known and loved (2009), http://www.billbuxton.com/multitouchOverview.html
3. Dourish, P.: Where the action is: The foundation of embodied interaction. MIT Press, Cambridge (2001)
4. Forina, M.: Arvus - an extendible package for data exploration, Classification and Correlation, http://www.radwin.org/michael/projects/learning/about-wine.html
5. Han, J.Y.: Low-cost multi-touch sensing through frustrated total internal reflection. In: UIST 2005: Proceedings of the 18th annual ACM symposium on User interface software and technology, pp. 115–118. ACM Press, New York (2005)
6. Fisher, R.A.: UCI Repository of Maschine Learning Databases – Iris Data Set (1999)
7. Fry, B., Reas, C.: Processing programminig environment (2001), http://processing.org/
8. Heidegger, M.: Sein und Zeit. Niemeyer, Halle a. d. S. (1927)
9. Hermann, T.: Taxonomy and definitions for sonification and auditory display. In: Katz, B. (ed.) Proc. Int. Conf. Auditory Display (ICAD 2008), France (2008)
10. Hermann, T., Ritter, H.: Listen to your data: Model-based sonification for data analysis. In: Lasker, G.E. (ed.) Advances in intelligent computing and multimedia systems. Int. Inst. for Advanced Studies in System research and cybernetics, Baden-Baden, Germany, pp. 189–194 (1999)
11. Hermann, T.: Sonification for exploratory data analysis. PhD thesis, Bielefeld University (February 2002), http://www.techfak.uni-bielefeld.de/ags/ni/publications/media/Hermann2002-SFE.pdf
12. Hermann, T., Krause, J., Ritter, H.: Real-time control of sonification models with an audio-haptic interface. In: Nakatsu, R., Kawahara, H. (eds.) Proc. Int. Conf. Auditory Display (ICAD 2002), Kyoto, Japan, pp. 82–86 (2002)
13. Hermann, T., Milczynski, M., Ritter, H.: A malleable device with applications to sonification-based data exploration. In: Stockman, T. (ed.) Proc. Int. Conf. Auditory Display (ICAD 2006), pp. 69–76. University of London, London, UK (2006)
14. Hunt, A., Hermann, T., Pauletto, S.: Interacting with sonification systems: Closing the loop. In: Banissi, E., Börner, K. (eds.) IV 2004: Proceedings of the Information Visualisation, Eighth International Conference on (IV 2004), Washington, DC, USA, pp. 879–884. IEEE Computer Society, Los Alamitos (2004)
15. Levenberg, K.: A method for the solution of certain non-linear problems in least squares. Quarterly of Applied Mathematics, 164–168 (1944)
16. Marquardt, D.: An algorithm for least-qquares estimation of nonlinear parameters. SIAM Journal on Applied Mathematics 11, 431–441 (1963)
17. Kramer, G. (ed.): Auditory display: sonification, audification, and auditory interfaces. Addison-Wesley, Reading (1994)
18. Kramer, G., Walker, B., Bonebright, T., Cook, P., Flower, J., Miner, N., Neuhoff, J., Bargar, R., Barrass, S., Berger, J., Evreinov, G., Fitch, W.T., Grohn, M., Handel, S., Kaper, H., Levkowitz, H., Lodha, S., Shinn-Cunningham, B., Simoni, M., Tipei, S.: Sonification report: Status of the field and research agenda (1997)
19. McCartney, J.: SuperCollider hub. (July 2004) http://supercollider.sourceforge.net/
20. Hermann, T., Ritter, H.: Neural gas sonification: Growing adaptive interfaces for interacting with data. In: Banissi, E., Börner, K. (eds.) IV 2004: Proceedings of the Information Visualisation, Eighth International Conference on (IV 2004), Washington, DC, USA, pp. 871–878. IEEE Computer Society, Los Alamitos (2004)

21. Fritzke, B.: A growing neural gas network learns topologies. In: Tesauro, G., Touretzky, D., Leen, T. (eds.) Advances in Neural Information Processing Systems, vol. 7, pp. 625–632. MIT Press, Cambridge (1995)
22. Kaltenbrunner, M., Bovermann, T., Bencina, R., Costanza, E.: TUIO: A protocol for table based tangible user interfaces. In: GW (2005)
23. Hansen, T., Hourcade, J.P., Virbel, M., Patali, S., Serra, T.: PyMT: A post-WIMP multi-touch user interface toolkit. In: Proceedings of the International Conference on Interactive Tabletops and Surfaces (ITS 2009), Banff, Canada (2009)
24. Zito, T., Wilbert, N., Wiskott, L., Berkes, P.: Modular toolkit for data processing (MDP): A python data processing frame work. Frontiers in Neuroinformics (2008), `http://mdp-toolkit.sourceforge.net`
25. Stinson, P.K.: SCOSC: SuperCollider OSC interface for python (October 2009), `http://trac2.assembla.com/pkaudio/wiki/SuperCollider`

Quantum Harmonic Oscillator Sonification

Anna Saranti, Gerhard Eckel, and David Pirrò

Institute of Electronic Music and Acoustics
University of Music and Dramatic Arts Graz
Inffeldg. 10/3, 8010 Graz, Austria
anna.saranti@student.kug.ac.at, eckel@iem.at, pirro@iem.at

Abstract. This work deals with the sonification of a quantum mechanical system and the processes that occur as a result of its quantum mechanical nature and interactions with other systems. The quantum harmonic oscillator is not only regarded as a system with sonifiable characteristics but also as a storage medium for quantum information. By representing sound information quantum mechanically and storing it in the system, every process that unfolds on this level is inherited and reflected by the sound. The main profit of this approach is that the sonification can be used as a first insight for two models: a quantum mechanical system model and a quantum computation model.

1 Introduction

The quantum harmonic oscillator is one of the most fundamental quantum mechanical systems. It describes, as in classical mechanics, the motion of an object subjected to a parabolic potential [1, pp. 54–63]. Like every other quantum mechanical system it is described by its Hamiltonian, which for this system is solvable with known eigenstates and eigenvalues. Any state of the system can be expressed as a superposition of its eigenstates. The quantum harmonic oscillator provides a physical realization of a quantum computer model [2, pp. 283–287] where quantum information is stored in the state of the quantum harmonic oscillator and then either processed through its intrinsic time evolution or through coupling with the environment. The sonification choices that were adopted in this work could also be associated with these information processing operations.

At a first step, sound information is stored quantum mechanically in the system's state. Letting the system evolve in time or interact with other systems affects the state and, thereby, the stored information. The deformation of the stored sound reflects the characteristics and properties of the system and the processes that occur. In those cases where the eigenvalues and eigenstates are affected, their sonification could also provide more insight into the phenomena.

The motivation for this approach is to gain a first insight into quantum computational storage operations through sound. Quantum mechanical memory has, in general, different properties from the classical [2, pp. 13–17], which can be highlighted through sonification. The impact of an external disturbance to the stored quantum information is a fairly complex procedure with interdependencies that

S. Ystad et al. (Eds.): CMMR/ICAD 2009, LNCS 5954, pp. 184–201, 2010.

can be perceived coherently through sound. The part of the stored quantum information which is classically accessible through quantum measurement and the impact of the measurement operations in the classically retrieved part can be also acoustically represented using this approach.

The best known model of a quantum mechanical memory unit is the qubit [2, pp. 13–17] which is abstract and unbounded from the properties of the physical system that realizes it. The harmonic oscillator quantum computer model bases on the features of the underlying system and therefore the representation of the quantum information is directly interconnected with the system properties.

Many other quantum mechanical problems, such as the single mode of an electromagnetic field in a one-dimensional cavity and the vibration spectra of diatomic molecules are based on the quantum harmonic oscillator [1, pp. 19–32]. Thus this sonification could begin to acquire knowledge not just about quantum mechanical systems of the same form but also quantum mechanical systems in general, because they follow the same principles.

This paper is organized as follows: The second section provides a very brief description of the system that is needed for the understanding of the sonification decisions. The third section concentrates on the sonification of the time evolution process of the quantum harmonic oscillator as a closed system, which is a derivation of the time-dependent Schrödinger equation. In the fourth section, the system is subjected to two types of disturbances where the influence of the interactions on several other systems is described with the help of perturbation theory. The fifth section provides some details of the implementation whereas the sixth section presents some future plans and ideas for future works.

2 Quantum Harmonic Oscillator

2.1 Description of the System

Every quantum mechanical system's total energy is described by its Hamiltonian $\hat{H}$. Leaving the time evolution of the system aside and concentrating on the description of the system for a specific time point, the time-independent Schrödinger equation is [1, pp. 19–32]:

$$\hat{H}\psi(x) = E\psi(x) \tag{1}$$

where $\hat{H}$ is the Hamiltonian of the system, $\psi(x)$ the wavefunction that represents the state of the system and E the eigenvalues of $\hat{H}$. The value of $|\psi(x)|^2$ expresses the probability density of finding the oscillating object at the position x [3, pp. 54–57]. The Hamiltonian $\hat{H}$ is mathematically represented by the equation [1, pp. 54–63]:

$$\hat{H} = K + V = \frac{\hat{p}^2}{2m} + \frac{m\omega^2}{2}\hat{x}^2 \tag{2}$$

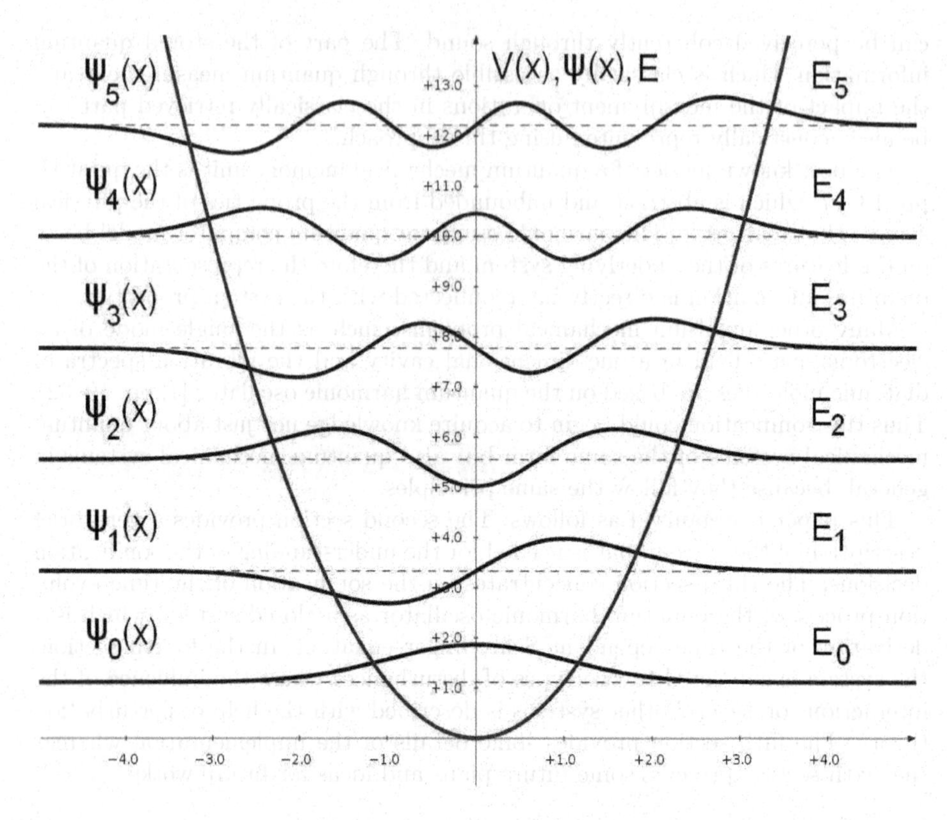

Fig. 1. The first six (from 0 to 5) Eigenenergies and Eigenfunctions of the Quantum Harmonic Oscillator. The Eigenergies are depicted with equally spaced vertical lines. The corresponding Eigenfunctions are shifted on the y-axis each one with offset the particular Eigenenergy. On the same plot is also the parabolic potential V.

Where K is the kinetic energy, V the potential energy, $\hat{p}$ the momentum operator, m the mass of the oscillating particle, ω the eigenfrequency and $\hat{x}$ the displacement operator. The eigenvalues that satisfy the equation (1) are quantized and represent the eigenenergies of the quantum harmonic oscillator:

$$E_n = \left(n + \frac{1}{2}\right) \hbar\omega, \quad n = 0, 1, 2... \tag{3}$$

The eigenstates that satisfy the equation (1) are mathematically expressed with the help of the Hermite Polynomials $H_n(x)$:

$$H_n(x) = (-1)^n e^{x^2} \frac{\mathrm{d}^n}{\mathrm{d}x^n} e^{-x^2}, \quad n = 0, 1, 2... \tag{4}$$

The eigenstates $\psi_n(x)$ which satisfy the equation (1) are weighted Hermite polynomials and represent the eigenfunctions of the quantum harmonic oscillator:

$$\psi_n(x) = \left(\frac{a}{\sqrt{\pi}\; n!\; 2^n}\right)^{\frac{1}{2}} H_n(ax)\; e^{-\frac{a^2}{2}x^2}, \quad a = \sqrt{\frac{m\omega}{\hbar}} \tag{5}$$

The eigenfunctions of the quantum harmonic oscillator constitute a complete and orthonormal basis. Therefore any state of the system which is represented by the wavefunction can be written as a linear combination of its eigenstates.

$$\psi(x,t) = \sum_{n=1}^{\infty} c_n \psi_n(x) \tag{6}$$

The sum of all the probabilities should sum up to one. The c_n coefficients are complex numbers that are called probability amplitudes [5] and fulfill the normalization condition:

$$\sum_{n=1}^{\infty} |\, c_n\,|^2 = 1 \tag{7}$$

2.2 Shapelet Basis Expansion Method

The description of the audio signals that is realized in this work is based on their decomposition onto the eigenfunctions of the quantum harmonic oscillator thus falling to the category of a non-parametric signal expansion method [6, pp. 9–21]. The signal can be expanded as a linear combination of the basis functions $\{\psi_n(x)\}$. The coefficients for a signal y can be obtained from the following equation:

$$y = \sum_n c_n \psi_n \;\Rightarrow\; c_n = B^{-1} y \tag{8}$$

Where B is the matrix that contains the eigenfunctions $\psi_n(x)$ of the quantum harmonic oscillator. The $\psi_n(x)$ functions are called Shapelets [7], [8], because they form a perturbed version of a Gaussian function as shown by the equation (5). Shapelets have a main difference to wavelets; namely the various shapes of the basis functions. The wavelet transform basis functions are the same up to a scaling factor. On the other hand, the Shapelet basis functions are of a different size and form.

The first step of the sonification procedure is to store an audio signal into the quantum harmonic oscillator using the overlap-add method [9, pp. 237–238]. The signal is multiplied by a sliding window of length N. The successive windowed signal frames are expanded as a linear combination of the eigenfunctions of the same quantum harmonic oscillator.

The number of eigenfunctions of the quantum harmonic oscillator is theoretically infinite but in this work only a finite number of eigenfunctions are implemented, depending on the needs for a good analysis and resynthesis. Throughout this work the number of the coefficients c_n used was also N for a windowed part of a signal with N number of samples. An extended testing of this basis for audio signal processing applications could be made in future work.

2.3 Harmonic Oscillator Quantum Computation Model

The computational building block of a quantum mechanical memory is the qubit, whereas the memory of a quantum computer consists of several qubits [2, pp. 13–17]. As with the classical bit, the qubit is realized on a physical system. The main difference is that this physical system is at a level where quantum mechanical phenomena are apparent and determine the properties of the storage.

We will now concentrate on the mathematical abstraction of the qubit. The state in which a quantum system lives is represented mathematically with the symbol $|\psi\rangle$ using the Bra-Ket or Dirac notation [2, pp. 61-62], [3, pp. 164–169], which is the standard notation for the description of vectors in vector spaces. In particular, the state of one qubit $|\psi\rangle$ can be described as a linear combination of the two states "0" and "1", which is called superposition.

$$|\psi\rangle = a\,|0\rangle + b\,|1\rangle, \quad a^2 + b^2 = 1, \quad a, b \in \mathbb{C} \tag{9}$$

The states "0" and "1" correspond to the states that the classical bit can have [2, pp. 81]. The essential difference between the qubit and the classical bit is that the latter can only be in one of the "0" and "1" states at once, whereas the qubit coexists in both of them. Attention needs to be drawn to the fact that the a, b numbers do not represent the proportion of time duration in which the qubit lives in each state (just like in the probabilistic bit case), but are an indication for the weighting and contribution each of the "0" or "1" states to the whole quantum mechanical state $|\psi\rangle$. The factors a and b are complex valued and continuous, for reasons that are declared in quantum theory [5], [4, pp. 29–31].

Apart from the Bra-Ket or Dirac notation, quantum mechanical states can be described with the help of vectors [4, pp. 4–5], [2, pp. 61–62]. The state $|\psi\rangle$ of the qubit can be represented as a unit column vector in the two dimensional complex vector space or Hilbert space $H = \mathbb{C}^2$. The complex values a and b are the elements of this vector:

$$|\psi\rangle = a\,|0\rangle + b\,|1\rangle = a\begin{bmatrix}1\\0\end{bmatrix} + b\begin{bmatrix}0\\1\end{bmatrix} = \begin{bmatrix}a\\b\end{bmatrix}, \quad a^2 + b^2 = 1, \quad a, b \in \mathbb{C} \tag{10}$$

Together the $|0\rangle$ and $|1\rangle$ states constitute an orthonormal basis for the two dimensional complex vector space. This base is called the computational basis [2, pp. 13], but Hilbert space also has other orthonormal bases. Therefore the expression of the quantum state could be made with respect to any other basis of the Hilbert space, as the basis created by the eigenfunctions of the quantum harmonic oscillator.

According to the fourth postulate of quantum mechanics, "The state space of a composite physical system is the tensor product of the state spaces of the component physical systems" [2, pp. 94]. Hence the state space or Hilbert space of the two-qubit system will be the $H = (\mathbb{C}^2)^2$. The computational basis for the two qubits system is constructed from all the possible combinations (expressed in tensor product relations) of the computational basis states of each qubit: $|0\rangle\otimes|0\rangle$, $|0\rangle\otimes|1\rangle$, $|1\rangle\otimes|0\rangle$ and $|1\rangle\otimes|1\rangle$ compactly written as $|00\rangle$, $|01\rangle$, $|10\rangle$, $|11\rangle$ accordingly [2, pp. 16–17]. The mathematical symbol $\otimes$ denotes the tensor product [4, pp. 26–27]. Thus, the computational basis has $2^2 = 4$ states and every possible state of the quantum system can be expressed again as a linear combination of the basis states:

$$|\psi\rangle = a\,|00\rangle + b\,|01\rangle + c\,|10\rangle + d\,|11\rangle, \quad |a|^2 + |b|^2 + |c|^2 + |d|^2 = 1, \quad a, b, c, d \in \mathbb{C} \tag{11}$$

The state of an N-qubit system $|\psi\rangle$ is a vector with magnitude one in the 2^N-dimensional complex vector space $(\mathbb{C}^2)^N$ [10, pp. 25–27]. The computational basis for the N-qubit system is constructed from all the possible combinations of the computational basis states of each qubit. The state can be represented in analogy with the one and two qubit states [4, pp. 54–55]:

$$|\psi\rangle = \sum_n c_n\,|d_0...d_{m-1}\rangle = \sum_n c_n\,|n\rangle, \quad \sum_n |c_n|^2 = 1 \tag{12}$$

Where c_n the probability amplitudes, $|n\rangle$ the computational basis vectors and $d_0, d_1...d_{m-1}$ binary numbers (either 0 or 1) that represent the $|n\rangle$. For example, a four qubit state it can be written as:

$$|\psi\rangle = c_0\,|0000\rangle + c_1\,|0001\rangle + c_2\,|0010\rangle + c_3\,|0011\rangle + ... + c_{15}\,|1111\rangle \tag{13}$$

Which could be also expressed as:

$$|\psi\rangle = c_0\,|0\rangle + c_1\,|1\rangle + c_2\,|2\rangle + c_3\,|3\rangle + ... + c_{15}\,|15\rangle \tag{14}$$

The state of the quantum harmonic oscillator is in correspondence with the state of a quantum mechanical memory created from qubits. In the quantum harmonic oscillator model one possible physical implementation of the qubits is made in such a way that the state of the whole memory can be expanded as a linear combination of the eigenfunctions $\psi_n(x)$ [2, pp. 283–287]. The analogy is for every 2^N eigenstates that are used for the expansion of the signal represent the quantum information storage capability of N qubits because they create an equivalent complex Hilbert space.

It is assumed that the system can be prepared in a desired state through an initialization procedure. Special attention needs to be drawn to the fact that the coefficients that are computed for the expansion of the stored audio signal not only need to fulfill the equation (8) but also the normalization condition (7). The normalization of the coefficients to probability amplitudes is also implemented.

3 Closed System Time Evolution

A quantum system that evolves without coupling to the environment is called closed or isolated [2, pp. 81–84]. The time - dependent Schrödinger equation (15) describes the evolution of the closed system in time [1, pp. 19–32].

$$i\hbar\frac{\partial\psi(x,t)}{\partial t} = \hat{H}\psi(x,t) \tag{15}$$

where $\hbar$ is the Planck's constant. The time evolution is a procedure that changes the state of the system but leaves the eigenenergies and eigenfunctions unaffected. If the wavefunction of the system $\psi(x,0)$ at time $t_0 = 0$ is described by the equation [2, pp. 13–17]:

$$\psi(x,0) = \sum_{n=1}^{\infty} c_n^{(0)}\psi_n(x) \tag{16}$$

where $c_n^{(0)}$ are the coefficients of the input sound according to the $\psi_n(x)$ basis at time $t_0 = 0$, then after time t each coefficient will be multiplied by a different complex exponential term:

$$c_n^{(t)} = c_n^{(0)} e^{\frac{-iE_n t}{\hbar}} \tag{17}$$

where E_n is the n-th Eigenenergy. The state of the system will change accordingly:

$$\psi(x,t) = \sum_{n=1}^{\infty} c_n^{(t)}\psi_n(x) = \sum_{n=1}^{\infty} c_n^{(0)} e^{\frac{-iE_n t}{\hbar}}\psi_n(x) \tag{18}$$

Every time a windowed sound segment is stored in the quantum oscillator the coefficients $c_n(0)$ for this window are computed with respect to the basis of the eigenfunctions $\psi_n(x)$. Each coefficient is then multiplied with its corresponding exponential term. The real and imaginary part of the time evolved coefficients are separately used for the resynthesis of the sound and produce two individual tracks that are merged in a stereo file. The phenomenon produces a modulation in the spectral domain which repeats itself after a specific period of time (Fig. 2). The period duration T_n of the evolution process is individual for each coefficient c_n and is the same for the real and imaginary part:

$$T_n = \frac{4\pi}{(2n+1)\omega} \tag{19}$$

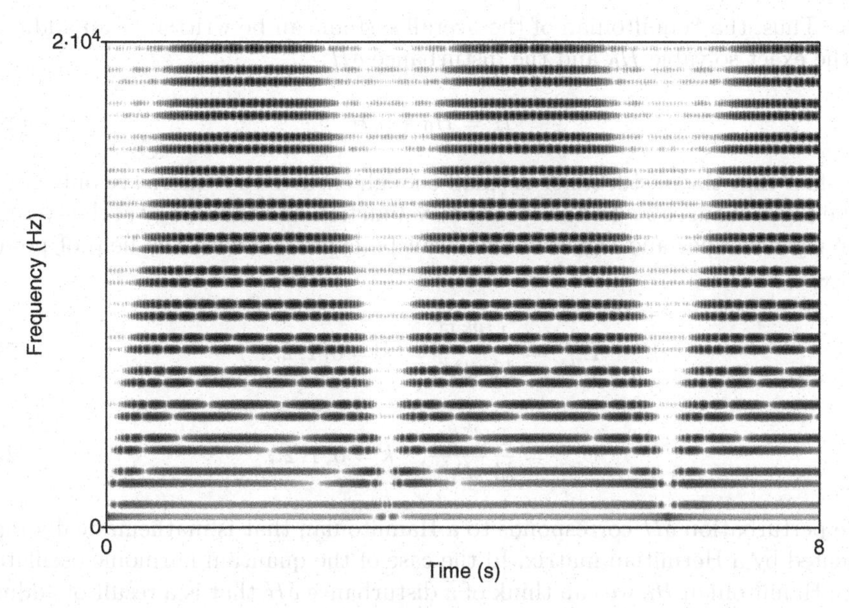

Fig. 2. Spectrum of a sinusoidal of frequency 440 Hz expanded over the Shapelet basis, then resynthesized and time-evolved. According to the time-evolution described by equation (18), the spectrum presents modulating frequency components that rise through the periodic change of the expansion coefficients.

In this implementation the time variable t doesn't flow continuously at each sample but increases by the value of the hop size that is used in the overlap-add for every window. The phenomenon can be heard at various speeds using a scaling factor on the time parameter.

The time evolution implements a unitary transformation and, therefore, is the main procedure that can be used for the realization of quantum gates in this computational model [2, pp. 283–287]. Information processing is achieved with the additional use of perturbation of the eigenfunctions, as described in the next chapter.

4 Open System

4.1 Overview of Perturbation Theory

When a quantum mechanical system has strong interactions with the environment it is called open [2, pp. 353–354]. Solving such systems, i.e. finding their eigenenergies and eigenfunctions, is a complex and difficult procedure. Therefore an approximation method needs to be used. The perturbation theory is one of them and can be applied when a system with a solvable Hamiltonian $\hat{H}_0$ is subjected to a relatively weak disturbance $\delta\hat{H}$ in regard to the value of $\hat{H}_0$ [1, pp.

133]. Thus, the Hamiltonian of the overall system can be written as an addition of the exact solvable $\hat{H}_0$ and the disturbance $\delta\hat{H}$:

$$\hat{H} = \hat{H}_0 + \delta\hat{H} \tag{20}$$

The fact that this disturbance is small enough ensures that there are only going to be slight changes $\delta\psi$ and δE to the wavefunction and the energy of the system. The eigenenergies and eigenfunctions can be expressed with the help of power series:

$$E_n^{(k)} = \frac{1}{k!}\frac{d^k E_k}{d\lambda^k}, \quad k = 0, 1, 2... \tag{21}$$

$$\psi_n^{(k)} = \frac{1}{k!}\frac{d^k \psi_n}{d\lambda^k}, \quad k = 0, 1, 2... \tag{22}$$

The perturbation $\delta\hat{H}$ corresponds to a Hamiltonian that is mathematically represented by a Hermitian matrix. In the case of the quantum harmonic oscillator with Hamiltonian $\hat{H}_0$ we can think of a disturbance $\delta\hat{H}$ that is a result of adding or removing some energy from the system. Throughout this work, the use of approximation approaches other than the perturbation theory are not addressed, but this could be a topic that could be further explored.

There are two types of perturbation approaches: the time-independent and the time-dependent. The time-independent procedure describes the system's behavior when the disturbance is constant, whereas the time-dependent deals with systems that are subjected to a time-varying disturbance.

4.2 Time-Independent or Rayleigh-Schrödinger Method

Description of the Process. The undisturbed or principal system will have an exact solution according to the time-independent Schrödinger equation [1, pp. 134–140]:

$$\hat{H}_0 \, \psi_n^{(0)}(x) = E_n^{(0)} \psi_n^{(0)}(x) \tag{23}$$

The zero at the superscript of $E_n^{(0)}$ denotes that the eigenenergies are from the undisturbed system, whereas the n at the subscript of the eigenenergies shows the correspondence of the n-th eigenenergy to the n-th eigenfunction. After the disturbance is applied, the Hamiltonian will, by means of the equation (20), change where the term $\delta\hat{H}$ is replaced by $\lambda\, V$ and thus the Schrödinger equation for this system will be:

$$\hat{H}\psi_n(x) = (\hat{H}_0 + \lambda\, V)\psi_n(x) = E_n\psi_n(x) \tag{24}$$

Where $\hat{H}$ is the Hamiltonian, $\psi_n(x)$ the eigenfunctions and E the eigenenergies of the disturbed system. The λ term is a factor that controls the disturbance intensity and can take values with a range from 0 to 1 which represent no perturbation to full perturbation accordingly. Just because the disturbance is weak, the eigenenergies E_n of the disturbed system will not deviate very much from the eigenenergies $E^{(0)}$ of the undisturbed. The same property holds for the eigenfunctions. The power series expansion will be in accordance with the equations (21) and (22).

$$E_n = E_n^{(0)} + \lambda E_n^{(1)} + \lambda^2 E_n^{(2)} + ... \tag{25}$$

$$\psi_n(x) = \psi_n^{(0)}(x) + \lambda \psi_n^{(1)}(x) + \lambda^2 \psi_n^{(2)}(x) + ... \tag{26}$$

The superscripts $0, 1, 2...$ denote the zero-th, first and second term of the power series. The zero superscript is the unperturbed one. The n at the subscript of the eigenenergies shows the correspondence of the n-th eigenenergy to the n-th eigenfunction.

The derivation of the solution occurs by inserting the equations (25) and (26) into (24). The expression of the first order term of the eigenfunction's correction as a linear combination with respect to the orthonormal basis that is formed from the eigenfunctions $\psi_n^{(0)}(x)$ of the unperturbed system, leads to the first and second correction to the energy of the system:

$$E_n^{(1)} = \psi_n^{(0)}(x)^\dagger V \psi_n^{(0)}(x), \; E_n^{(2)} = \sum_m{}' \frac{| \psi_m^{(0)}(x)^\dagger V \psi_n^{(0)}(x) |^2}{E_n^{(0)} - E_m^{(0)}} \tag{27}$$

The first term of the eigenfunction correction is expressed by the following equation:

$$\psi_n^{(1)}(x) = \sum_m{}' \frac{\psi_m^{(0)}(x)^\dagger V \psi_n^{(0)}(x)}{E_n^{(0)} - E_m^{(0)}} \psi_m^{(0)}(x) \tag{28}$$

Where the acute in the summation denotes that the sum is made over all n eigenfunctions except the m. Higher terms can be obtained iteratively but are not used for the implementation to reduce the computational complexity of the implementation.

Audification Choices. Various disturbance types that correspond to different Hermitian matrices V were used for the audification of this perturbation type. One example of a perturbation used corresponds to a constant electrical field with a potential that has linear dependency from the displacement x which is added to the parabolic potential. The λ factor can also be used to control how intense the development of the disturbance phenomena will be.

By applying the same disturbance type V many times consecutively, a slow deformation of the shape of each of the eigenfunctions can be examined at first. The eigenenergie's values also slightly deviate from their initial value, each one

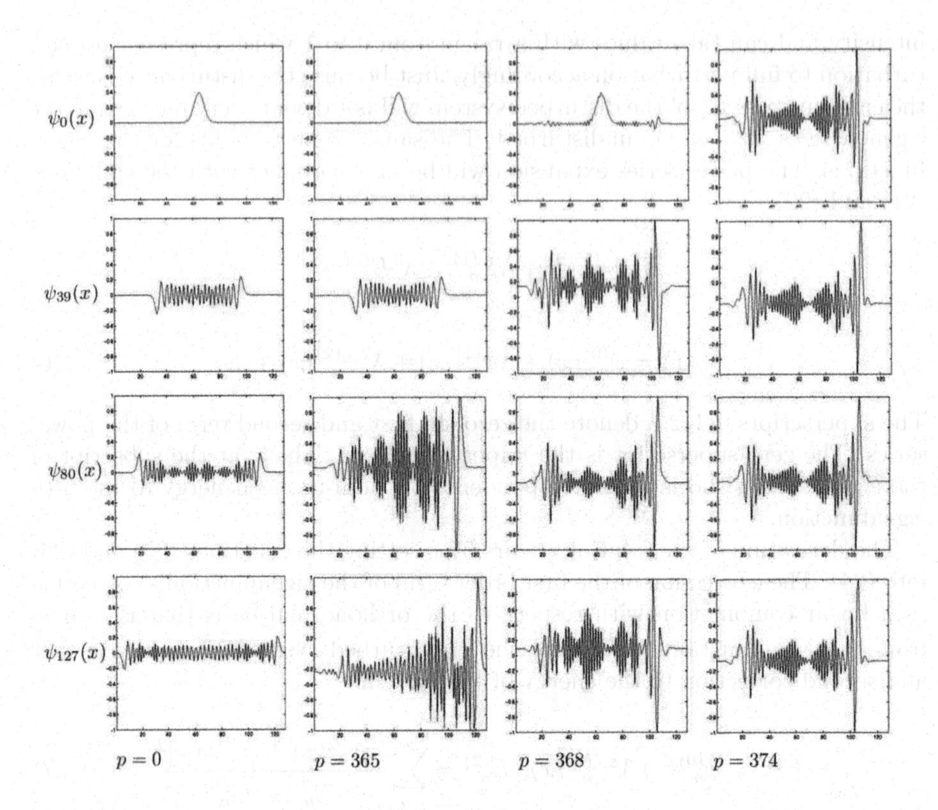

Fig. 3. The deformation of the Eigenfunctions $\psi_0, \psi_{39}, \psi_{80}$ and ψ_{127} after perturbing successively several times. In the leftmost part of the figure the Eigenfunctions are unperturbed and at the rightmost they are one step before collapse. The value of p denotes the number of time-independent perturbations that are already applied to the system. The analogy between the amplitude of the perturbed Eigenfunctions is not in direct proportion with their computed values after the dissolving starts, due to normalizing conditions.

differently but consistently as a whole. Each one of the perturbation types produces a characteristic change which is clearly recognizable.

A phenomenon that occurs in every tested disturbance type is a deformation of the eigenfrequencies and eigenvalues after the application of sufficient many consecutive time-independent perturbations. Suddenly the system starts to decompose and, after a while, it collapses. The eigenenergies E_n value range grow and the simulation eventually stops. The eigenfunctions $\psi_n(x)$ are also greatly deformed at the same time because as the eigenenergie's and eigenfunction's changes are closely linked, as expressed also by equations (27) and (28).

The alteration of the eigenfunctions can be made independently hearable by a direct mapping of the eigenfunctions in the time axis, where each channel holds one eigenfunction. Figure 3 shows the deformation of the eigenfrequencies in subsequent perturbations. One crucial point in which audification and sonification are profitable over a visual representation is the fact that the eye cannot grasp the interconnections and the consistency of the changes of all eigenfunctions as an integrated entity.

As previously mentioned, the eigenfunction's transformations can be also made recognizable by analyzing the windowed part of the audio signal as a linear combination of the eigenbasis. In every step of the overlap-add procedure a time-independent perturbation is applied which alters the eigenfunctions in such a way that they may no longer constitute a complete orthogonal basis. Despite this fact, the coefficients are computed as if the underlying basis was orthonormal. By this means the deformation of the sound is an indication for the decomposition of the eigenfunctions and their orthonormality.

Perturbations of small intensity have no recognizable audible effects on the stored sound. The effects start to take place only a little before the collapse occurs. Because of the rich content of the eigenfunction's alterations, a sonification procedure that would be more reflective of the phenomenon could be addressed.

Sonification Choices. Because the eigenenergies of the unperturbed system are equally spaced, as presented in (3), the idea of a projection of their values on the frequency plane has arisen. With an appropriate scaling factor the eigenenergies can be seen as the frequencies of sinusoidals that create a harmonic sound before any perturbation. Each time the perturbation is applied, the values of the frequencies of the sinusoidals are slightly changed. To make the sound effect more recognizable, the amplitude of all the sinusoidal components of the spectrum was set to the same value and then was filtered with the spectral envelope of a vowel of small duration with the help of cepstrum analysis [9, pp. 319–321].

As can also be seen in Fig. 4, the first times the perturbation is applied the spectrum of the sound has a characteristic development. After a critical number of perturbations, the decomposition of the system begins and an inharmonic sound is produced.

4.3 Time-Dependent or Dirac Method

Description of the Process. In this case the perturbation is denoted with the $V(t)$ operator which is assumed to be small with regard to the Hamiltonian $\hat{H}_0$ of the undisturbed system and the time duration of the disturbance reaction to the system is considered to be small enough. The eigenenergies and eigenfunctions of the system will also change with time. The Hamiltonian will be the addition of the unperturbed solvable and the time-dependent term [1, pp. 149–153]:

$$\hat{H}(t) = \hat{H}_0 + V(t) \tag{29}$$

The time-dependent Schrödinger equation is in this case:

$$i\hbar \frac{\partial \psi(x,t)}{\partial t} = \hat{H}(t)\psi(x,t) = (\hat{H}_0 + V(t))\psi(x,t) \tag{30}$$

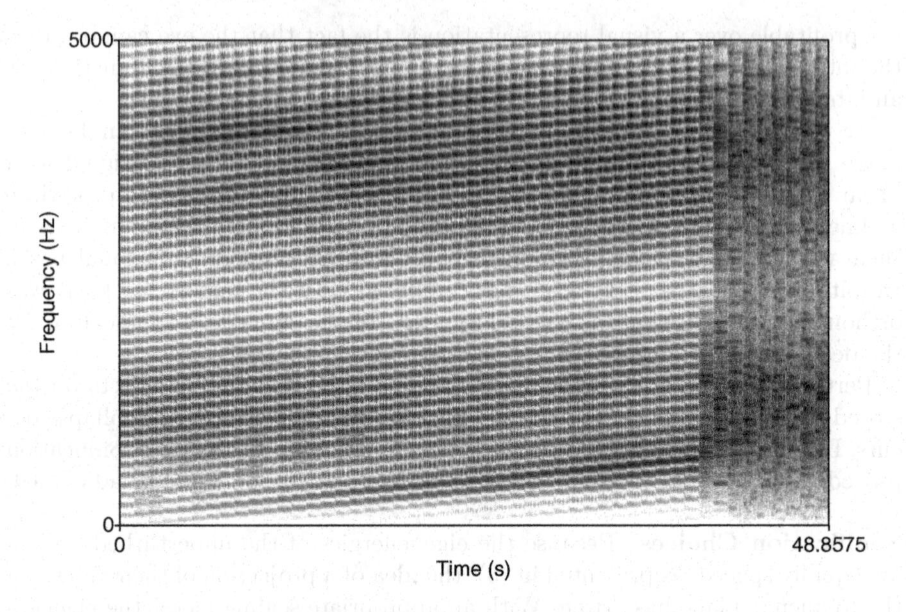

Fig. 4. Spectrum of the sonified Eigenenergies after perturbing successively several times. The new Eigenenergies are computed at each frame, easily seen in the figure by the vertical characteristic. The spectrum maintains a recognizable structure for the first perturbations and the dissolution of the spectral structure is apparent after enough perturbations are applied.

It cannot be solved by separating the spacial and temporal parts with the use of variable separation. That is the reason why in this case the solution cannot be implemented with the approach of the time-independent case. In analogy with the methodology of the time-independent case, the wavefunction of the system will be expanded as a linear combination of the basis of the unperturbed system's eigenfunctions, whereas the solution involves the detection of the expansion coefficients.

$$\psi(x,t) = \sum_{m=1}^{\infty} c_m(t)\psi_m^{(0)}(x) \tag{31}$$

The coefficients $c_m(t)$ are represented by a mathematical expression that includes both the time-evolution term that is caused by the unperturbed Hamiltonian $\hat{H}_0$ combined with the time-dependent transformation $a_m(t)$ that is generated from $V(t)$:

$$c_m(t) = a_m(t)e^{\frac{-iE_m^{(0)}t}{\hbar}} \tag{32}$$

The $a_m(t)$ terms are expanded with the help of power series. The equation (31) is solved to:

$$\psi(x,t) = \sum_{m=1}^{\infty} a_m(t) e^{\frac{-iE_m^{(0)} t}{\hbar}} \psi_m^{(0)}(x) \tag{33}$$

Where the $a_m^{(1)}(t)$ is the first correction term of the $a_m(t)$ expansion:

$$a_m^{(1)}(t) = -\frac{i}{\hbar} \int_0^t V_{nm}(t') e^{\frac{-i(E_n - E_m)t'}{\hbar}} \mathrm{d}t' \tag{34}$$

$V_{nm}(t)$ expresses the term:

$$V_{nm}(t) = \int_x \psi_m^{(0)}(x)^{\dagger} V(t) \psi_n^{(0)}(x) \mathrm{d}x \tag{35}$$

The further higher terms are computed iteratively but are not used in the implementation of this work due to their computational complexity.

The term $a_m^{(1)}(t)$ in equation (34) represents the fact that the system possesses a kind of memory. The integration is always computed from the time point where the perturbation started. Even if the disturbance stops its action the effects of the interaction are "remembered" and maintained in the system. This phenomenon is inherited by the stored sound.

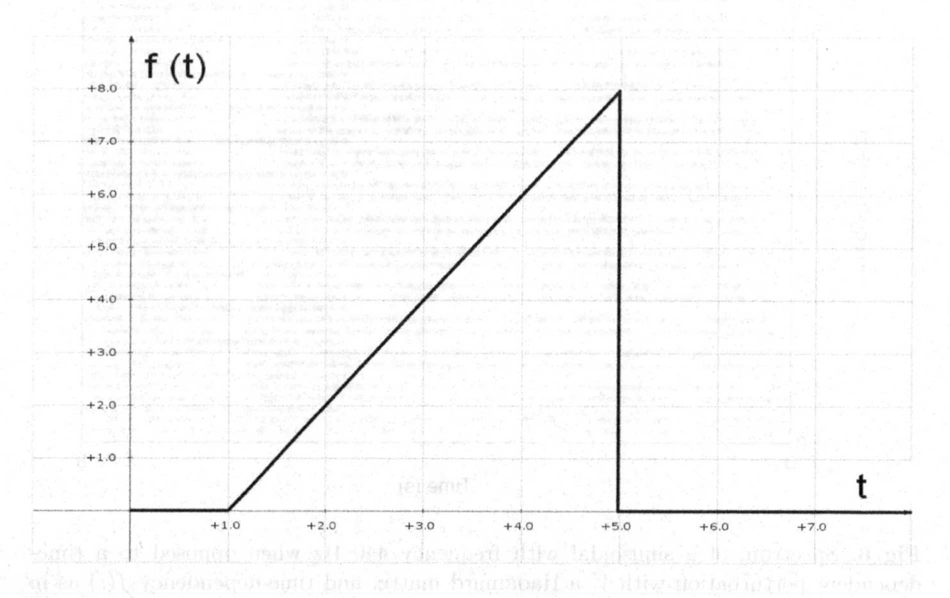

Fig. 5. The time-dependency $f(t)$ of the perturbation that was used for the creation sound in Fig. 6

Audification Choices. The time-dependent perturbation only affects the state of the system. Therefore an insight into the influence of the disturbance can be only made through the audification of the stored sound. More specifically, the first order correction term is computed and added for every windowed segment coefficient, as shown in (29). The resynthesized sound with respect to the basis of the unperturbed eigenfunctions contains the changes that are implied by the disturbance.

So far the type of perturbations $V(t)$ that were used could be decomposed as a product of a constant Hermitian matrix V and a function of time $f(t)$. The V term contains the spatial dependency and is in analogy with the systems that were used in the time-independent perturbation and the $f(t)$ which expresses the time dependency and contains combinations of linear, step and sinusoidal functions.

In the signals treated with a time-dependent perturbation there is always an existing component that evolves in time according to the unperturbed Hamiltonian as seen in (32) and a component that evolves under the influence of the perturbation. These two evolutions interfere with each other and create recognizable interference patterns in the spectral domain. Especially in the case where $f(t)$ is almost constant for a specific duration, a periodic component which is acoustically clearly separated from the evolution modulation appears, as shown in Fig. 6.

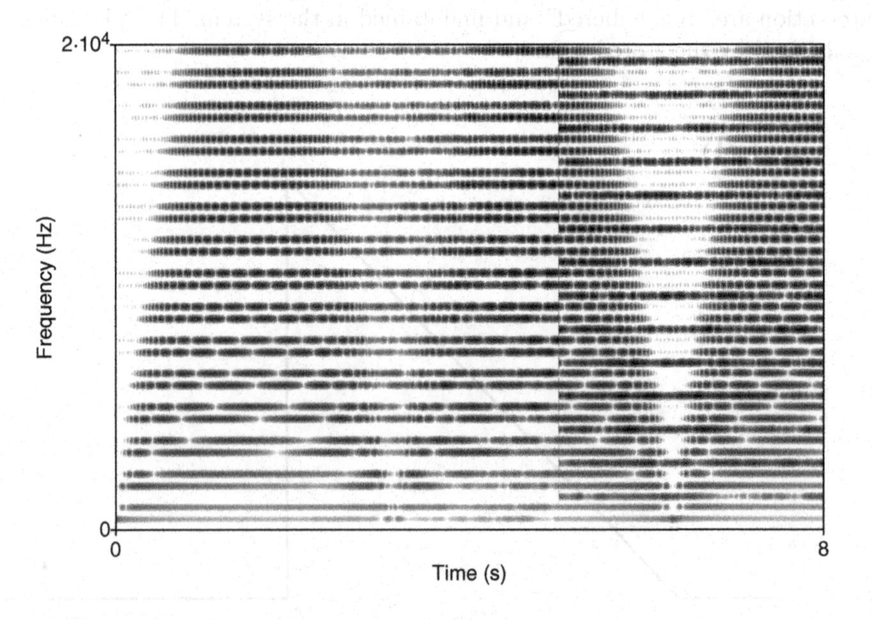

Fig. 6. Spectrum of a sinusoidal with frequency 440 Hz when opposed to a time-dependent perturbation with V a Hadamard matrix and time-dependency $f(t)$ as in the Fig. 5. The development of the sound has similarities with the time-evolution spectrum of Fig. 2 in the beginning but the perturbation effects gradually grow. At the time point $t = 5$ sec the perturbation stops but its effects remain as a constant periodical component.

By using perturbations with different types of V parts and the same time dependency $f(t)$ it became apparent that the developed sounds more reflect the time dependency than the V component.

5 Implementation Details

Two different software packages that implement the functionality mentioned above have been programmed, one in Matlab and one in C for a PD external. The QLib Matlab package [11] was used for the computation of probability amplitudes and exploration of the quantum measurement process. The GNU Multiple Precision Arithmetic Library (GMP) [12] provided an efficient solution to the computation of the factorial, which is the main obstacle for the implementation of the eigenfunctions $\psi_n(x)$ for large n. For the efficient computation of the externals, specially in the analysis-resynthesis stage, the CBLAS [13] and CLAPACK [14] were included.

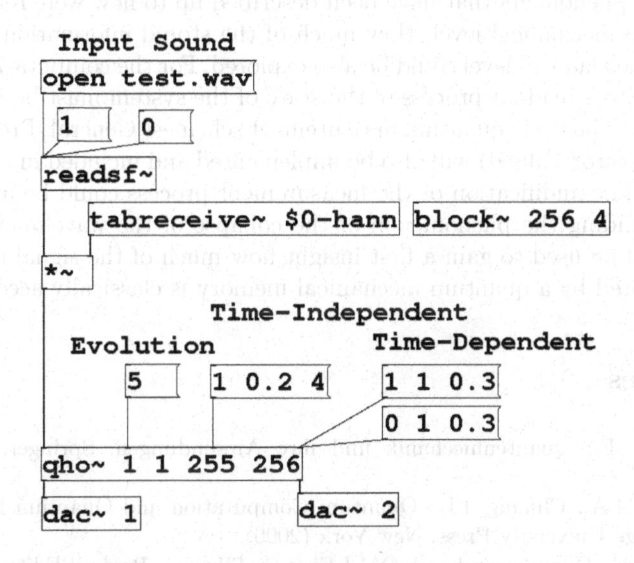

Fig. 7. Pure Data External *qho* $\sim$ that implements the impact that the time evolution, time-independent and time-dependent perturbation procedures have to a stored test sound

6 Future Work

More sonification decisions need to be undertaken to enable further comprehension of the system's behavior . As seen in the eigenfunction's deformation from time-independent perturbation case, appropriate methods that include psycho acoustical principles and are more strongly interconnected to the nature of some of the phenomena, should be further explored.

Apart from the perturbation method, other disturbance approaches can be explored, such as the variational method [1, pp. 147–149]. These approaches are applied in the cases where the decomposition of the disturbed Hamiltonian cannot be made by means of an approximative method because the system cannot be described through a small disturbance on a solvable system.

According to the correspondence principle of Bohr, for relatively big values of n the behavior of the quantum harmonic oscillator should be consistent with its classical counterpart. In particular, the larger the n becomes, the more we approach the classical harmonic oscillator [3, pp. 54–57]. On the other hand, the behavior of the oscillating object deviates the most from the classical one for the lowest possible value of n. Quantum mechanical phenomena obey the uncertainty principle. Therefore the comparison can only be made in terms of probability densities of finding the oscillating object in a specific position x. A sonification scheme that concentrates on the transition from the quantum to the classical equivalent harmonic oscillator could be one of the possible ways for the sonification of the correspondence principle.

All of the phenomena that have been described up to now were realized in/at the quantum mechanical level. How much of the stored information will be accessible in the classical level could be also explored. For the computational model to be complete a readout process of the state of the system must be included [2, pp. 283–287]. The three quantum measurement schemes(General, Projective and Positive Operator Valued) will also be implemented and included in this work [2, pp. 84–93]. The audification of the measurement process could be an approach for understanding the phenomenon of the collapse of the wavefunction. Audio signals could be used to gain a first insight how much of the signal information that was stored by a quantum mechanical memory is classically accessible.

References

1. Ehlotzky, F.: Quantenmechanik und ihre Anwendungen. Springer, Heidelberg (2005)
2. Nielsen, M.A., Chuang, I.L.: Quantum Computation and Quantum Information. Cambridge University Press, New York (2000)
3. Schwabl, F.: Quantenmechanik QM I Eine Einführung, Revised Edition. Springer-Verlag, Heidelberg (2005)
4. Nakahara, M., Ohmi, T.: Quantum Computing, From Linear Algebra to Physical Realizations. CRC Press, Taylor and Francis Group, Boca Raton (2008)
5. Deutsch, D., Ekert, A., Lupacchini, R.: Machines, Logic and Quantum Physics. Bull. Symbolic Logic 6(3), 265–283 (2000)
6. Goodwin, M.M.: Adaptive Signal Models - Theory, Algorithms and Audio Applications. Kluwer Academic Publishers, Dordrecht (1998)
7. Refregier, A.: Shapelets: I. A Method for Image Analysis. Mon. Not. Roy. Astron. Soc 338(1), 35–47 (2003)
8. Coffey, M.W.: Properties and possibilities of quantum shapelets. J. Phys. A: Math. Gen. 39(4), 877–887 (2006)
9. Zölzer, U.: Digital Audio Effects. John Wiley & Sons, Chichester (Reprinted 2005)

10. Ömer, B.: Quantum Programming in QCL, Master Thesis, Institute of Information Systems, Technical University of Vienna, (2007), http://tph.tuwien.ac.at/~oemer/doc/quprog.pdf(last accessed October 23, 2009)
11. QLib Matlab Package, http://www.tau.ac.il/~quantum/qlib/qlib.html (last accessed October 23, 2009)
12. GNU Multiple Precision Arithmetic Library, http://gmplib.org/ (last accessed October 23, 2009)
13. CLAPACK (f2c'ed version of LAPACK), http://www.netlib.org/clapack/ (last accessed October 23, 2009)
14. CBLAS Library, http://www.gnu.org/software/gsl/manual/html_node/GSL-CBLAS-Library.html (last accessed October 23, 2009)

Using Sound to Identify Correlations in Market Data

David Worrall

Sonic Communications Researh Group, University of Canberra
worrall@avatar.com.au

Abstract. Despite intensive study, a comprehensive understanding of the structure of capital market trading data remains elusive. The one known application of audification to market price data reported in 1990 that it was difficult to interpret the results probably because the market does not resonate according to acoustic laws. This paper reports on some techniques for transforming the data so it *does* resonate; so audification can be used as a means of identifying autocorrelation in capital market trading data. Also reported are some experiments in which the data is sonified using a homomorphic modulation technique. The results obtained indicate that the technique may have a wider application to other similarly structured time-series data.

Keywords: audification, sonification, homomorphic modulation, stock market, trading data, time-series, statistics.

1 Introduction

The statistical analysis of trading data, and stochastic modelling using that analysis, is an area of ongoing quantitative research in finance. Two principal concerns are to find techniques to accurately describe the way prices are distributed statistically and to what extent or auto-correlation exists and can be detected, even pre-empted, as market prices evolve. Understanding the first is important for the risk analysis of various trading instruments in the longer term, and understanding the second is important in attempts to predict future, especially catastrophic, events.

The power of visual representation to enhance and deepen an understanding of phenomena through their data abstractions is undisputed. Yet, as with many time-domain processes, a visual representation does not always reveal the structure of the data. Mandelbrot, arguably the most well known Quant[1], argues that is not possible to *visually* distinguish graphs of real market data and those of Brownian-motion-generated data[2] [30]. This leads to a data sonification question, which the research

[1] The term 'Quant' is used in the field to identify those who quantitatively analyze capital market data, or use such analysis to construct investment portfolios with a specific risk profile. See §2.2.

[2] Brownian motion is an independent (that is uncorrelated) random walk in which the size and direction of the next (price) move is independent of the previous move(s). A statistical analysis of time series data is concerned with the distribution of values without taking into account their sequence in time.

S. Ystad et al. (Eds.): CMMR/ICAD 2009, LNCS 5954, pp. 202–218, 2010.

reported in this paper seeks to answer: Can trading data be presented in a way that permit its autocorrelation to be aurally distinguished from statistically similar, but uncorrelated, data?[3]

This investigation was conducted during the development of software solutions in the *SoniPy* environment for the sonification of large multidimensional datasets [44], so the primary emphasis was tool development and perceptual testing was informal.

2 The Data

The importance of understanding the data itself before attempting to sonify it has been long emphasised [5]. As will be discussed later, it is important in this particular circumstance, to distinguish between real trading data, simulated trading data and financial data such as general economic indicators, as they usually, arise out of different processes and have quite different structural characteristics.

2.1 Context

Capital markets are (increasingly virtual) places where companies and traders converge around an exchange to raise new investment capital, and investors and speculators trade exchange-registered securities such as stocks (shares), bonds, currencies, futures and other derivatives contracts. These exchanges have strict government-regulated mechanisms for such activity and the community of freely-participating individuals around them communicate more-or-less informally with each other and formally through exchange-registered brokers who themselves provide information to their clients about specific trading activity as well as about other more general environmental (financial, political, meteorological etc) conditions that may affect an individual's trading decisions. Such decisions, enacted by the brokers, cause excitations of the trading system, known colloquially as a 'trading engine', which in turn produces data records of its activities. Some of that data, and various summaries of it, are fed back for the information of market participants. In turn, these marketplaces operate as systems within national and global economies and international companies may be listed on more than one exchange. Each exchange's trading system is designed to be acephalously appropriate for the types of securities that are traded on it.

Trading engines need to be fast, efficient and accurate[4]. They generate large quantities of data, reflecting the moment-to-moment shifting situation of the order book of each of their trading securities as potential buyers and sellers adjust their declared

[3] Decorrelation can be achieved by changing the sequence of values in a time series. In so doing, any spectral information in the series is destroyed, while its statistical properties remain invariant.

[4] It is somewhat ironic that, in an enterprise that relies on 'clean' data, financial data often requires considerable 'washing' before its sonification can be undertaken. This situation is exacerbated by the trait that, with datasets over a certain size, the use of metadata tagging is uncommon, principally because it significantly increases the overall size of the dataset, even though the omission increases the likelihood of error. In any event, any cleaning has to be undertaken algorithmically and so it is expedient to have the tools for doing so integrated with the sonification software being used.

positions, and then sometimes undertake trades. Security Trading datasets are sets of time-ordered trading events having a number of empirical dimensions, such as price and volume[5], depending on the particular type of security being traded (share, futures, options etc) and from which other data may be derived. A medium-sized exchange such as the Australian Securities Exchange (ASX) processes approximately two million trades a month: an average of 100,000 trades a day[6].

2.2 Quantitative Analysis

As a discipline in finance, quantitative analysis begins with Bachelier's speculation that the market is efficient and thus price movement must be a random walk [1]. At the time of his speculation in 1900, the mathematics of random walks was well known and an analysis of markets in terms of it enabled the construction of portfolios of stocks with defined risk profiles. Benoît Mandelbrot's study of price action in cotton led him to question this received wisdom, and to develop another mathematics, which he called "fractal" to accommodate his analytic findings. Fractal mathematics has become a widely applicable tool in many fields, both analytic and generative, and continues to be the basis of contemporary quantitative analysis.

Quantitative analysts generally use statistical analysis and stochastics to model the risk profiles of market indices, segments and individual securities as accurately as possible so as to assist in the construction of investment portfolios with certain characteristics, such as risk exposure, for example. To underestimate the risk of a portfolio is to court calamity, while overestimating it invites lower returns than might have otherwise been possible. Quantitative analysis, especially of high-frequency data, remains an area of active research [4] [14] and readable introductions are available in the popular science press for those less mathematically inclined [30][37][38][39].

3 Previous Work

3.1 Survey of the Sonification of Financial Data

The first users of technology-enabled financial market data sonification[7] were probably the bucket-shop traders in the early years of the twentieth century, who were

[5] The term ‘volume’ is used throughout to mean ‘trading volume’ not as a psychoacoustic parameter.

[6] A breakdown of recent ASX trading volumes is available from their website: www.asx.com.au/asx/statistics/TradingVolumes.jsp

[7] The presence of auditing (hearing of accounts from the Latin *auditus*) has been inferred from records of Mesopotamian civilizations going back as early as 3500 BCE. To ensure that the Pharaoh was not being cheated, auditors compared the ‘soundness’ of strictly independently scribed accounts of commodities moving in, out and remaining in warehouses [7]. In the alternating intoning of such lists, differences can be easily identified aurally. A faster and more secure method that eliminates any ‘copy-cat’ syndrome in such alternation, is to have the scribes read the records simultaneously–a type of modulation differencing technique. While we have no evidence that this specific technique was practiced in ancient times, such a suggestion does not seem unreasonable, and would represent possibly the earliest form of data sonification.

reputed to be able to differentiate the sounds of stock codes, and prices that followed, from the sounds made by their stock-ticker machines as they punched recent trading information, telegraphed from an exchange, into a strip of rolling paper tape [28]. Janata and Childs suggest that Richard Voss may have been the first to experiment with the sonification of historical financial data: stock prices of the IBM corporation [22]. This is possible, as Voss and Mandelbrot were research collaborators in fractal mathematics at IBM's Thomas J. Watson Research Center and Voss played an early seminal role in the visualisation of fractal structures and in the analysis of the fractal dimensions of music and speech [41][42].

Kramer and Ellison used financial data in the early 1990's to demonstrate multivariate sonification mapping techniques [24]. This work was later summarized and published with sound examples [25]. The trading data used included four–and–a–half years of the weekly closing prices of a US stock index, a commodity futures index, a government T-bond index, the US federal funds interest rates, and value of the US dollar. Mappings were to pitch, amplitude and frequency modulation (pulsing and detuning), filter coefficients (brightness) and onset time (attack). Mapping concepts included redundant mapping and datum highlighting (beaconing).

Ben-Tal et al. sonified up to a year's end–of–day data from two stocks simultaneously by mapping them to perceptually distinct vowel-like sounds of about one second duration [3]. A single trading day was represented as a single sound burst. The closing price for the day was mapped to the center frequency, and the volume of trade to the bandwidth. These values were scaled such that the parameters for the last day of trade in each period corresponded to a reference vowel. Closing price was mapped to the number of sound bursts and volume (the number of trades) to duration. They informally observed that they could categorise high volume, high price trading days as loud, dense sounds, while low volume, low price days were heard as pulsed rhythmic sounds.

Brewster and Murray tested the idea that traders could use sounds instead of line-graphs to keep track of stock trends when they are away from the trading floor [6]. Using Personal Digital Assistants with limited screen space over a wireless network, one month of (presumably intraday) price data for a single share was mapped to pitch via MIDI note numbers. Participants, all students whose previous trading experience was unreported, were required to try to make a profit by buying and selling shares while monitoring price movement using either line or sound graphs. As trade transaction costs appear not to have been factored into the calculations, profits or losses were presumably gross. The experimental results showed no difference in performance between the two modes, but participants reported a significant decrease in workload when they used the sonification, as it enabled them to monitor the price aurally while simultaneously using the visual display to execute trades.

Nesbitt and Barrass also undertook a multimodal sonification and visualisation study, this of market depth,[8] to test whether subjects could predict the price direction of the next trade [34]. They used real data from a single security's order book. The visualisation used a landscape metaphor in which bid and ask orders (to buy and sell), were 'banked' on either side of a river, the width of which thus represented the size of

[8] *Market depth* is a term used to denote the structure of potential buy and sell orders clustered around the most recently traded price.

price gap between the highest bid and the lowest ask, known as the 'bid–ask spread'. A wider river implied slower flow (fewer trades) and so on. The sonification employed the metaphor of an open-outcry market. A sampled male 'buy' and a female 'sell' voice displaying a discretely partitioned dataset (price, volume, price-divergence) was mapped into a discretely partitioned three-dimensional 'importance space' (pitch, loudness, stereo-location). This experimental design illustrates how sonification can be used to assist the apprehension of data segmentation such as where the trajectory of a parameter under focus changes.

Janata and Childs developed *Marketbuzz* as an add-on to conventional trader's terminals, such as those by Bloomberg, for the sonification of real-time financial data [23]. They used it to evaluate tasks involving the monitoring of changes in the direction of real-time price movements, with and without auditory or visual displays. A significant increase in accuracy using auditory displays was reported, especially when traders were visually distracted by a simultaneous diversionary "number-matching" task. Further, Childs details the use of sonification to highlight significant price movements relative to opening price, as well as continually changing features of Stock Options [8].

Mezrich, Frysinger and Slivjanovski developed a dynamic representation, employing both auditory and visual components, for redundantly displaying multiple multivariate time-series [32]. Each variable was represented by a particular timbre. The values of the variable were mapped to pitch. The analyst could focus on a subset of the data by interactively brightening or muting individual variables and could play the data both forwards and backwards. Subsets of the data could be saved and juxtaposed next to each other in order to compare areas where the data might be similar. In almost all cases, the sonified data performed as well as or better than the static displays.

Two other types of sonifications of securities data demonstrate different motivations but are mentioned here for completeness. The first is Ciardi's *sMax*, a toolkit for the auditory display of parallel internet-distributed stock-market data [9]. *sMax* uses a set of Java and Max modules to enable the mapping and monitoring of real time stock market information into recognizable musical timbres and patterns. The second is Mauney and Walker's rendering of dynamic data specifically for peripheral auditory monitoring. The system reads and parses simulated real-time stock market data that it processes through various gates and limiters to produce a changing soundscape of complementary ecological sounds [31].

There are a number of studies, principally those whose purpose was the study of parameter-mapping and auditory graphs, which have been omitted from this survey because it is not clear that there is anything in the findings specific to the structure of financial data: Unless it is generated using advanced modelling techniques, fictional data is unlikely to exhibit the same structural characteristics as real financial time series data.

3.2 Audification

In the 1990's Frysinger experimented with playing back market price data directly as a sound waveform. He reported that he found that *the results proved difficult to interpret, probably because the stock market does not follow physical-acoustic resonance laws* resulting in natural or 'ecological' sounds that can be understood from everyday listening experience [16][17][18]. There appears to be no further reports of security data audification prior to the work reported in this paper. Hayward also suggested that another reason audification fails for arbitrary data such as stock market figures or daily temperatures is that the amount of data required: even at low sampling rates, it is difficult to make a sound with a duration long enough to reveal valuable information to the listener [21].

In summarizing previous work on seismic audification, Hayward reported both Speeth's original experiment on discriminating the seismic sounds of earthquakes from atomic explosions and Frantti's repeat of it with a larger number of participants with less training and data from diverse locations. Frantti found a lower average accuracy and a wider variance in participant performance, which was also critically affected by the audification's time-compression ratio and the number of repeat audits. Neither study used trained seismologists, nor did participants have any interactive control [15][11]. Concentrated on single wavelett and quantitative questions, Hayward indicated a number of solutions to the difficulties encountered as well as some strategic extensions to planetary seismology in general. Dombois reported that he could hear seismological station-specific characteristics in his time-compressed audifications [12]. He informally found that, over time, overall information of a dynamic state was better comprehended with audification, whereas visualization was more effective when a detailed analysis of a single wavelett was required. He developed a unified acceleration method to make records taken under different meteorological and seismic conditions more compatible and in a later report on the state of research in auditory seismology, documented several other investigations in the field in the 1990s and much earlier [13]. He reported an increase in interest among seismologists, and this is also evidenced by the recent reporting in the popular media of the audification of stellar seismology [19].

Using data from a helicopter flight recorder, Pauletto and Hunt showed that audification can be used as an equally effective alternative to spectrograms for the discernment of complex time-series data attributes such as noise, repetitive elements, regular oscillations, discontinuities, and signal power [35]. However, another empirical experiment found that the use of audification to represent data related to the rubbing of knee-joint surfaces was less effective at showing the difference between normal and abnormal signals than other sonification techniques [27].

3.3 Sonification of Stochastic Functions

Aside from their use in algorithmic music composition, stochastic functions have received little attention in sonification research. Perhaps the first was a study of the use of parameter-mapping and physical model sonification in a series of experiments monitoring the performance of Markov chain Monte-Carlo simulations for generating statistical data from higher dimensional probability density functions [22]. The inclusion of some sound-generating tools in the statistical package *R* has the potential to generate wider interest, as exemplified by its use in tuning a parameter in the Hybrid Monte-Carlo algorithm [20]. Informal auditing of a technique to sonify, using amplitude modulation, cross-correlations in irregularly spiking sequences that resemble a Poisson process led to the postulation that the use of sonification for time series analysis is superior to visualisation in cases where the intrinsic non-stationarity of an experiment cannot be ruled out [2]. Time series data was generated by shaping a uniform distribution (white noise) with a cumulative probability density function, (similar to that used by Xenakis for his *ST* series of compositions [45]), in a differentiation study of the perceptualisation of some statistical properties of time series data generated using a Lévy skew alpha-stable distribution.This distribution is of interest to modellers of financial time series [38]. The study found no evidence that skewness in their data was perceivable, but participants were able to distinguish differences in kurtosis, which correlated with roughness or sharpness of the sound. This research provided empirical support for a part of the earlier initial findings of the experiments outlined below [44].

4 Informal Experiments

The experiments described here sought to (a) discover a way to directly audify a capital market trading dataset that preserved its autocorrelation characteristics and (b) ascertain informally whether such a dataset can be aurally discriminated from an audification of a statistically equivalent uncorrelated dataset. The null hypothesis in each case was that no distinction could reliably be made between the audifications.

4.1 The Dataset

The dataset chosen consists of twenty-two years of the daily closing price of All Ordinaries Index (ticker XAO) of the Australian Securities Exchange (ASX)[9], as illustrated by the plot in Fig. 1.

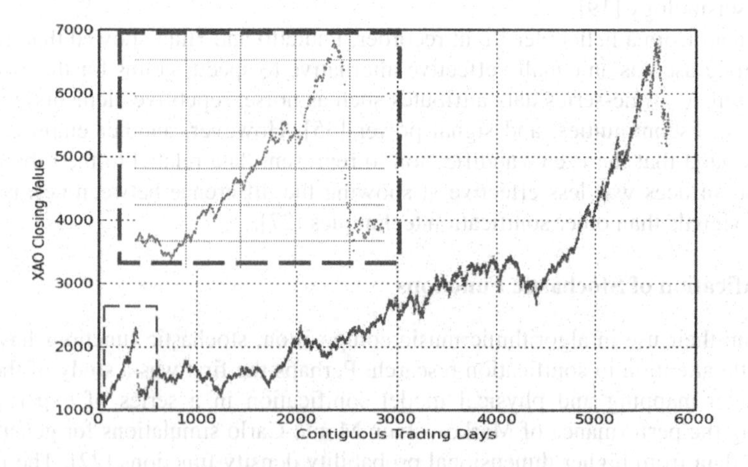

Fig. 1. A plot of 22 years of daily closing values of the ASX's All-Ordinaries Index (XAO). The inserted enlargement shows the activity in the region of "black" Tuesday (20 October 1987), the largest one-day percentage decline in the history of the market.

[9] The XAO is the broad Australian market indicator, a composite of the 500 largest companies, weighted by capitalisation, which are listed on the exchange. Contextual details are available at http://www.asx.com.au/research/indices/description.htm

The first task was to find a way to overcome the non-resonance problem referred to earlier, as discussed by Hayward [21]; one that transformed the dataset to be suitably oscillatory while preserving its correlational integrity. An equivalent problem is to be found in quantitative analysis, as observed by Stony Brook computer scientist Steven Skiena:

> *The price of an asset as a function of time is perhaps the most natural financial time series, but it is not the best way to manipulate the data mathematically. The price of any reasonable asset will increase exponentially with time, but most of our mathematical tools (e.g. correlation, regression) work most naturally with linear functions. The mean value of an exponentially-increasing time series has no obvious meaning. The derivative of an exponential function is exponential, so day-to-day changes in price have the same unfortunate properties.* [38]

The Net Return, or simply, the Return, is a complete and scale-free summary of investment performance that oscillates from positive values (increase) around zero (no change). \So the XAO dataset was converted to such market returns. For an asset whose price changed from p_t at time t to $p_{t+\partial t}$ at time $t+\partial t$, the simple linear return R_{lin} is defined as

$$R_{lin} = p_{t+dt} - p_t \,. \tag{1}$$

Because prices tend to move exponentially over longer timeframes, that is, in percentage terms, a better measure than R_{lin} is the ratio of successive price differences to the initial prices. These are known as net linear returns [3]:

$$R_{net} = \frac{(p_{t+dt} - p_t)}{p_t} \,. \tag{2}$$

Table 1 summarises the statistical properties of these returns for the XAO dataset, clearly showing that they are not a normally distributed.

Table 1. Basic statistics for XAO net returns

number of samples	5725
minimum sample	0.333204213
maximum sample	0.05886207483
arithmetic mean	2.1748845e04
variance	9.5685881e-05
skewness	7.6491182
Kurtosis	241.72988

Fig. 2 is a plot of net returns of the XAO dataset. The insert is of the first 500 samples, similarly to the insert in Fig. 1.

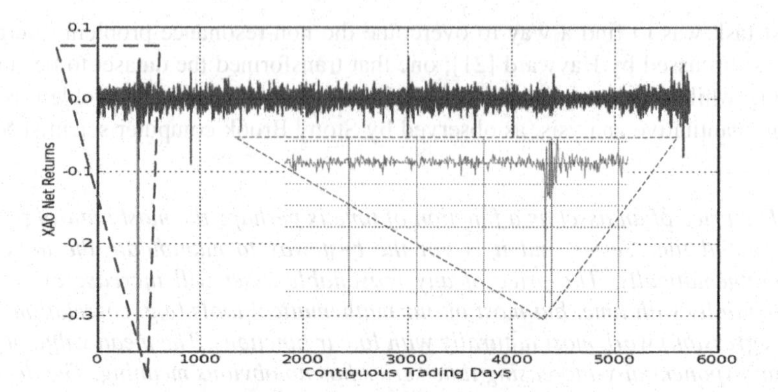

Fig. 2. A plot of XAO net returns. The insert is of t500 samples from the period leading to and from "black" Tuesday.

The single largest Net Return, clearly visible in Fig.1 and Fig. 2, was on 20 October 1987 ("black" Tuesday) the largest one-day percentage decline in stock market history. The difference between this largest minimum and the second-largest minimum is 62% of the total returns space. This is shown in the histogram of Fig. 3, which illustrates the frequency of net returns. The single minimum and second-largest minimum are circled, but barely visible at this scale.

So, despite its anecdotal interest, an 'audacious' clipping of the largest minimum sample to that of the second-largest minimum was performed and the resulting returns plotted in Fig. 4. Its histogram is show in Fig. 5, which illustrates both the skewness and kurtosis of the dataset, when compared to a normal distribution.

The size of the sample-bins of this histogram is kept constant to those of the Fig. 4 histogram by decreasing the number of bins. Of interest is the asymmetry of the outliers (the data at the extremities): there are more of the negative variety than positive,

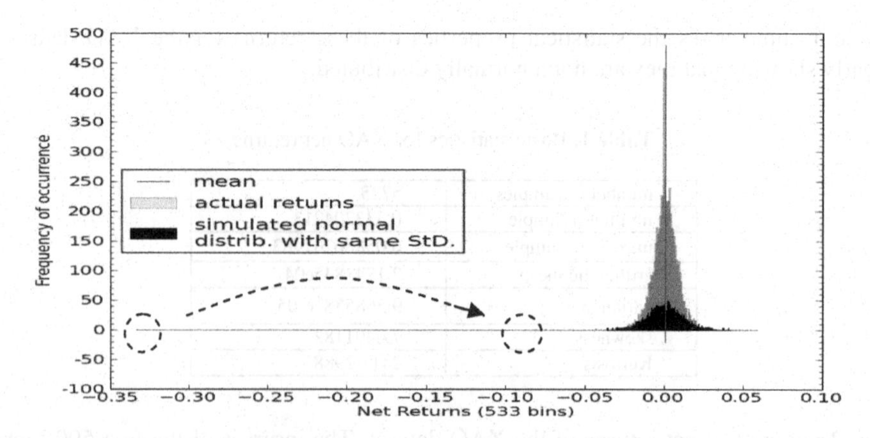

Fig. 3. Histogram of XAO net returns that illustrates the proportion of dataspace allocated to a single negative outlier without clipping

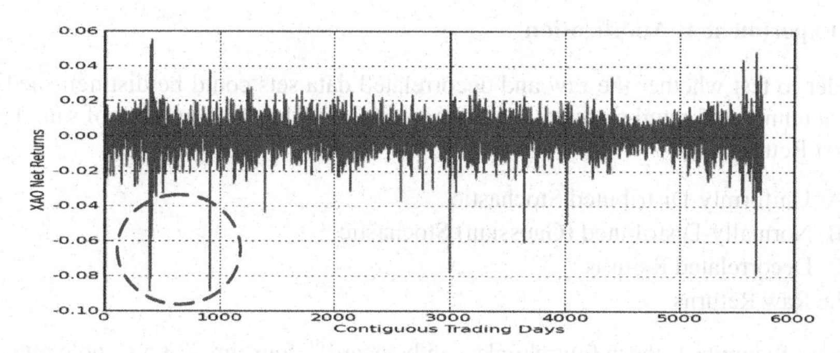

Fig. 4. Plot of XAO net returns, clipped so as to be suitable for audification

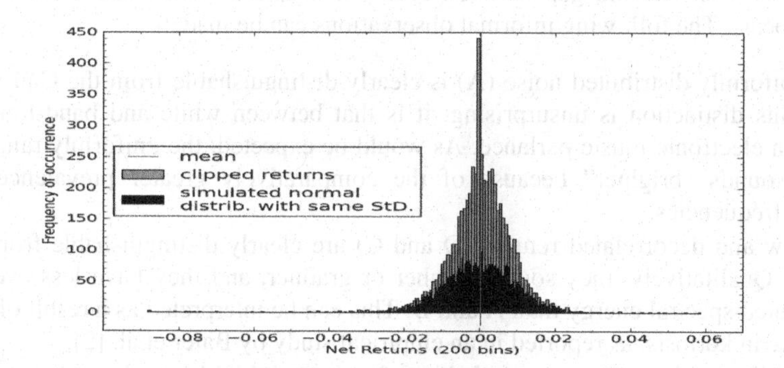

Fig. 5. Histogram of the clipped net returns overlaid with a simulated normal distribution with the same standard deviation and number of datum for comparison

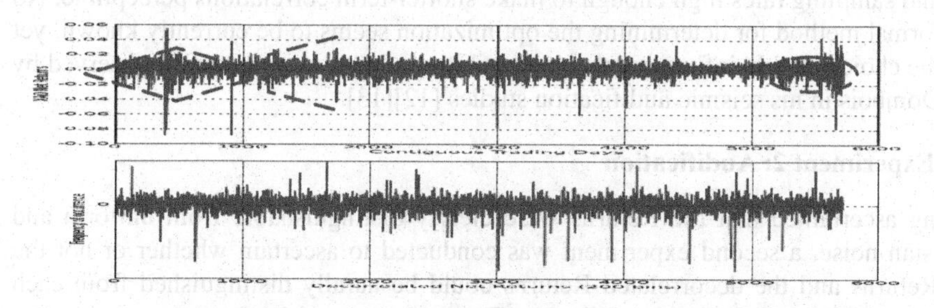

Fig. 6. A plot of the correlated (top) and decorrelated net returns

and negative ones exist further from the mean; even more so when considering this is the clipped dataset.

For comparison, the net returns were decorrelated. Fig. 6 shows the plots of both the correlated and decorrelated datasets. A number of features are visually apparent. Both have long tails but they appear more evenly distributed throughout the decorrelated dataset, contributing to its more chaotic visual appearance, whilst the correlated dataset appears to have periods of increasing (trapezoid), low (circle), and ramped (diamond) volatility.

4.2 Experiment 1: Audification

In order to test whether the raw and decorrelated data sets could be distinguished aurally, a number of chunks of audio were prepared with the same number of samples as the Net Returns.

- A. Uniformly-Distributed Stochastic
- B. Normally-Distributed (Gaussian) Stochastic
- C. Decorrelated Returns
- D. Raw Returns

In Audio Example 1, these four chunks can be heard[10] four times at a sample rate of 8 kHz in the order A-B-C-D. Each audio chunk is of approximately 0.7 seconds duration. There is a one-second gap between each chunk and a further one-second gap between repeats. The following informal observations can be made:

- The uniformly distributed noise (A) is clearly distinguishable from the Gaussian (B). This distinction is unsurprising: it is that between white and band-limited noise in electronic music parlance. As would be expected, the uniformly random noise sounds "brighter" because of the comparatively greater prevalence of higher frequencies.
- The raw and decorrelated returns (D and C) are clearly distinguishable from A and B: Qualitatively, they sound rougher or grainier, and they have less evenly distributed spectral energy than A and B. This can be interpreted as a result of the increase in kurtosis, as reported in an empirical study by Baier et al. [2].
- The 8 kHz sampling rate was settled on after some initial heuristic experimentation with higher and lower values. There appears to be an optimal compromise between durations long enough for possible temporal patterns to be perceptible and sampling rates high enough to make shorter-term correlations perceptible. No formal method for determining the optimization seems to be currently known, yet the choice clearly influences the perceptibility of pattern, as was also observed by Dombois in his seismic audification studies [12][13].

4.3 Experiment 2: Audification

Having ascertained that the Returns were clearly distinguishable from uniform and Gaussian noise, a second experiment was conducted to ascertain whether or not the raw Returns and the decorrelated Returns could be aurally distinguished from each other. An additional decorrelated Return (E) was generated in the same manner described for C, in Experiment 1, and three files were prepared with the following sequences in which the original raw returns (D) was placed in first second and third place respectively:

- Audio Example 2. D-C-E
- Audio Example 3. C-D-E
- Audio Example 4. C-E D

[10] All the Audio Examples referred to can be directly downloaded from http://www.sonification.com.au/springer09/

The listening task, on multiple random presentations of these audio files, was to try to determine, in each case, which one of the three chunks sounded different from the other two. The informal findings of several listeners, all of who had musical training, can be summarised as follows:

- The task was a more cognitively demanding than those in Experiment 1.
- Distinguishability was dependent on a narrower band of sampling rates. Above 8 kHz the characteristics described earlier seem to disappear. Below 3-4 kHz the roughness created by the individuation of large-valued samples meant that the principal means of identifying the raw returns was probably more by its invariance across all chunk presentations than by direct chunk comparison.
- Between 4 kHz and 8 kHz sampling rate, a distinct, though subtle, amplitude modulation was observable in the Net Return chunks that seems not to be present in the decorrelated ones. This amplitude modulation effect required attentive listening, probably, in part, due the relatively short duration of the audio chunks (less that 700 ms).

This last observation pointed to the need for more data to enable longer durations or the application of a technique other than audification that enables a slower sample presentation rate. As no intraday data was available for the dataset in question, the latter approach was chosen in Experiment 3.

4.4 Experiment 3: Homomorphic Modulation Sonification

This experiment was designed to test a simple proposition: That the four datasets A, B, C and D of Experiment 1 could be distinctly identified under homomorphic mapping into a pitch-time auditory space. A homomorphic mapping is one in which the changes in a dimension of the auditory space track changes in a variable in the dataset, with only as few mediating translations as are necessary for comprehension [26]. A narrow interpretation, called Homomorphic Modulation Sonification is used, in which time in the dataset was mapped to time in the auditory display, and sample value was mapped to pitch deviation (both positive and negative) from a centre frequency.

There is a subtle but important distinction between Homomorphic Modulation Sonification and the type of parametric mapping in which each datum is played as, or contributes to, a separate tone with its own amplitude envelope. In the separate-tones case, the audio-amplitude profile of the resulting audible stream fluctuates from–and–to zero, resulting in a sequence of auditory objects individuated by more–or–less rapid onset transients. With modulation however, a single continuous pulsed waveform results, affording the opportunity for the amplitude formant to be held relatively constant, resulting in a lower perceptual loading [36][4].

A csound [9] instrument, illustrated in Fig. 7, was constructed to produce this homomorphic mapping. Structurally, this is a basic frequency modulator in which an ADSR[11] for controlling modulation index is replaced by a sample buffer of the AIFF

[11] Computer music parlance: An ADSR is an Attach-Delay-Sustain-Release envelope shaper, a common tool for synthetically controlling the amplitude evolution of computed sounds.

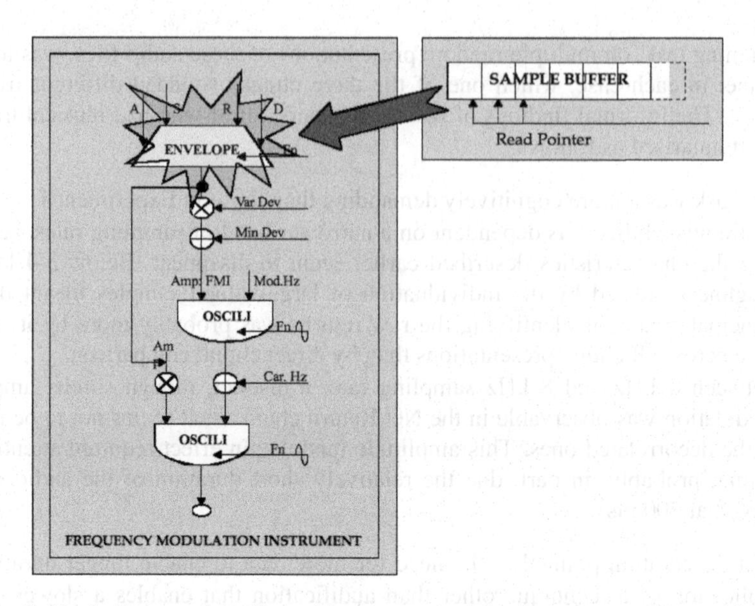

Fig. 7. Block diagram of the Csound instrument used for the homomorphic mappings

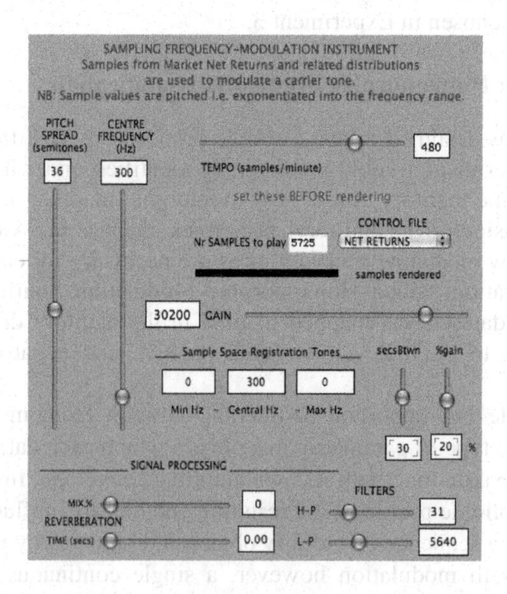

Fig. 8. User interface to the sampling frequency modulation instrument

samples. These samples, which can be read directly from an audio file, are then used in sequence to control the frequency deviation from a user-defined centre 'reference' carrier frequency.

Control was implemented to facilitate heuristic investigation of the perceptual space in *SoniPy* (via MIDI and Csound's *Python* API)[10] and in MacCsound. Fig. 8 shows the MacCsound controller interface [29]. The setting shown is for frequency-modulating a 300Hz tone according to the values of successive samples within a range of three octaves at the rate of 480 modulations per minute (8 Hz) from the Net Returns distribution. With this controller, it is possible to dynamically adjust the pitch spread and centre frequency during audition.

In addition to the sample modulator, the sonification also uses has a simple audio 'tick' registration tone generator that acts as an auditory reminder of the upper, lower and central limits of the current render. Both its frequency-of-occurrence (*secsBtwn*) and relative loudness (*%gain*) is adjustable.

Audio Examples 5–9 provide various sonic realisations of the homomorphic mappings of the Net Returns samples generated for Experiments 1 and 2. For consistency, all are rendered with the settings illustrated in Figure 8.

- Audio Example 5 is a series of twenty-second 'snapshots' of each of the four sample sets A, B C and D.
- Audio Example 6 is of A.
- Audio Example 7 is of B
- Audio Example 8 is of C.
- Audio Example 9 is of D.

The informal findings of Experiment 3 can be summarised as follows:

- The difference between homomorphic mapping sonifications of A, and B is easily noticeable, at least to a musically trained listener, as it was in the audifications of Experiment 2. The homomorphic mapping of A can be observed to be evenly spread across the pitch gamut, while the Normal distribution of B can be observed to be more closely clustered around the centre of the gamut, the mean.
- Again, C and D are noticeably different to A and B. Whilst both C and D appear to have short trending auto-correlative sequences, those of C (the Net Returns) appear more consistently and when they do they appear to last for longer periods of time. This is particularly noticeable in frequency of sequences of consecutive zero or small Net Returns, a characteristic consistent with the observation by chartists and technical analysts that securities prices frequently shift to a new price 'zone' quite quickly interspersed with longer times consolidating those zones before moving again [33].

5 Conclusions and Suggestions for Further Work

This investigation was conducted during the development of software solutions in the *SoniPy* environment for the sonification of large multidimensional datasets [43]. As such, the primary emphasis was tool development and perceptual testing was informal.[12] It is apparent from these experiments that the simple technique of using net returns is applicable to the sonification of capital market trading data and so it may be possible to apply the same techniques to other similarly structured time-series datasets

[12] A heavily commented version of the script developed for these experiments, along with the audio examples, is available for download from http://www.sonification.com.au/springer09/

from such applications as electoencephalography, trans-synaptic chemical transmitters, and the hierarchical networks arising from social affiliations. A further interesting extension study would be the application of the techniques to the functional simulations of financial-market-like time-series, an active field of econometric investigation in which large datasets can be generated when needed.

Although other techniques can be applied, the directness of audification makes it appealing and controlled empirical experiments to determine which features of the dataset are perceivable under those conditions would be worthwhile. The observation of amplitude modulation in the raw returns in Experiment 2 suggests that an empirical study to isolate the aural characteristics of cross-correlation, such as the spectral modulation suggested by the study, may be useful. This would require the preparation of additional, unrelated, raw returns datasets of the same sample size.

The choice of sampling rate clearly influences pattern perceptibility, as was also observed in seismic audification studies [40] but, apart from limiting the resulting frequency band imposed, no reliable formal optimization method is known and this deserves empirical attention, perhaps by using the using the fractal dimension of the dataset as a potential correlation index.

The effect of the size of the dataset on the sampling rate also needs to be isolated, as, whether or not higher sampling rates on larger datasets reveal other distinguishing features is currently not known.

Acknowledgements

This work was supported by funding from The University of Canberra, The Capital Markets Cooperative Research Centre and The Australian Music and Psychology Society.

References

1. Bachelier, L.: Théorie de la Speculation. Annales de l'Ecole Normale Superieure 17, pp. 21–86; Translated into English by Boness, A.J. In: Cootner, P.H. (ed.) The random character of stock market prices, pp. 17–78. MIT Press, Cambridge (1967)
2. Baier, G.T., Hermann, T., Lara, O.M., Müller, M.: Using Sonification to Detect Weak Cross-Correlations in Coupled Excitable Systems. In: Proceedings of The Eleventh Meeting of the International Conference on Auditory Display, Limerick, Ireland, July 6-9 (2005)
3. Ben-Tal, O., Berger, J., Cook, B., Daniels, M., Scavone, G., Cook, R.: "SONART: The Sonification Application Research Toolbox. In: Proceedings of the 2002 International Conference on Auditory Display, Kyoto, Japan, July 2-5 (2002)
4. Best, V., Ozmeral, E.J., Kopco, N., Shinn-Cunningham, B.G.: Object Continuity Enhances Eelective Auditory Attention. Proceedings of the National Academy of Sciences of the United States of America 105(35), 13174–13178 (2008)
5. Bly, S.: Multivariate Data Mappings. In: Kramer, G. (ed.) Auditory Display: Sonification, Audification, and Auditory Interfaces. Santa Fe Institute Studies in the Sciences of Complexity, Proceedings, vol. XVIII. Addison Wesley Publishing Company, Reading (1994)
6. Brewster, S., Murray, R.: Presenting Dynamic Information on Mobile Computers. Personal Technologies 4(2), 209–212
7. Boyd, E.: History of auditing. In: Brown, R. (ed.) A History of Accounting and Accountants. T.L & E.C. Jack, Edinburgh (1905)

8. Childs, E.: Auditory Graphs of Real-Time Data. In: Proceedings of ICAD 2005-Eleventh Meeting of the International Conference on Auditory Display, Limerick, Ireland, July 6-9 (2005)
9. Ciardi, F.C.: sMAX. A Multimodal Toolkit for Stock Market Data Sonification. In: Proceedings of ICAD 2004-Tenth Meeting of the International Conference on Auditory Display, Sydney, Australia, July 6-9 (2004)
10. http://www.csounds.com (Last viewed, March 2009)
11. Dombois, F.: Using Audification in Planetary Seismology. In: Proceedings of Eighth International Conference on Auditory Display, Espoo, Finland, July 29-August 1 (2001)
12. Dombois, F.: Auditory Seismology on free Oscillations, Focal Mechanisms, Explosions and Synthetic Seismograms. In: Proceedings of the 2002 International Conference on Auditory Display, Kyoto, Japan, July 2-5 (2002)
13. Dombois, F.: Underground Sounds. An Approach to Earthquake Prediction by Auditory Seismology. In: Geophysical Research Abstracts, vol. 4 (2002) EGS02-A-02741
14. Farmer, J.D., Patelli, P.: Zovko. I.I. The Predictive Power of Zero Intelligence in Financial Markets. Proceedings of the National Academy of Sciences of the United States of America 102(6), 2254–2259 (2005), http://www.pnas.org/content/102/6/2254.full.pdf+html (Includes links to extensive range of supplementary material) (last viewed, october 2007)
15. Frantti, G.E., Leverault, L.A.: Auditory Discrimination of Seismic Signals from Earthquakes and Explosions. Bulletin of the Seismic Society of America 55, 1–25 (1965)
16. Frauenberger, C., de Campo, A., Eckel. G. : Analysing Time Series Data. In: Proceedings of the 13th International Conference on Auditory Display, Montreal, Canada, June 26-29 (2007)
17. Frysinger, S.P.: Applied Research in Auditory Data Representation. In: Extracting Meaning from Complex Data: Processing, Display, Interaction, Santa Clara, California. SPIE-The International Society for Optical Engineering (1990)
18. Frysinger, S.P.: A Brief History of Auditory Data Representation to the 1980s. In: First Symposium on Auditory Graphs, Limerick, Ireland, July 10 (2005)
19. Gosh, P.: Team Records 'Music' from Stars. BBC news website (2008), http://news.bbc.co.uk/2/hi/science/nature/7687286.stm (Last viewed March/October, 2008)
20. Heymann, M., Hansen, M.: A New Set of Sound Commands for R; Sonification of the HMC algorithm. In: ASA Proceedings, Statistical Computing Section, pp. 1439–1443. American Statistical Association, Alexandria (2002)
21. Hayward, C.: Listening to the Earth sing. In: Kramer, G. (ed.) Auditory Display: Sonification, Audification, and Auditory Interfaces. Santa Fe Institute Studies in the Sciences of Complexity, Proceedings, vol. XVIII, pp. 369–404. Addison Wesley Publishing Company, Reading (1994)
22. Hermann, T.M., Hansen, H., Ritter, H.: Sonification of Markov Chain Monte Carlo simulations. In: Proceedings of the 7th International Conference on Auditory Display. Helsinki University of Technology, pp. 208–216 (2001)
23. Janata, P.: Childs. E. Marketbuzz: Sonification of Real-Time Financial Data. In: Proceedings of the 2004 International Conference on Auditory Display, Sydney, Australia, July 5-9 (2004)
24. Kramer, G., Ellison, S.: Audification: The Use of Sound to Display Multivariate Data. In: Proceedings of the International Computer Music Conference, San Francisco, C.A. ICMA, pp. 214–221 (1991)
25. Kramer, G.: Some Organising Principles for Representing Data with Sound. In: Kramer, G. (ed.) Auditory Display: Sonification, Audification, and Auditory Interfaces. Santa Fe Institute Studies in the Sciences of Complexity, Proceedings, vol. XVIII. Addison Wesley Publishing Company, Reading (1994)

26. Kramer, G.: An Introduction to Auditory Display. In: Kramer, G. (ed.) Auditory display: Sonification, Audification, and Auditory Interfaces. Santa Fe Institute Studies in the Sciences of Complexity, Proceedings, vol. XVIII, pp. 1–77. Addison Wesley Publishing Company, Reading (1994)
27. Krishnan, S., Rangayyan, R.M., Douglas, B.G., Frank, C.B.: Auditory Display of Knee-Joint Vibration Signals. Journal of the Acoustical Society of America 110(6), 3292–3304 (2001)
28. Lefèvre, E.: Reminiscences of a Stock Operator. John Wiley & Sons, NJ (1923/1994)
29. Ingalls, M.: http://www.csounds.com/matt/MacCsound/ (Last viewed March 2009)
30. Mandelbrot, B.B., Hudson, R.L.: The (Mis)behaviour of Markets. Basic Books, NY, USA (2004)
31. Mauney, B.S., And Walker, B.N.: Creating Functional and Livable Soundscapes for Peripheral Monitoring of Dynamic Data. In: Proceedings of ICAD 2004 - The Tenth International Conference on Auditory Display, Sydney, Australia, July 6-9 (2004)
32. Mezrich, J.J., Frysinger, S.P., Slivjanovski, R.: Dynamic Representation of Multivariate Time-Series Data. Journal of the American Statistical Association 79, 34–40 (1984)
33. Murphy, J.: Technical Analysis of the Financial Markets. New York Institute of Finance, New York (1999)
34. Nesbitt, K.V., Barrass, S.: Evaluation of a Multimodal Sonification and Visualisation of Depth of Market Stock Data. In: Proceedings of the 2002 International Conference on Auditory Display, Kyoto, Japan, July 2-5 (2002)
35. Pauletto, S., Hunt, A.: A Comparison of Audio and Visual Analysis of Complex Time-Series Data Sets. In: Proceedings of ICAD 2005-Eleventh Meeting of the International Conference on Auditory Display, Limerick, Ireland, July 6-9 (2005)
36. Patterson, R.D.: Guidelines for the Design of Auditory Warning Sounds. In: Proceeding of the Institute of Acoustics, Spring Conference, vol. 11(5), pp. 17–24 (1989)
37. Peters, E.E.: Chaos and Order in the Capital Markets. John Wiley & Sons, Inc, NY (1991)
38. Skiena, S.S.: Financial Time-Series Data (2004), http://www.cs.sunysb.edu/skiena/691/lectures/lecture6/lecture6.html (Last viewed January 2008)
39. Sornette, D.: Why Stock Markets Crash. In: Critical Events in Complex Financial Systems. Princeton University Press, Princeton (2003)
40. Speeth, S.D.: Seismometer Sounds. Journal of the Acoustic Society of America 33, 909–916 (1961)
41. Voss, R.F., Clarke, J.: 1/f Noise in Music and Speech. Nature 258, 317–318 (1975)
42. Voss, R.F., Clarke, J.: 1/f Moise in Music: Music from 1/f Noise. Journal of the Acoustical Society of America 63(1), 258–263 (1978)
43. Worrall, D.R.: Audification Experiments using XAO Returns in Auditory Display as a Tool for Exploring Emergent Forms in Exchange-Trading Data (2004). Report to the Sonic Communication Research Group, University of Canberra, (October 2004), http://www.avatar.com.au/sonify/research/sonseml/4audExperi.html (Last viewed March 2009)
44. Worrall, D.R., Bylstra, M., Barrass, S., Dean, R.T.: SoniPy: The Design of an Extendable Software Framework for Sonification Research and Auditory Display. In: Proceedings of the 13th International Conference on Auditory Display, Montréal, Canada, June 26-29 (2007)
45. Xenakis, I.: Formalized Music: Thought and Mathematics in Music, pp. 136–143. Indiana University Press, Indiana (1971)

Intelligibility of HE-AAC Coded Japanese Words with Various Stereo Coding Modes in Virtual 3D Audio Space

Yosuke Kobayashi, Kazuhiro Kondo, and Kiyoshi Nakagawa

Graduate School of Science and Engineering, Yamagata University,
4-3-16 Zyonan, Yonezawa, Yamagata, 992-8510, Japan
yosuke_kobayashi@m.ieice.org,
{kkondo,nakagawa}@yz.yamagata-u.ac.jp

Abstract. In this paper, we investigated the influence of stereo coding on Japanese speech localized in virtual 3-D space. We encoded localized speech using joint stereo and parametric stereo modes within the HE-AAC encoder. First, we tested subjective quality of localized speech at various azimuths on the horizontal plane relative to the listener using the standard MUSHRA tests. We compared the encoded localized speech quality with various stereo encoding modes. The joint stereo mode showed significantly higher MUSHRA scores than the parametric stereo mode at azimuths of ±45 degrees. Next, the Japanese word intelligibility tests were conducted using the Japanese Diagnostic Rhyme Tests. Test speech was first localized at 0 and ±45 degrees and compared with localized speech with no coding. Parametric stereo-coded speech showed lower scores when localized at -45 degrees, but all other speech showed no difference between speech samples with no coding. Next, test speech was localized in front, while competing noise was localized at various angles. The two stereo coding modes with bit rates of 56, 32, and 24 kbps were tested. In most cases, these conditions show just as good intelligibility as speech with no encoding at all noise azimuths. This shows that stereo coding has almost no effect on the intelligibility in the bit rate range tested.

Keywords: Head-Related Transfer Function (HRTF), High-Efficiency Advanced Audio Coding (HE-AAC), stereo coding, joint stereo, parametric stereo, Japanese Diagnostic Rhyme Tests (JDRT).

1 Introduction

In this research, we are aiming to use sound localization to separate the primary speaker speech from the other speakers in a multi-party virtual 3D audio conferencing environment. We are using HRTF (Head-Related Transfer Function) to separate sound sources. Vocal Village [1] and Voiscape [2] are examples of such systems. These systems integrate both audio and images (still and moving) in a virtual 3D environment. Avatars indicating participants and sound-generating objects are placed at arbitrary locations in this virtual space. Each participant's speech and sound objects

S. Ystad et al. (Eds.): CMMR/ICAD 2009, LNCS 5954, pp. 219–238, 2010.

are localized atcorresponding locations. The user is free to move around in this space, and the sound image locations are altered according to relative position changes.

The focus of the Voiscape system is in the creation of a 3-D multimedia "chat" environment. However, we are focusing more on the communication networking aspects of a similar system. When speech is localized within virtual space, they require multi-channel representation, most likely stereo if to be presented over a headphone. Thus, stereo signal processing and transmission is required for localized speech. Stereo coding is known to influence stereo sound image. For example, sound image is known to be broadened with the Twin VQ coding [3]. Mid-Side Stereo coding and Parametric Stereo coding [4] were shown to have different sound localization azimuth dependency in terms of quality and perceived localization accuracy.

In this paper, we study the effect of stereo audio coding on localized stereo speech, specifically on sound quality and Japanese word intelligibility. Although we are aware of attempts to assess conversional (bidirectional) quality in a similar setup [5], we will only deal with single-directional listening quality here. We used HE-AAC (High-Efficiency Advanced Audio Coding) [6] [7] which is the latest audio encoder currently available, and compared Joint Stereo [7] (which adaptively switches between Simple Stereo and Mid-Side Stereo) with Parametric Stereo [7] [8], which are both part of the HE-AAC standard codec. We only vary the stereo coding mode. The same single-channel audio coding was used (*i.e.* the AAC framework of the HE-AAC codec). Sampling rates supported by the standard are 32, 44.1 and 48 kHz. We used 32 kHz in our experiments since we are mainly dealing with wideband speech, which typically has a bandwidth of 7 kHz, and generally requires sampling rates at or above 16 kHz.

In previous work [9], we have shown that the speech intelligibility of target speech when competing noise is present can be kept above 70 % if the competing noise is placed at azimuths of more than 45 degrees away from the target speech on the horizontal plane. The sound localization in this case was achieved by convolving HRTFs of KEMAR (Knowles Electronics Mannequin for Acoustic Research) [10] (to be noted KEMAR-HRTFs) with the individual mono sources. We will attempt the same test with stereo-coded speech in this paper.

This paper is organized as follows. In the next chapter, subjective quality listening tests as well as its results are given for the two stereo coding methods. This is followed by speech intelligibility tests when localized speech is presented both with and without competing noise. Finally, conclusions and discussions are given.

2 Subjective Quality Tests

In this chapter, we investigated the subjective audio quality of localized speech using the MUSHRA (MUlti Stimulus test with Hidden Reference and Anchors) method [11]. The coding rate as well as the speech localization azimuth was varied. The number of subjects of this test was 20 males and 1 female. The tests were run semi-automatically on a Windows computer using a GUI-based Perl script.

2.1 Sound Sources

These speech samples were selected from the ASJ Continuous Speech Database for Research (ASJ-JIPDEC) [12], which is one of the widely-used Japanese speech databases. All were read ATR phonetically-balanced sentences. Read sentences used in the MUSHRA tests are summarized in Table 1. One of the male speakers had relatively high tone, while the other had a low voice. For all sound sources, the sampling frequency was 16 kHz, and the quantization bits were 16 bits linear. The sampling rate was up-sampled to 32 kHz.

Table 1. Sound sources used in the MUSHRA tests

Speaker	**Spoken sentence**
Male 1	bunsyoha nennen hueteiku
Male 2	mou yugureno youdearu
Female	anosakaonoboreba yuhiga mieru

2.2 Audio Codecs Used in the MUSHRA Listening Test

Table 2 lists the codecs used in this test. The HE-AAC coding and decoding were done at 24, 32 and 56 kbps with both joint stereo and parametric stereo coding using the aacPlus™ Encoder ver.1.28 [13]. The labels shown in the Table will be used in the figures in later chapters as well.

We used not only the recommended 3.5 kHz low-pass filtered audio, but also a 2.0 kHz low-pass filtered speech to use as anchors in the MUSHRA test. The reason for this is that since we are using speech, most of the spectral components are below 3.5 kHz, and thus this filter will not filter out enough components to serve as anchors. The references were localized speech samples which were not encoded. The listener always had access to the reference speech. The listeners were also given a hidden reference (*i.e.* the listeners were not told that there was a reference), and the listeners were instructed to give 100 points to samples which they thought were references in the tests.

Table 2. Audio codecs used in this research

<table>
<tr><th>Label</th><th>Stereo coding</th><th>Data rate (kbps)</th><th>Codec</th></tr>
<tr><td>Ref.</td><td rowspan="3">None
(Simple Stereo)</td><td rowspan="3">1024</td><td rowspan="3">None</td></tr>
<tr><td>LPF3.5k</td></tr>
<tr><td>LPF2.0k</td></tr>
<tr><td>JS24</td><td rowspan="3">Joint
Stereo</td><td>24</td><td rowspan="6">HE-AAC</td></tr>
<tr><td>JS32</td><td>32</td></tr>
<tr><td>JS56</td><td>56</td></tr>
<tr><td>Pa24</td><td rowspan="3">Parametric
Stereo</td><td>24</td></tr>
<tr><td>Pa32</td><td>32</td></tr>
<tr><td>Pa56</td><td>56</td></tr>
</table>

2.3 Sound Localized in Virtual Audio Space

The test speech was localized by convolving monaural sources with the KEMAR-HRIR [10]. As recommended in the accompanying documentation, the HRTFs for the left ear are the mirror images of the HRTFs for the right ear. All speech sources were localized at an azimuth of 0 and ±45 degrees. The localization procedure is shown in Fig.1, and the localized positions are shown in Fig.2. The sound sources were normalized to approximately the same sound pressure by a fixed scaling factor α_1.

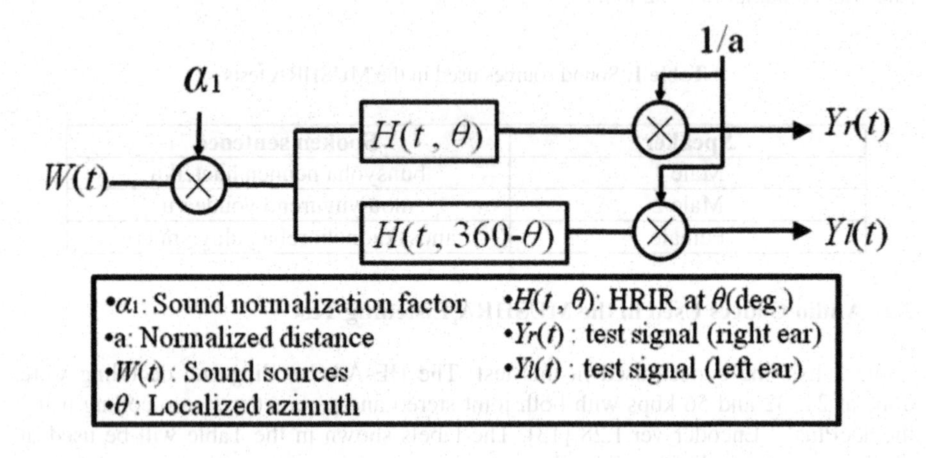

Fig. 1. Test signal generation procedure (chapters 2 and 3)

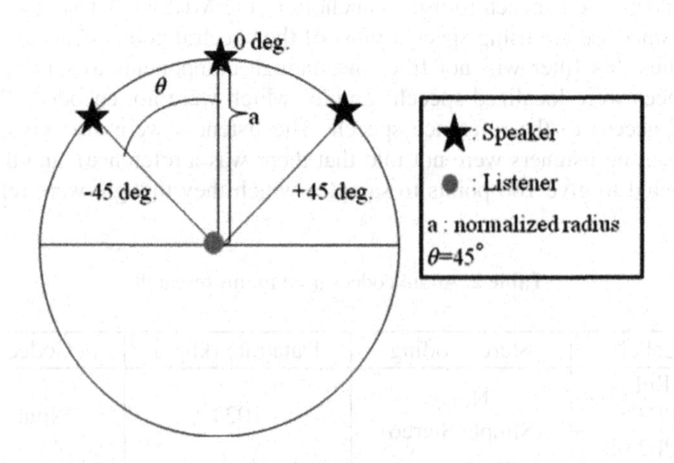

Fig. 2. Location of the sound sources (chapters 2 and 3)

Standard MUSHRA listening tests were conducted with these localized speech as well as the reference and the two anchors. The listeners listened to the localized speech for one speaker, and rated the subjective quality on a 100-point scale. The subjects rated all test sources from the same speaker, and moved on to the next speaker. The presentation order of the speakers was randomized. The test speech was played out from stereo headphones (Sennheiser HD-25-1 II).

2.4 Results

Fig.3. (a) to (c) shows the results of the MUSHRA listening test for the 3 azimuths. The scores are averaged across the three tested speakers. The error bars shown are the 95 % confidence intervals.

First, for speech localized at 0 degree azimuth, subjective quality for both joint stereo and parametric stereo were relatively independent of the data rate. All scores were in the range of 80-100, which refers to a quality rating of "excellent," and these sources are essentially indistinguishable from the reference.

Next, for sound sources localized at ±45 degree azimuth, all joint stereo-coded speech was higher than parametric stereo at same bit rates. Moreover, joint stereo was higher at -45 degree azimuth by about 10 points than +45 degrees except at 24 kbps. While joint stereo-coded quality is rate-dependent, the parametric stereo quality is relatively independent of the bit rate, and remains constantly below joint stereo at 24 kbps.

Joint stereo-coded speech at 56 and 32 kbps, *i.e.* JS56 and JS32, show significantly higher scores at -45 degrees than +45 degrees. This difference is not limited to a few listeners, but is seen in at least some samples for almost all listeners. The cause for this difference is still unclear, and will be investigated.

To summarize, in terms of subjective quality, joint stereo coding generally gives superior quality speech than parametric stereo coding, especially when speech is localized to the sides.

2.5 Results of ANOVA

We examined these results for statistical significance using the analysis of variance (ANOVA) test to validate the quality differences between encoding methods (including the reference, and excluding anchors). Significant difference was seen between the reference and each encoded sample with various stereo mode and bit rate with a significance level of 1% at all azimuths ((a): $F(6, 140) = 5.87$, $p < 0.001$ (b): $F(6, 140) = 24.11$, $p < 0.001$, (c): $F(6, 140) = 22.33$, $p < 0.001$). However, the results for parametric stereo were not significantly different by bit rate at all azimuth ((a): $F(2, 60) = 0.14$, $p = 0.873$, (b): $F(2, 60) = 0.13$, $p = 0.878$, (c): $F(2, 60) = 0.19$, $p = 0.824$). Joint stereo in (a) was significantly different by a significance level of 5 % ((a): $F(2, 60) = 3.90$, $p = 0.026$). Moreover, joint stereo in (b) and (c) were significantly different by a significance level of 1% ((b): $F(2, 60) = 10.60$, $p < 0.001$, (c): $F(2, 60) = 7.88$, $p < 0.001$).

Finally, we analyzed the results for speech coded at 24 kbps which showed mutually similar values in Fig. 3 (a) to (c) with the t-test. Significant difference was not seen between joint and parametric stereo coding in (a) ((a): $t(20) = -0.70$, $p = 0.493$). However, significance difference was seen at a significance level of 5 % in (b) and (c) ((b): $t(20) = 2.62$, $p = 0.0164$, (c): $t(20) = 2.44$, $p = 0.0242$).

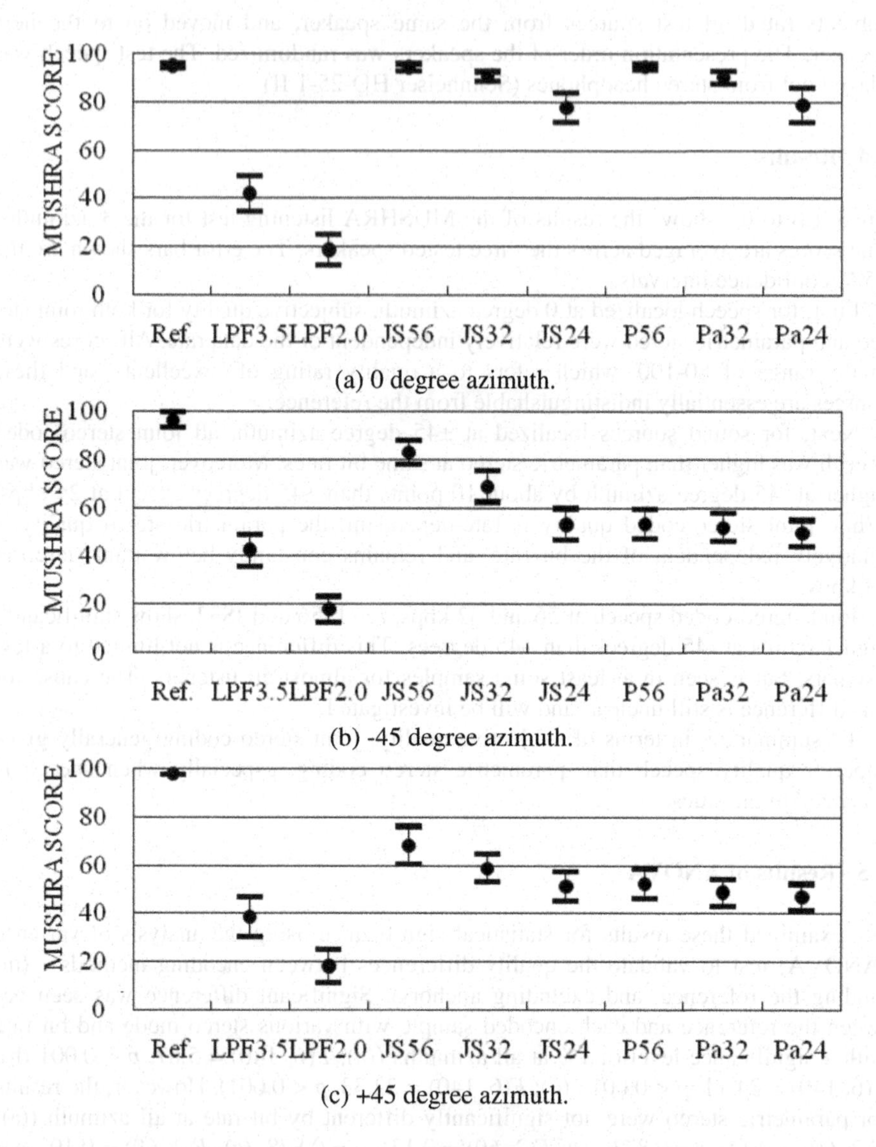

Fig. 3. Results of the MUSHRA listening tests

3 Speech Intelligibility Test with No Competing Noise

We conducted the Japanese Diagnostic Rhyme Tests (JDRT) [14] to measure intelligibility of localized speech. In this section, we will describe intelligibility tests without competing noise to measure the effect of localization and stereo coding on the speech signal itself. Intelligibility tests with competing noise to measure the combined effect of localization, coding and noise on intelligibility will be described in the next section.

3.1 The Japanese Speech Intelligibility Test

The Japanese Diagnostic Rhyme Test (JDRT) uses word-pairs that are different only by one initial phoneme as stimulus words to measure the speech intelligibility. Most Japanese syllables are formed by only a vowel, or a consonant and a vowel. (The geminate consonant, the Japanese syllabic nasal and the contracted sound are excluded.) In the JDRT, we did not use words that start with a vowel. Therefore, changing one initial phoneme means changing the consonant. Consonants are categorized into six attributes, and intelligibility is measured by attributes. We chose a word pair list consisting of 120 words, or 60 word pairs, 10 word pairs per attribute. Table 3 shows the consonant taxonomy used in the JDRT.

Table 3. Consonant taxonomy of the JDRT

Phonetic Feature	Classification	Example
Voicing	Vocalic and Nonvocalic.	Zai - Sai
Nasality	Nasal and oral.	Man - Ban
Sustention	Continuant and Interrupted.	Hashi -Kashi
Sibilation	Strident and Mellow.	Jyamu - Gamu
Graveness	Grave and Acute.	Waku - Raku
Compactness	Compact and Diffuse.	Yaku - Waku

During the test, the subject listens to the sound of one word. Both words in the word pair are then presented visually on the screen, and the subject selects one of the words as the correct word. At this point, the subject can repeatedly hear the same sound. When the next button is selected, the following sound is presented. This procedure is repeated until the predetermined numbers of words are tested. The words are presented in random order. The selected word is recorded and processed with the personal computer automatically. The percentage of correct response is adjusted for chance, noted CACR, and is evaluated using the following expression (1).

$$CACR = \frac{Correct\ responses - Incorrect\ responses}{Total\ responses} \times 100\ [\%] \quad (1)$$

3.2 Experimental Conditions

In the following tests, we used joint stereo coding and parametric stereo coding, both at 24 kbps (JS24, Pa24), as well as the reference signal (Ref., no coding), which are partial selections from Table 2. We chose 24 kbps as the common coding bit rate since we have shown in the previous section that speech quality at this rate is almost as good as speech at higher rates, especially with parametric coding. The encoder used was the aacPlus™ encoder ver.1.28. In the JDRT, we used read speech of one female speaker. Speech samples were originally sampled 16 kHz, and the quantization bits were 16 bits linear. These samples were up-sampled to 32 kHz.

The test speech was localized using the same method as described in the previous chapter for the MUSHRA listening tests using the KEMAR-HRIR [10]. All speech sources were localized at an azimuth of 0 degree and ±45 degrees. Thus, the sound localization process is the same as shown in Fig.1, and the sound image locations are the same as shown in Fig.2. These localized speech samples were then coded with the two stereo coding methods mentioned above.

The number of subjects of this test was 7 males and 3 females, and the test was run semi-automatically on a Windows PC.

3.3 Results

Fig.4 (a) to (c) shows the results of JDRT with no competing noise. All figures show the CACR by phonetic feature, as well as the average over all features. Overall, JS24, Pa24 and Ref. show similar intelligibility (CACR) for most cases (azimuth and phonetic feature). However, at -45 degrees, parametric stereo coding seems to result in lower CACR. This is especially true for graveness and compactness, where parametric stereo coding is lower than joint stereo coding by 12 % and 10 %, respectively. Interestingly, these two phonetic features are also known to be affected most by competing noise compared to other features [14].

It is also interesting to note that Pa24 actually gives about the same intelligibility at both the left and the right azimuths, *i.e.* symmetrical intelligibility, although lower than JS24 and the reference. JS24 shows significantly lower intelligibility at +45 degrees compared to -45, and so is not symmetrical. The reason for this is unclear, and we would like to further investigate the cause of this lack of symmetry with joint stereo coding.

The conclusion from these results can be said that in most cases, stereo coding will not significantly affect intelligibility of the target speech itself, but it does have some azimuth dependency, especially for the joint stereo coding.

3.4 Results of ANOVA

We examined these results using ANOVA to validate the difference between stereo mode as well as with the reference. Only graveness in Fig. 4 (b) was shown to be significantly different by stereo coding modes with a significance level of 5% ($F(2, 27) = 4.36$, $p = 0.0478$). This was somewhat surprising since compactness in Fig. 4 (b) with parametric coding was much lower than the other two. However, this does back up our initial observation that in most cases, stereo coding does not affect intelligibility.

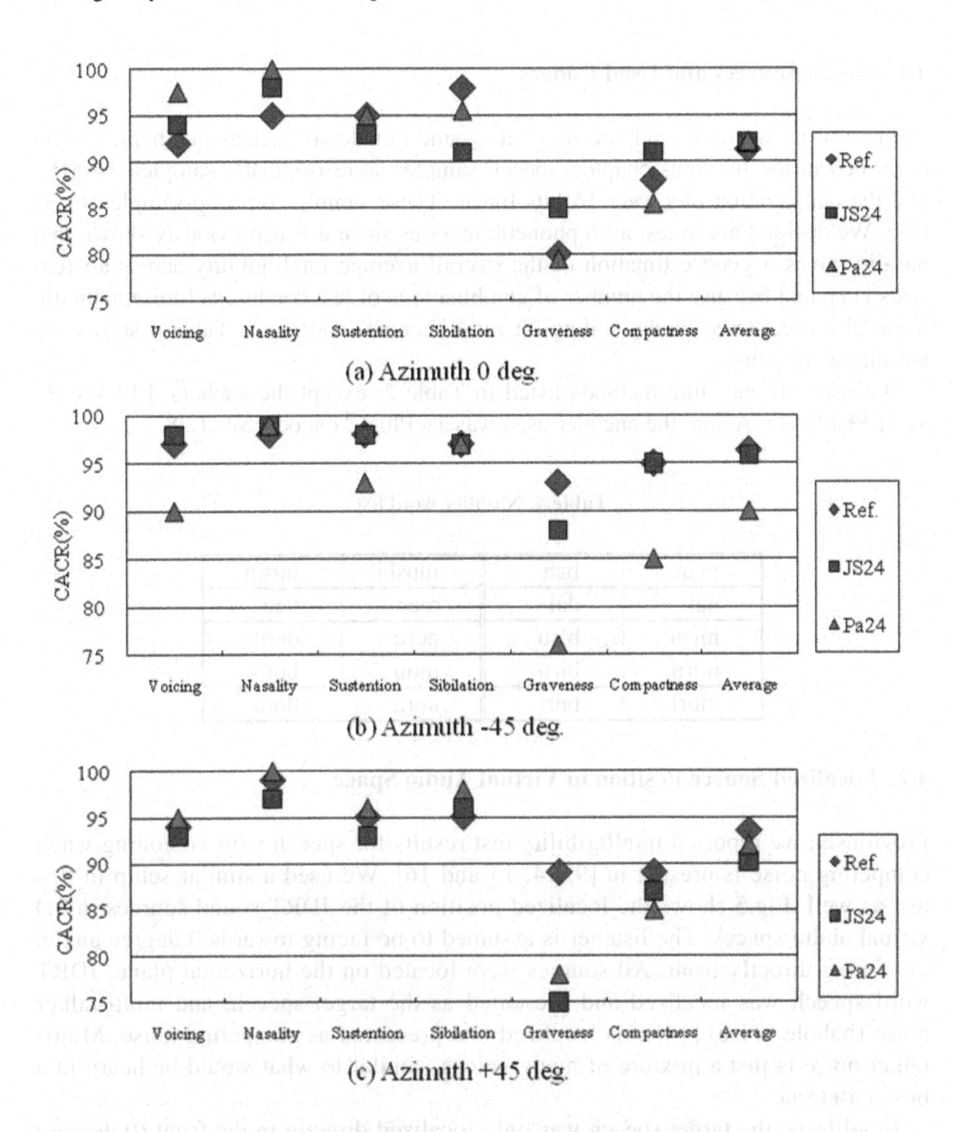

Fig. 4. Results of intelligibility tests with no competing noise (CACR by phonetic feature)

4 Speech Intelligibility Test with Competing Noise

In this chapter, we tested the Japanese speech intelligibility with the two stereo coding modes at various bit rates when competing noise was present. We used the Japanese Diagnostic Rhyme Tests (JDRT) again for localized speech in 3-D virtual space, as has been investigated in [9, 15, 16, and 17]. All tests were again run semi-automatically on a Windows PC.

4.1 Sound Sources and Used Codecs

In these tests, we used read speech of the same female speaker as in Chapter 3. As described in the previous chapter, speech samples were originally sampled 16 kHz, and the quantization bits were 16 bits linear. These samples were up-sampled to 32 kHz. We decided not to test all 6 phonetic features since it was previously shown that nasality gives a good estimation of the overall average intelligibility across all features [17], and because the number of combinations of test conditions (noise azimuth, noise distance, stereo mode, coding bit rate) becomes enormous. Table 4 shows the nasality word pairs.

We used all encoding methods listed in Table 2, except the anchors, LPF3.5kHz and LPF2.0kHz. Again, the encoder used was aacPlus^{TM} encoder ver.1.28.

Table 4. Nasality word list

man	ban	mushi	bushi
nai	dai	men	ben
misu	bisu	neru	deru
miru	biru	mon	bon
muri	buri	nora	dora

4.2 Localized Source Position in Virtual Audio Space

Previously, we reported intelligibility test results for speech with no coding when competing noise is present in [9, 14, 15 and 16]. We used a similar setup in this test as well. Fig.5 shows the localized position of the JDRT sound sources in 3D virtual audio spaces. The listener is assumed to be facing towards 0 degree angle, *i.e.* facing directly front. All sources were located on the horizontal plane. JDRT word speech was localized and presented as the target speech, and multi-talker noise (babble noise) [14] was localized and presented as competing noise. Multi-talker noise is just a mixture of many voices, similar to what would be heard in a busy cafeteria.

In all tests, the target speech was only localized directly in the front (0 degree) this time. We localized the noise at various azimuths in 15 degree increments in the frontal region between ±45 degrees, and in 45 degree increments outside this region. We located the noise on a radius relative to the distance between the target speaker and the listener.

Denoting as "a" the normalized speaker-listener distance, noise was located on a radius with the same distance (“1.00a”), twice the distance (“2.00a”), half the distance (“0.50a”), and quarter the distance (“0.25a”). The location of the target speech and the competing noise is shown in Fig. 5, and the test signal generation procedure is shown in Fig. 6.

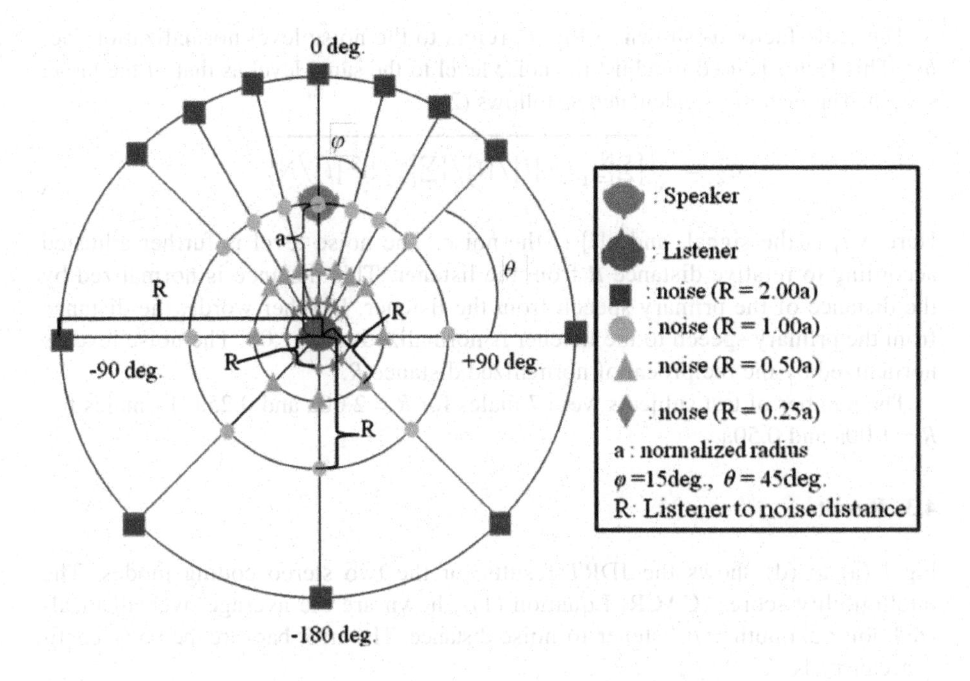

Fig. 5. Location of the sound sources (chapter 4)

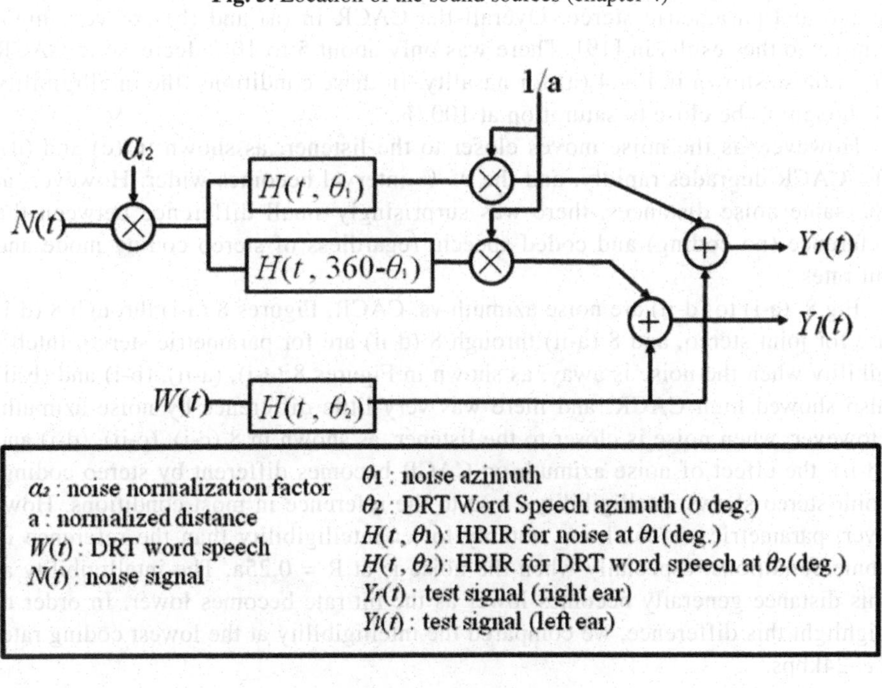

Fig. 6. Test signal generation procedure (chapter 4)

The scale factor α_2 shown in Fig. 6, refers to the noise level normalization factor. This factor is used to adjust the noise level to the same level as that of the target speech. The factorα_2 is calculated as follows (2).

$$\alpha_2 = \sqrt{\{\sum_{i=1}^{N} s^2[i]/N\}/\{\sum_{i=1}^{N} n^2[i]/N\}} \tag{2}$$

Here, $s[i]$ is the signal, and $n[i]$ is the noise. The noise level is further adjusted according to relative distance R from the listener. This distance is normalized by the distance of the primary speech from the listener. In other words, the distance from the primary speech to the listener is normalized to be 1.00. The noise level is normalized by the reciprocal of normalized distance R.

The number of test subjects were 7 males for R = 2.00a and 0.25a, 14 males for R = 1.00a and 0.50a.

4.3 Results

Fig.7 (a) to (d) shows the JDRT results for the two stereo coding modes. The intelligibility scores (CACR, Equation (1)) shown are the average over all localized noise azimuth and listener-to-noise distance. The error bars are the 95 % confidenceintervals.

The CACR shown in (a) and (b) were not significantly different for both joint stereo and parametric stereo. Overall the CACR in (a) and (b) are very high, similar to the results in [19]. There was only about 5 to 10% decrease in CACR from those shown in Fig.4 (a) for nasality. In these conditions, the intelligibility is thought to be close to saturation at 100 %.

However, as the noise moves closer to the listener, as shown in (c) and (d), the CACR degrades rapidly, and the 95 % interval becomes wider. However, at the same noise distances, there was surprisingly small difference between the reference (no coding) and coded speech, regardless of stereo coding mode and bit rates.

Fig.8. (a-i) to (d-ii) are noise azimuth vs. CACR. Figures 8 (a-i) through 8 (d-i) are for joint stereo, and 8 (a-ii) through 8 (d-ii) are for parametric stereo. Intelligibility when the noise is away, as shown in Figures 8 (a-i), (a-ii), (b-i) and (b-ii) also showed high CACR, and there was very little difference by noise azimuth. However, when noise is closer to the listener, as shown in 8 (c-i), (c-ii), (d-i) and (d-ii), the effect of noise azimuth on CACR becomes different by stereo coding. Joint stereo shows intelligibility close to the reference in most conditions. However, parametric stereo shows notably lower intelligibility than the reference in some conditions, especially when the noise is at R = 0.25a. The intelligibility at this distance generally becomes lower as the bit rate becomes lower. In order to highlight this difference, we compared the intelligibility at the lowest coding rate, *i.e.* 24kbps.

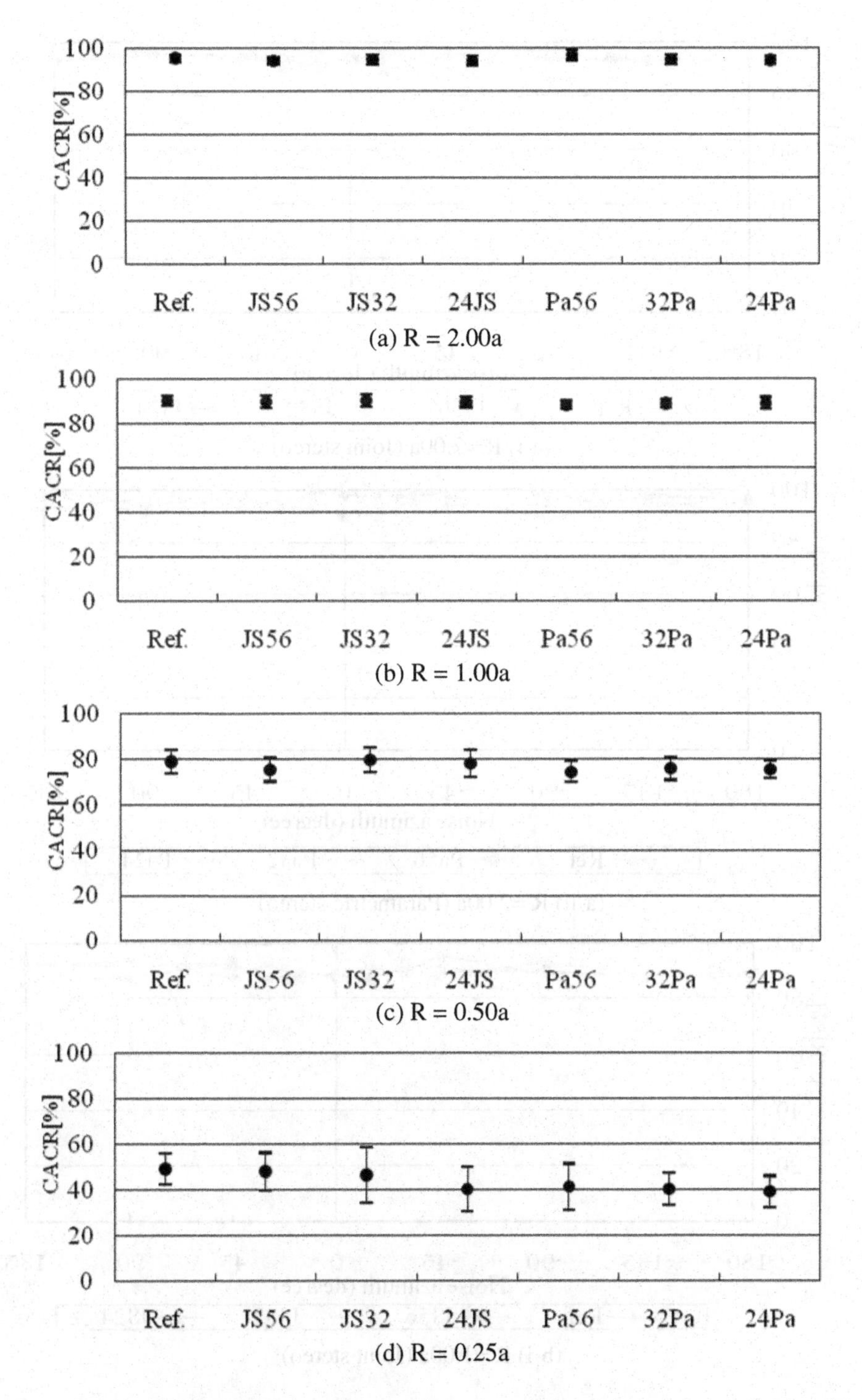

Fig. 7. Intelligibility (CACR) vs. stereo coding mode and bit rates(average over all azimuths)

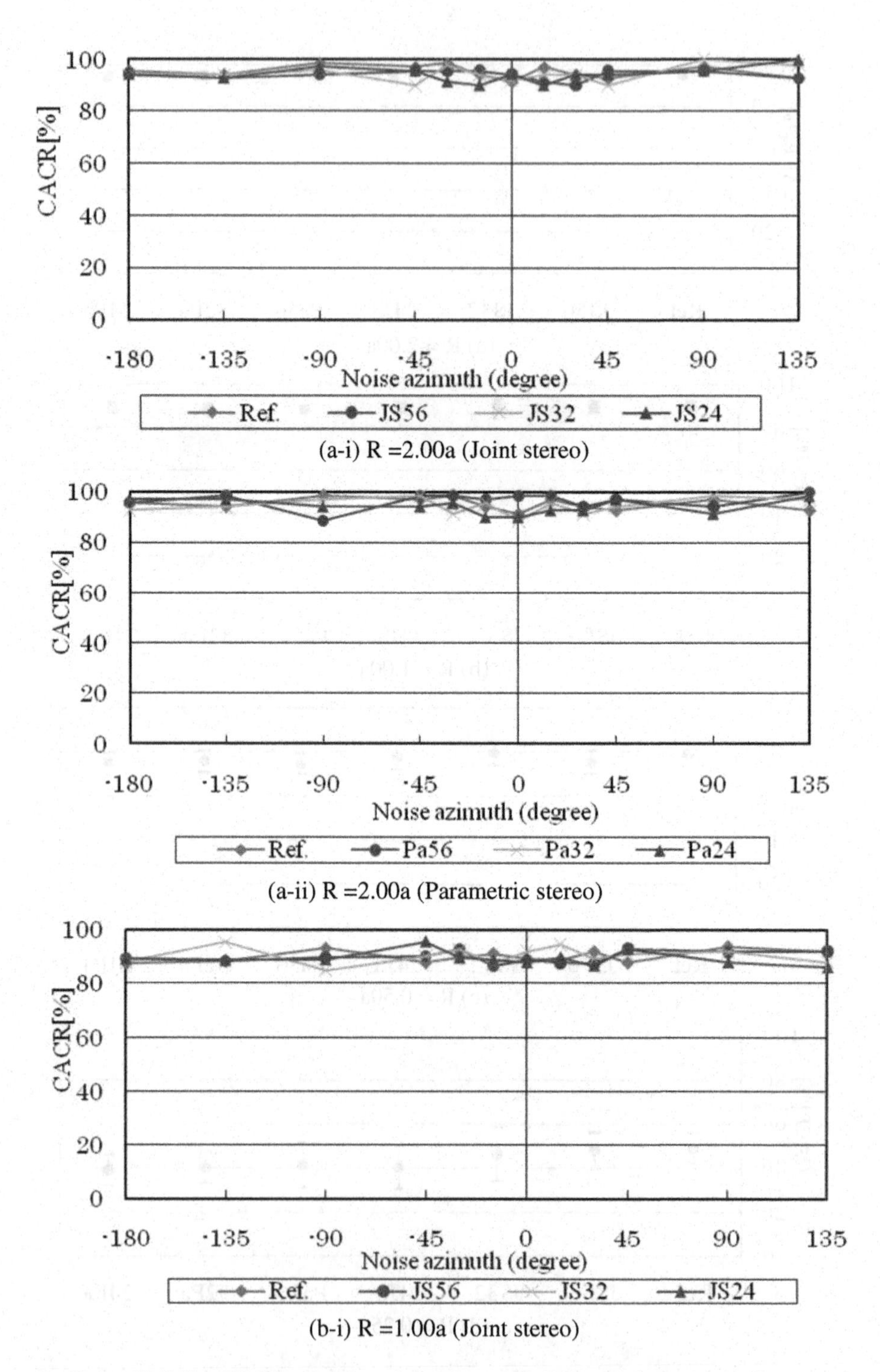

Fig. 8. Intelligibility (CACR) vs. noise azimuth by stereo coding modes and noise distance

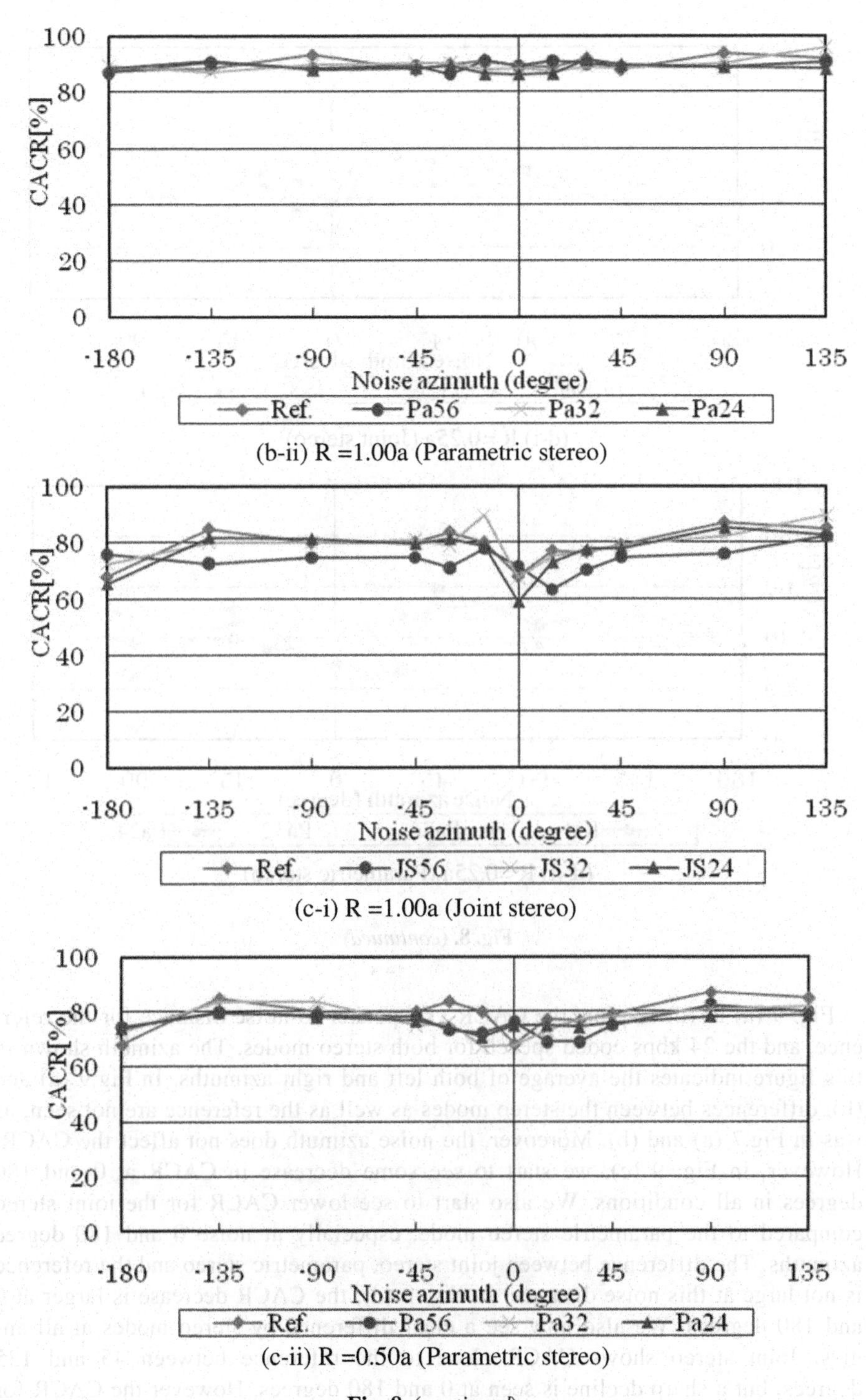

(b-ii) R =1.00a (Parametric stereo)

(c-i) R =1.00a (Joint stereo)

(c-ii) R =0.50a (Parametric stereo)

Fig. 8. *(continued)*

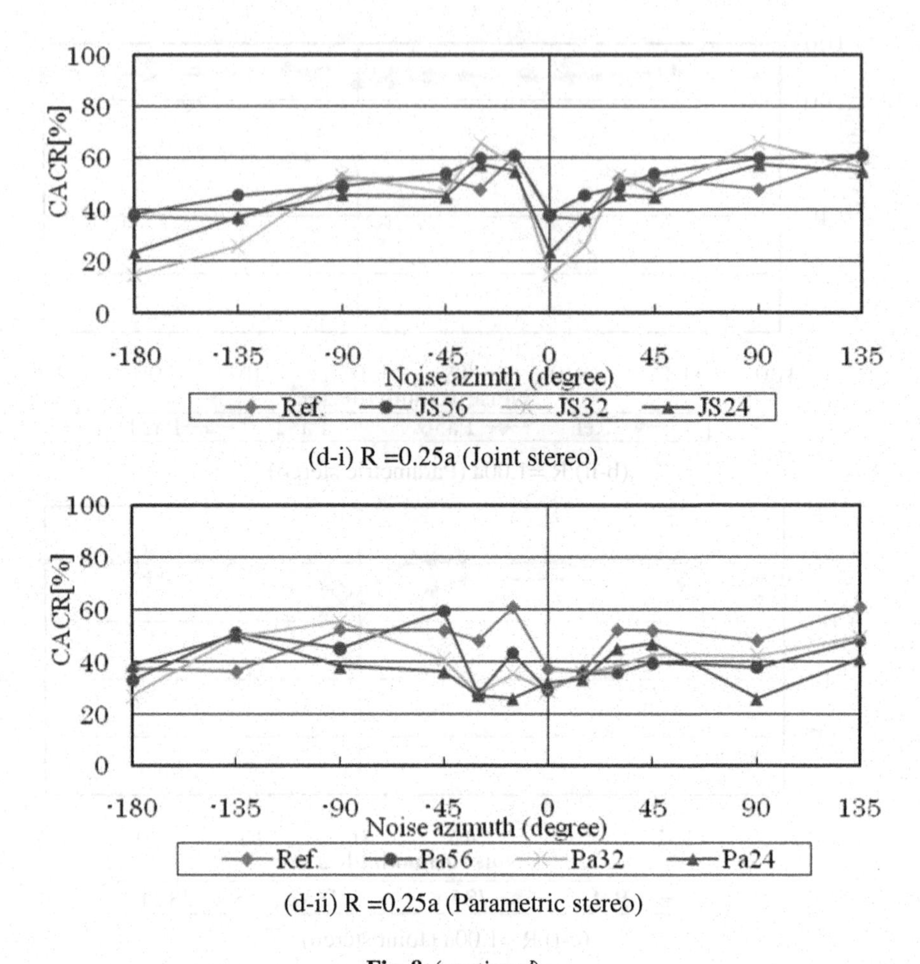

(d-i) R =0.25a (Joint stereo)

(d-ii) R =0.25a (Parametric stereo)

Fig. 8. *(continued)*

Fig. 9 (a) to (d) re-plots the CACR vs. speaker-to-noise distance for the reference, and the 24 kbps coded speech for both stereo modes. The azimuth shown in this figure indicates the average of both left and right azimuths. In Fig.9 (a) and (b), differences between the stereo modes as well as the reference are not seen, as was in Fig.7 (a) and (b). Moreover, the noise azimuth does not affect the CACR. However, in Fig. 9 (c), we start to see some decrease in CACR at 0 and 180 degrees in all conditions. We also start to see lower CACR for the joint stereo compared to the parametric stereo mode, especially at noise 0 and 180 degree azimuths. The difference between joint stereo, parametric stereo and the reference is not large at this noise distance. In Fig. 9 (d), the CACR decrease is larger at 0 and 180 degrees. We also now see a clear difference by stereo modes at all angles. Joint stereo shows CACR close to the reference between 45 and 135 degrees, but a sharp decline is seen at 0 and 180 degrees. However the CACR for the parametric stereo mode was relatively independent of noise azimuths, even at 0 and 180 degrees, and mostly lower than joint stereo; most notably about 5 to 20% lower between 0 to 45 degrees. This may be because parametric stereo cannot code multiple sound sources from different azimuths as effectively as joint stereo. More research is needed to confirm this observation.

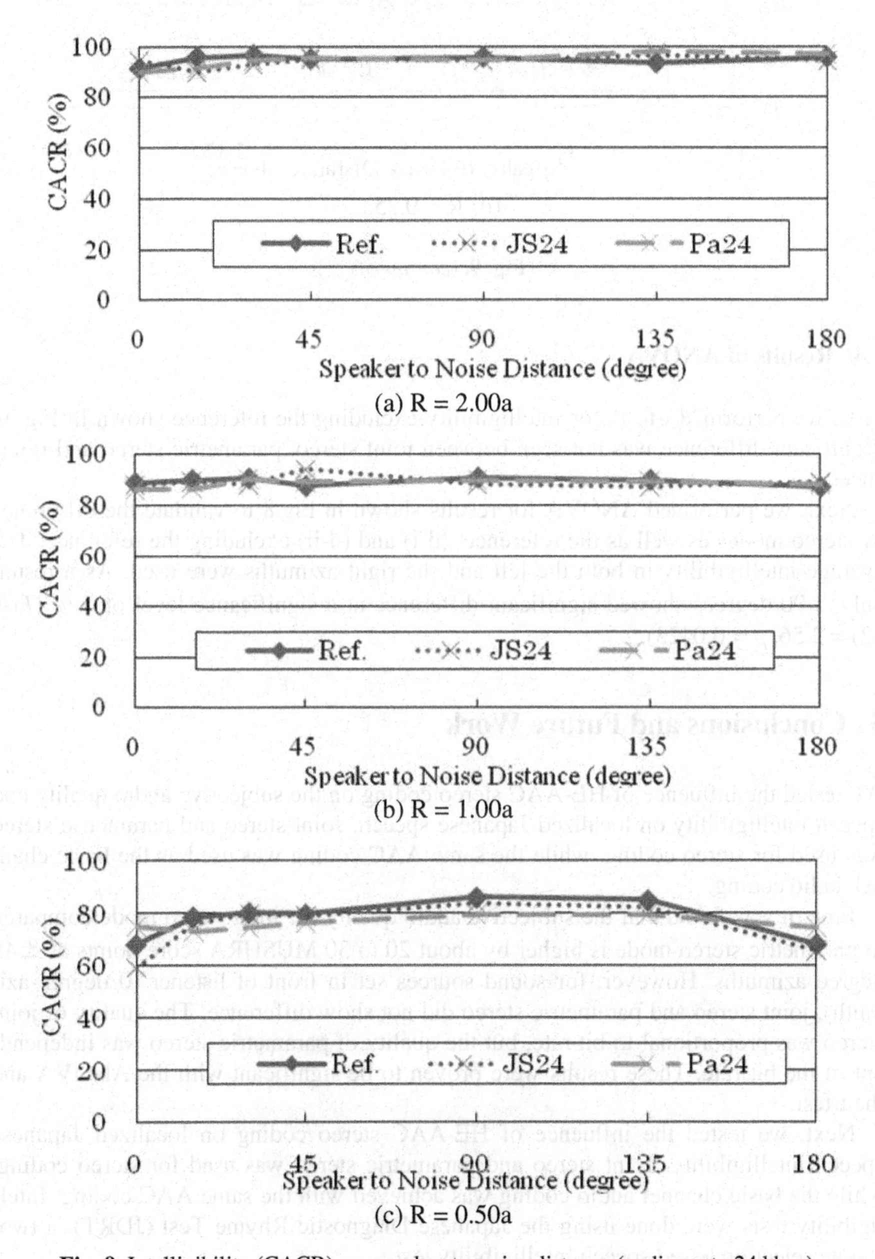

Fig. 9. Intelligibility (CACR) vs. speaker-to-noise distance (coding rate 24 kbps)

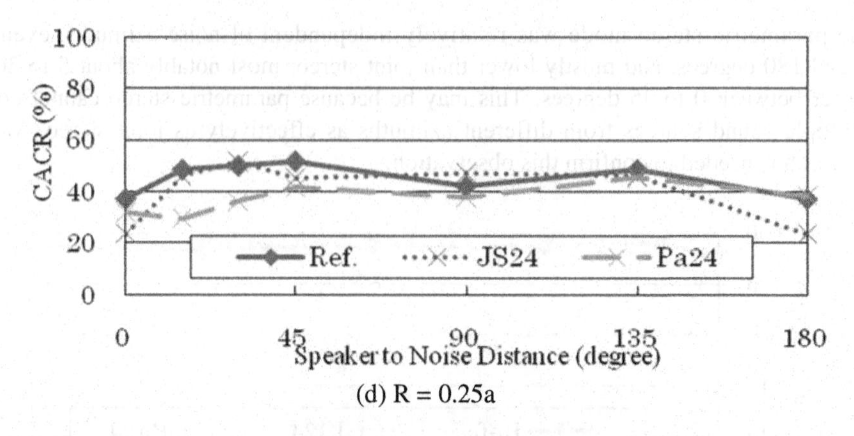

(d) R = 0.25a

Fig. 9. *(continued)*

4.4 Results of ANOVA

First, we performed t-tests for intelligibility excluding the reference shown in Fig. 9. Significant difference was not seen between joint stereo, parametric stereo and reference.

Next, we performed ANOVA for results shown in Fig.8 to validate the difference by stereo modes as well as the reference. (d-i) and (d-ii) excluding the reference. The average intelligibility in both the left and the right azimuths were used. As a result, only ±90 degrees showed significant difference at a significance level of 5 % ($F(6, 42) = 2.56$, $p = 0.0423$).

5 Conclusions and Future Work

We tested the influence of HE-AAC stereo coding on the subjective audio quality and speech intelligibility on localized Japanese speech. Joint stereo and parametric stereo was used for stereo coding, while the same AAC coding was used as the basic channel audio coding.

First, it was found that the subjective audio quality for joint stereo mode compared to parametric stereo mode is higher by about 20 to 30 MUSHRA score points at ±45 degree azimuths. However, for sound sources set in front of listener (0 degree azimuth), joint stereo and parametric stereo did not show difference. The quality of joint stereo was proportional to bit rate, but the quality of parametric stereo was independent of the bit rate. These results were proven to be significant with the ANOVA and the t-test.

Next, we tested the influence of HE-AAC stereo coding on localized Japanese speech intelligibility. Joint stereo and parametric stereo was used for stereo coding, while the basic channel audio coding was achieved with the same AAC coding. Intelligibility tests were done using the Japanese Diagnostic Rhyme Test (JDRT), a two-to-one selection based speech intelligibility test.

The speech intelligibility was tested first with target speech only (no competing noise). The influence of stereo coding on intelligibility was shown to differ by phonetic feature. In particular, graveness was shown to degrade significantly by stereo coding. However, intelligibility for other phonetic features, especially nasality, does not seem to be affected by coding.

The speech intelligibility was then tested when target speech was localized in front of the listener at a distance *a*, and the competing noise was localized on the horizontal plane at various azimuths and at a radius of 2.00a, 1.00a, 0.50a and 0.25a. Only words in the nasality list were tested in this case. Joint stereo and parametric stereo at a coding rate of 24 kbps, and when noise was located at R = 2.00a, R = 1.00a and R = 0.50a did not show significant difference. However, at R = 0.25a, when the noise is located between +45 and -45 degrees, significant difference in intelligibility was shown; intelligibility of parametric stereo-coded speech in this range was about 5% to 20% lower than joint stereo or the reference (non-coded) speech. Intelligibility outside of this azimuths range was similar. Thus, overall, stereo coding modes and coding bit rate do not have clear effect on speech intelligibility. Intelligibility was rather shown to be highly related to speaker-to-noise azimuth.

In this paper, we examined coding at 24kbps closely. This is because we concluded that this rate was reasonably low bit rate and achieves relatively high speech quality which is required in our application. However, there are lower coding rates available, both in parametric and joint stereo coding modes, in the standard. We are currently investigating the performance with these lower rates, and will report these results in a separate paper [20].

From these results, when encoding speech for 3-D audio displays, coding bit rates and stereo coding modes were shown not to be the most critical factor. Care should be rather taken to localize sources so that they may be located well away from competing sources. These results suggest some valuable guidelines when designing a 3-D audio conferencing systems using stereo audio coding in the future.

In this research, we only tested with the HE-AAC implementation from Coding Technologies, who proposed the main parts of the standard. Thus, we believe their implementation is fully compliant with their standard, including the inter-channel phase processing. Nonetheless, we would like to test with other standard implementations as well. We also would like to expand our experiments to include other locations of the target speech, including azimuths and radii.

Acknowledgments

Financial support was provided by the NEC C&C Foundation and the Yonezawa Industrial Association. This work was also supported in part by the Cooperative Project Program of the Research Institute of Electrical Communication at the Tohoku University. Finally, we also wish to express our gratitude to the ICAD'09/CMMR'09 coordinators for the chance to contribute to these post-conference manuscripts.

References

1. Kilgore, R., Chignell, M., Smith, P.: The Vocal Village: Enhancing Collaboration with Spatialized Audioconferencing. In: Proc. World Conference on E-Learning in Corporate, Government, Healthcare, and Higher Education (2004)
2. Kaneda, Y.: Subjective Evaluation of Voiscape - A Virtual "Sound Room" Based Communication-Medium. Tech. Rep. of the IEICE EA 2007-42 (2007)
3. Junichi, N., Kenji, O.: Effects of Reproduction Methods and High-Efficiency Audio Coding on Word Intelligibility with Competing Talkers. J. IEICE A J88-A, 1026–1034 (2005)
4. Kobayashi, Y., Kondo, K., Nakagawa, N., Takano, K.: The Influence of Stereo Coding on the 3D Sound Localization Accuracy. Tech. Rep. of the IEICE EA2008-56 (2008)
5. Alexander, R., Claudia, S.: Auditory Assessment of Conversational Speech Quality of Traditional and Spatialized Teleconferences. In: Proc. 8. ITG-Fachtagung Sprachkommunikation, pp. 8–10 (2008)
6. ISO/IEC 14496-3:2003/Amd.1
7. ISO/IEC 14496-3:2005/Amd.2
8. Breebaart, J., Par, S.v.d., Kohlrausch, A., Schuijers, E.: Parametric Coding of Stereo Audio. EURASIP J. on Applied Signal Processing 9, 1305–1322 (2004)
9. Kitashima, Y., Kondo, K., Terada, H., Chiba, T., Nakagawa, K.: Intelligibility of read Japanese words with competing noise in virtual acoustic space. J. Acoustical Science and Technology 29(1), 74–81 (2008)
10. HRTF Measurements of a KEMAR Dummy-Head Microphone, `http://sound.media.mit.edu/resources/KEMAR.html`
11. Recommendation ITU-R BS.1534-1: Method for the subjective assessment of intermediate quality level coding system (2001-2003)
12. ASJ Continuous Speech Database for Research (ASJ-JIPDEC), `http://www.jipdec.jp/chosa/public/report/onseidb/`
13. Coding Technologies Acquired by Dolby Laboratories, `http://www.aacplus.net/`
14. Kondo, K., Izumi, R., Fujimori, R., Rui, K., Nakahawa, K.: On a Two-to-One Selection Based Japanese Speech Intelligibility Test. J. Acoust. Soc. Jpn. 63, 196–205 (2007)
15. Chiba, T., Kitashima, Y., Yano, N., Kondo, K., Nakagawa, K.: On the influence of localized position of interference noise on the intelligibility of read Japanese words in remote conference systems, Inter-noise 2008, PO-2-0294 (2008)
16. Kobayashi, Y., Kondo, K., Nakagawa, K.: Intelligibility of Low Bit rate MPEG-coded Japanese Speech in Virtual 3D audio space. In: Proc. 15th International Conference on Auditory Display, pp. 99–102 (2009)
17. Yano, N., Kondo, K., Nakagawa, K., Takano, K.: The Effect of Localized Speech and Noise Distance on the Speech Intelligibility. IPSJ-Tohoku B-2-3 (2008)
18. Rice University: Signal Processing Information Base (SPIB), `http://spib.rice.edu/spib/select_noise.html`
19. Stoll, G., Kozamernik, F.: EBU Report on the Subjective Listening Tests of Some Commercial in EBU Technical Review, no. 283 (2000)
20. Kobayashi, Y., Kondo, K., Nakagawa, K.: Influence of Various Stereo Coding Modes on Encoded Japanese Speech Intelligibility with Competing Noise. In: Proc. International Work-shop on the Principles and Applications of Spatial Hearing, P19(Poster) (2009)

Navigation Performance Effects of Render Method and Head-Turn Latency in Mobile Audio Augmented Reality

Nicholas Mariette

LIMSI-CNRS
Audio and Acoustics Group
Orsay, France
nicholas.mariette@limsi.fr
http://www.limsi.fr/Scientifique/aa/

Abstract. This study assessed participant performance of an outdoor navigation task using a mobile audio augmented reality system. Several quantitative performance measures and one subjective measure were used to compare the perceptual efficacy of Ambisonic and VBAP binaural rendering techniques, and a range of head-turn latencies. The study extends existing indoors research on the effects of head-turn latency for seated listeners.

The pilot experiment found that a source *capture radius* of 2 meters significantly affected the sole participant's navigation distance efficiency compared to other radii. The main experiment, using 8 participants, found that render method significantly affected all performance measures except subjective stability rating, while head-turn latency only affected mean track curvature and subjective stability. Results also showed an interaction in which the choice of rendering method mitigated or potentiated the effects of head-turn latency on perceived source stability.

Keywords: Mobile audio augmented reality, navigation, binaural rendering, VBAP, Ambisonic, head-turn latency.

1 Introduction

This paper describes a pilot study and main experiment that attempt to measure variations of navigation performance supported by a mobile audio audio reality (AR) system. Examined here are two common technological limitations of such systems: binaural rendering resolution, and latency between head-turns and corresponding rendered audio changes.

The experiments also contribute a new perceptual evaluation paradigm, in which participants perform an outdoor navigation task using personal, location-aware spatial audio. The task is to navigate from a central base position to the location of a simulated, world-stationary sound source. For each stimulus, system parameters are varied, providing the experimental factors under examination. Simultaneously, body position/orientation and head-orientation sensor data are

S. Ystad et al. (Eds.): CMMR/ICAD 2009, LNCS 5954, pp. 239–265, 2010.

recorded for later analysis. Objective performance measures enable assessment of the degradation of participants' navigation performance from the ideal, due to tested system parameter values.

This participant task was designed as a generalisation of real-world mobile audio AR applications. For instance, any navigation task can be generalised as a series of point-to-point navigations (A-B, A-C, ...) like those in the present experiment, whether it involves a closed circuit (e.g. A-B-C-A) or an open, continuous series of way-points (e.g. A-B-C-D-...). This experiment thus provides a method of examining a given system implementation's effectiveness at supporting any navigation activity. The same experimental paradigm could also be used to evaluate factors other than rendering technique and latency. For example, it could be used to compare generalised and individualised head-related impulse response (HRIR) filters, or the presence or absence of sophisticated acoustics simulation.

2 Background

To date, few mobile audio AR systems have been reported, particularly those with unrestricted outdoor functionality. There are even fewer published performance evaluations of such systems, possibly due to extensive existing research on *static* spatial audio perception. However, several technical and perceptual factors are unique to the experience of interacting with simulated sound sources that are fixed in the world reference frame. Thus, new experiments are required to understand the performance supported by different system designs.

The most relevant prior research is by Loomis et al. [3], who created a "simple virtual sound display" with analogue hardware controlled by a 12MHz 80286 computer, video position tracking using a light source on the user's head, and head-orientation tracking using a fluxgate compass. Video position measurements occurred at a rate of 6 Hz, and covered 15×15 m with a maximum error "on the order of 7 cm throughout the workspace". Head tracking occurred at a sampling rate of 72 Hz, with 35 ms measurement latency, 1.4° resolution and 2° or 3° accuracy in normal conditions, although it occasionally exceeded 20° with poor calibration or head tilts. Spatial audio rendering consisted of azimuth simulation by synthesis of inter-aural time and intensity differences. Distance simulation was provided through several cues: the "first power law for stimulus pressure"; atmospheric attenuation; a constant-level artificial reverberation (also for externalisation cues); and the naturally occurring cue of "absolute motion parallax" – i.e. changing source azimuth due to body position.

The experiment required participants to "home" into real (hidden) or virtual sounds placed at the same 18 locations arranged in a circle around the origin, with azimuth and radius slightly randomised. Results measured the time to localise; time for participant satisfaction of successful localisation; distance error at the terminus; and absolute angular change from near the start of a participant's path to near its end. ANOVA testing showed significant differences between participants for some performance measures, however the real/virtual condition had no significant effect on any measure. Mean results for most measures were also quite similar across conditions, except the path angular change, which was much larger for virtual stimuli (33°) than real stimuli (14°), despite no significant effect shown by the ANOVA ($p<0.05$). The researchers concluded that a simple virtual display could be "effective in creating the impression of external sounds to which subjects can readily locomote", but that more sophisticated displays may improve space perception and navigation performance.

That research has many similarities to the present experiment, although it focused on *verification* of navigation ability using simple spatialised stimuli, even though, as the authors stated: "homing can be accomplished merely by keeping the sound in the median plane until maximum sound intensity is achieved". In other words, the closed perception-action feedback loop enables the correct source azimuth to be determined with the aid of head-turns, after which body translation and intensity distance cues can be used to find the sound position.

The present experiment assumes that a basic system will support navigation, then examines the hypothesis that navigation performance might be affected by system parameters such as latency and rendering technique. The present focus is to determine how system technical design and performance affects participant navigation performance.

Another relevant study by Walker and Lindsay [12] investigated navigation in a *virtual* audio-only environment with potential application to navigation aids for visually-impaired people. This study focused on the performance effect of way-point "capture radius", which describes the proximity at which the system considers an auditory beacon to be successfully reached. The simulation environment provided head-orientation interaction but participants navigated while sitting in a chair using buttons on a joystick to translate forwards or backwards in the virtual world, using orientation to steer. Thus, the participants' proprioception and other motion-related senses were not engaged, even though auditory navigation was possible. The present pilot study investigated the same factor, but in augmented reality (that mixed synthetic spatial audio with *real* vision/motion).

Walker and Lindsay's results showed successful navigation for "almost all" participants, with relatively direct navigation between way-points, and some individual performance differences. The main result identified a varied effect of capture radius on navigation speed and path distance efficiency, with a performance trade-off between the two. For the medium capture radius, which had the best distance efficiency, navigation speed was worst. For the small and large capture radii, distance efficiency was lower, but speed was higher. It appears that with a medium capture radius, participants were able to navigate a straighter path through the course, and they took more time to do so. This study informed the present pilot study, which sought to find an optimal capture radius that would *accentuate* navigation performance differences due to other system parameters, without making the navigation task excessively difficult.

3 Method

3.1 Setup

Experiment trials were performed in daylight, during fine weather conditions, on university sport fields, which provided a flat, open, grassy space. The pilot study was conducted with only the author as a participant. The main experiment was performed by eight male volunteers, of unknown exact age, in their 20s or 30s, with one in his early 50s, all with no known hearing problems.

Participants used a mobile hardware system comprised of a Sony Vaio VGN-U71 handheld computer, interfaced (using a Digiboat 2-port USB-Serial converter) to a Honeywell DRM-III position tracker and Intersense InertiaCube3, with audio output to Sennheiser HD-485 headphones. The experiment was controlled using custom software written in C# .NET 2.0, which received and logged data from the DRM-III at one sample per 250 ms, and from the InertiaCube3 at one sample per 6 ms. The software also provided a graphical user interface for the participant to enter responses and control their progress, and it controlled the real-time stimulus playback and binaural rendering, developed in Pure Data [9], interfaced using the Open Sound Control (OSC) protocol [14].

3.2 Procedure

The experiment was self-paced by the participant using the custom software. For each stimulus, the participant was required to begin by facing in a given direction at a marked *base position* in the center of a clear space with a radius of at least 35 meters. The accuracy of the base position and starting direction was not critical because the software re-zeroed its position and direction just prior to starting each stimulus.

When ready, the participant presses a software button to start the current stimulus, which is synthesised to simulate a stationary sound source at a chosen distance and azimuth. The participant task is to walk to the virtual source position and stop moving when they reach a given "capture radius" from the precise source location, indicated when the source stops playing. If the participant fails to locate the stimulus within 60 seconds, the stimulus also stops and a time-out message is displayed. In either case, the software prompts the participant to rate the perceived stability of the source position on a scale of 1 (least stable) to 5 (most stable). In the pilot study, the participant rated "perceived latency" (to head-turns), but this was considered potentially too esoteric for the main study. Finally, the participant must walk back to the central base position, face in the start direction and repeat the process for each subsequent stimulus, until the trial is completed. Figure 1 shows a photograph of a participant walking to locate a stimulus sound source during an experiment trial.

Participants were given the following guidance to prepare them for the experiment:

1. As per previous experiments using the DRM-III position tracker [5,6], for optimal tracking, participants were asked to walk at a steady, medium pace,

Fig. 1. Participant navigating to target during an experiment trial

only in the direction their torso was facing – i.e. to walk forward and avoid side-stepping or walking backwards.
2. Participants were reminded that the sound is intended to be stationary in the world, positioned at head height (since elevation was not simulated).
3. To help imagine a real sound source, participants were encouraged to look ahead (not at the ground) while they walked, and use head-turns to find the correct source direction.

3.3 Stimuli and Experimental Factors

Stimuli consisted of a continuous train of noise-bursts, spatialised in real time. The raw noise-burst-train itself was a continuously-looped, ten-second sample of Matlab-generated Gaussian white noise enveloped by a rectangular wave with duty cycle of 50 ms on, 100 ms off. The raw stimulus sound pressure level at the minimum simulated source distance of one meter was set to 75dBA in each headphone channel, with the Vaio sound output at full volume. Thus, the level was repeatable and would never exceed 75dBA even if the participant was at the same position as a sound source.

Both pilot study and main experiment employed factorial designs, with head-turn latency as one of the experimental factors. Latency is expressed as the amount of additional latency over the baseline total system latency (TSL), which was found to be 176 ms (±28.9 ms s.d.), using a method based on [8].

Pilot study stimuli were rendered using a technique of vector-based amplitude panning to six virtual speakers (denoted *VBAP6*) [10]. Each speaker was simulated binaurally by convolving its signal with the appropriate pair of head related impulse responses (HRIRs) from subject three, chosen arbitrarily from the CIPIC database [1]. Distance was simulated only by controlling stimulus level in proportion to the inverse square of source distance.

The pilot used 40 factor combinations of five capture radii (1, 2, 3, 4, 5 meters) and eight values of additional head-turn latency (0, 25, 50, 100, 200, 300, 400 and 500 ms). Two repetitions of each factor combination resulted in 80 stimuli, each spatialised to a random direction at a distance of 20 meters.

The main experiment used only 40 stimuli in total, comprised of four repetitions of two render methods and five latencies. The first render method (denoted *Ambi-B*) was first-order Ambisonic decoded to binaural via a six-speaker virtual array, using an *energy decoder* [2]. The second render method (denoted *VBAP12-G*), used a twelve-speaker virtual VBAP array, with a single, distance-variable ground reflection, with low-pass filtering to simulate a grass surface. Additional latency was set to 0, 100, 200, 400 or 800 milliseconds. Stimuli were spatialised to a random distance between 15 and 25 meters at a random azimuth within five sectors of the circle (to avoid direction clustering). Source capture radius was set to 2 meters due to the pilot study results (reported in Section 4).

3.4 Results Analysis

For each stimulus, all raw data available from the DRM-III position tracker and IntertiaCube3 orientation tracker was logged four times a second. This included position easting and northing, the number of steps taken, body heading and head yaw, pitch and roll. Another file recorded all stimulus factor values (latency, render method, azimuth and range), as chosen randomly for each trial. Lastly, a results file recorded the stimulus factors again, with the source position, participant's rating response, time elapsed, and distance walked (integrated along the participant's path). These data were then imported into Matlab and analysed with respect to the stimulus factors.

Results were analysed using several performance measures based on the participant's response time, position tracks and head-orientation movements during navigation. These performance measures were designed in terms of the participant-source geometry shown in Figure 2, and they were analysed as described below.

The first performance measure, *distance efficiency (DE)* is calculated as the ratio of the ideal, direct path length ($d+c$) to the actual, walked path length ($a+c$), as shown in Equation 1. Both distances are measured to the center of the stimulus capture circle of radius c, assuming the last section of the actual walk path would be a straight line if it were to be completed.

$$DE = \frac{d+c}{a+c} \tag{1}$$

The second is the *inter-participant-normalised mean velocity (MV)* to the edge of the capture radius (Equation 2), used as a measure related to the navigation

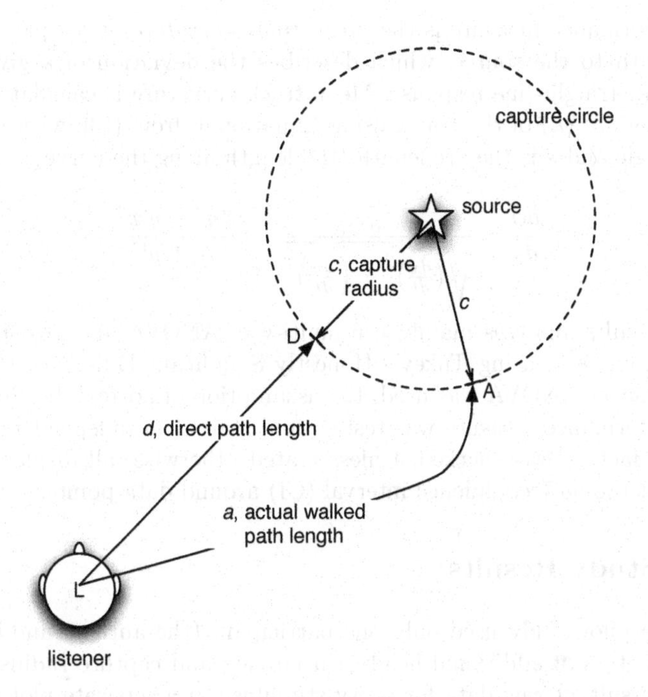

Fig. 2. Participant and source geometry, showing source capture radius, direct path and participant's actual walked path to the edge of the source capture circle

time [12]. Note that the navigation time itself is not used because it is influenced by the randomised source distance for each stimulus and by participants' different walking speeds. Instead, mean velocity results are used, since they account for distance and they are normalised across all participants' results, with respect to the overall mean.

$$MV = \frac{d}{t} \tag{2}$$

A third performance measure is the *root mean square head-yaw deviation (HYD_{RMS})* from the body heading (Equation 3), which represents the average amount of head-turn activity during a stimulus. In calculating this measure, the yaw records for the first four meters of walk are discarded, since this early response is strongly affected by the initial source azimuth, which is not a pertinent factor in this study. This correlation of initial head-yaw and initial source azimuth occurs because the participant often turns their head around to find the source direction before moving very far from the base position.

$$HYD_{RMS} = \sqrt{\frac{1}{n}\sum_{i=1}^{N}(\theta_{head_i} - \theta_{body_i})^2} \tag{3}$$

A fourth performance measure is the *mean track curvature* of the participant's navigation path to the source, which describes the deviation of a given path from the ideal straight-line response. Mean track curvature is calculated as the *extrinsic curvature* (κ) of the track, using Equation 4, from [13], where ϕ is the tangential angle and s is the arc length (the length along the curve).

$$\kappa \equiv \frac{d\phi}{ds} = \frac{\frac{d\phi}{dt}}{\sqrt{\left(\frac{dx}{dt}\right)^2 + \left(\frac{dy}{dt}\right)^2}} = \frac{x'y'' - y'x''}{x'^2 + y'^2} \tag{4}$$

Subsequent results analyses mainly use multi-way ANOVA and post-hoc multiple comparison tests using Tukey's Honestly Significant Difference (HSD) at $p<0.05$. Whenever ANOVA was used, the assumption of normal distribution of the given performance measure was tested, and statistical independence of the experimental factors was checked. Unless stated otherwise, all displayed error bars represent the 95% confidence interval (CI) around data points.

4 Pilot Study Results

As noted, the pilot study used only one participant (the author) and only examined the factors of additional head-turn latency and capture radius. Figure 3 shows the resultant raw data for every stimulus, on a separate plot for each capture radius setting. The plots show the position tracks with a marker at every footstep and a short vector displaying the recorded head yaw data. They also show a straight line from the base position to the source position at the center of the capture circle.

On inspecting the stimulus position tracks, it's clear that some are straighter than others. Ideally, the participant would perfectly localise the sound source before they move, then walk directly towards the point source, stopping at the capture radius, where the target is considered reached. In reality, this does not occur. The participant isn't able to perfectly localise the stimulus from the beginning, although thanks to head-turn interaction, the initial source bearing determination is usually fairly accurate. Then, while the participant walks in the initial chosen direction, any perceived azimuth error increases as the participant nears the source, which forces continual reassessment of the momentary source direction. Errors are potentially exacerbated by head-turn latency, especially if the participant keeps walking towards a source rendered using a delayed head-yaw reading. Thus, added latency was expected to degrade the participant's navigation path from the ideal straight line into a curved or piecewise path with progressive corrections.

Initial visual comparison of participant tracks reveals they are straighter for the largest (5 meter) capture circle when compared to the smallest capture circles of 1 or 2 meters. Another basic feature is the apparent tendency for this participant to curve anti-clockwise rather than clockwise as he nears the source position. However, close inspection indicates that anti-clockwise curves are not ubiquitous, for example, see the source at approximately 30° clockwise from north on the 3 m capture radius plot (Figure 3c). Also, some tracks curve clockwise at the beginning and anti-clockwise at the end.

Another feature is the tendency for occasional use of larger head-turns, presumably to reassess the source azimuth, e.g. the track just anti-clockwise from north on the 3 m capture radius plot. Larger head-turns mostly occur at the beginning and towards the end of the navigation. At the beginning, the task resembles a traditional stationary localisation experiment, since the participant usually doesn't move away from the base position until they've judged the source direction. Head-turns here enable almost complete mitigation of front-back localisation errors, evidenced by the participant moving away from the base in almost the correct direction in most cases. Greater head movement again becomes useful to adjusts navigation direction along the way to the target.

In general, participant tracks show that navigation to the target was usually very successful, even for the most difficult 1 m capture radius. Further conclusions require derivation of performance measures and statistical analysis with respect to experimental factors. Thus, ANOVA tests were performed to look for significant effects of source capture radius and added head-turn latency on distance efficiency.

For the pilot study, a visual test[1] of the ANOVA requirement for normally distributed independent variables was passed for distance efficiency, but not for latency rating, so that was not analysed. Results show significant effects of capture radius on distance efficiency ($F(4,40) = 6.3$, $p = 0.00048$). Post-hoc multiple comparisons of capture radius values were then performed using Tukey's HSD ($p<0.05$).

Figure 4 shows capture radius versus distance efficiency, with significant differences between the 2 m radius and 3, 4 and 5 m radii, but insignificant differences between other value pairs. This insignificant effect of the 1 m radius is unexpected, but might be explained by a weak effect of capture radius, and the small number of repetitions using only one participant. Nevertheless, there appears to be a general trend of increasing distance efficiency with increasing capture radius. This is expected since most path curvature occurs close to the source position, so a larger circle allows less opportunity for such navigation perturbations to be caused by azimuth localisation errors and head-turn latency. Conversely, smaller capture circles include a greater proportion of the critical final stage of homing-in to the source. A similar conclusion was reached in [12], using a virtual audio-only environment.

5 Main Results

5.1 Raw Position and Head-Yaw Tracks

Figure 5 shows the raw data for every stimulus, on a separate plot for each participant (with the author as participant 1). As for the pilot study, navigation tracks show each footstep, with head-yaw data at that position, and a straight

[1] Using the Matlab function `normplot`.

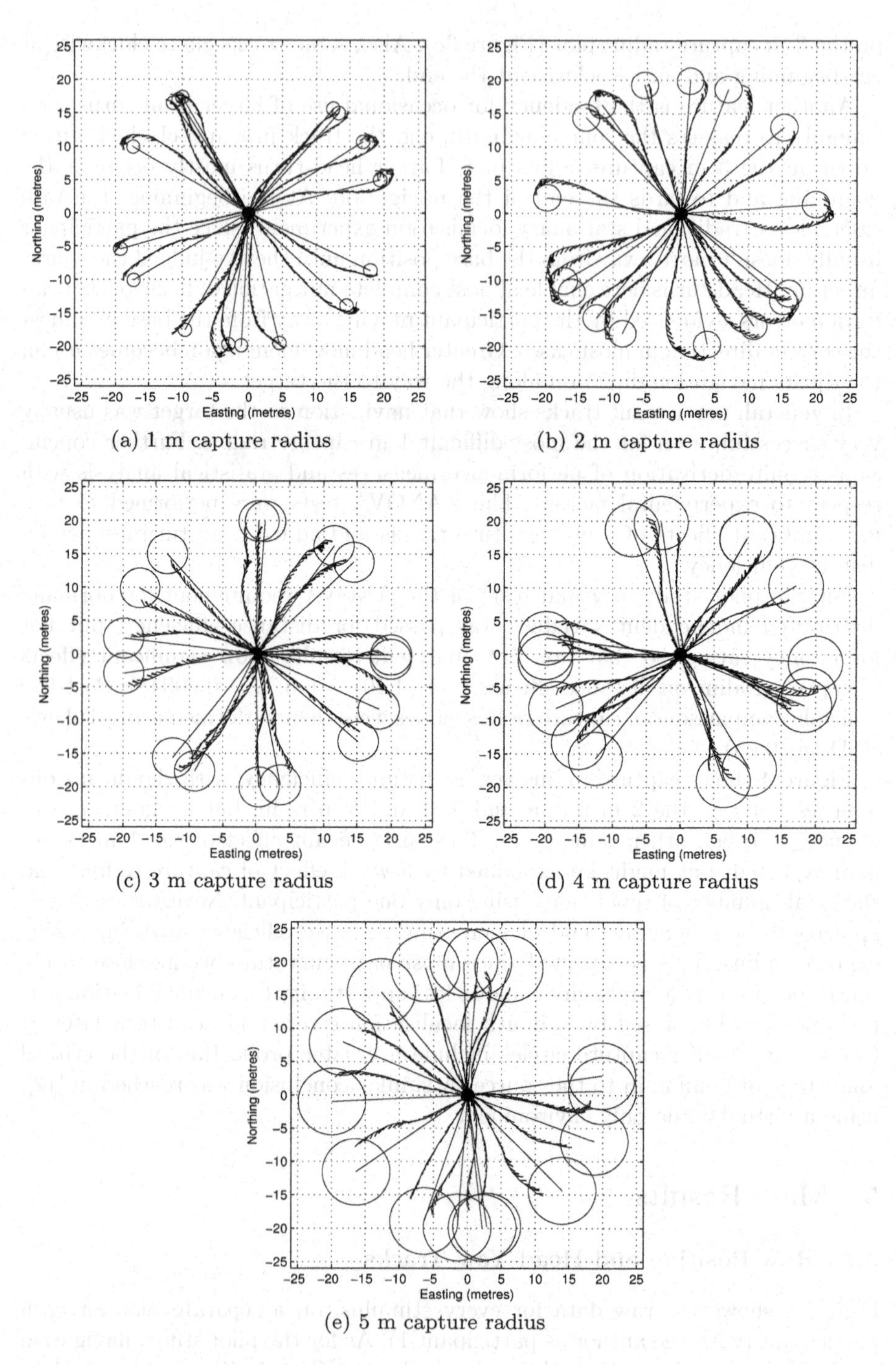

Fig. 3. Pilot study tracks, showing participant body position, head orientation, source position, bearing and capture radius

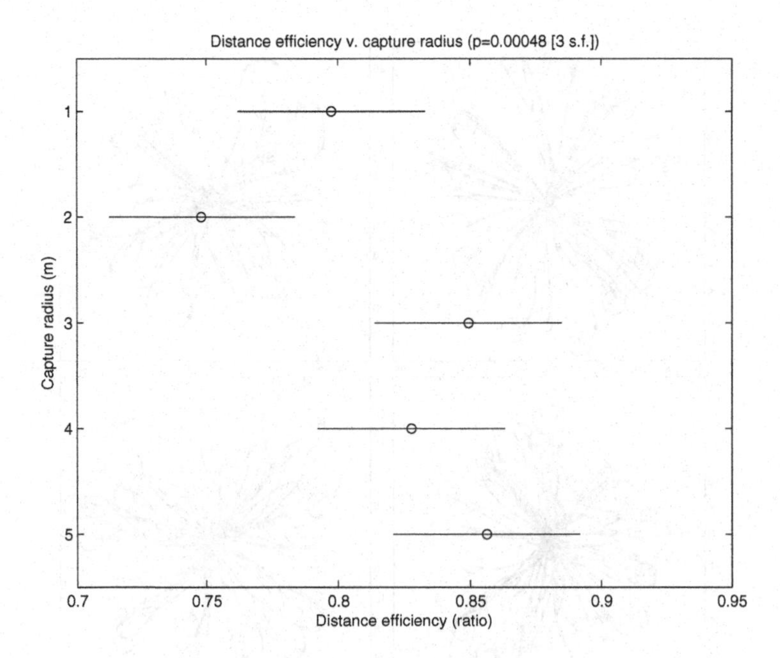

Fig. 4. Post-hoc Tukey HSD multiple comparison of capture radius vs. distance efficiency

line to the target. The capture radius was set to 2 meters for all stimuli as a result of the pilot study, since this was the largest radius that significantly affected distance efficiency.

Participants varied significantly in their ability to navigate directly to the target and individual navigation performance also varied between stimuli. For all participants, some stimuli were reached almost directly, while others were overshot or reached via a curved path that often became more curved towards the source position. Participants usually headed away from the base position in approximately the right direction, rarely in the opposite direction. Then, during navigation, static localisation errors and possibly head-turn latency affected the momentarily-perceived target azimuth. Resulting momentary heading errors then increase due to the source-listener geometry as the participant nears the source. This prompts more drastic path corrections, or pauses with head-turns, as participants proceed towards the target. Effectively, the participant's ability to walk an ideal straight path towards the source is reduced depending on their navigation strategy in combination with their perceptual ability, the inherent rendering resolution and system latency.

5.2 Head-Yaw and Distance from Base

Figure 6 displays head-yaw data and distance from the base over time in a separate plot for each participant. Head yaw values are plotted to the left y-axis,

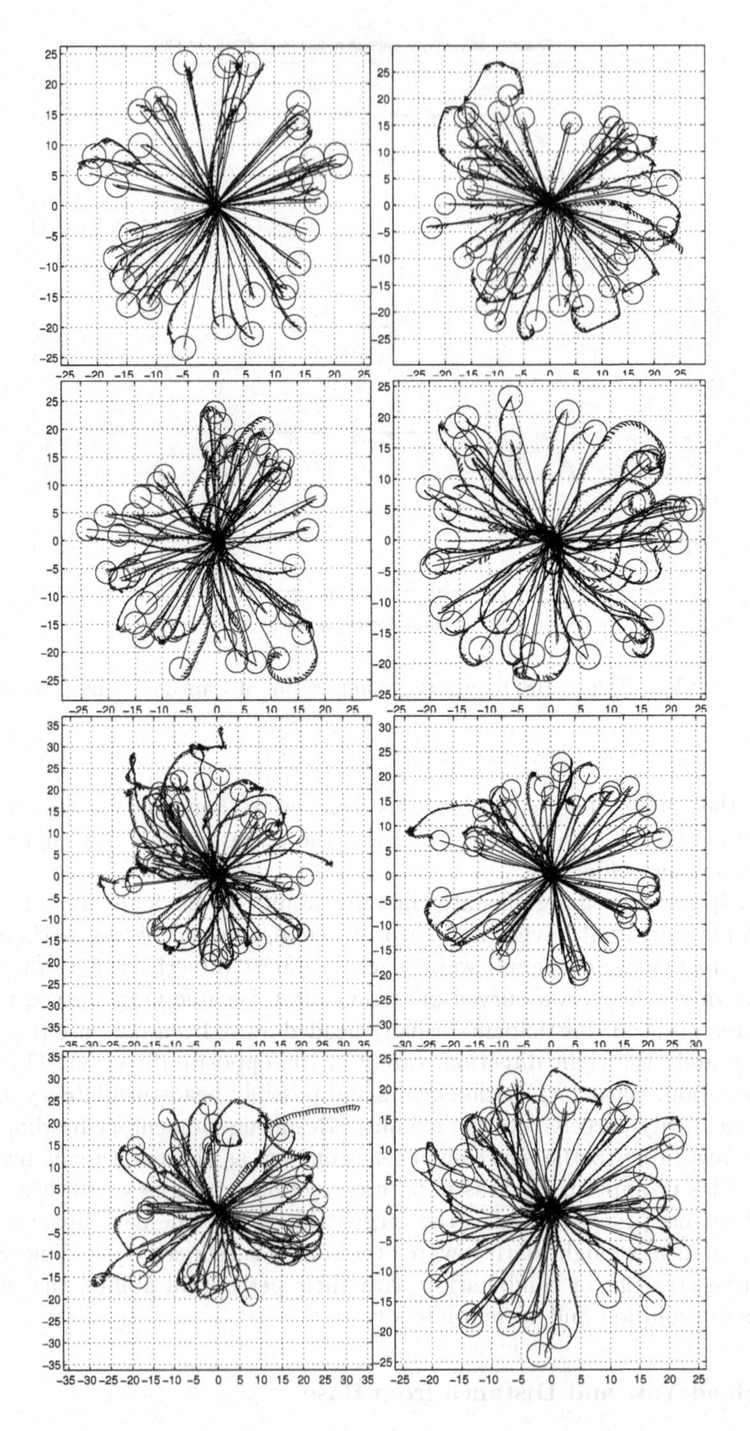

Fig. 5. Experiment record of all participant tracks (participants 1-8, in order: left to right, top to bottom)

while distance-from-base is plotted to the right y-axis. Each stimulus response is represented by one trace of each kind, both ending at the same time. Any pair of traces that ends before the 60 second limit indicates a successfully located source, while traces that reach 60 seconds were not located and thus triggered the time-out message. Timeouts were also explicitly recorded, with a total of 10 occurring over all participants and all stimuli, 6 of which occurred during the first 3 stimuli.

Head yaw represents the participant's momentary head-orientation with respect to their body orientation at each step. Some processing is required to get head-yaw data that center around zero degrees. First, the head and body orientation data must be phase-unwrapped (since they are reported in a modulo form between 0° and 360°), then their initial difference is subtracted, which assumes the participant kept both their head and body facing forward at the start of each trial. As an automated process, the occasional problem occurs, which potentially explains a single odd result for participant 7 (Figure 6g) in which the head-yaw seems to center around approximately 100° instead of zero, with some dramatic deviations in each direction. This track is also apparent in Figure 5 (participant 7 is bottom-left) as the most prominent track in the top-right of the plot, where the head-yaw vectors are offset by approximately 90° from the track bearing for some distance.

Participant 5 also exhibits some larger head-turns than other participants, although they are centered much closer to zero than the unusual track from participant 7. It is possible the problem is related to tracking inaccuracies during some stimuli when participant 5 appears to stop and rotate on-the-spot (possibly more than 360°), or other tracks that show one or more loops. Several of participant 5's tracks which exhibit larger head yaw also lasted 60 seconds, indicating the source was not located within the time limit. Note however that even these two participants successfully navigated to the source within the time limit in most cases, often via a reasonably direct path.

Apart from these few unusual results, most participants' head-yaw results lie within reasonable angular bounds (around ±90° maximum). The plots also reveal that participants generally followed the expected strategy to perform lots of head-rotations early during each stimulus, usually before they moved far from the base position. This can be observed as larger head yaws before the distance-from-base traces rise far above zero. This observation is somewhat prominent for every participant, but most evident for participants 1 and 8. The early localisation head rotations are correlated to the (randomised) initial source azimuth, since the participant usually rotates all the way around to face the source, sometimes overshooting the correct bearing, before moving away from the base in the perceived source direction. This observation informed the development of the RMS head-yaw performance measure, which discards the head-yaw trace for the first four meters to avoid correlation with initial source azimuth, whilst measuring any remaining variations that may be due to render method or head-turn latency.

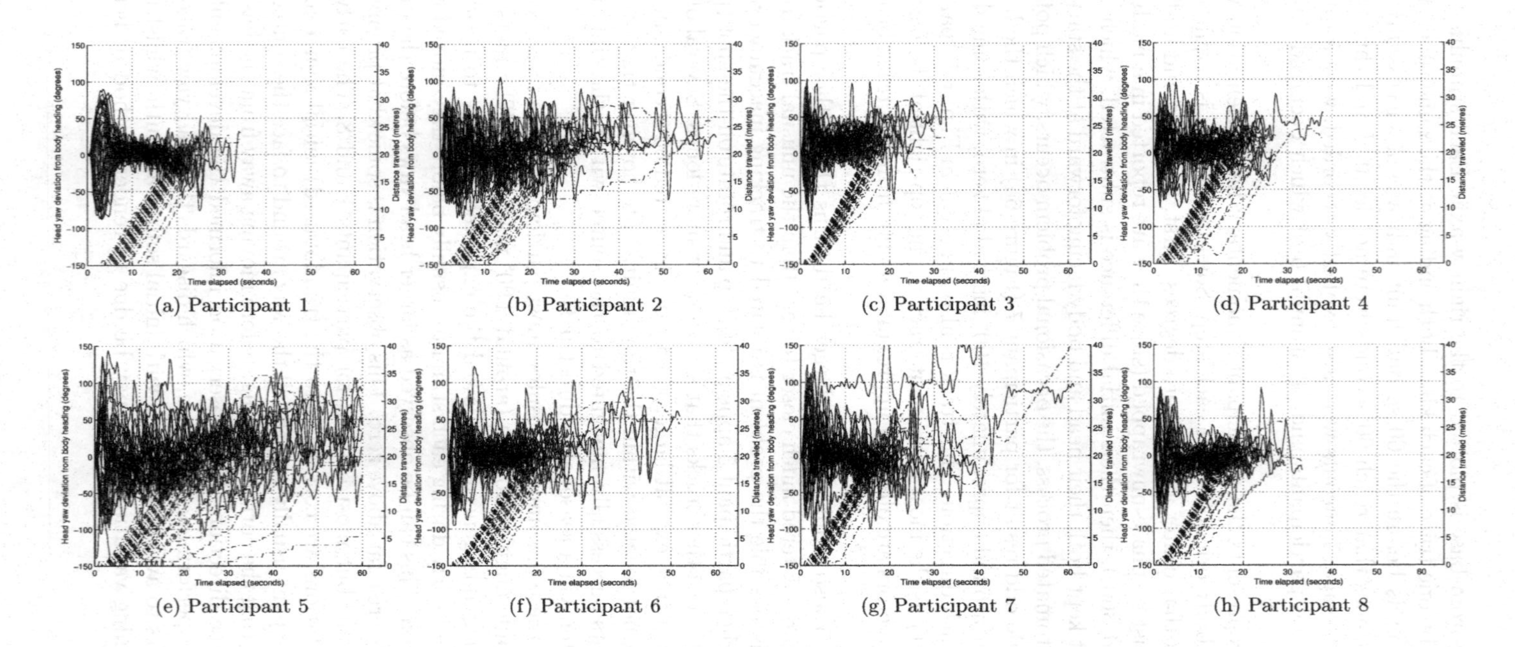

Fig. 6. Experiment records of participant head yaw and distance from base

The traces of travelled distance in Figure 6 reveal another aspect of participants' results. Most participants travel at a quite constant radial velocity (indicated by the slope of the radial distance trace) over the majority of each stimulus response. They make their initial move away from the base position sometime during the first ten seconds and often reach the source capture circle within about 30 seconds. Most radial velocity variation occurs towards the end of each individual navigation response, in the form of reversals in the distance-from-base (i.e. a curved track looping back towards the base position), non-radial movements, slower walking, or some combination of the above. These movement variations are then quantified in one or more of the performance measures: distance efficiency, mean velocity or track curvature.

5.3 Analysis of Variance

ANOVA tests were run for all performance measures against the factors of latency, render method and participant, after testing all assumptions. After initial visual confirmation of the data distributions, transformations were determined to be necessary for distance efficiency and RMS head yaw data. Satisfactory results were achieved for distance efficiency by taking the arcsine value – a transformation considered suitable for proportion data [7]. RMS head yaw was transformed satisfactorily by taking the base ten logarithm.

Individual participant differences were highly significant for all performance measures except stability ratings and mean velocity, which had been normalised between participants to enable comparison of the *unanchored* stability scale and to account for variable partipant walking speeds. Individual differences were expected, so further significant effects of the participant factor are not mentioned unless they are remarkable.

Distance efficiency was significantly affected by render method ($F(1{,}258) = 41$, $p = 8.7 \times 10^{-10}$) and the interaction between head-turn latency and participant ($F(28{,}258) = 1.5$, $p = 0.048$). Inter-participant normalised mean velocity, was significantly affected by render method ($F(1{,}258) = 45.8$, $p = 8.6 \times 10^{-11}$) alone. Inter-participant normalised stability ratings were significantly affected by head-turn latency ($F(4{,}258) = 8.2$, $p = 3.2 \times 10^{-6}$), the interaction of latency with render method ($F(4{,}258) = 5.8$, $p = 0.00018$), and the interaction of render method with participant ($F(7{,}258) = 3.8$, $p = 0.00059$). Mean track curvature was significantly affected by latency ($F(4{,}258) = 12.7$, $p = 1.84 \times 10^{-9}$), render method ($F(1{,}258) = 10.2$, $p = 0.00155$) and participant ($F(7{,}258) = 5.74$, $p = 3.5 \times 10^{-6}$). Finally, other than individual differences, RMS head-yaw was significantly affected only by render method ($F(1{,}258) = 10.6$, $p = 0.0013$).

Interestingly, the factor of render method alone significantly affected every performance measure except for the stability rating. However, stability rating *was* affected by the *interactions* of render method with head-turn latency and participant. The interaction with latency is discussed later in Section 5.7. The intereaction with participant is interesting in itself because stability rating was normalised between participants, effectively cancelling out individual differences on average. Thus, some participants must have rated stability significantly differently, depending on the stimulus render method.

Head-turn latency alone only had a significant effect on the stability rating and mean track curvature, although it almost had a significant effect on RMS head yaw (p=0.065), which might become significant given a more powerful experiment with more participants. The significant effects of each experimental factor are further investigated below, using post-hoc multiple comparison tests (Tukey HSD, $p<0.05$), with error bars on all plots representing 95% confidence intervals.

5.4 Post-Hoc Analysis: Participant Differences

Individual differences (Figure 7) were prominent, as expected, for all performance measures except mean velocity and stability ratings, which were normalised between participants. That said, for distance efficiency, only participant 1 (the author) was significantly different, with remaining participants averaging around 0.75 efficiency. This suggests a possible learning effect, given that the author had more experience with the experimental task, the system, and the binaural rendering used.

For the mean track curvature results (Figure 7c), participants' mean results ranged from mildly positive (anti-clockwise) to significantly negative (clockwise), although many were insignificantly different to zero. Thus, while an inspection of participant tracks in 5 might draw the conclusion that most tracks exhibit clockwise curves, this is not evidenced by the post-hoc analysis.

5.5 Post-Hoc Analysis: Effects of Render Method

Render method significantly affected every performance measure except for stability rating. In all cases, VBAP12-G rendering significantly outperformed AMBI-B. In detail, mean distance efficiency was approximately 0.72 for Ambisonic, 0.81 for VBAP12-G (Figure 8a); inter-participant normalised mean velocity was 0.8 m/s for Ambisonic, 0.93 m/s for VBAP12-G (Figure 8b); RMS head-yaw was 18.5° for Ambisonic, 15.5° for VBAP12-G (Figure 9a); and lastly, mean track curvature was -0.02 m^{-1} for Ambisonic, -0.002 m^{-1} for VBAP12-G (Figure 9b).

These results might be explained by the different angular resolution or localisation blur resulting from the differences between rendering methods. VBAP12-G used 12 virtual speakers (with 12 HRIRs, assuming head-symmetry), and a panning method that only interpolates between adjacent virtual speakers. In contrast, Ambi-B used first-order, 2D Ambisonic rendering with energy decoding to 6 virtual speakers (using 6 filters by combining the 2D Ambisonic speaker decode with the HRIRs). A separate, unpublished static localisation experiment using these rendering methods showed better mean localisation errors for the VBAP12-G method (20.5°) compared to Ambi-B (23.1°) [4]. Similar relative results were also found in another localisation performance comparison between amplitude panning and Ambisonic rendering to a real 6-speaker array [11].

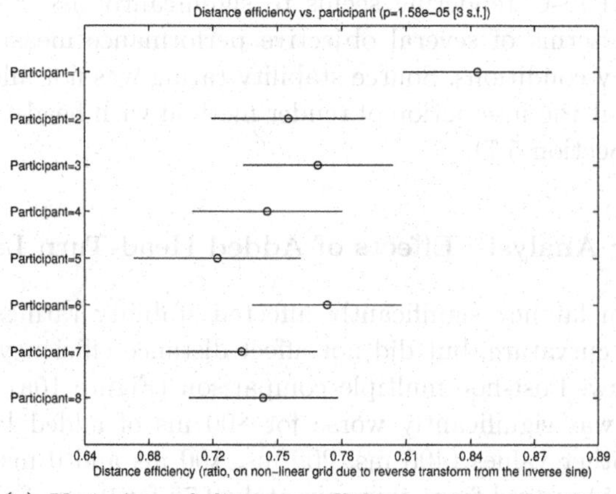

(a) Variation of distance efficiency across participants

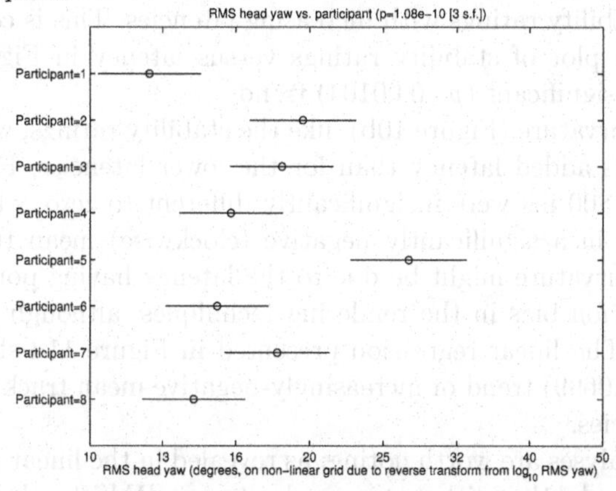

(b) Variation of RMS head yaw across participants

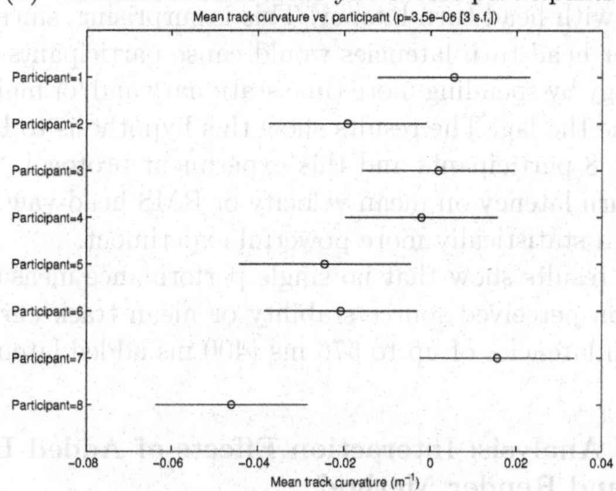

(c) Variation of mean track curvature across participants

Fig. 7. Post-hoc Tukey HSD multiple comparisons of participant differences

Overall, VBAP12-G rendering seems to significantly assist navigation to a sound source in terms of several objective performance measures, under any head-turn latency conditions. Source stability rating was not affected by render method alone, but the interaction of render method with head-turn latency was significant (see Section 5.7).

5.6 Post-Hoc Analysis: Effects of Added Head-Turn Latency

Added head-turn latency significantly affected stability ratings (as expected) and mean track curvature, but did not affect distance efficiency, mean velocity or RMS head-yaw. Post-hoc multiple comparison (Figure 10a) shows that the stability rating was significantly worse for 800 ms of added latency (976 ms TSL) than for other values (400 ms, 200 ms, 100 ms and 0 ms). The stability ratings (scale: 1-5) ranged from approximately 2.75 for 0 ms of added latency to below 2.1 for 800 ms of added latency. There appears to be a monotonic trend of decreasing stability ratings with increasing latencies. This is confirmed in the linear regression plot of stability ratings versus latency in Figure 11c, which reveals a highly significant (p=0.00104) trend.

Mean track curvature (Figure 10b), like the stability ratings, was significantly worse for 800 ms added latency than for the lower latencies. Furthermore, all latencies except 800 ms were insignificantly different to zero, while the 800 ms latency resulted in a significantly negative (clockwise) mean track curvature. The clockwise curvature might be due to the latency having potentiated a pre-existing localisation bias in the rendering techniques, although the evidence is not conclusive. The linear regression presented in Figure 11d shows an almost significant (p=0.0699) trend of increasingly-negative mean track curvature with increasing latencies.

Two other analyses are worth noting: as revealed in the linear regression plots in Figures 11a and 11b, neither mean velocity nor RMS head-yaw exhibited a significant trend with head-turn latency. This is surprising, since it was hypothesised that higher head-turn latencies would cause participants to modify their navigation strategy by spending more time stationary and/or making more head-turns to overcome the lag. The results show this hypothesis to be unsupported, at least for these 8 participants and this experiment protocol. To discover any effects of head-turn latency on mean velocity or RMS head-yaw might require a modified task or a statistically more powerful experiment.

Overall, these results show that no single performance measure had any significant impact on perceived source stability or mean track curvature for total system head-turn latencies of up to 576 ms (400 ms added latency).

5.7 Post-Hoc Analysis: Interaction Effects of Added Head-Turn Latency and Render Method

ANOVA results (Section 5.3) revealed that stability rating was significantly affected by the interaction of added head-turn latency and render method

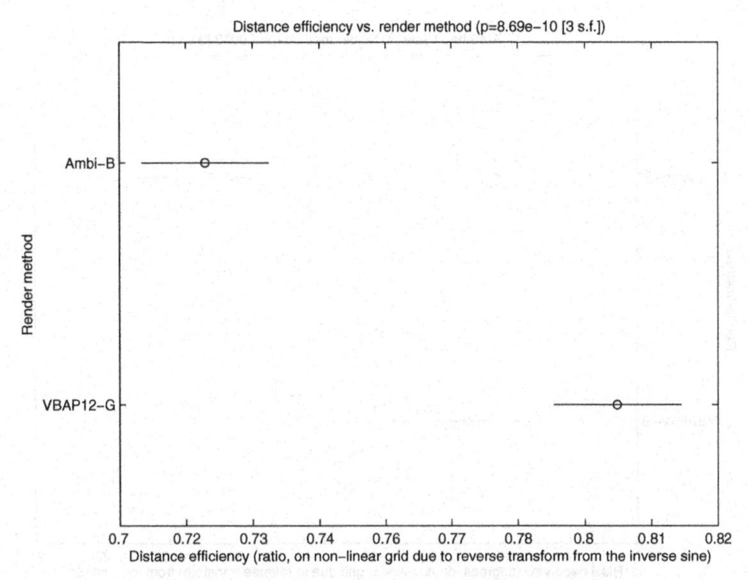

(a) Effect of render method on distance efficiency

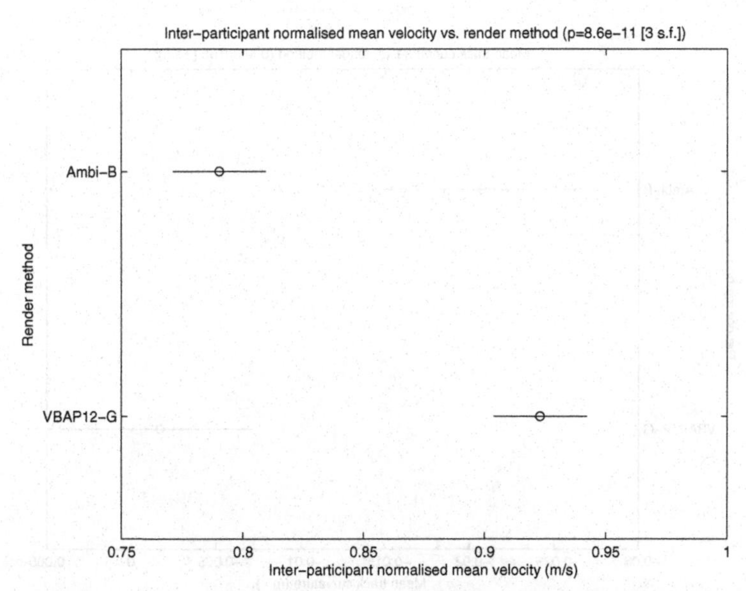

(b) Effect of render method on inter-participant normalised mean velocity

Fig. 8. Post-hoc Tukey HSD multiple comparison of render method effects on distance efficiency and mean velocity

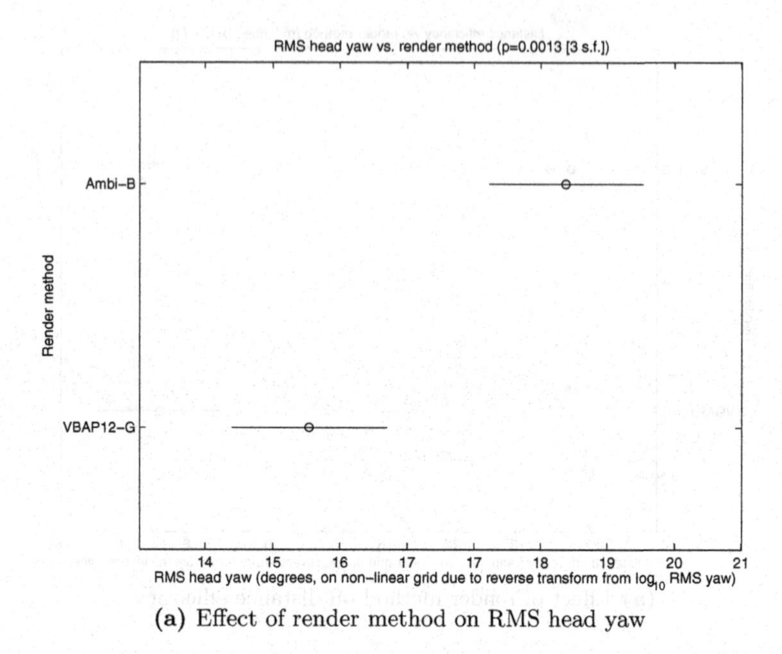

(a) Effect of render method on RMS head yaw

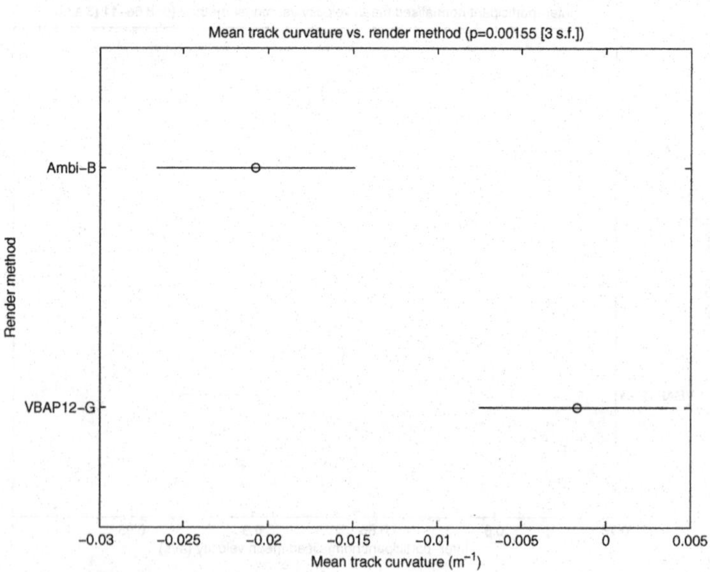

(b) Effect of render method on mean track curvature

Fig. 9. Post-hoc Tukey HSD multiple comparisons of render method effects on RMS head-yaw and mean track curvature

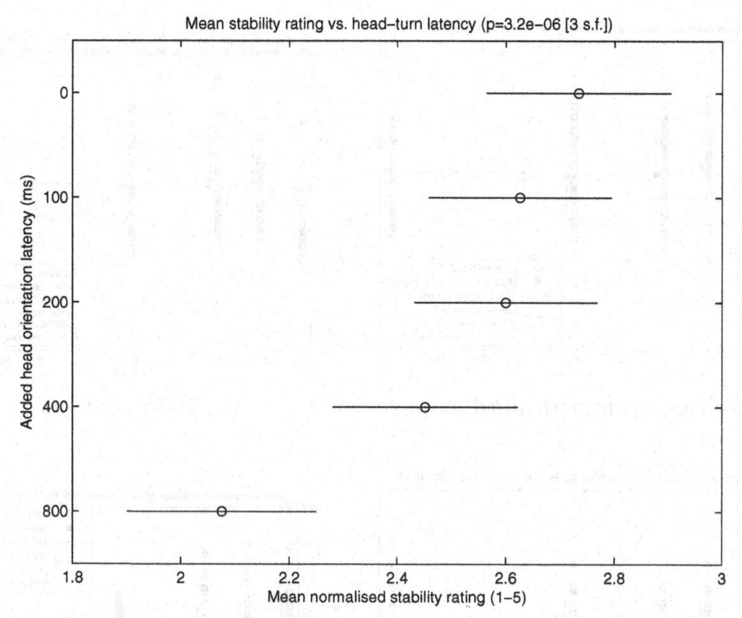

(a) Effect of added head-turn latency on inter-participant normalised stability ratings

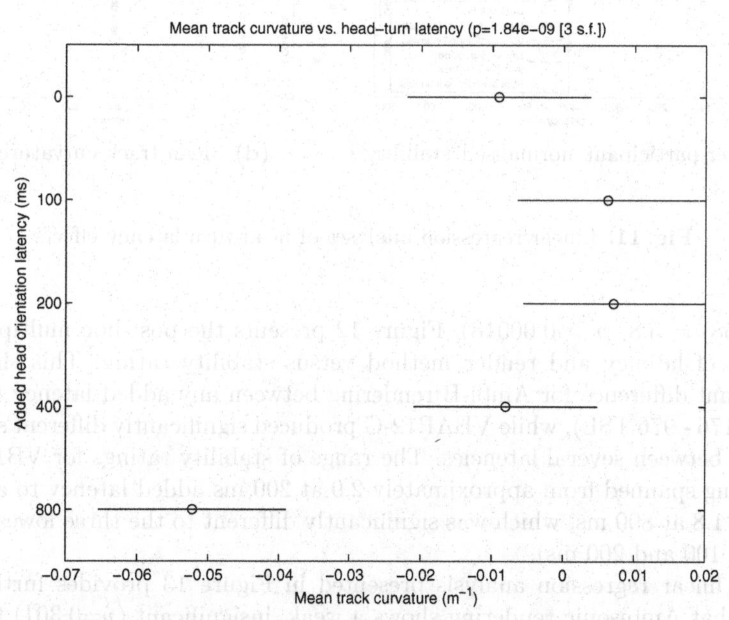

(b) Effect of added head-turn latency on mean track curvature

Fig. 10. Post-hoc Tukey HSD multiple comparisons of head-turn latency effects on stability ratings and mean track curvature

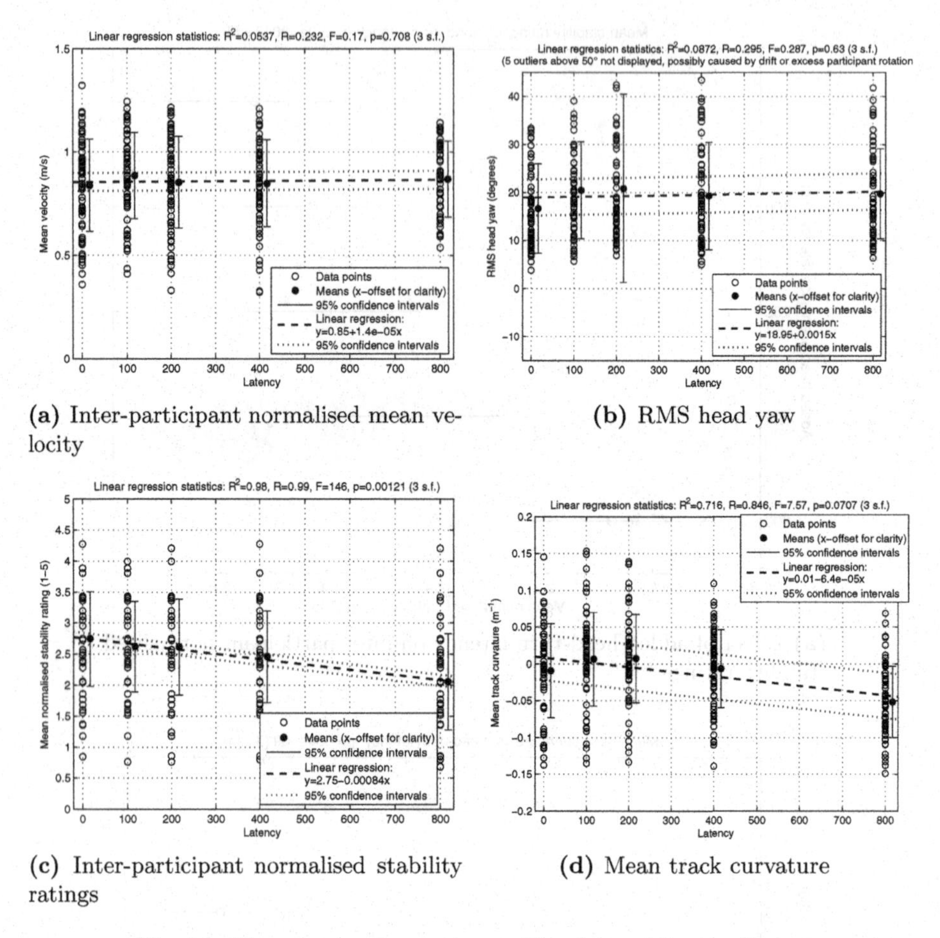

(a) Inter-participant normalised mean velocity

(b) RMS head yaw

(c) Inter-participant normalised stability ratings

(d) Mean track curvature

Fig. 11. Linear regression analyses of head-turn latency effects

(F(4,258) = 5.8, p = 0.00018). Figure 12 presents the post-hoc multiple comparison of latency and render method versus stability rating. This shows no significant difference for Ambi-B rendering between any added latency (0 - 800 ms, or 176 - 976 TSL), while VBAP12-G produced significantly different stability ratings between several latencies. The range of stability ratings for VBAP12-G rendering spanned from approximately 2.9 at 200 ms added latency to approximately 1.8 at 800 ms, which was significantly different to the three lowest latencies (0, 100 and 200 ms).

The linear regression analysis presented in Figure 13 provides further evidence that Ambisonic rendering shows a weak, insignificant (p=0.301) trend of decreasing stability rating with increased head-turn latency, while VBAP12-G shows a stronger, significant (p=0.0269) trend. Apparently the VBAP12-G stimuli account for most of the significant trend of subjective stability rating versus latency (Figure 11c).

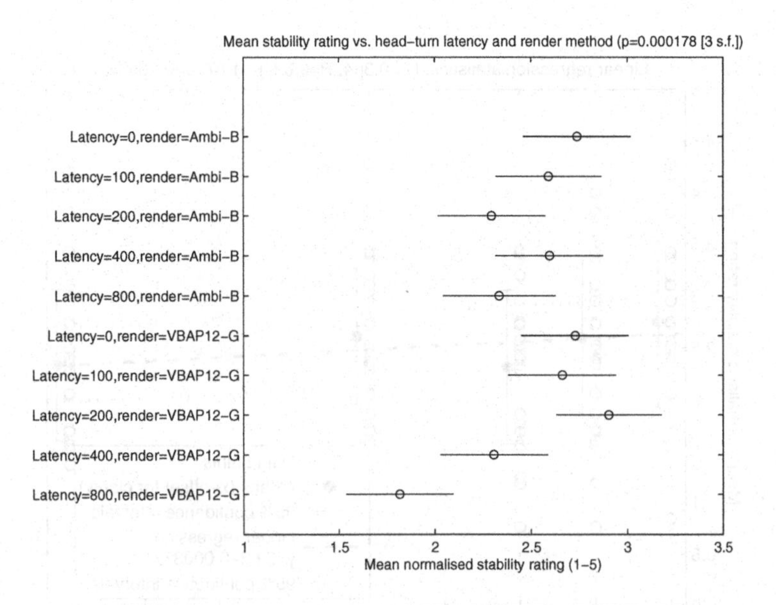

Fig. 12. Post-hoc Tukey HSD multiple comparisons of added head-turn latency and render method combinations vs. inter-participant normalised stability ratings

This interesting result could be interpreted as showing that the lower-resolution, higher-blur Ambisonic rendering mitigates the impact of increasing head-turn latency on subjective source stability, which is evident for VBAP12-G rendering. By reducing the azimuth resolution and increasing localisation blur, head-turn latency becomes less perceptible, effectively increasing the upper threshold of head-turn latency before measurable performance degradation. Conversely, higher resolution, lower-blur rendering requires lower head-turn latencies to avoid a perceptible lag.

6 Summary, Discussion and Conclusions

This study investigated the navigation performance of users of a mobile audio AR system and how this may be measurably affected by the binaural rendering method and total system latency to head-turns. These two system parameters are critical to good performance and are often limited by other aspects of the system design. Rendering is limited by mobile computing power, battery life, rendering technique, and generalised HRIRs. Head-turn latency is potentially limited by orientation sensor communication delays, the mobile device operating system, available computing power, rendering software design and sound output buffer size.

The experiment task was designed to generalise any potential navigation task using positional auditory beacons. It required participants to navigate from a single, central base position to the location of multiple virtual sound sources.

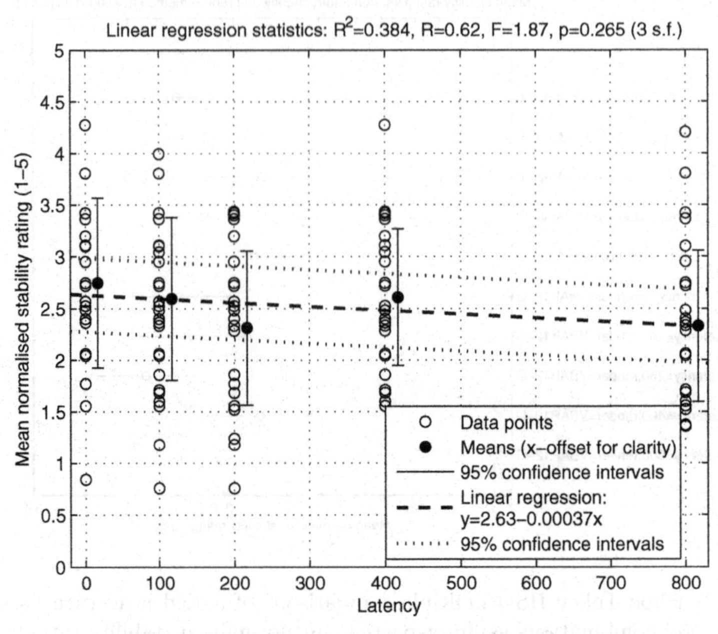

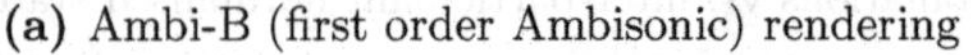
(a) Ambi-B (first order Ambisonic) rendering

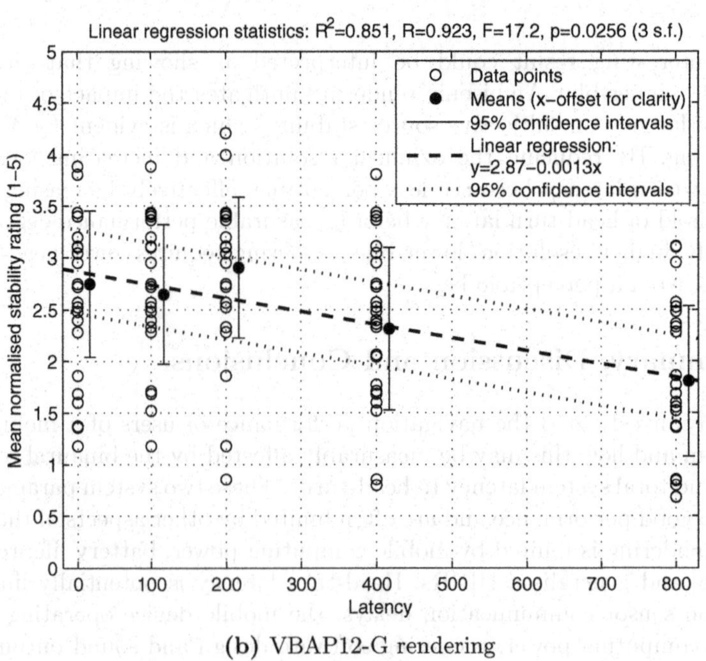

(b) VBAP12-G rendering

Fig. 13. Linear regression of inter-participant normalised stability rating vs. added head-turn latency, across render methods

The pilot study examined the impact of source *capture circle* radius and head-turn latency on performance measures of distance efficiency and head-turn latency rating. Decreasing capture radius significantly reduced distance efficiency ($p<0.05$), showing that straighter navigation paths were supported by larger capture circles. Pilot study results informed the main experiment, in which capture radius was set to 2 m, the largest circle with significant detrimental impact on distance efficiency. Therefore, for better navigation performance in mobile AR applications, capture radius should be 3 meters or more for a system with similar binaural rendering techniques, tracking accuracy, and latency specifications to the experiment system.

The main experiment examined the performance impact of head-turn latency and render method (Ambi-B or VBAP12-G), which differ mainly in terms of inherent azimuth resolution and localisation blur. Performance measures were distance efficiency, mean velocity, mean track curvature, RMS head-yaw and subjective source stability rating.

Overall results showed that regardless of sometimes severe system performance degradation, all eight participants successfully navigated to most source positions within 60 seconds. In comparison to static (non-interactive) binaural audio displays, improved source localisation performance (i.e. navigation) is made possible by the perceptual feedback provided by system interactivity to user position movements and head-turns. For instance, participants mostly began walking towards a sound source in approximately the correct direction, in contrast to the high front-back confusion rates and azimuth errors common in static experiments.

In greater detail, VBAP12-G rendering performed significantly better than Ambi-B on four out of five performance measures, with no significant effects on stability rating ($p<0.05$). In contrast, added head-turn latency only showed significant effects on stability rating and mean track curvature, which were both significantly worse for 800 ms added latency (976 ms TSL). No significant effects were observed on any performance measures for head-turn latency values of 200 ms (376 ms TSL) or less.

The most interesting result was the significant effect on subjective stability of the interaction of render method and latency, despite render method alone having no effect. Significant stability degradation due to increased head-turn latency occurred only for VBAP12-G rendering, not for Ambi-B rendering. The lower resolution, higher blur of Ambi-B rendering apparently mitigated the detrimental effect of high head-turn latency on perceived stability. Conversely, VBAP12-G rendering apparently exacerbated latency's effect on stability.

In conclusion, the study made several key observations useful for the implementation of mobile audio AR systems. First, even systems with high head-turn latency or relatively low resolution (high blur) rendering can afford successful user navigation to positional sound sources, but degradation of both specifications does damage objective and subjective participant performance.

Lower resolution, higher-blur Ambisonic rendering decreased navigation distance efficiency, but reduced the detrimental effects of latency on source stability rating. Higher resolution, lower-blur VBAP rendering improved all qualitative performance measures but enabled high latencies to degrade the source stability ratings.

Improved navigation performance is best supported by improving both system specifications at the same time (as might be expected). However, within the parameters of this study, a greater performance benefits were achievable by increasing rendering resolution than by reducing system latency. Mid-range latencies up to 376 ms TSL can be tolerated, regardless of rendering method.

Acknowledgements

This research took place in 2007 and 2008, during the author's PhD candidacy at University of New South Wales, Sydney, Australia. It occurred within the AudioNomad project lead by Dr. Daniel Woo, Dr. Nigel Helyer and Prof. Chris Rizos. AudioNomad was supported by the Australian Research Council as a Linkage Project (LP0348394) with the Australia Council for the Arts, under the Synapse Initiative. Writing of this paper was supported by LIMSI-CNRS, through the ANR SoundDelta project.

References

1. Algazi, V.R., Duda, R.O., Thompson, D.M., Avendano, C.: The CIPIC HRTF Database. in: Proc. IEEE Workshop on Applications of Signal Processing to Audio and Electroacoustics. pp. 99–102, Mohonk Mountain House, New Paltz, NY (October 21-24 2001)
2. Benjamin, E., Lee, R., Aficionado, L., Heller, A.: Localization in Horizontal-Only Ambisonic Systems. 131st AES Convention, preprint 6967, 5–8 (October 8 2006)
3. Loomis, J., Hebert, C., Cicinelli, J.: Active localization of virtual sounds. J. Acoust. Soc. Am. 88(4), 1757–1764 (October 1990)
4. Mariette, N.: Perceptual Evaluation of Personal, Location Aware Spatial Audio. Ph.D. thesis, School of Computer Science and Engineering, University of New South Wales (2009 [in examination])
5. Mariette, N.: A Novel Sound Localization Experiment for Mobile Audio Augmented Reality Applications. in: al, Z.P.e. (ed.) 16th Int. Conf. on Artificial Reality and Tele-existence. pp. 132–142, Springer, Berlin/Heidelberg, Hangzhou, China (2006)
6. Mariette, N.: Mitigation of binaural front-back confusions by body motion in audio augmented reality. in: Int. Conf. on Auditory Display, Montreal, Canada (June 26-29 2007)
7. McDonald, J.: Data Transformations (2007), `http://udel.edu/~mcdonald/stattransform.html`
8. Miller, J.D., Anderson, M.R., Wenzel, E.M., McClain, B.U.: Latency measurement of a real-time virtual acoustic environment rendering system. in: Int. Conf. on Auditory Display, Boston, MA, USA (6-9 July 2003)
9. Puckette, M.: Pure Data. in: Int. Computer Music Conf. pp. 269–272, Int. Computer Music Association, San Francisco (1996)
10. Pulkki, V.: Virtual Sound Source Positioning Using Vector Base Amplitude Panning. J. Audio Eng. Soc. 46(6), 456–466 (1997)

11. Strauss, H., Buchholz, J.: Comparison of Virtual Sound Source Positioning with Amplitude Panning and Ambisonic Reproduction. in: The 137th regular Meeting of the Acoustical Society of America (1999)
12. Walker, B.N., Lindsay, J.: Auditory Navigation Performance is Affected by Waypoint Capture Radius. in: The 10th Int. Conf. on Auditory Display, Sydney, Australia (July 6-9 2004)
13. Weisstein, E.W.: Curvature (2007), `http://mathworld.wolfram.com/Curvature.html`
14. Wright, M., Freed, A.: Open Sound Control: A New Protocol for Communicating with Sound Synthesizers. in: Int. Computer Music Conf., Thessaloniki, Greece (September 25-30 1997)

Evaluating the Utility of Auditory Perspective-Taking in Robot Speech Presentations

Derek Brock, Brian McClimens, Christina Wasylyshyn, J. Gregory Trafton, and Malcolm McCurry

Navy Center for Applied Research in Artificial Intelligence, Naval Research Laboratory, 4555 Overlook Ave., S.W. Washington, DC, 20375 USA
{Derek.Brock,Brian.McClimens,Christina.Wasylyshyn, Gregory.Trafton,Malcolm.McCurry}@nrl.navy.mil

Abstract. In speech interactions, people routinely reason about each other's auditory perspective and change their manner of speaking accordingly, by adjusting their voice to overcome noise or distance, or by pausing for especially loud sounds and resuming when conditions are more favorable for the listener. In this paper we report the findings of a listening study motivated both by this observation and a prototype auditory interface for a mobile robot that monitors the aural parameters of its environment and infers its user's listening requirements. The results provide significant empirical evidence of the utility of simulated auditory perspective taking and the inferred use of loudness and/or pauses to overcome the potential of ambient noise to mask synthetic speech.

Keywords: auditory perspective-taking, adaptive auditory display, synthetic speech, human-robot interaction, auditory interaction, listening performance.

1 Introduction

The identification and application of human factors that promote utility and usability is an overarching concern in the design of auditory displays [1]. The importance of this tenet is especially relevant for robotic platforms that are intended to be actors in social settings. People naturally want to interact with robots in ways that are already familiar to them, and aural communication is arguably the medium that many would expect to be the most intuitive and efficient for this purpose.

Implementing an auditory interface for a robot requires the integration of complementary machine audition and auditory display systems. These are ideally multifaceted functions and consequently pose a variety of interdisciplinary challenges for roboticists and researchers with related concerns. Audition, for instance, requires not only an effective scheme for raw listening, but also signal processing and analysis stages that can organize and extract various kinds of information from the auditory input. Important tasks for a robot's listening system include speech recognition and understanding, source location, and ultimately, a range of auditory scene analysis skills. The auditory display system, in contrast, should be capable of presenting

S. Ystad et al. (Eds.): CMMR/ICAD 2009, LNCS 5954, pp. 266–286, 2010.

speech and any other sounds that are called for by the robot's specific application. To support aurally based interactions with users and the environment—and thus be useful for more than just simplistic displays of information in auditory form—these complementary listening and display systems must be informed by each other (as well as by other systems) and coordinated by an agent function designed to implement the robot's auditory interaction goals.

In practice, the current ability of robots to flexibly exercise interactive behaviors informed by the interpretation and production of sound-based information remains far behind the broad and mostly transparent skills of human beings. The computational challenges of auditory scene analysis and certain aspects of natural language dialogue are two of the primary reasons for this, but it is surprising that little attention has been given to some of the practical kinds of situational reasoning robots will need for successful auditory interactions in everyday, sound-rich environments.

For example, in speech and auditory interactions with each other, people typically account for factors that affect how well they can be heard from their listener's point of view and modify their presentations accordingly. In effect, they reason about their addressee's auditory perspective, and in most situations, their exercise of this skill markedly improves communication and reduces shared interactional effort. Talkers learn from experience that an addressee's ability to successfully hear speech and other sorts of sound information depends on a range of factors—some personal and others contextual. They form an idea of what a listener can easily hear and usually try not to adjust their manner of speaking much beyond what is needed to be effective. Certainly, one of the most common accommodations talkers make is to raise or lower their voice in response to ambient noise or to compensate for distance or changes in a listener's proximity. If an ambient source of noise becomes too loud, talkers will often enunciate their words more carefully or move closer to their listener or pause until the noise abates, and then will sometimes repeat or rephrase what they were saying just before they stopped.

All together, these observations show that achieving effectiveness in aural interactions often involves more than just presenting and listening. Given this perspective, it is not difficult to imagine that people are likely to find speech and other forms of auditory information an unreliable medium for human-robot interaction if the robot is unable to sense and compensate for routine difficulties in aural communication. Listeners count on talkers to appreciate their needs when circumstances undermine their ability to hear what is being said. And if this expectation is not met, they must redouble their listening effort, or ask talkers to speak louder, and so on. Giving an auditory user interface the ability to diagnose and adapt to its listeners needs, then, is a practical imperative if the underlying platform is targeted for social roles in everyday environments or noisy operational settings.

Motivated by this insight, the first author and a colleague recently demonstrated a prototype computational auditory perspective-taking scheme for a mobile robot that monitors both its user's proximity and the status of the auditory scene, and inferentially alters the level and/or progress of its speech to accommodate its user's listening needs [2]. The hardware and software framework for this system is primarily a proof of concept rather than a full solution. In particular, system parameters must be tuned for specific environments and there is limited integration with non-auditory sensors and functions that can play important roles in sound-related behaviors. The prototype's conduct involving auditory perspective taking is demonstrated in the context of an interactive auditory display that might be used as a mobile information kiosk in a lobby or in a museum or exhibit hall where groups of people and other sources of noise are expected to be present on an intermittent but frequent basis (cf. [3]). Speech-based user interactions are limited to a few fixed phrases, and the auditory display is essentially a text-to-speech system that reads selected paragraphs of information with a synthetic voice. The system develops a map of auditory sources in its immediate surroundings, detects and localizes its user's voice, faces and follows the user visually, and monitors the user's proximity and the varying levels of ambient noise at its location. It then judges how loudly it needs to speak to be easily heard, pauses if necessary, and can even propose moving to a quieter location. Further details about the system and its implementation are described in [4], and a more thorough development of the idea of auditory perspective taking is given in [5].

Although casual experience with this prototype interface in demonstration runs confirms that it functions as intended, it is nevertheless important to formally show whether adaptive auditory display techniques in human-robot interaction, or in other user interaction paradigms, can, in fact, meet listeners' expectations and improve their listening performance in difficult auditory situations. This paper describes the objectives and method of an initial empirical evaluation of this question and presents the findings of the resulting experiment. Additionally, implications for the design of auditory interfaces for robotic platforms and future adaptive auditory display research are discussed.

A study of the effectiveness of an automated auditory perspective-taking scheme could be approached in a number of ways, the most obvious being an *in situ* evaluation. Consideration of both the number of interrelated parameters used by the system outlined in [4] and the range of its adaptive actions, however, argued here for the design of a smaller, more constrained initial experiment. Moreover, it was recognized that the system's key auditory behaviors, namely, its ability to make changes in the level and progress of presented speech, were essentially the most important actions to evaluate in terms of usability and impact on users' listening performance. Consequently, several of the interactions the prototype addresses were not incorporated in the present study, particularly, changes in listener proximity (cf. [6]) and the role and utility of speech-based user controls.

Focusing solely on the utility of changes in auditory level and the use of pauses made it unnecessary to employ the robotic implementation in the experiment. All of the sound materials and adaptive actions could be simulated in a studio setting where participants could comfortably perform the response tasks used to measure their listening performance while seated. Similarly, to avoid the artificial manipulation and seemingly arbitrary selection of one set of noisy real-world environments over another (e.g., urban traffic, factory floor, busy theatre lobby, stadium crowd, etc.), a small number of broadband noise types was used for maskers.

Last, the expository materials and techniques employed here to measure participants' listening performance were adapted from, and are largely the same as, those developed by the authors for a previous but unrelated study involving a somewhat similar set of issues [7]. Here, though, the spoken information used in the earlier study—short segments of public radio commentaries—has been converted to "robot" speech with a commercial speech-to-text engine. Synthetic voices (of both genders) are now in relatively wide use, but they are substantially varied and are known to be more difficult for listeners to process than natural speech (see e.g., [8][9]). Hence, to remove voice type as a factor, a single, "standard" synthetic male voice was used for all of the information presented to listeners.

2 Method and Apparatus

Fourteen participants, five female and nine male, all personnel at the authors' institution, and all claiming to have normal hearing, took part in the experiment. A within-subjects design was employed. The timing and display of all sounds and response materials were coordinated by software, coded in Java by one of the authors, running on a laboratory PC. The auditory component was rendered with three Yamaha MSP5 powered studio monitors placed directly left, right, and in front of the listener, all at a distance of approximately 1.32 m (this layout is shown below in Fig. 1). Sound was limited to a maximum of 85 dB SPL. The response tasks were presented visually on a 0.61m (diagonal) Samsung SyncMaster 243T flat-panel monitor.

2.1 Listening Materials and Experimental Manipulations

A corpus of expository materials derived from an archive of editorials on topics of general interest originally broadcast on public radio was developed for the study. Ten commentaries were transcribed, and in some cases edited for length, and then re-recorded as synthesized "robot" speech using the Cepstral text-to-speech engine [10] and a standard male synthetic voice named "Dave." The resulting audio materials were randomly assigned to three training sessions, which allowed participants to become familiar with the listening and response tasks, and to seven formal listening exercises that made up the body of the experiment. The assignments were the same for all listeners. Additionally, a test for uniformity among the commentaries assigned to the listening exercises showed no significant differences between a number of lexical parameters (number of sentences, words, and syllables, etc.). The training sessions were each about a minute in length and the listening exercises lasted between 2.5 and 3.5 minutes, depending on the particular manipulation (see below).

Most real-world noise environments have notably different and variable time-frequency characteristics, which in turn make their effectiveness as maskers difficult to systematize in a controlled experiment. To avoid this potential confound, four types of broadband noise were selected to simulate the occurrence of ambient, potentially speech-masking noise events in the study: brown noise (used only in the training sessions), and white, pink, and "Fastl" noise [11]. The latter of these is white noise filtered and modulated to simulate the average spectral distribution and fluctuating temporal envelope of an individual's speech. A digital audio editing tool was used to normalize, and create a matrix of four masking events for, each kind of noise. For white, pink, and Fastl noise, two short events (5 sec.)—one "forte" (-26 dB) and the other "fortissimo" (-19 dB)—and two long events (30 sec.) differing in loudness in the same manner were created. (The terms "forte" (loud) and "fortissimo" (very loud) are used here for expository purposes and are borrowed from the lexicon of musical dynamics for indicating relative loudness in performance.) Onset and offset ramps were linear fades lasting 0.51 sec. for short events and 7.56 sec. for long events. Six of the experimental conditions featured noise, and listeners heard each of the four kinds of masking events in these manipulations twice in random order (for a total of eight events). A slightly different matrix of brown noise events was created for use in two of the training sessions.

Design. The experiment's scheme of manipulations involved a Baseline listening exercise and a two factor, two-level by three-level (2x3) design with repeated measures. Participants heard the Baseline manipulation first and then each of the remaining six manipulations in counterbalanced order. In the Baseline condition, participants simply listened to one of the commentaries and carried out the associated response tasks (see Section 2.2 below). In the other six conditions, they performed equivalent listening and response tasks with the addition of eight intermittent noise events. As is shown in Fig. 1, the commentaries were rendered by the audio monitor in front of the listener, and instances of broadband noise were rendered by the monitors on the listener's left and right.

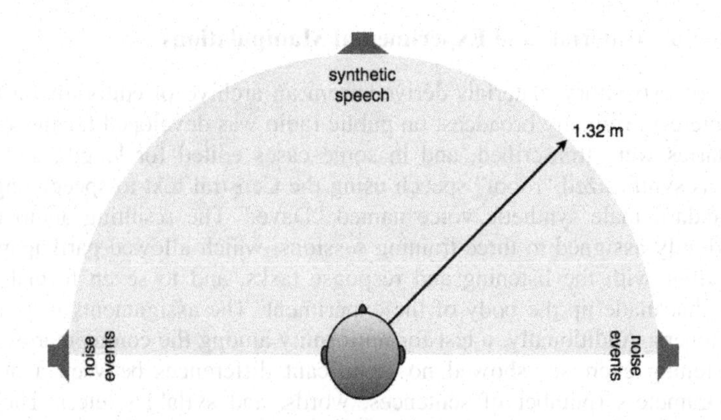

Fig. 1. Layout of the experimental listening environment, showing the location and purpose of each audio monitor and its distance from the listener

The main intent of the experiment was to evaluate the combined utility of automated pauses and level changes when ambient noise with the potential to mask the auditory display arises; interest in the comparative effects of one type of noise versus another was secondary. Accordingly, the two levels of the first factor in the design of the non-baseline manipulations entailed the respective non-use and use of the automated adaptive presentation strategies, and the second factor (three levels) involved the respective use of white, pink, and Fastl noise events. Consistent with this focus on the utility of adaptivity, the three training sessions highlighted only the first factor by introducing abbreviated versions of the "baseline" manipulation and then the contrast between "non-adaptive" and "adaptive" presentations of synthetic speech during episodes of brown noise rather than any of the three types of noise used in the formal study. A summary of the seven listening exercises participants carried out in the body of the experiment is given in Table 1.

Table 1. A summary of the seven experimental conditions and their coded designations. Participants heard all seven conditions in counter-balanced order.

Condition	Description
Baseline	**Baseline** synthetic speech, no noise events
NA-white	**Non-adaptive** synthetic speech and **white** noise events
NA-pink	**Non-adaptive** synthetic speech and **pink** noise events
NA-Fastl	**Non-adaptive** synthetic speech and **Fastl** noise events
A-white	**Adaptive** synthetic speech and **white** noise events
A-pink	**Adaptive** synthetic speech and **pink** noise events
A-Fastl	**Adaptive** synthetic speech and **Fastl** noise events

Predictions and planned comparisons. The seven conditions chosen for the study were motivated by a specific set of anticipated outcomes. First, it was expected that measures of listening performance (see Section 2.2) in the Baseline condition would be the best in the study, but would fail to approach perfect performance due to the use of a synthetic voice. In contrast, listening performance in the three Non-Adaptive conditions (those in which broadband noise events were allowed to mask portions of the spoken commentary: NA-white, NA-pink, and NA-Fastl) was expected to be poorest in the study, both collectively and individually. More importantly, and the focus of the experiment, listening performance in the three Adaptive auditory display conditions (A-white, A-pink, and A-Fastl) was expected to be nearly as good as the Baseline and substantially better than in the non-adaptive conditions.

Since the prototype auditory perspective-taking system makes no distinction between one type of noise and another, and only tries to infer listening needs on the basis of amplitude, it was unclear how each of the broadband noise manipulations would affect participants' comparative performance, particularly in the three Adaptive conditions. White noise and pink noise are both continuous at a given volume and are both effective auditory maskers. But white noise, with equal energy in all frequencies, is the more comprehensive masker of the two and, for many individuals, it may also be the more attentionally and cognitively disruptive under any circumstances, but especially when it is very loud. Fastl noise, on the other hand, because of the shape of its underlying spectral power density and fluctuating amplitude envelope, provides the least comprehensive coverage as a masker. However, if auditory cognition is perceptually tuned to attend to voices, Fastl noise may be more distracting than either white or pink noise due to its speech-like properties. Nevertheless, all three types of noise should be good maskers of speech. Because of these qualified differences and the difficulty of predicting how broadband noise events may interact with auditory concentration in various circumstances, planned comparisons (contrasts) are used below to evaluate how performance in the two presentation strategy manipulations differ from performance in the Baseline condition across the three manipulations of noise-type.

Adaptive auditory display behaviors. To approximate the prototype auditory perspective-taking system's response to different levels of ambient noise in the Adaptive auditory display manipulations (i.e., A-white, A-pink, and A-Fastl), the three commentaries respectively assigned to these conditions were modified in the following ways. First, as in the Non-Adaptive conditions, they were appropriately aligned with noise events on separate tracks in a sound editor. Next, using linear onset and offset ramps, the amplitude envelope of each commentary was modulated (increased by 6 dB) to compete in parallel with the eight randomly ordered noise events in its particular manipulation. The resulting modulations were then appropriately delayed (i.e., shifted forward in running time) to simulate the time it takes for the onset of a noise event to cross the system's response threshold. Thus for noise events at the forte level (-26 dB), the synthetic speech starts to become louder 1.0 sec. after the short event begins and 3.0 sec. after the long one begins, the difference being due to the more gradual onset ramp of long events (see Section 2.1 above). The short and long episodes of noise at the fortissimo level (-19.dB) have correspondingly steeper onset ramps, so the response for these events begins at 0.8 and 2.0 sec., respectively. Fortissimo events, though, are intended to trigger the prototype's pause response. To mimic this effect, corresponding periods of silence were inserted in the commentaries with the sound editor (thus increasing their length). During short episodes of fortissimo noise, pauses begin at the first word boundary following 1.2 sec. of the loudness response; during long episodes they begin similarly at or beyond the 5.0 sec. mark. The commentaries were then edited to resume at the point where the noise event drops below the pause threshold by re-uttering the interrupted sentence or phrase. Long pauses, however, first resume with the words, "As I was saying…" The idea of resuming interrupted synthetic speech in this manner arose during the development of the prototype and was found to be consistent with listeners' intuitions about verbal pauses in piloting for the study.

To summarize, eight noise events (two of each of the four kinds outlined in Section 2.1) occurred in each of the three Adaptive conditions, and the auditory display took the following actions to overcome the potential for its presentation of synthetic speech to be masked from its listener's perspective. When the respective short- and long-forte events occurred, the level of the speech rose by 6 dB to be easy to hear over the level of the noise and then fell to its previous level as the noise abated. When the respective short- and long-fortissimo events occurred, the speech became louder to a point and then paused. After the noise abated, the auditory display resumed from the beginning of the phrase or sentence it interrupted, but in the case of the long-fortissimo event prefaced its resumption with the words, "As I was saying." A schematic of the auditory display's four adaptive behaviors showing level changes and pauses is given in Fig. 2.

Auditory Examples. Edited examples of the sound materials used in the study are given in the binaural recordings listed below, which are available by email from the first author as .wav or .mp3 files. NADAPT presents an instance of each of the four noise event types in the Non-Adaptive manipulations: long-fortissimo/NA-white, long-forte/NA-pink, short-fortissimo/NA-Fastl, and short-forte/NA-Fastl. ADAPT presents an instance of each of the four noise event types in the Adaptive manipulations: long-fortissimo/A-white, long-forte/A-pink, short-fortissimo/A-Fastl, and short-forte/A-Fastl.

NADAPT: example speech and noise events in Non-Adaptive conditions
ADAPT: example speech and noise events in Adaptive conditions

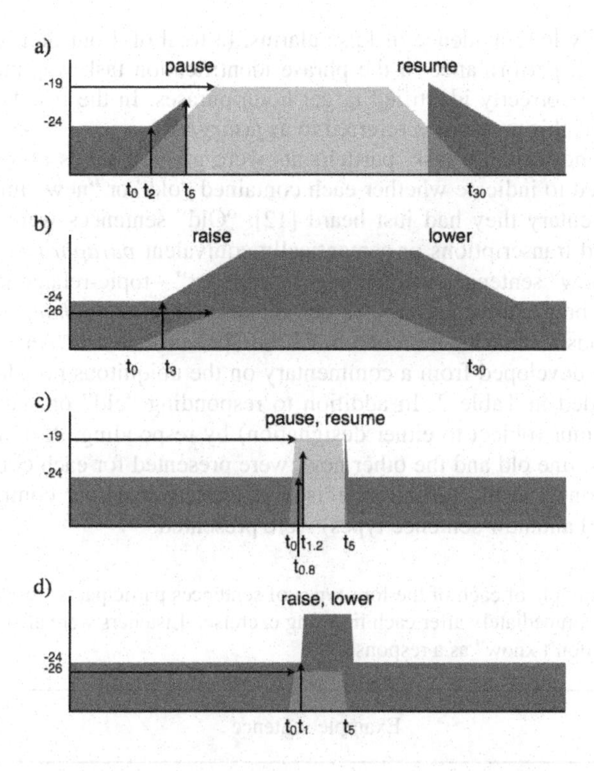

Fig. 2. Schematic diagrams showing actions taken by the auditory display in the experiment's Adaptive conditions to counter noise events with the potential to mask speech from the listener's perspective: (a) long-fortissimo, (b) long-forte, (c) short-fortissimo, and (d) short-forte. Time in seconds is shown on the horizontal axis, and level in dB is shown on the vertical axis. Noise event envelopes are shown as semi-transparent light gray trapezoids. Envelopes of continuous speech are shown in dark gray. See the text for additional details.

2.2 Response Tasks and Dependent Measures

In both the training sessions and the listening exercises, participants carried out two response tasks, one while listening and the other immediately after. After each training session and listening exercise, participants were also asked to rate their preference for the way the synthetic speech was presented.

The first response task involved listening for noun phrases in the spoken material and marking them off in an onscreen list. Each list contained both the targeted noun phrases and foils in equal numbers (eight targets in each of the training sessions and twenty targets per manipulation in the formal listening exercises). Targets were listed in the order of their aural occurrence and were randomly interleaved with no more than three intervening foils; foils were selected from commentaries on similar but not identical topics.

Participants proved to be quite good at discriminating between target phrases and foils on the basis of the speech materials, and only rarely mistook foils for utterances in any of the commentaries, regardless of their ability to verify targets. Thus, because of an extremely low incidence of false alarms, (a total of 4 out of 1960 possible correct rejections), performance in the phrase identification task was measured only as the percentage correctly identified target noun phrases. In the results and discussion sections below, this measure is referred to as *p*(*targets*).

In the second response task, participants were given a series of sentences to read and were asked to indicate whether each contained "old" or "new" information based on the commentary they had just heard [12]. "Old" sentences were either *original*, word-for-word transcriptions or semantically equivalent *paraphrases* of commentary sentences. "New" sentences were either "*distractors*"—topic-related sentences asserting novel or bogus information—or commentary sentences *changed to make their meaning* inconsistent with the content of the spoken material. An example of each sentence type developed from a commentary on the ubiquitous popularity of baseball caps is provided in Table 2. In addition to responding "old" or "new," participants could also demur (object to either designation) by responding, "I don't know." Only two sentences, one old and the other new, were presented for each commentary in the training sessions. In the formal exercises, eight sentences per commentary (two of each of the old and new sentence types) were presented.

Table 2. An example of each of the four types of sentences participants were asked to judge as "old" or "new" immediately after each listening exercise. Listeners were also allowed to demur by selecting "I don't know" as a response.

Sentence type	Example sentence	Designation
Original	Baseball caps are now bigger than baseball.	Old
Paraphrase	Baseball caps have become more popular than the game of baseball.	Old
Meaning change	Baseball caps are now bigger than football.	New
Distractor	Most baseball caps are now made in China.	New

Two measures were calculated from the participants' sentence judgments in each condition. The primary measure, denoted *p*(*sentences*), is the proportion of sentences correctly judged as old or new. The second measure, denoted *p*(*demurs*), is the proportion of "I don't know" responses. Both measures are calculated as a percentage of the eight sentences presented for verification in each condition.

Last, to gage participants' subjective impressions, after completing the sentence judgment task in the training sessions and in each of the experimental conditions, they were asked to rate their preference for the auditory display. They did this by indicating their agreement with the statement, "I prefer the way the synthetic speech was presented in this listening exercise," on a seven point Likert scale, with 1 = "strongly disagree" and 7 = "strongly agree."

3 Results

The performance measures for both response tasks were largely consistent with the pattern of listening performance that was expected to arise between the noise-free Baseline condition and the six conditions featuring noise events and the non-use or use of the adaptive auditory display. In particular, participants' abilities to correctly recognize targeted noun phrases, *p*(*targets*), and judge sentences as old or new, *p*(*sentences*), were both highest in the Baseline condition and lowest in the three Non-Adaptive conditions (NA-white, NA-pink, and NA-Fastl). Mean scores for the target phrase recognition response task were only slightly lower than Baseline in the three Adaptive conditions (A-white, A-pink, and A-Fastl), as predicted. Scores for the sentence judgment task in the Adaptive conditions, however, were not as high as expected (see Section 2.1.2), and fell in a more intermediate position between the scores for the respective Baseline and Non-Adaptive conditions. Even so, the correlation between *p*(*targets*) and *p*(*sentences*) is significant (Pearson's $r = 0.573$, $p = 0.05$ (2-tailed)). Plots of the mean proportions of correctly identified target noun phrases, *p*(*targets*), and sentences correctly judged as "old" or "new," *p*(*sentences*), in all seven conditions are respectively shown in Fig. 3 and Fig. 4.

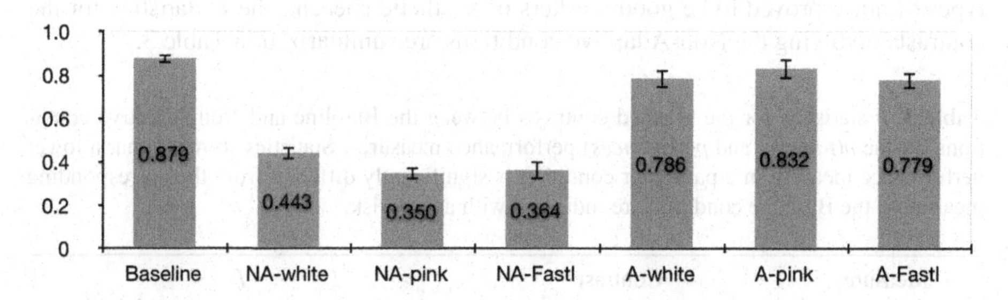

Fig. 3. Plot of the mean proportion of correctly identified target noun phrases, p(targets), in each condition. The y-axis shows proportion. Error bars show the standard error of the mean.

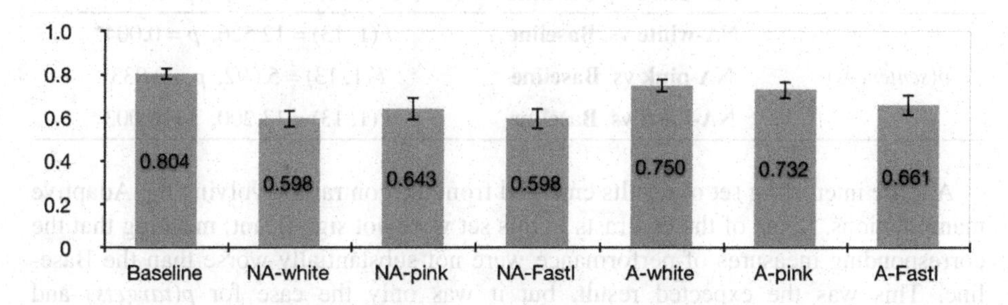

Fig. 4. Plot of the mean proportion of sentences correctly judged as "old" or "new," p(sentences), in each condition. The y-axis shows proportion. Error bars show the standard error of the mean.

To evaluate effects of presentation strategy—non-adaptive vs. adaptive—and the three types of noise on listening performance, the six conditions involving noise events were construed as a factorial design, and a two-level by three-level, repeated measures analysis of variance was performed for each of the dependent measures. In these analyses, there was a main effect for presentation strategy but not for noise type. Specifically, the 2x3 ANOVA for *p*(*targets*) showed that participants were significantly better at the target phrase task when Adaptive presentations were used to counter noise events ($F(1, 13) = 190.7, p < 0.001$). The corresponding ANOVA for *p*(*sentences*) showed, similarly, that performance of the sentence judgment task was significantly better in the conditions involving Adaptive presentations ($F(1, 13) = 5.077, p = 0.042$). Additionally, there was a significant interaction between presentation strategy and noise type in the analysis for *p*(*targets*) ($F(2, 26) = 4.518, p = 0.021$), but not in the analysis for *p*(*sentences*).

Because it was unclear how each type of noise might impact listening performance when the respective Non-Adaptive and Adaptive auditory display strategies were used, planned contrasts were carried out to evaluate how the dependent measures in these manipulations differed with performance in the Baseline condition. All of these comparisons involving the Non-Adaptive manipulations were significant, meaning that both performance measures, *p*(*targets*) and *p*(*sentences*), were meaningfully hurt by the noise events in these conditions. In other words, as was expected, all three types of noise proved to be good maskers of synthetic speech. The *F* statistics for the contrasts involving the Non-Adaptive conditions are summarized in Table 3.

Table 3. *F* statistics for the planned contrasts between the Baseline and Non-Adaptive conditions for the *p*(*targets*) and *p*(*sentences*) performance measures. Statistics showing that a lower performance measure in a particular condition is significantly different from the corresponding measure in the Baseline condition are indicated with an asterisk.

Measure	Contrast	*F*
p(*targets*)	**NA-white** vs. **Baseline**	$F(1, 13) = 200.718, p < 0.001$*
	NA-pink vs. **Baseline**	$F(1, 13) = 354.169, p < 0.001$*
	NA- Fastl vs. **Baseline**	$F(1, 13) = 232.386, p < 0.001$*
p(*sentences*)	**NA-white** vs. **Baseline**	$F(1, 13) = 12.526, p = 0.004$*
	NA-pink vs. **Baseline**	$F(1, 13) = 5.692, p = 0.033$*
	NA-Fastl vs. **Baseline**	$F(1, 13) = 13.200, p = 0.003$*

A more interesting set of results emerged from the contrasts involving the Adaptive manipulations. Some of the contrasts in this set were not significant, meaning that the corresponding measures of performance were not substantially worse than the Baseline. This was the expected result, but it was only the case for *p*(*targets*) and *p*(*sentences*) with pink noise events and for *p*(*sentences*) with white noise events. The other three contrasts were all significant: in spite of the Adaptive presentation strategies, both white noise and Fastl noise had a meaningful impact on listeners' ability to perform the target phrase recognition task, and Fastl noise significantly hurt their corresponding ability to perform the sentence judgment task. The F statistics for the contrasts involving Adaptive speech are summarized in Table 4.

Table 4. F statistics for the planned contrasts between the Baseline and Adaptive conditions for the *p*(*targets*) and *p*(*sentences*) performance measures. Statistics showing that a lower performance measure in a particular condition is significantly different from the corresponding measure in the Baseline condition are indicated with an asterisk.

Measure	Contrast	F
	A-white vs. **Baseline**	$F(1, 13) = 10.876, p = 0.006$*
p(*targets*)	**A-pink** vs. **Baseline**	$F(1, 13) = 1.441,\ p = 0.251$
	A- Fastl vs. **Baseline**	$F(1, 13) = 7.280,\ p = 0.018$*
	A-white vs. **Baseline**	$F(1, 13) = 1.918,\ p = 0.189$
p(*sentences*)	**A-pink** vs. **Baseline**	$F(1, 13) = 2.537,\ p = 0.135$
	A-Fastl vs. **Baseline**	$F(1, 13) = 5.438,\ p = 0.036$*

Since the study, as it was primarily conceived, can be characterized as a test of the merit of adaptively improving the signal-to-noise ratio (SNR) of an auditory display, the outcome of these latter contrasts raises an important concern about the underlying uniformity of the experimental treatments. In particular, because of the fact that the auditory signals corresponding to both of the primary response measures—target noun phrases and sentences—were synthetic speech, it was not feasible to tightly equate the SNRs of the stimuli across the manipulations involving noise events. Noun phrases, such as "hammy lines," "hour-long interview," and "the whole point," for example, have different signal powers due to their differing lengths and differing patterns of phonemes, and the same can be said for each of the sentences that formed the basis of the sentence judgment task. A fair degree of preparatory attention was given to this matter, but as can be seen by examining the dark bars in Figs. 5 and 6, this key dimension of the stimuli was not distributed in a strictly uniform manner. (The underlying SNRs in these figures were calculated with a noise floor of -96 dB; the plots show the mean SNR of signals occurring both with and without noise that required a response in each condition.) To determine whether or not there were significant differences between the stimulus SNRs in each of the six conditions involving noise events, a 2x3 ANOVA (two presentation strategies by three types of noise) was carried out for a) the SNRs of the target phrases and b) the SNRs of the commentary sentences that were selected for listeners to evaluate as "old" or "new." As expected, the SNRs in the adaptive presentations were higher than in the non-adaptive presentations, (targets: $F(1, 114) = 21.236$, $p < 0.001$; sentences: $F(1, 30) = 7.44$, $p =$ .011. Note that this was by design. Critically, there were no differences in the SNRs between noise types (targets: $F(2, 114) < 1$, n.s.; sentences: $F(2, 30) < 1$, n.s.) and no interactions between the two factors (presentation and noise) (targets: $F(2, 114) < 1$, n.s.; sentences: $F(2, 30) < 1$, n.s.).

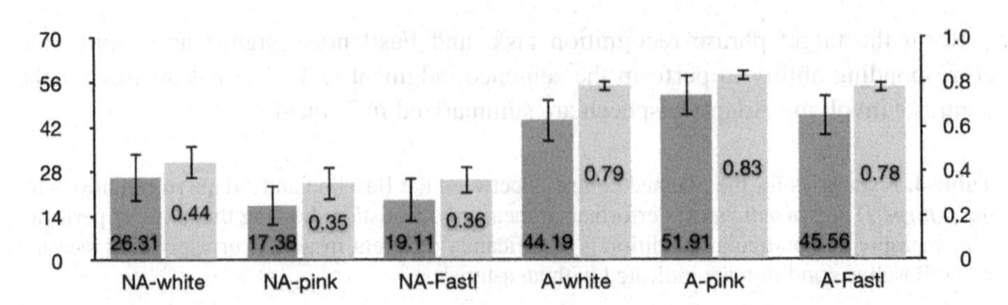

Fig. 5. Comparative plots of a) the mean signal-to-noise ratio (SNR) of the 20 target noun phrases in each of the manipulations involving noise events (dark bars) and b) the mean proportion of correctly identified target noun phrases (*p*(*targets*), light bars) in each condition. The y-axis on the left shows SNR in dB; the y-axis on the right shows proportion. Error bars show the standard error of the mean.

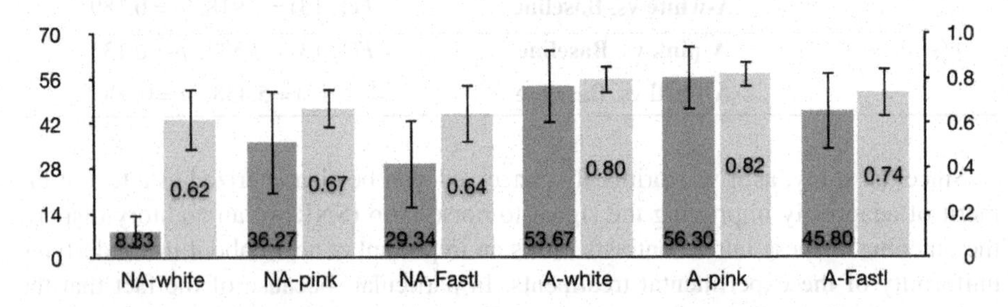

Fig. 6. Comparative plots of a) the mean signal-to-noise ratio (SNR) of the six sentences that corresponded to judgments entailing "original" or "paraphrased" content or a "meaning change" in each of the manipulations involving noise events (dark bars) and b) proportion of just these six sentences correctly judged as "old" or "new," (different from p(sentences), light bars) in each condition. (Note that the "distractor" sentences participants were asked to judge as old or new were entirely made up and did not correspond to specific commentary sentences.) The y-axis on the left shows SNR in dB; the y-axis on the right shows proportion. Error bars show the standard error of the mean.

Also plotted in Figs. 5 and 6, however, are the comparative mean proportions of correct responses for each type of stimuli (light bars). In Fig. 5, these are simply the values of *p*(*targets*) reported in Fig. 3. The mean proportional scores shown in Fig. 6 differ from *p*(*sentences*) in that they reflect only judgments involving sentences that were actually presented and potentially heard. (Only six of the eight sentences participants were asked to judge in each condition—specifically, those that were either "original" (verbatim), a "paraphrase," or that involved a conspicuous "meaning change"—meet this criterion (see Table 2). The other two sentences—the "distractors"—were topically-related fabrications that were not present in the commentaries; participants were expected to recognize that they had not heard distractors and, thus, mark them as "new.") A consistent correspondence between the response measures and the SNRs in these plots is readily apparent and the respective correlations are significant (targets: $R^2(118) = 0.191, p < 0.001$); sentences: $R^2(34) = 0.31, p < 0.001$).

Overall, then, 19% of the variability in *p(targets)* and, in Fig. 6, 31% of the variability in the proportion of correct, stimuli-based sentence judgments can be explained by the corresponding SNRs of the stimuli, and these correlations should be taken into account in the material below assessing differences in listening performance relative to the three types of noise used in the study.

In addition to the primary performance measures, an additional measure was associated with the sentence judgment response task, specifically, the proportion of "I don't know" responses participants made in each condition, denoted *p(demurs)*. Giving participants the option to make this response allowed them to indicate they felt they had no basis to judge a particular sentence as old or new information. Intuitively, a greater percentage of demurs should be expected in the Non-Adaptive manipulations because of the masking effects of noise. This proved to be the case, and a plot of the overall mean proportion of demurs in all seven experimental conditions, shown in Fig. 7 (columns with error bars), exhibits, inversely, the same broad pattern as that seen for both *p(targets)* and *p(sentences)* in Figs. 2 and 3. A 2x3 ANOVA of the six non-Baseline conditions for *p(demurs)*, however, showed no main effect for either factor and no interaction. Out of the six planned contrasts, only NA-Fastl vs. Baseline was significant ($F(1, 13) = 7.495$, $p = 0.017$), meaning that the number of demurs in each of the other five conditions was not meaningfully greater than in the Baseline condition.

Also shown in Fig. 7 are the corresponding counts of participants in each condition who chose to demur one or more times (inverted triangles), and the mean proportion of demurs made by just these individuals (square bracketed values; the lower bound of this proportion is 1/8 or 0.125). Not surprisingly, both of these series are significantly correlated with the overall mean *p(demurs)* values (respectively, Pearson's $r = 0.864$, $p < 0.02$ (2-tailed) and Pearson's $r = 0.851$, $p < 0.02$ (2-tailed)). These augmenting data are given to provide additional perspective on the nature of the comprehension task as a measure of human listening performance. In particular, the values in the more aurally favorable Baseline and Adaptive conditions show that a number of listeners (though less than half) did not recall what was said well enough, to various degrees, to be confident in their judgment of sentences as old or new information. In contrast, the corresponding numbers in the Non-Adaptive conditions reveal that several listeners (though again, less than half) chose not to demur in spite of noise events that masked sentences they were asked to judge.

Finally, a plot of the listeners' mean level of subjective agreement in each condition with the statement, "I prefer the way the synthetic speech was presented in this listening exercise," is shown in Fig. 8. As mentioned above, the range of this data corresponds to a seven point Likert scale. The resulting pattern of ratings across manipulations is somewhat similar to the patterns seen in Figs. 3 and 4. However, there is an interesting difference here in that while participants' mean preference for the Baseline presentation is greater than their preference for any of the Non-Adaptive presentations, it is not greater than their preference for any of the Adaptive presentations. Planned contrasts with the Baseline condition were not significant, but a two factor ANOVA of this data in the non-Baseline conditions showed a main effect for presentation strategy ($F(1, 13) = 10.538$, $p = 0.006$).

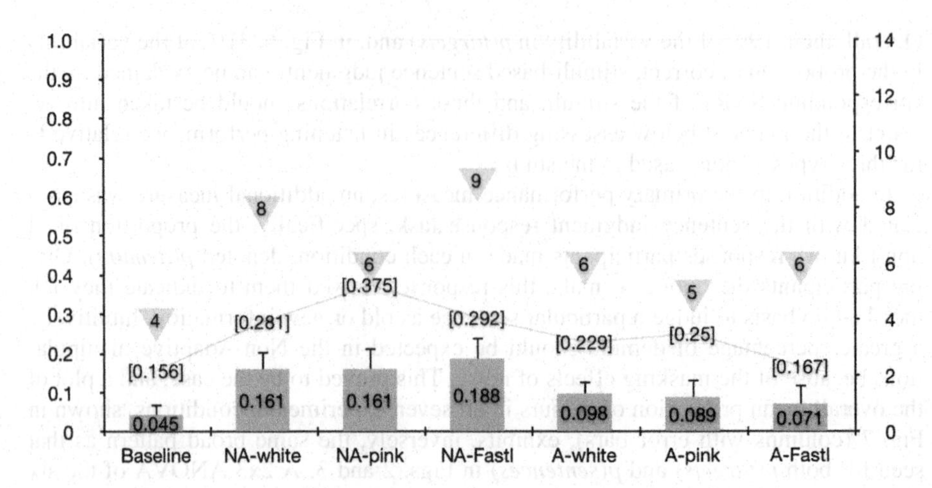

Fig. 7. Combined plot showing a) p(demurs), the mean proportion of "I don't know" responses across all listeners in each condition, (gray columns with error bars (showing standard error of the mean; negative error bars not shown to reduce clutter) corresponding to y-axis on left), b) number of listeners in each condition who demurred one or more times (inverted triangles corresponding to y-axis on right), and c) the mean proportion of "I don't know" responses in each condition made by listeners who demurred one or more times (square bracketed values corresponding to y-axis on left). See text for additional information.

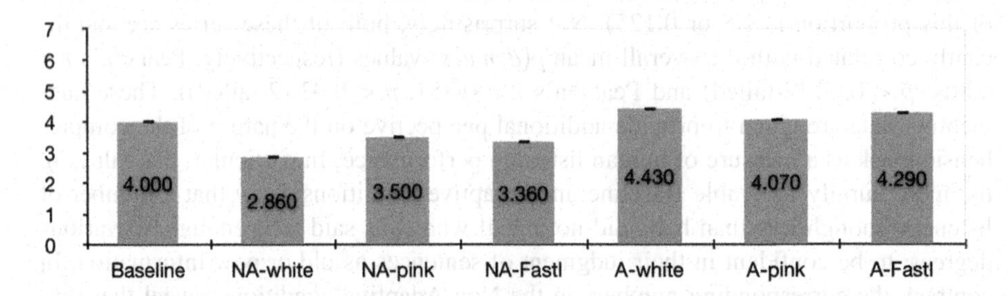

Fig. 8. Plot showing the mean level of participants' agreement with the statement, "I prefer the way the synthetic speech was presented in this listening exercise," in each condition. The y-axis in this plot reflects a seven-point Likert scale ranging from 1 = "strongly disagree" to 7 = "strongly agree." Error bars show the standard error of the mean.

4 Discussion

The chief motivation for this experiment was to evaluate the combined utility of two adaptive auditory display techniques for individual listeners in noisy settings, namely automated changes in loudness and the use of pauses. In the application context of the study—human-robot interaction involving synthetic speech—both of these flexible presentation strategies are intended to anticipate listening requirements from the user's auditory perspective and improve the overall effectiveness of his or her listening experience. To test these ideas, participants were asked to listen to seven short commentaries spoken by a synthetic voice and, for each commentary, carry out two response tasks designed to measure a) their ability to attend to the content while listening and b) the consistency of their understanding of the content afterwards. The commentaries were randomly assigned to a set of experimental conditions that elicited a noise-free baseline of listening performance and, in six additional manipulations, tested how the non-use and combined use of the two adaptive aural presentation techniques affected listening performance in the presence of eight coordinated episodes of three types of broadband noise.

Collectively, the results of the study provide significant empirical evidence of the utility of simulated auditory perspective taking and the inferred use of loudness and/or pauses to overcome the potential of noise to mask synthetic speech. In particular, while measures of listening performance aided by the adaptive techniques in the presence of noise were not as robust as listening in the absence of noise, they were demonstrably better than unaided listening in the presence of noise. Additionally, when asked, listeners indicated a significant subjective preference for the adaptive style of synthetic speech over the non-adaptive style.

Overall, this finding has implications for the design of auditory interfaces for robots and, more generally, for adaptive auditory display research, some of which will be covered below. Certain aspects of the study, however, warrant further consideration and/or critique. Among these are how Baseline performance in the study compares to listening performance involving human speech, the impact of noise type on listening performance in the Adaptive conditions, and listeners' subjective preferences.

4.1 Listening to Synthetic and Human Speech

Although listening performance in the Baseline condition, as measured by *p*(*targets*) and *p*(*sentences*), was expected to be the best in the study, it was also expected to fail to approach perfect performance due to the use of a synthetic voice. No test of this conjecture was made here, but a specific manipulation in the concurrent vs. serial talker experiment by Brock et al. in [7] offers a useful, if imperfect means for comparison.

In the cited experimental condition, a different group of participants from those in the present study listened to a serial presentation of four commentaries that were drawn from the same source as those used here. The commentaries were spoken by human talkers and were rendered with headphones at separate locations in a virtual listening space using a non-individualized head-related transfer function. During the listening exercise, the same target phrase and sentence judgment methods used in the present study were employed to measure listening performance, but all four commentaries were presented before the corresponding sentence judgment tasks were given to listeners.

The resulting mean proportion of correctly identified target phrases was 0.91, and the corresponding mean proportion of correctly judged sentences was 0.87. When these numbers are compared with their counterparts in the present Baseline condition (respectively, 0.88 and 0.80), it can be seen that listening performance, in spite of a number of experimental differences, was somewhat poorer when the information medium involved synthetic speech.

The purpose in making this rough comparison is not to claim significance, which has been shown elsewhere (e.g., [8][9]), but rather to stress the aurally anomalous properties of current synthetic voice technology and thus point to a further motivation for accommodating users' listening requirements when this technology is used in noisy settings. Canned human speech can be used for limited purposes, but there is little alternative to synthetic speech in the less constrained aural interaction models that are called for if robots are to be accepted as credible actors in social settings.

4.2 The Impact of Noise Type on Listening Performance

Although the use of three different types of broadband noise, as surrogates for real-world noise capable of masking synthetic speech, was a secondary consideration in the design of this study, its outcome, with respect to maskers, suggests that one type of noise may be more difficult to adapt for than another.

Taken together, the significant interaction between the presentation and noise factors in the *p(targets)* data and the pattern of significant performance differences among the contrasts reported in Table 4 provide a degree of evidence that some forms of noise—presumably, because of their particular characteristics (see Section 2.1.2)—can potentially undermine a listener's auditory concentration. First, note that the *p(targets)* interaction, which can be seen in both Fig. 3 and Fig. 5, arises primarily from the fact that listening performance in the NA-pink and A-pink manipulations are respectively lower and higher than listening performance in the other Non-Adaptive and Adaptive conditions. Although this pattern is correlated with the SNRs of the stimuli, the latter does not reflect an interaction and the observed variance in *p(targets)* is much smaller, particularly in the adaptive manipulations. Thus, pink noise was a good masker of synthetic speech, but it was also the best type of noise to successfully adapt for. Add to this the pattern of significant contrasts in the *p(targets)* portion of Table 4, and it is plausible that in spite of the adaptive auditory display, both white and Fastl noise impacted listeners in a way that pink noise did not.

Did these effects happen for similar reasons? Arguably not, when the contrasts in the *p(sentences)* portion of Table 4 are also taken into consideration. These contrasts show that listeners demonstrated a good understanding of the commentaries in the Adaptive manipulations involving white and pink noise events—relative to their performance in the Baseline condition—but were unable to do so in the A-Fastl condition. Given the notable descriptive differences between white and Fastl noise, the one being continuous and spectrally uniform and the other having fluctuating, speech-like properties, it would appear that competing ambient noise with speech-like qualities may be a challenging type of masker to consistently overcome.

In seeming opposition to this interpretation, though, is the pattern of *p(demurs)* data shown in Fig. 5 and the associated data for participants electing to respond in this way in the post-listening sentence judgment task. If Fastl noise does impair auditory concentration, observing a substantial number of demurs would be good supporting evidence. However, the value of *p(demurs)* in the A-Fastl condition is essentially no different from this measure in the other two Adaptive conditions. Instead, the largest number of demurs occurs in the NA-Fastl condition, and furthermore, only this contrast with the Baseline value of *p(demurs)* is significant. In conjunction with the low value of *p(sentences)* in this condition, Fastl noise is the most successful masker of non-adapting synthetic speech in the context of post listening measures.

But is this pattern in *p(demurs)* data inconsistent with the premise that some forms of noise can substantially undermine a listener's auditory concentration? If an aural masker has this additional cognitive effect (as opposed to simply overwhelming a target sound energetically), then it could be an even better masker than, say, unvarying continuous noise. Certainly, more participants in the NA-Fastl condition than in any other appear to have decided that they had a poor understanding of the commentary they had just heard, and thus responded appropriately. So in the A-Fastl condition, it may only be the case that listeners were unaware of the extent of their impaired understanding because the adaptive auditory display ensured that none of the commentary was fully masked. If this is so, then there should be a greater mean proportion of sentence judgment errors relative to the other Adaptive conditions, and this turns out to be the case (see Fig. 4). In fact, the mean proportion of sentence judgment errors in the A-Fastl condition (calculated as the remainder when *p(sentences)* and *p(demurs)* are deducted from a perfect score) is greater, at 0.268, than the corresponding proportion of errors in any of the other conditions in the study.

Thus, more so than white or pink noise, Fastl noise appears to be a good masker of synthetic speech, both during and after listening and even when adaptive changes in loudness and the use of pauses are employed. If this conclusion is right, it has implications for the design of auditory human-robot interaction in social settings because Fastl noise is taken to be a type of analog for speech noise. However, because the underlying pattern of stimulus SNRs in the study is significantly correlated with the primary performance measures (see Section 3), this result will require additional study.

4.3 Listeners' Subjective Preferences

The purpose of asking participants after each listening exercise to rate their agreement with the sentence, "I prefer the way the synthetic speech was presented in this listening exercise," was to determine, in a relatively unbiased way, how much they liked or disliked the particular auditory display they had just worked with. Ratings of this sort are inherently subjective, but can nevertheless provide useful insights and/or reveal unanticipated issues.

The mean preference data shown in Fig. 6 shows a significant main effect in favor of the Adaptive auditory display, and it also reveals a consistently greater preference for the Adaptive manipulations over the Baseline condition. The contrasts are not significant, but the trend is conspicuous and unexpected: it seems counter-intuitive that an uninterrupted presentation in the quiet would be less preferable than adaptively modified presentations accompanied by multiple noise events.

The rating for the Baseline condition, though, turns out to be exactly midway between the two ends of the Likert scale used for this measure. On balance, then, listeners seem to have been indifferent to the use of synthetic speech by itself. But two factors may have contributed to this outcome. First, this condition was always heard first by listeners in the full experiment and, second, nothing of consequence (i.e., no interruptions, etc.) occurs in this manipulation, which, after having gone through the training exercises, would make the response tasks seem relatively straightforward.

Not knowing what might be coming next and having little additional basis for expressing a preference may well be the best explanation for this neutral outcome.

In the Adaptive manipulations, on the other hand, substantial impediments to listening arise and the auditory display responds to the intruding noise effectively and with dispatch. More importantly, it does this in ways that are modeled on human solutions. Without corroborating data to specifically indicate why participants rated each manipulation as they did, it can only be speculated that their agreement with the preference statement was somewhat higher in the Adaptive conditions because noise events were stimulating and the synthetic voice acted on their listening needs transparently and in ways that met their expectations or, at the very least, facilitated their performance of the response tasks. If this interpretation is correct, it shows that simulated perspective taking in this type of auditory interaction design has important collaborative utility and merits further development.

4.4 Implications for Design and Research

The outcome of the study supports the idea that auditory interaction designs for robotic platforms can and should account for their users' listening requirements, especially in operational settings where ambient noise is likely to be an issue. This idea also extends to situations in which the proximity between the robot and its user is likely to vary with any frequency. The small but measurably different impact that Fastl noise had on listening performance in the study suggests that additional adaptive strategies such as enunciation and repair may be needed in some circumstances to cope with the distracting and informational masking effects of extraneous speech. Recognizing the need for enunciation could perhaps be informed by machine classification of the ambient noise environment. Yet, another aspect of auditory perspective-taking that will need to be addressed in future research involves inferences made on the basis of users' privacy concerns and other socially motivated considerations.

It is also possible to imagine a range of non-speech applications for robot auditory interfaces such as aural monitoring and playback and sonification of process or sensor data. Auditory displays of this sort on robots or in other formats may be even harder to use in the presence of ambient noise than speech displays precisely because of the way they represent information. Real-world noise is likely to be a good informational masker of non-speech sounds in much the same way that speech and speech-like noise can be an informational masker of speech. Ambient speech may also have masking effects on non-speech auditory displays, especially if sonifications are involved, because of the nature of their information content and the sustained auditory attention they require. Effective adaptive presentation strategies in these circumstances will require additional research and may prove to be different from the techniques evaluated here.

5 Conclusions

The notion that robots will eventually assume collaborative roles involving aural interactions in social settings has already materialized in the form of self-serve check out registers at stores, automated telephone support, and toys that talk and respond to voice commands. In the relatively near future, it is widely expected that mobile robotic platforms capable of far greater autonomy than is technically feasible today will be deployed for a wealth of interactive societal purposes ranging from service and caretaking to military and logistical applications. Soon, people will not only expect to be able to interact with robots in much the same way they interact with each other in face-to-face activities, but they will also expect these advanced systems to understand their communicative needs. The idea of auditory perspective taking—inferring what an addressee's listening requirements are on the basis of ambient sound, proximity, and, ultimately, social constraints—is just one element of this understanding, albeit an important one, that will eventually be joined with other communication skills users will expect robots and other systems to be capable of, such as gaze following, contextual awareness, and implied goal recognition. The success of the adaptive auditory display strategies evaluated in the present study confirms the importance of this emerging direction in user interface design.

Acknowledgments. This research was supported by the Office of Naval Research under work order number N0001409WX30013. The authors would like to thank Hesham Fouad for technical help with the production of the accompanying audio files.

References

1. Peres, S.C., Best, V., Brock, D., Frauenberger, C., Hermann, T., Neuhoff, J., Nickerson, L.V., Shinn-Cunningham, B., Stockman, A.: Auditory Interfaces. In: Kortum, P. (ed.) HCI Beyond the GUI, pp. 145–195. Morgan Kaufman, San Francisco (2008)
2. Brock, D., Martinson, E.: Exploring the Utility of Giving Robots Auditory Perspective-Taking Abilities. In: Proceedings of the 12th International Conference on Auditory Display (ICAD), London (2006)
3. Thrun, S., Beetz, M., Bennewitz, M., Burgard, W., Cremers, A.B., Dellaert, F., Fox, D., Hähnel, D., Rosenberg, C., Roy, N., Schulte, J., Schulz, D.: Probabilistic Algorithms and the Interactive Museum Tour-guide Robot Minerva. Intl. J. Robotics Res. 19, 972–999 (2000)
4. Martinson, E., Brock, D.: Improving Human-Robot Interaction through Adaptation to the Auditory Scene. In: HRI 2007: Proceedings of the ACM/IEEE International Conference on Human-Robot Interaction, Arlington, VA (2007)
5. Brock, D., Martinson, E.: Using The Concept of Auditory Perspective Taking to Improve Robotic Speech Presentations for Individual Human Listeners. In: AAAI 2006 Fall Symposium Technical Report: Aurally Informed Performance: Integrating Machine Listening and Auditory Presentation in Robotic Systems, Washington, DC (2006)
6. Kagami, S., Sasaki, Y., Thompson, S., Fujihara, T., Enomoto, T., Mizoguchi, H.: Loudness Measurement of Human Utterance to a Robot in Noisy Environment. In: HRI 2008: Proceedings of the 3rd ACM/IEEE International Conference on Human-Robot Interaction, Amsterdam (2008)
7. Brock, D., McClimens, B., Trafton, J.G., McCurry, M., Perzanowski, D.: Evaluating Listeners' Attention to and Comprehension of Spatialized Concurrent and Serial Talkers at Normal and a Synthetically Faster Rate of Speech. In: Proceedings of the 14th International Conference on Auditory Display (ICAD), Paris (2008)

8. Hardee, J.B., Mayhorn, C.B.: Reexamining Synthetic Speech: Intelligibility and the Effect of Age, Task, and Speech Type on Recall. In: Proceedings of the Human Factors and Egonomics Society 51st Annual Meeting, Baltimore, MD, pp. 1143–1147 (2007)
9. Stevens, C., Lees, N., Vonwiller, J., Burnham, D.: Online Experimental Methods to Evaluate Text-to-speech (TTS) Synthesis: Effects of Voice Gender and Signal Quality on Intelligibility, Naturalness, and Preference. Computer Speech and Language 19, 129–146 (2005)
10. Cepstral, http://cepstral.com
11. Fastl, H., Zwicker, E.: Psychoacoustics: Facts and Models, 3rd edn. Springer, Berlin (2007)
12. Royer, J.M., Hastings, C.N., Hook, C.: A Sentence Verification Technique for Measuring Reading Comprehension. J. Reading Behavior 11, 355–363 (1979)

Simulator Sickness in Mobile Spatial Sound Spaces

Christina Dicke[1], Viljakaisa Aaltonen[2], and Mark Billinghurst[1]

[1] The Human Interface Technology Laboratory New Zealand
University of Canterbury
Christchurch, New Zealand
firstname.lastname@hitlabnz.org
[2] Nokia Research Center
Immersive Communication Team
Tampere, Finland
viljakaisa.aaltonen@nokia.com

Abstract. In this paper we summarize, evaluate, and discuss the effect of movement patterns in a spatial sound space on the perceived amount of simulator sickness, the pleasantness of the experience, and the perceived workload. During our user study nearly 48 percent of all participants showed mild to moderate symptoms of simulator sickness, with a trend towards stronger symptoms for those experiencing left to right movements. We found evidence for predictable left to right movements leading to a perceived unpleasantness that is significantly higher than for unpredictable or no movement at all. However none of the movement patterns had a noticable effect on the perceived cognitive load for simple tasks. We also found some differences in the perception of the sound space between men and women. Women tended to have a stronger dislike for the sound space and found the task to be more difficult.

Keywords: Illusory self-motion, vection, spatial sound, auditory display, motion sickness, simulator sickness.

1 Introduction

Simulator sickness is a form of motion sickness, in which users of simulators or virtual environments develop symptoms such as dizziness, fatigue, and nausea [1] [2]. Both simulator sickness and the related phenomenon of motion sickness are difficult to measure. They are polysymptomatic and many of the symptoms are internal, non-observable, and subjective.

One of the most popular theories for explaining simulator sickness is the sensory conflict theory of Reason and Brand [3]. They believe that motion sickness occurs if there is a conflict between visual, vestibular, and proprioceptive signals in response to a motion stimulus. This disconcordance between the different cues leads the brain to conclude that the conflict is a result of some kind of poisoning [4]. To protect the physical health of the individuum, the brain reacts by inducing sickness and even vomiting to clear the supposed toxin.

S. Ystad et al. (Eds.): CMMR/ICAD 2009, LNCS 5954, pp. 287–305, 2010.

We are interested in researching the connection between audio cues and simulator sickness. Virtual auditory environments and 3D auditory interfaces can evoke the convincing illusion of one or more sound sources positioned around a listener [5] [6]. Trends in consumer audio show a shift from stereo to multi-channel audio content, as well as a shift from stationary to mobile devices, providing a rich array of mobile audio experiences. Auditory stimuli, often in combination with visual stimuli, are often rendered so realistically, that users perceive a high degree of presence within social or recreational virtual environments [7]. This means that if there is connection between audio cues and simulator sickness, then there may be an increased risk of sickness occuring.

Feeling present and immersed in a virtual environment can result in perceiving vection, the illusion of self-motion. Several investigations have shown a correlation between spatial presence/immersion and vection [8] [9]. Vection can also occur in many real life situations - usually when an observer is not moving, but is exposed to a moving visual pattern, such as when watching a moving train through the windows of a stationary train, or seeing a film in the front rows of the movie theatre. Vection has been attributed by Hettinger and Riccio [10] and McCauley and Sharkey [11] to be one of the major candidates for causing simulator sickness. Studies concerning vection often assume a link between the vection measured and the potential for the device or environment to cause sickness.

In our research we are interested in the connection between spatial audio, vection and simulator sickness. In the rest of this paper, we will first give a brief overview of the most important findings on the effects of vection in relation to spatial audio. While vection is a thoroughly researched phenomenon, very few researchers have investigated simulator sickness in audio only simulators or interfaces. In the main part of the paper, we present an experiment in which we explore the influence of movement patterns within a sound space on the perceived pleasantness of the experience and occurance of simulator sickness. We also studied the influence of the perceived pleasantness/simulator sickness of a spatial sound space on the cognitive load generated by a simple task. The design rationale of the conducted experiment is described in detail, followed by a summary of the findings. To conclude, the results of our study are discussed and directions for future research are given.

2 Related Work

Vection is well researched for visual stimuli. Vection occurs for all motion directions and along all motion axes. In a typical vection experiment, participants are seated inside a optokinetic drum and are asked to report on their perception of self motion, and discomfort level. Most participants quickly perceive vection in the direction opposite to the drum's true rotation. Depending on the type of simulator used, over 60% of participants can experience motion sickness-like symptoms [12] [13] [14].

As found by Brandt et al. [15] and Pausch et al. [16], visual stimuli covering a large part of the field of view will usually induce stronger circular vection with shorter onset latencies. Stimulation of the entire field of view will result in strongest vection.

Auditory vection has been less thoroughly researched. Although initial research was conducted almost 90 years ago [17], only recently has there been an increased interest in the phenomenon. See [18] [19] [20] and a review by Väljamäe [21] for a comprehensive overview of auditorily-induced vection research.

Lackner [22] found that a rotating sound field generated by either an array of six loudspeakers or dichotic stimulation can induce illusionary self-rotation, if the subject has their eyes shut or covered. See also Riecke et al. [23] for vection induced by moving sounds. If the subject has their eyes open and a stable visual field is given, vection does not occur. This and other research [24] suggest that visual cues dominate auditory cues in determining apparent body orientation and sensory localization.

Al'tman et al. [25] found that faster sound source movement was associated with an increase in the illusion of head rotation. In their study the subject was seated on a rotating platform and had their eyes closed. The subject's head was fixed in an immobile position, while an impulse series was played to them binaurally via headphones. The moving sound image affected the subject's postural reactions and the illusion of head rotation. When there were changes in the sound source movement, vection effects, such as the perceived rotation speed, were particularly strong.

Larsson et al. [18] found that in a rotating sound field, sound sources associated with immovable objects (such as church bells) are more likely to induce vection than both moving (e.g. cars) and artificial sound sources. In addition they found that a realistically rendered environment may increase perception of self-motion. The playback of multiple sound sources also induces significantly more vection responses than playing only a single sound source.

To summarize: a rotating sound field or moving sounds, higher rotation speeds, changes in speed, sounds representing immovable sound sources, and a realistic sound scene are all factors intensifying the perception of vection. A stable visual field on the other hand decreases or impedes the perception of vection.

Although several studies have shown that vection can be evoked by auditory stimuli, it is important to keep in mind that vection is only one possible cause for simulator sickness and that often symptoms are non-observable, subjective, and temporal.

The experiment depicted in this paper did not aim to reproduce the findings summarized above. Our primary intention was to investigate the general effects, including effects similar to simulator sickness, on a listener when exposed to a binaural listening experience. In particular we investigated the effects of predictable and unpredictable movement patterns of different speeds in an audio scene consisting of multiple sound sources.

3 Experiment

3.1 Design Rationale

Our experiment was designed to induce motion sickness or a certain degree of unpleasantness in participants through playback of binaural recordings of movements between several competing sound sources. Having a mobile user in mind, we consider the applicable scenarios for spatial sound to be:

- Mobile sound spaces, that move relative to a user, as in navigation support systems [26] [27] [28].
- Binaural media consumption such as listening to binaural recordings of concerts or audio books, etc.
- Spatial mobile conferencing with attendants located in a spatial, navigatable sound space [29].
- Spatial auditory interfaces that support navigation between and interaction with different sound items [30] [31] [32].

As we are explicitly interested in the effects on mobile users, after careful consideration we refrained from blindfolding or restricting the participants' body positions to a special pose. In a mobile setting users will always have visual stimuli, unless they are visually impaired, and it is unlikely that they cannot freely determine their body positions. However, if under these "realistic" conditions effects of simulator sickness arise, we assume that closing the eyes will intensify, but not drastically change these effects.

In our experiment we compared the following conditions:

Condition 1 (left-right): Left-right audio movements, simulating predictable movements while navigating, or interacting with a spatial audio interface.
Condition 2 (random): Random audio movements of objects within a sound space which may occur during media consumption or live feeds from other users.
Condition 3 (control): No audio movements; a control condition.

To study the effects of movement patterns on the perceived workload, participants were asked to identify random, nonsensical numbers in a text read to them (see below for a detailed description). This task was designed to create a cognitive workload similar to the challenges of orientation or navigating while focusing on a primary task.

Based on the summarized earlier research, our main hypotheses were:

H1: Participants would feel more discomfort when listening to random, unpredictable audio movements.

We assume that, given the findings by Lackner [22] and Al'tman et al. [25], random and unpredictable movements would intensify the perception of vection and hence the perceived discomfort. Predictable movements, on the other hand, would allow the subject to form a mental model which may potentially destroy the illusion of self-motion [21] and therefore the experience of simulator sickness, assuming that vection is one of the main causes for simulator sickness.

H2: The distraction generated by random audio movements affects the cognitive load and decreases task performance.

As unpredictable movements impede the formation of a mental model of the scene and its movement pattern, we hypothesize that in the condition with random movements more cognitive load is generated by the continuous necessity to orientate within the scene. Therefore the subject may not be able to focus their attention fully on the task, which may have a negative effect on task performance.

3.2 Participants

Eighty-two participants volunteered for the experiment ranging in age from 15 to 54 years old (M = 33 years), and were recruited within the Nokia community and several sport clubs. Forty-nine of the participants were male, thirty-three female. All participants were native Finnish speakers. This was a between subjects study with participants randomly allocated to the three conditions: left-right movements (N = 28), random movements (N = 25), and control, no movements (N = 27). Three participants reported minor hearing problems.

3.3 Audio Material

We used twenty minutes of binaurally recorded sound for the experiment. The recording was produced by an experimenter wearing an Augmented Reality Audio (ARA) headset which consists of binaural microphones, an amplifier/mixer, and in-ear headphones [30] (see Figure 1). The ARA headset is a "hear through" headset which allows users to hear audio cues superimposed over the audio from the surrounding real world. We chose to use the ARA headset instead of a manikin as it allowed the experimenter to move freely during the recording, which was especially important for recordings of random, 3-DOF movements. We decided on using binaural recordings with the ARA headset instead of binaural synthesis to ensure the reproduction of authentic head and body movements.

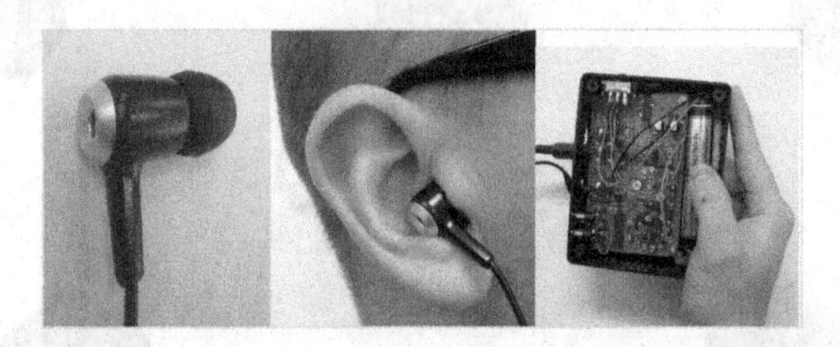

Fig. 1. ARA headset used in the study. Left: In-ear headphones equipped with microphones. Middle: Headset fitted in ear. Right: ARA mixer (Picture taken from [33]).

During the recording the experimenter sat on a swivel chair and was surrounded by five Genelec 6020A bi-amplified active loudspeakers fixed at face level. The recording was made in a soundproof studio with room acoustics. As can be seen in Figure 2, the loudspeakers were set up in a circular layout with a diameter of approx. 3 meters. The sound field created by the loudspeakers playing the following:

- Music, easy listening (Loud Speaker 1)
- Male speaker reading the text (Finnish) for the task (LS 2)
- Street noise, including cars passing by (LS 3)
- Podcast (Finnish), male and female speakers (LS 4)
- Environmental noise, birds, river (LS 5)

For the recording of condition 1 (left-right), the sound source the participants were asked to concentrate on was played from the back on speaker number 2 (conf. fig 2.). The experimenter moved her head from left to right through an angle of 80 degrees over approximately 0.8 seconds as illustrated in Figure 3. Preliminary testing indicated that having the target sound source appearing in the back is perceived as less natural and hence more annoying than facing it.

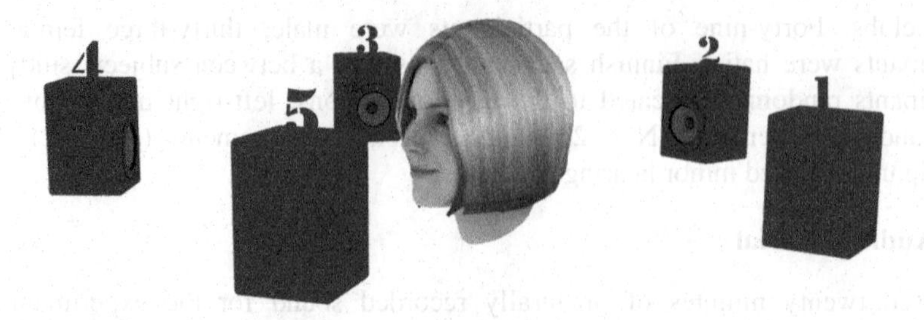

Fig. 2. Setup used for binaural recordings with the experimenter surrounded by five loudspeakers

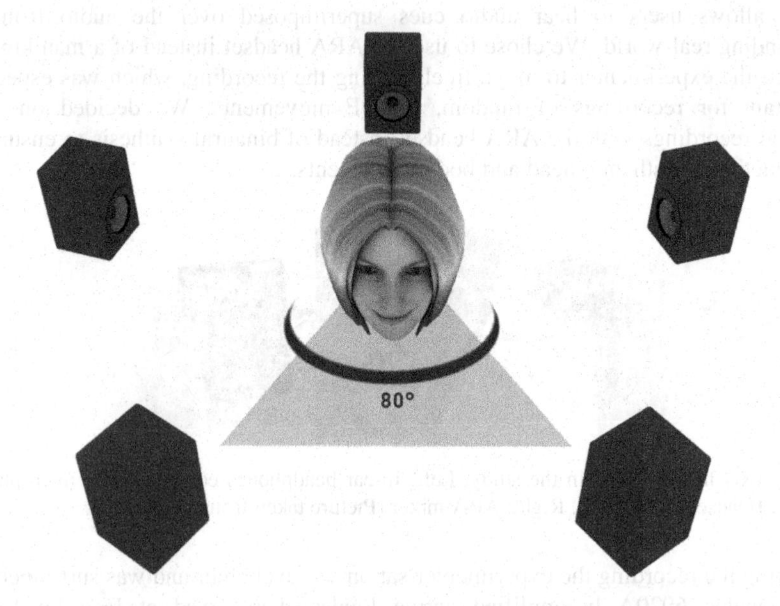

Fig. 3. Illustration of head orientation movements made for the recording of the left-right condition

For condition 2 (random) the experimenter moved her head in random, unpredictable movements. This included approaching or withdrawing from a sound source, rotations about her x- ,y-, and z-axis, and rapid changes of acceleration during movements. For the control condition the experimenter did not move at all, and was always facing the target sound source.

3.4 Experiment Task

In the experiment subjects wore headphones in a sound proof booth and were asked to listen to the recordings previously made in the different conditions. In all conditions participants were asked to concentrate on one of the sound sources, a male voice talking about dogs and horses. The script was read by a professional male speaker and consisted of adaptations of the Wikipedia Finland [34] entries on dogs and horses. At random positions in the text 33 numbers between 1 and 120 were placed. These numbers were out of context and we made sure that they did not make sense in the given context. Participants were asked to identify and chronologically write down numbers they identified as nonsensical.

The task required participants to concentrate and stay focussed on only one of the sound sources. They had to process the received information and identify numbers out of context. The task was designed to investigate the differences in cognitive load placed upon the participants over the three conditions. Since it required participants to focus their attention on one fixed spot in space, we assumed that the perceived changes of the target source's position might add an additional challenge.

3.5 Procedure

Before their first trial, participants were familiarized with the sound proof listening booths and were instructed on how to put on and adjust the Sennheiser HD580 headphones. After these instructions they were asked to fill the Simulator Sickness Questionnaire (SSQ) introduced by Kennedy et al. [1]. They were then given an oral and written explanation of the task and were encouraged to talk to the experimenter if they had questions or needed further clarification. After the trial participants were asked to fill out the SSQ again, followed by a second questionnaire asking about their perception of various aspects of the experiment. After completing the questionnaire, participants were debriefed, compensated, and dismissed.

The following dependent measures taken were: the pleasantness of the experience (including simulator sickness), the perception of the sound space, and the perceived cognitive load. The data from the various dependent measures were mostly analyzed using a one-way analysis of variance (ANOVA) with a fixed confidence level (p-value = .05). A seven-point Likert scale was used in the questionnaire handed to participants after the trial (1 = "I totally agree" and 7 = "I totally disagree").

4 Results

The results presented in the following paragraphs are deduced from the data gathered through the SSQ, the error rate of the task, and the second questionnaire.

4.1 Simulator Sickness Questionnaire (SSQ)

The SSQ was used as a measure in this experiment. The symptoms used, and their weightings, are given in table 1.

Subscales of the SSQ are: nausea, oculomotor, and disorientation. Participants reported the extend to which they experienced each of the symptoms shown in table 1 as one of "None", "Slight", "Moderate" and "Severe" before and after the trial. These are scored respectively as 0, 1, 2, and 3. The subscales of the SSQ were computed by summing the scores for the component items of each subscale.

Weighted scale scores for each column individually were computed by multiplying the nausea scale score by 9.54 and the disorientation subscale by 13.92. The total SSQ score was obtained by adding nausea and disorientation values and multiplying by 3.74 [1].

Table 1. The Simulator Sickness Questionnaire (SSQ) used as a measure in this experiment, the symptoms used, and their weightings

Symptom	Severity			
General discomfort	None	Slight	Moderate	Severe
Fatigue	None	Slight	Moderate	Severe
Headache	None	Slight	Moderate	Severe
Eye strain	None	Slight	Moderate	Severe
Difficulty focusing	None	Slight	Moderate	Severe
Increased salivation	None	Slight	Moderate	Severe
Sweating	None	Slight	Moderate	Severe
Nausea	None	Slight	Moderate	Severe
Difficulty concentrating	None	Slight	Moderate	Severe
"Fullness of the head"	None	Slight	Moderate	Severe
Blurred vision	None	Slight	Moderate	Severe
Dizzy (eyes open)	None	Slight	Moderate	Severe
Dizzy (eyes closed)	None	Slight	Moderate	Severe
Vertigo (Giddiness)	None	Slight	Moderate	Severe
Stomach awareness	None	Slight	Moderate	Severe
Burping	None	Slight	Moderate	Severe

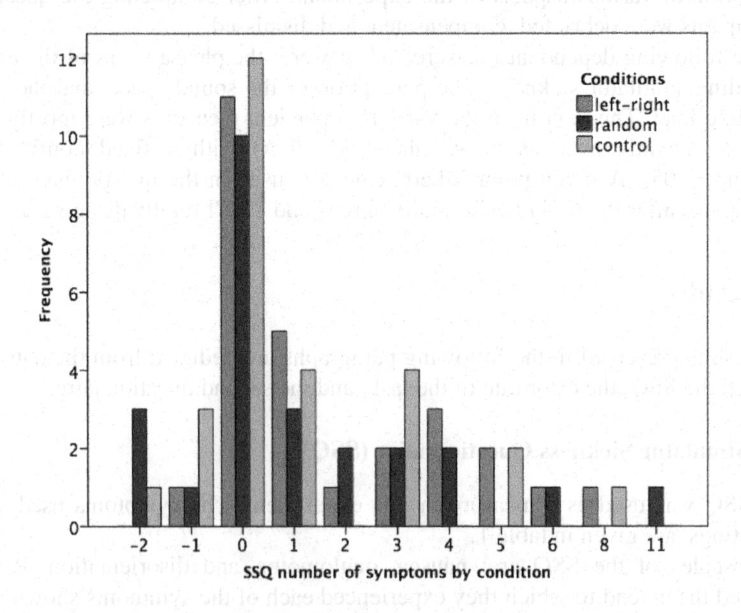

Fig. 4. Frequencies for SSQ Total (not weighted) for all participants (N = 81) per condition

As can be seen in Figure 4, 51.9 percent of all participants had a score of zero or below for the SSQ Total, indicating that they did not show any symptoms of simulator sickness. However, approx. 48 percent of all participants showed slight to moderate symptoms.

Tables 2 and 3 show pre-exposure scores, post-exposure scores and differences between the post- and pre-scores. To adapt the results to the requirements of measuring simulator sickness induced by a purely auditory stimulus, we neglected the scores for oculomotor problems.

Table 2. Mean pre- and post exposure SSQ scores for nausea over all three conditions

Condition	Nausea Pre (Mean)	SD	Nausea Post (Mean)	SD	Nausea Post-Pre (Mean, weighted)	SD
Left-right	1.26	1.56	2.56	2.12	12.37	15.16
Random	1.08	1.23	1.6	1.58	4.96	13.53
Control	.81	.8	1.69	1.95	8.44	14.92

Table 3. Pre- and post exposure SSQ scores for disorientation over all three conditions

Condition	Disorient. Pre (Mean)	SD	Disorient. Post (Mean)	SD	Disorientation Post-Pre (Mean)	SD
Left-right	1.04	1.67	1.52	2.65	6.7	18.66
Random	.36	.57	1.12	1.81	10.58	23.18
Control	.42	.81	.69	1.34	3.61	13.15

4.1.1 Nausea

A paired t-test showed a significant difference ($t(26) = -4.24$, $p < .001$) between pre (M = 1.26, SD = 1.56) and post (M = 2.56, SD = 2.13) exposure scores for nausea in the left-right condition. A paired t-test showed a near significant difference ($t(24) = -1.83$, $p = .079$) between pre (M = 1.08, SD = 1.29) and post (M = 1.6, SD = 1.58) exposure scores for nausea in the random condition. A paired t-test showed a significant difference ($t(26) = -2.76$, $p = .01$) between pre (M = .081, SD = .8) and post (M = 1.69, SD = 1.95) exposure scores for nausea in the control condition. However, the results from an analysis of variance on the scores for each condition shown in table 2 did not indicate significant differences in perceived nausea between the conditions.

4.1.2 Disorientation

A paired t-test showed a significant difference ($t(24) = -2.28$, $p = .032$) between pre (M = 0.36, SD = .57) and post (M = 1.12, SD = 1.81) exposure scores for disorientation in the random condition. However, the results from an analysis of variance on the mean scores for each condition shown in table 3 did not indicate significant differences in perceived disorientation between the conditions.

4.1.3 SSQ Total

For the left-right condition the SSQ total is 6.65 (weighted), for the random condition it is 4.79 (weighted) and for the control condition 4.02 (weighted) (see Figure 5).

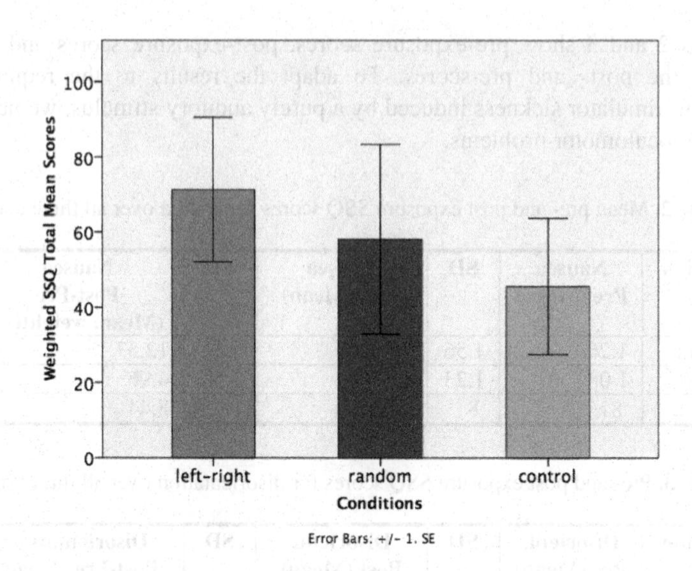

Fig. 5. Mean scores for SSQ total over all conditions

No significant difference (ANOVA F(2,77) = .58, p=.56) for SSQ Total could be found between the three conditions, but there is a trend towards a higher mean score for the left-right condition.

4.2 Pleasantness

In the post-study questionnaire we asked participants to agree or disagree to statements about the general pleasantness of the experience. This was done on a Likert scale of 1 (totally agree) to 7 (totally disagree). The included statements were:

- The task was pleasant.
- The task was boring.
- The listening experience was good.
- I could have continued to listen to this for a longer period of time.
- I would have liked to quit the test before the end.
- The sound volume was just right.

Number of participants, mean scores, and standard deviations are summarized in table 4.

Participants in the control group were on average indifferent about the pleasantness of the task. Participants from the left-right and the random group found the task to be significantly more unpleasant (ANOVA F(2,77) = 5.39, p=.006, confirmed by a post hoc Bonferroni test with p=.022 for left-right and .014 for random) compared to the control group (see Figure 6).

Table 4. Results from the post-study questionnaire on single items concerning the pleasantness of the experience

Condition / Statement	N	Mean Score	SD
"The task was pleasant."			
Left-right	28	4.89	1.6
Random	25	5.0	1.36
Control	27	3.74	1.5
"The listening experience was nice/good."			
Left-right	28	5.25	1.65
Random	25	4.68	1.91
Control	27	4.04	1.74
"I could have continued to listen to this for a longer period of time."			
Left-right	28	6.29	1.15
Random	25	5.85	1.28
Control	27	5.15	1.82
"I would have liked to quit the test before the end."			
Left-right	28	4.32	1.87
Random	25	4.8	2.1
Control	27	4.85	1.9
"The sound volume was just right."			
Left-right	28	2.25	1.18
Random	25	1.72	.74
Control	27	2.22	1.25

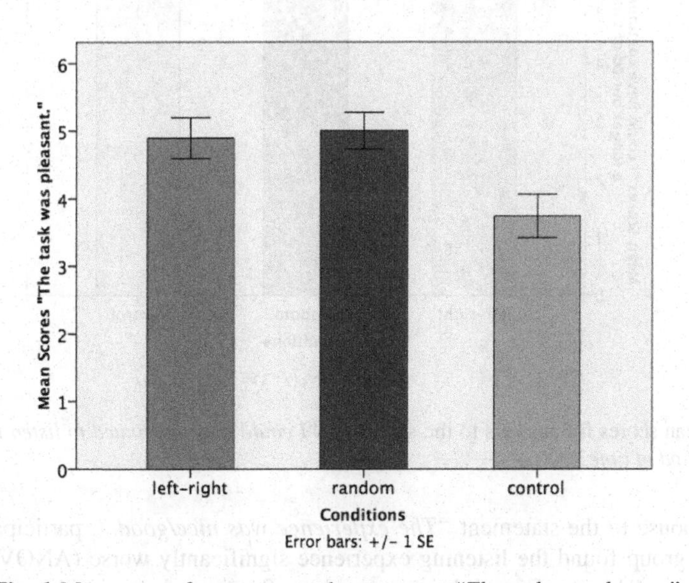

Fig. 6. Mean scores for answers to the statement *"The task was pleasant"*

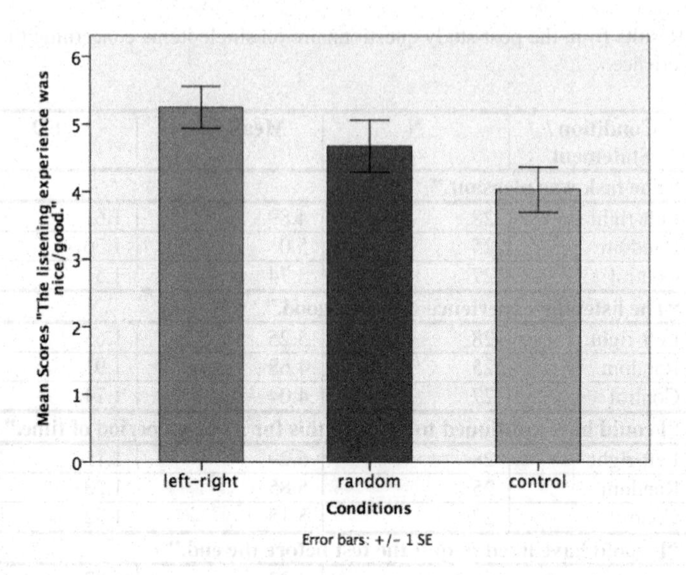

Fig. 7. Mean scores for answers to the statement *"The experience was nice/good"*

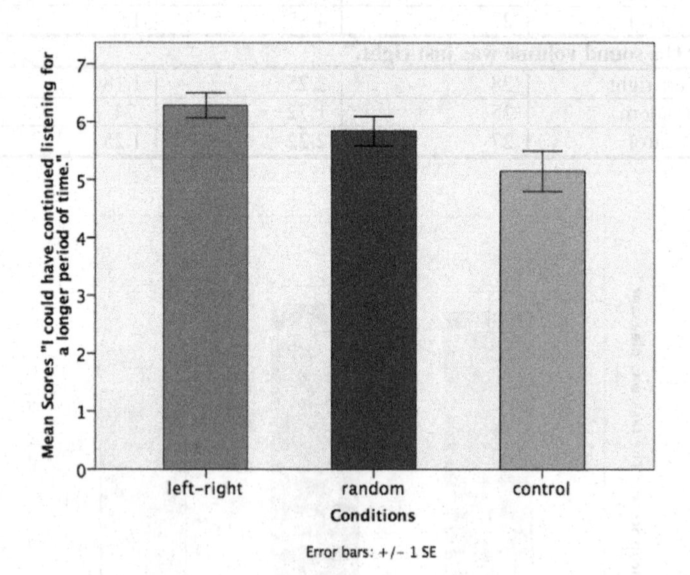

Fig. 8. Mean scores for answers to the statement *"I could have continued to listen to this for a longer period of time"*

In response to the statement *"The experience was nice/good."*, participants in the left-right group found the listening experience significantly worse (ANOVA $F(2,77) = 3.251$, $p=.044$, confirmed by a post hoc Bonferroni test with $p=.038$) than participants in the control group (see Figure 7).

In response to "I could have continued to listen to this for a longer period of time.", over all conditions participants felt like they would not want to listen to the sound space for a longer period of time. Though participants from the left-right group had significantly higher scores (ANOVA F(2,77) = 4.32, p=.017, confirmed by a post hoc Bonferroni test with p=.014) (see Figure 8) than participants from the control group.

4.3 Perception of the Sound Space

We also asked participants if they perceived the sound space to be chaotic. As can be seen in Figure 9, participants in the control group (N = 27, M = 3.37, SD = 1.85) found the sound space to be significantly less chaotic (ANOVA F(2,77) = 6.67, p=.002, confirmed by a post hoc Bonferroni test with p=.03 for random and .002 for left-right) than participants in the left-right (N = 27, Mean = 2.07, SD = 1.12) and random (N = 25, Mean = 2.36, SD = .95) groups.

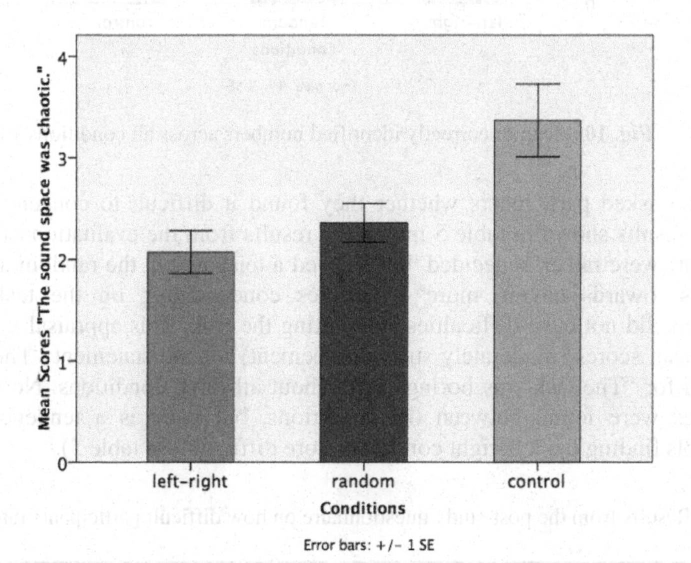

Fig. 9. Mean scores for answers to the question about whether the sound space was perceived to be chaotic

4.4 Cognitive Load

To measure the cognitive load of participants during the trial we evaluated the results from the listening task. Thirty-three nonsensical numbers were randomly inserted into the text and subjects were asked to record the numbers. When we compared the amount of numbers recorded across conditions we could not identify a difference between the conditions (ANOVA F(2,74) = .072, p=.931), in fact, as depicted in Figure 10, the results are almost identical. For control (N = 25) the mean of detected nonsensical numbers is 31 (SD = 2.4), for left-right (N = 27) the mean is 30.7 (SD = 3.1) and for random (N = 25) the mean is 30.9 (SD = 4.0).

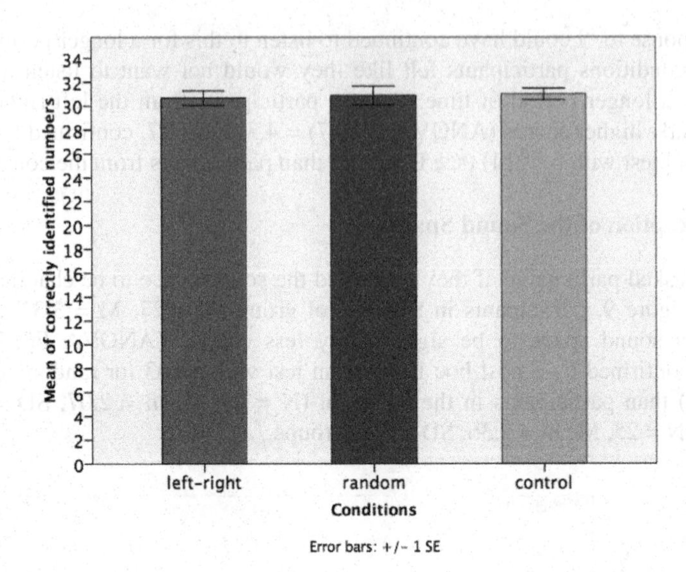

Fig. 10. Mean of correctly identified numbers across all conditions

We also asked participants whether they found it difficult to concentrate on the task. The results shown in table 5 mirror the results from the evaluation of the task – participants were rather undecided, but showed a tendency in the random and control conditions towards having more difficulties concentrating on the task. Overall participants did not have difficulties completing the task. This appraisal is supported by low mean scores (moderately strong agreement) for the statement "The task was easy" and for "The task was boring" throughout all three conditions. No significant differences were found between the conditions, but there is a tendency towards participants finding the left-right condition more difficult (see table 5).

Table 5. Results from the post-study questionnaire on how difficult participants rated the task

Condition / Statement	N	Mean Score	SD
"It was difficult to concentrate on the task."			
Left-right	28	4.04	1.67
Random	25	4.4	1.5
Control	27	4.41	1.82
"The task was easy."			
Left-right	27	3.22	1.55
Random	25	3.08	1.58
Control	27	2.44	1.5
"The task was boring."			
Left-right	28	3.71	2.12
Random	25	3.08	1.55
Control	27	2.85	1.38

4.5 Gender Differences

We found evidence for different perceptions of the task and the sound space between men and women in this study. Women (N = 33, M = 3.85, SD = 1.72) found it significantly more difficult (t(80) = -2.04, p = .04) to concentrate on the task than the men did (N = 49, M = 4.59, SD = 1.55). Both women and men did not want to listen to the sound space for a longer period of time. However women (M = 6.27, SD = .94) disagreed significantly stronger (t(80) = 2.6, p = .05) with the statement "I could have continued to listen to this for a longer period of time.". Men and women found the sound volume to be very good, nevertheless men (M = 1.84, SD = .8) perceived it to be significantly better (t(80) = 2.33, p = .04) than the women (M = 2.39, SD = 1.36).

Figure 11 illustrates the differences in weighted mean values for disorientation pre-study, post-study, and in total. It also shows the differences between men and women for the SSQ Total score (including nausea).

Results from an independent t-test show a significant difference (t(79) = 1.31, p = .017) between men (M = 4.54, SD = 14.87) and women (M = 10, SD = 22.7) in perceived disorientation (weighted).

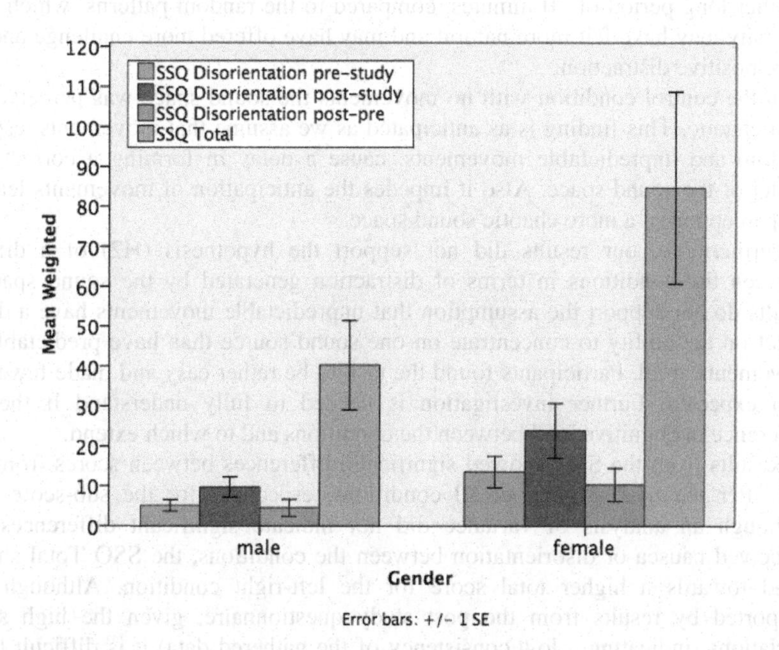

Fig. 11. Weighted mean scores of men and women for disorientation measured before and after the experiment, including the overall score for disorientation (post-pre score) and the cumulative SSQ Total including nausea scores

An examination whether these differences in perceived disorientation were already existent before the study affirmed our assumption: indeed there was a significant difference (t(79) = 1.31, p = .017) between men (M = 4.54, SD = 14.87) and women (M = 10, SD = 22.7) in the perceived disorientation before the experiment. Results from an independent t-test also show a near significant difference (t(78) = -1.87, p = .065) between men (M = 40.3, SD = 78.37) and women (M = 84.66, SD = 133.74) in perceived overall simulator sickness (SSQ Total weighted). It is likely that this difference is also due to the preexisting difference in disorientation.

5 Discussion

Our study showed that different movement patterns of spatial sound sources do indeed affect the perceived pleasantness of the listening experience. We found that predictable left to right movements make the listening experience less pleasant and generate stronger irritations compared to random movements or no movements at all.

Our first hypothesis (H1) that random, unpredictable spatial sound movements would make the experience more unpleasant could hence not be supported. The left-right movements were in sequence the most unnatural pattern. Humans use these movements to orientate when crossing a street or when turning their heads towards a sound source, but not repeatedly for a period of 20 minutes. Participants may have also found it particularly annoying and/or boring to listen to left-right movements for a rather long period of 20 minutes, compared to the random patterns, which in their diversity may have felt more natural and may have offered more challenge and hence more positive distraction.

In the control condition with no movements the sound space was perceived to be least chaotic. This finding is as anticipated as we assume that movements, especially random and unpredictable movements, cause a delay in forming a correct mental model of the sound space. Also it impedes the anticipation of movements leading to the perception of a more chaotic sound space.

Furthermore, our results did not support the hypothesis (H2) of a difference between the conditions in terms of distraction generated by the sound space. Our results do not support the assumption that unpredictable movements have a different effect on the ability to concentrate on one sound source than have predictable or no movements at all. Participants found the task to be rather easy and made fewer errors than expected. Further investigation is needed to fully understand if there is a difference in cognitive load between the conditions and to which extend.

Results from the SSQ showed significant differences between scores from before and after the trial throughout all conditions, especially for the sub-score nausea. Although an analysis of variance did not indicate significant differences in the perceived nausea or disorientation between the conditions, the SSQ Total showed a trend towards a higher total score for the left-right condition. Although this is supported by results from the post study questionnaire, given the high standard deviations (indicating a low consistency of the gathered data) it is difficult to make viable assumptions. We observed that some participants reacted strongly and showed sever symptoms of simulator sickness whereas others did not show any symptoms at all and actually felt better after the experiment than they did before.

We also found differences in the perception of the sound space between men and women. Women found it more difficult to concentrate on the task and they had a stronger dislike for the sound space. This perception was not confirmed by their task performance as there were no significant differences found between men and women. Although earlier findings by Kennedy et al. [35] and Biocca [36] indicate that women are more susceptible to simulator sickness, we came to believe that the differences in simulator sickness and disorientation scores are likely due to an already existing difference before the experiment.

6 Conclusions and Future Work

In this paper we have shown that predictable left to right movements lead to a perceived unpleasantness that is significantly higher than the unpleasantness experienced for unpredictable or no movements at all. Approximately 48 percent of all participants showed mild to moderate symptoms of simulator sickness, with a trend towards stronger symptoms for the left to right movements. Although the consistency of the data gathered using the SSQ was very low, data from the post-study questionnaire and our observations indicate that spatial audio and especially regular movement patterns repeated for a longer period of time will have a negative impact on the perception of the sound space.

Considering that the experiment was designed with the intend to evoke symptoms of simulator sickness, the current data suggests that even under the extreme conditions we created for the study the perceived unpleasantness did not exceed an amount that would have lead to stopping the trials.

Taking into account that unpredictable movements of sound sources in the sound space do not seem to reduce the listening experience to a critical degree, and that we could not provide evidence of a negative effect on cognitive load for simple tasks, we are rather optimistic about the use of spatial audio in mobile applications, such as navigation support systems, spatial auditory interfaces or entertainment applications. Based on these results we feel that use of spatial audio cues in mobile applications should not easily lead to simulator sickness.

With our experiment we investigated whether there is an influence of movement patterns within a spatial sound space on the perceived pleasantness of the experience and the perceived cognitive load on a listener. There are several directions for future research. In general, raising the consistency of the SSQ dataset should be aimed for. Varying the difficulty of the task used in the study as well as its realism would certainly be worth investigating.

Although we tried to acknowledge some criteria of a mobile usage scenario, (eyes open, no fixed posture) we also neglected others. For this study we have not been able to include a realistic mobile setting, for example an outdoors navigation task, or a task that forces the participants to react to an unpredictable environment. It would also be very interesting to study the effects of a more consistent and realistic spatial sound space on a listener, like for example in a navigation support system or a spatialised conference call. Furthermore it would be interesting to investigate the user perception of spatial augmented reality audio applications, where a real sound environment is extended with virtual auditory environments.

Acknowledgments

We thank Tim Bell and Andy Cockburn for early discussion and feedback, Miikka Vilermo, Anssi Rämö, Henri Toukomaa and Julia Turku for helping with the recordings, Martin Schrader, Monika Polonen, Markus Virta, and Perttu Kivelä for their help running the experiment.

References

[1] Kennedy, R.S., Lane, N.E., Berbaum, K.S., Lilienthal, M.G.: Simulator sickness questionnaire: an enhanced method for quantifying simulator sickness. Int. J. of Aviation Psychology 3(3), 203–220 (1993)

[2] Kolasinski, E.M.: Simulator Sickness in Virtual Environments. U.S. Army Research Institute for the Behavioral and Social Sciences, Technical Report No. 1027, Alexandria, VA (1995)

[3] Reason, J.T., Brand, J.J.: Motion Sickness. Academic Press, London (1975)

[4] Treisman, M.: Motion sickness: an evolutionary hypothesis. Science 197(4302), 493–495 (1977)

[5] Wenzel, E.: Localization in virtual acoustic displays. Presence 1(1), 80–107 (1992)

[6] Rumsey, F.: Spatial audio. Focal Press, London (2001)

[7] Hendrix, C., Barfield, W.: The sense of presence within auditory virtual environments Presence, 5th edn., pp. 290–301 (1996)

[8] IJsselsteijn, W.A., de Ridder, H., Freeman, J., Avons, S.E.: Presence: Concept, determinants and measurement. Proc. of the SPIE 3959, 520–529 (2000)

[9] Riecke, B.E., Schulte-Pelkum, J., Avraamides, M., von Der Heyde, M., Bülthoff, H.H.: Cognitive Factors can Influence Self-Motion Perception (Vection) in Virtual Reality. TAP 3(3), 94–216 (2006)

[10] Hettinger, L.J., Riccio, G.E.: Visually-induced motion sickness in virtual environments. Presence 1, 306–310 (1992)

[11] McCauley, M.E., Sharkey, T.J.: Cybersickness: Perception of Self-Motion in Virtual Environments. Presence 1(3), 311–318 (1992)

[12] Kennedy, R.S., Hettinger, L.J., Lilienthal, M.G.: Simulator sickness. In: Crampton, G.H. (ed.) Motion and space sickness, pp. 317–341. CRC Press, Boca Raton (1990)

[13] Regan, E.C., Price, K.R.: The frequency of occurrence and severity of side-effects of immersion virtual reality. Aviation, Space, and Environmental Medicine 65(6), 527–530 (1994)

[14] Riecke, B.E., Feuereissen, D., Rieser, J.J.: Auditory self-motion illusions ("circular vection") can be facilitated by vibrations and the potential for actual motion. APGV, pp. 147-154 (2008)

[15] Brandt, T., Dichgans, J., Koenig, E.: Differential effects of central versus peripheral vision on egocentric and exocentric motion perception. Experimental Brain Research 16, 476–491 (1975)

[16] Pausch, R., Crea, T., Conway, M.: A Literature Survey for Virtual Environments: Military Flight Simulator Visual Systems and Simulator Sickness. Presence 1(3), 344–363 (1992)

[17] Dodge, R.: Thresholds of rotation. J. Exp. Psychology 6, 107–137 (1923)

[18] Larsson, P., Västfjäll, D., Kleiner, M.: Perception of Self-motion and Presence in Auditory Virtual Environments. Presence, 252–258 (2004)

[19] Sakamoto, S., Osada, Y., Suzuki, Y., Gyoba, J.: The effects of linearly moving sound images on selfmotion perception. Acoustical Science and Technology 25, 100–102 (2004)

[20] Väljamäe, A., Larsson, P., Västfjäll, D., Kleiner, M.: Auditory presence, individualized head-related transfer functions, and illusory ego-motion in virtual environments. Presence, 141–147 (2004)

[21] Väljamäe, A.: Auditorily-induced illusory self-motion: A review. Brain Research Reviews 61(2), 240–255 (2009)

[22] Lackner, J.R.: Induction of illusory self-rotation and nystagmus by a rotating sound-field. Aviation, Space and Environmental Medicine 48(2), 129–131 (1977)

[23] Riecke, B.E., Väljamäe, A., Schulte-Pelkum, J.: Moving Sounds Enhance the Visually-Induced Self-Motion Illusion (Circular Vection) in Virtual Reality. Applied Perception, vol 6(2) article 7 (2009)

[24] Miller, N.R., Walsh, F.B., Hoyt, W.F., Newman, N.J., Biousse, V., Kerrison, J.B.: Walsh and Hoyt's Clinical Neuro-Ophthalmology, 2nd edn. Lippincott Williams & Wilkins, Baltimore (1998)

[25] Al'tman, Y.A., Varyagina, O.V., Gurfinkel, V.S., Levik, Y.S.: The Effects of Moving Sound Images on Postural Responses and the Head Rotation Illusion in Humans. Neuroscience and Behavioral Physiology 35(1), 103–106 (2005)

[26] Holland, S., Morse, D.R., Gedenryd, H.: AudioGPS: Spatial Audio in a Minimal Attention Interface. Human Computer Interaction with Mobile Devices (2001)

[27] Mariette, N.: A Novel Sound Localization Experiment for Mobile Audio Augmented Reality Applications. In: Pan, Z., et al. (eds.) Advances in Artificial Reality and Tele-Existence, pp. 132–142 (2006)

[28] McGookin, D., Brewster, S.A., Priego, P.: Audio Bubbles: Employing Non-speech Audio to Support Tourist Wayfinding. In: Altinsoy, M.E., Jekosch, U., Brewster, S. (eds.) HAID 2009. LNCS, vol. 5763, pp. 41–50. Springer, Heidelberg (2009)

[29] Dicke, C., Deo, S., Billinghurst, M., Lehikoinen, J.: Experiments in Mobile Spatial Audio-Conferencing: Key-based and Gesture-based interaction. In: Mobile HCI, pp. 91–100 (2008)

[30] Sawhney, N., Schmandt, C.: Nomadic Radio: Speech and Audio Interaction for Contextual Messaging in Nomadic Environments. CHI 7(3), 353–383 (2002)

[31] Brewster, S., Lumsden, J., Bell, M., Hall, M., Tasker, S.: Multimodal Eyes-Free Interaction Techniques for Wearable Devices. CHI 5(1), 473–480 (2003)

[32] Sodnik, J., Dicke, C., Tomazic, S., Billinghurst, M.: A user study of auditory versus visual interfaces for use while driving. Int. J. of Human-Computer Studies 66(5), 318–332 (2008)

[33] Tikander, M., Karjalainen, M., Riikonen, V.: An Augmented Reality Audio Headset. DAFX (2008)

[34] Wikipedia Finnland: `http://fi.wikipedia.org/wiki/Wikipedia:Etusivu`

[35] Kennedy, R.S., Lanham, S.D., Massey, C.J., Drexler, J.M.: Gender differences in simulator sickness incidence: Implications for military virtual reality systems. Safe Journal 25(1), 69–76 (1995)

[36] Biocca, F.: Will simulation sickness slow down the diffusion of virtual environment technology? Presence 1(3), 334–343 (1992)

From Signal to Substance and Back: Insights from Environmental Sound Research to Auditory Display Design

Brian Gygi[1] and Valeriy Shafiro[2]

[1] Veterans Affairs Northern California Health Care System
Martinez, CA USA 94553
bgygi@ebire.org
[2] Communications Disorders and Sciences
Rush University Medical Center
Valeriy_Shafiro@rush.edu

Abstract. A persistent concern in the field of auditory display design has been how to effectively use environmental sounds, which are naturally occurring familiar non-speech, non-musical sounds. Environmental sounds represent physical events in the everyday world, and thus they have a semantic content that enables learning and recognition. However, unless used appropriately, their functions in auditory displays may cause problems. One of the main considerations in using environmental sounds as auditory icons is how to ensure the identifiability of the sound sources. The identifiability of an auditory icon depends on both the intrinsic acoustic properties of the sound it represents, and on the semantic fit of the sound to its context, i.e., whether the context is one in which the sound naturally occurs or would be unlikely to occur. Relatively recent research has yielded some insights into both of these factors. A second major consideration is how to use the source properties to represent events in the auditory display. This entails parameterizing the environmental sounds so the acoustics will both relate to source properties familiar to the user and convey meaningful new information to the user. Finally, particular considerations come into play when designing auditory displays for special populations, such as hearing impaired listeners who may not have access to all the acoustic information available to a normal hearing listener, or to elderly or other individuals whose cognitive resources may be diminished. Some guidelines for designing displays for these populations will be outlined.

Keywords: Environmental Sounds, Auditory Display, Auditory Icons, Special Populations.

1 Introduction

Compared to the other major classes of familiar sounds, speech and music, environmental sounds are often considered the least important and have certainly been least studied. This is partially because the main function of environmental sounds is

S. Ystad et al. (Eds.): CMMR/ICAD 2009, LNCS 5954, pp. 306–329, 2010.

conveying information about the physical sources of sound-producing events, which in modern living is less valued than the communicative content of speech or the aesthetic qualities of music.

Nevertheless, environmental sounds are an important component of our everyday listening experience. One of the main benefits listeners with cochlear implants report is an increased ability to perceive environmental sounds [3]. Although appeals to evolutionary aspects of hearing are difficult to prove, it is hard to argue against the notion that the ability to identify sound sources preceded that of listening to speech or music [4]. It would be a more difficult, dangerous and less meaningful world without the ability to recognize environmental sounds.

For the auditory display designer, environmental sounds are useful precisely because of their representational value. While it is difficult to hear speech and not concentrate on the linguistic message, and it is compelling simply to hear music as music, environmental sounds can convey a variety of messages and take on disparate functions: as a warning signal; representing data auditorally; as an icon for carrying out commands on the computer or notifying the user of changes of state in the computer system; or, as themselves, as part of a virtual scene. This paper will focus on their use as auditory icons in computer interfaces, but some other uses, specifically as alarms, will be discussed in the final section.

However, while there has been a great deal of research and standardization in the use of visual icons in computer interfaces, analogous efforts with respect to auditory icons have been somewhat scattered and not well connected. This paper will attempt to incorporate some of the research into auditory icons with established findings from more basic auditory research to further develop the effective use of environmental sounds in auditory displays.

2 Background

The theoretical bases for auditory icons were laid out fairly early on in [5]. In it the author, William Gaver, described the different ways information can be mapped to representations, which he labeled "symbolic", in which the mappings were essentially arbitrary and had to be learned (such as a siren for a warning sign); nomic, where the meaning depended on the physics of the situation, which in the auditory case that would mean inferring a sound-producing source from the acoustics generated in the event; and in between the two are metaphorical mappings, which make use of similarities between the thing to be represented and the representing system. One example would be to use a descending pitch to denote a falling object. There is a continuum between these levels of representation, and a given sound-meaning mapping can change the level with usage. Nomic mappings are the most common way we listen to sounds in the world, namely in terms of the sources that produce the sounds, which he termed everyday listening, as opposed to musical listening, which is more concerned with more symbolic aspects of the sounds. Identifying more environmental sounds is a process of recovering the nomic mappings.

Gaver suggested that auditory displays could be more powerful and useful by taking advantage of everyday listening. The power of nomic mappings comes from the fact that they can convey more multi-dimensional data than simple symbolic associations because of their spectral-temporal complexity. When recognizing an environmental sound, we not only can tell with a great deal of specificity, what objects interacting in what way caused the sound (such as wooden hammer striking a plate) but we can also infer properties of the objects, such as the shape of a struck plate [7], the length of a falling rod [8], and in the case of footsteps, the gender of the walker [9]. Since we are sensitive to change in source properties, Gaver proposed that dimensional information can be conveyed in auditory displays by manipulating a sound in terms of its source properties. For example, if the sound of something dropping into a mailbox is used to inform the use of incoming email, the loudness of the sound can be used to indicate the size of the incoming mail (the louder the sound, the larger the incoming mail) and the timbre of the sound could indicate something of type of mail – if the sound was like a crackling paper, it would indicate a text file.

Gaver pointed out a concurrent property of nomic mappings in auditory displays: they are easier to learn and retain than arbitrary mappings, because they rely on highly learned associations, acquired through our life experience. This was demonstrated in [10]. They used different types of relations in assessing learnability of auditory icons, and found that direct relations, which included both ecological (their term for nomic) and metaphorical relations, were much more learnable than random ones, i.e. symbolic ones, but somewhat surprisingly there was no difference between ecological and metaphorical relations in learnability.

Although the use of environmental sounds was given a solid theoretical foundation, the practical applications were slower to develop, for several reasons. Gaver designed an auditory accompaniment to the Finder feature in Macs, called SonicFinder[6], which had a nicely thought-out interface using a variant of auditory icons in useful ways (a portion of which is detailed in Table 1.). For instance, selecting a file was mapped to the sound of an object being tapped, with the type of object indicated by the object material, and size of file represented by the size of the struck object. Although the SonicFinder was often cited as a good example of a sound-based interface, the project was never implemented by Apple.

The end result is that although most operating systems use environmental sounds for a few specialized purposes, such as deleting an object generates a crunching metallic sound, the full functionality of auditory interfaces has not been generally implemented, despite the potential to have a profound impact (e.g. facilitating computer use for the blind). There are several reasons for this, some commercial (the SonicFinder used too much memory in the days of limited memory), but some are due to disadvantages inherent in auditory interfaces, which have been well documented [11, 12]. Sound is not well suited for representing absolute data (as opposed to relative values): the spatial resolution for sounds is not as fine-grained as it is for vision, and simultaneous sounds occlude and mask each other more than visual objects, and sound is not a static medium; it unfolds in time, so it cannot represent data instantaneously and thus is problematic for continuous data displays.

However, many of hindrances to full usability of auditory icons come from the relative lack of research into environmental sounds, compared to vision or even speech and music. Although some basic principles for using sounds based on auditory research were outlined in [13], and [14] developed some good heuristics for auditory icon usage, basic research in environmental sounds has so far not been much

Table 1. A partial listing of events in the SonicFinder interface and associated auditory icons adapted from [6]

Finder Events	Auditory Icons
Objects	
Selection	Hitting Sound
Type (file, application, folder, disk, trash	Sound source (wood, metal, etc.)
Size	Frequency
Opening	Whooshing sound
Size of opened object	Frequency
Dragging	Scraping sound
Size	Frequency
Where (windows or desk)	Sound type (bandwidth)
....	....
Windows	
Selection	Clink
Dragging	Scraping
Growing	Clink
Window size	Frequency

applied to auditory display design. This paper will discuss some major issues involved in using environmental sounds in auditory displays, and how some recent knowledge gained from the hearing sciences can help to resolve some of these issues.

One major issue is identifiability of the sounds used, that is, how can the designer ensure a sound is actually recognized as intended. The identifiability of the sound depends on both the acoustic properties of the sound, but also, when a sound is presented in an auditory scene, the relationship of the sound to that scene: does the sound "fit" in a semantic sense? Thirdly, what are the best usages of environmental sounds? Which sounds are well-suited for which types of functions? In order to fully utilize the informative capabilities of environmental sounds, it is necessary to enable them to portray changes in states, as [15] suggested. This requires parameterizing the sounds so that certain aspects of the acoustics of the sounds will change, reflecting events in the interface. Finally, designing auditory displays for special populations, such as hearing impaired listeners who may not have access to all the acoustic information available to a normal hearing listener, or to elderly or other individuals whose cognitive resources may be diminished, poses some special considerations.

3 Identifiability

3.1 Baseline Identifiability in the Clear

The identifiability of the environmental sounds used as auditory icons is a major component in their successful use, as noted by [11, 14, 16-18]. Part of that is due to the automaticity of sound recognition. Identifying the source of a sound is one of the basic functions of the auditory system [4] which may also involve a series of cognitive inferences following identification [17], which give the auditory icon its significance. If the sound is unable to be identified, then the usefulness of the icon is greatly diminished[14].

Although humans can identify an immense number of environmental sounds quickly and accurately, studies attempting to determine the baseline identifiability of environmental sounds have yielded mixed results. Often these differences are due to methodological differences. Several early studies [19-22] were hampered by poor quality of sound reproduction or lack of controlled testing conditions. Even among more recent experiments there are differences in the procedures (method of constant stimulus, method of limits), presentation (in the clear, in noise, filtered), and limits on the durations of the sounds presented [14, 23-28]. So, not surprisingly, Ballas, in [17] compared five studies and found quite variable results. However, in [1] identification data were obtained for a group of 52 sounds which had been previously used in three studies of a large corpus of environmental sounds identification [25, 26, 28]. Multiple (3-5) tokens of each sound were tested, for a total of 195 different tokens, and the stimuli were all presented in the clear at an audible level (75 dB SPL) with little or no filtering or editing for length. Listeners identified the sounds using three-letter codes from a list of 90 possible codes. In addition the listeners rated the typicality of the sounds, that is, how good an example of the sound the token they heard was. The results for the best token of each sound, based on 75 young normal hearing listeners (YNH), are shown in Table 2. Forty five of the 52 tokens listed were all recognized at p(c) = 0.9 or greater. In addition, these tokens all had a mean typicality rating of 4 or better on a scale of 1-7 where 1=not typical at all. In addition, the mean p(c) for all tokens for a particular sound is listed. When there is a large discrepancy between the two it means that there were some tokens for a sound that were not well recognized at all.

So there does seem to be a group of sounds that can be recognized reliably in isolation in the clear, as well as a few (e.g., Shovel) that were not well recognized. Further, the correlations between the recognition performance in this test and the data for the same tokens from [26, 28, 29] were quite high, $r = 0.88$ and 0.83, respectively, indicating that the variance in identification between studies noted in [17] is likely due to the different tokens used in each study.[1]

3.2 Sound Duration and Identifiability

As noted above, sounds unfold in time and so the time required to identify an environmental sound needs to be considered in the design of the auditory display. In [14] the author recommended using for auditory icons sounds that were short, with " a wide bandwidth and where length, intensity and sound quality are roughly equal." While that might make sense from a programmatic standpoint, in practice it might lead to choosing icons that all sound like short noise bursts and are thus not easily discriminable. Since environmental sounds represent events in the real world, they will by necessity have different lengths. As mentioned, most of the sounds listed in the previous section were presented with minimal editing, so they were fairly long, with a mean duration of 2240 ms (max 4812 ms, min 224 ms). The correlation between duration and p(c) was essentially zero, so the shorter sounds were as well recognized

[1] Unfortunately, most of the sounds tested were taken from commercial sound effects CDs, which meant they were copyrighted and are not freely distributable. The authors are currently involved in a project of collecting and norming freely distributable high-quality environmental sounds for research purposes.

Table 2. List of sounds tested in [1], the mean p(c) for the best token of that sound, and for all the tokens of that sound as a group

	Mean p(c)			Mean p(c)	
Label	Token	Sound	Label	Token	Sound
Baby	1.00	1.00	Ice drop	0.97	0.95
Cough	1.00	0.96	Match	0.97	0.93
Dog	1.00	1.00	Bubbles	0.96	0.91
Drums	1.00	0.97	Car accel.	0.96	0.78
Glass	1.00	0.98	Rooster	0.96	0.96
Gun	1.00	0.98	Gargle	0.96	0.95
Laugh	1.00	0.98	Thunder	0.96	0.91
Phone	1.00	1.00	Crash	0.96	0.87
Siren	1.00	0.97	Bells	0.95	0.92
Sneeze	1.00	0.94	Rain	0.94	0.87
Toilet	1.00	0.82	Scissor	0.94	0.81
Whistle	1.00	0.98	B-ball	0.93	0.86
Door	0.99	0.99	Train	0.93	0.88
Clock	0.99	0.93	Claps	0.93	0.80
Helicopter	0.99	0.91	Sheep	0.93	0.89
Gallop	0.99	0.95	Crickets	0.92	0.86
Zipper	0.99	0.99	Waves	0.91	0.80
Cat	0.99	0.88	Pour	0.91	0.87
Cow	0.99	0.98	Tennis	0.90	0.84
Ping-pong	0.99	0.96	Splash	0.89	0.81
Neigh	0.99	0.98	Footstep	0.88	0.82
Bowling	0.99	0.96	Stapler	0.85	0.78
Bird	0.99	0.84	Hammer	0.85	0.62
Typewriter	0.99	0.72	Harp	0.81	0.81
Airplane	0.97	0.83	Cymbal	0.79	0.76
Car start	0.97	0.97	Shovel	0.65	0.53

as the longer sounds. However, this does not mean that sounds can be edited for length with impunity. In a seminal study of environmental sounds perception, [23], the sounds were all edited to be no more than 625 ms. While some of the sounds preserved their identifiability in this way, for a large number of them the identification accuracy was essentially at chance.

So how much of an environmental sound is necessary to be heard in order to identify it? A study of 117 environmental sounds using a gated paradigm [30] in which the length of the sounds were increased on each presentation by 50 ms found that half of the sounds were identified within 150 ms, which is an amazingly brief span of time. However, the gating paradigm allows listeners multiple exposures to a sound, each time gaining more information about the sound, so the results do not necessarily indicate how listeners would perform on the first presentation of a sound.

Certainly there are some sounds whose identity is immediately evident. Harmonic sounds can be identified on the basis of their steady-state spectra alone, even when the temporal information is misleading [31]. These include whistles, phones, sirens, musical instruments and animal vocalizations, all of which were among the quickly-recognized stimuli in [30]. So these sounds can be edited for length and still be recognizable. However, many short impulsive sounds tend to have similar envelopes and bandwidths (such as a basketball bouncing, gun, hand claps, door knock, and chopping wood) and the only way to distinguish is in their temporal structure, which includes the periodicity and damping (which was demonstrated in the case of bouncing and breaking bottles in [32]) so a longer sample is necessary in those cases.

Some environmental sounds are actual composite events made up of simpler, more basic events (see [33] for a taxonomy of basic and complex events, discussed below), and in these cases it is necessary to perceive all or nearly of the constituent events to accurately recognize the sound. In an extreme case, the spectrogram of a bowling ball rolling down an alley is plotted in Figure 1. The bowling ball does not strike the pins until 2.3 seconds after the start of the sound. Although this token had a very high identifiability, if it was edited to exclude collision with the pins it would be almost unidentifiable.

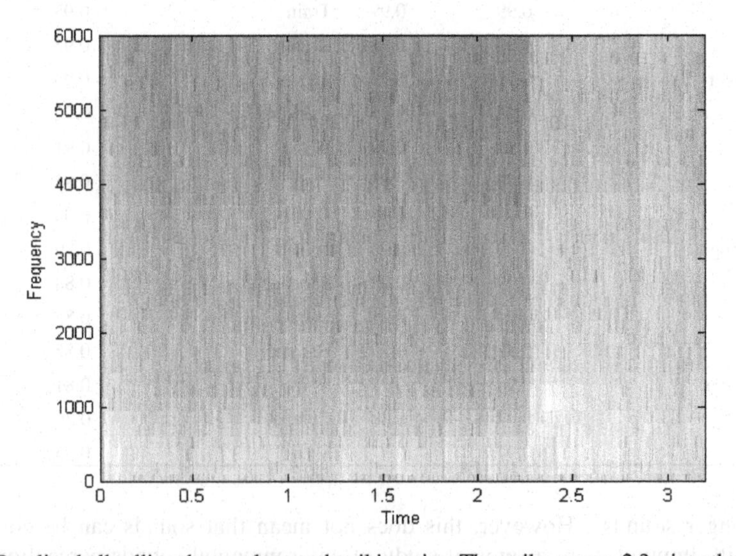

Fig. 1. Bowling ball rolling down a lane and striking pins. The strike occurs 2.3 s into the sound.

3.3 Effects of Filtering on Identification

There may be times when the auditory display designer will wish to filter the sounds used in auditory icons, to avoid masking other important sounds in the display (such as music), or to accommodate a narrow transmission bandwidth, or for special populations who may be using the display. For example [34] described an auditory GUI in which auditory icons were high- or lowpass filtered to indicate whether they were highlighted or deselected, respectively (see section 6 on auditory icons for special populations). Since environmental sounds are a result of the whole catalog of possible sound-producing events in the world, there is an extreme of spectral-temporal variation among environmental sounds, with some sounds with a strong harmonicity (vocalizations), others more like broad band noises (water sounds), some are relatively steady state (air sounds) and some are transient (impact sounds).

As a result, the effects of filtering on environmental sounds are not uniform, as was demonstrated in [25]. Figure 2 is adapted from that paper and shows the effects of low- and high-pass filtering on a selected group of environmental sounds. There are some sounds, such as weather sounds, like thunder and waves that are extremely adversely affected by low-pass filtering but are resistant to high-pass filtering. Some sounds, like an axe chopping, carry all the information in the envelope and so are robust to most filtering.

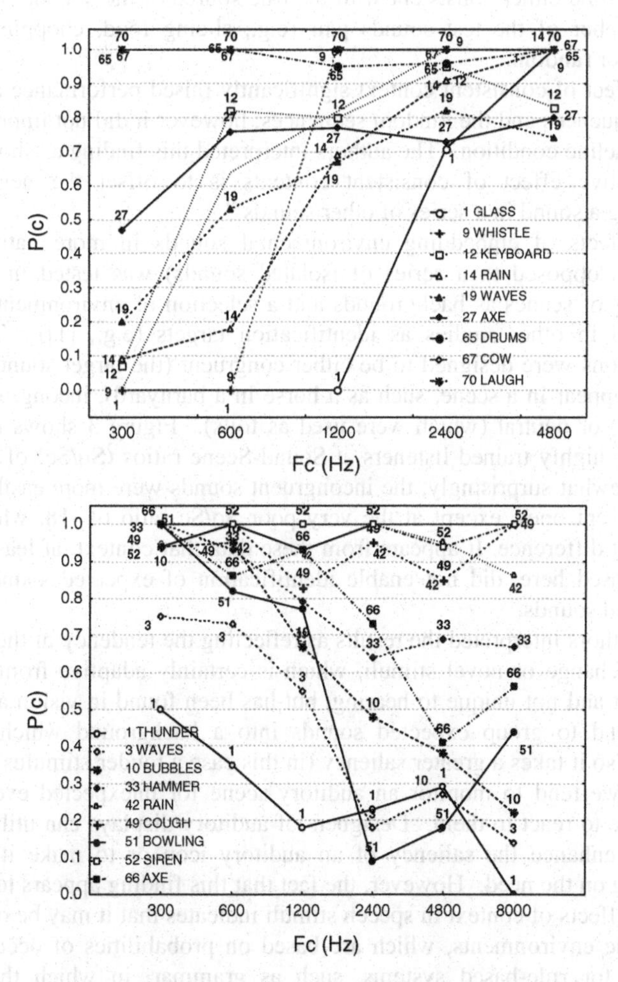

Fig. 2. Effects of lowpass (top) and highpass (bottom) filtering on a group of environmental sounds reprinted from [25] with the authors' permission. The cutoff frequency for the filters are shown on the abscissa.

3.4 Effects of Context on Environmental Sounds

Sounds in the world do not happen in isolation, but rather occur as part of an auditory scene concurrently with other related events. In any given setting there are sounds that are more or less likely to occur, i.e. that are congruent or incongruent with that setting. Although context has been found to be helpful in perceiving speech sounds [35] and in various psychophysical tasks [36-38], the effects of context on identification of environmental sounds have not been extensively researched.

It is often assumed that it is more difficult to recognize sounds out of context (or in no context at all) [14]. However [39] tested the identifiability of pairs of confusable sounds which were embedded in a sequence of other sounds that provided a scenario, such as a match lighting, a fuse burning, and an explosion. The sequences were designed to be either consistent with the true source of the sound, biased towards the other member of the test sounds pair (e.g., slicing food, chopping food and fuse burning), or random.

The effect of consistent context significantly raised performance above that of the biased sequences and the random sequences; however it did not improve performance over a baseline condition. The authors interpreted this finding as showing that "…the only positive effect of consistent contexts is to offset the negative effects of embedding a sound in a series of other sounds."

The effects of embedding environmental sounds in more naturalistic auditory scenes (as opposed to a series of isolated sounds) was tested in [40] using field recordings of scenes as backgrounds and a selection of environmental sounds which were used in other studies as identification targets (e.g., [1]). The sound-scene combinations were designed to be either congruent (the target sound was considered likely to appear in a scene, such as a horse in a barnyard), incongruent (a horse in a restaurant) or neutral (which were used as foils). Figure 3 shows the identification results for highly trained listeners at Sound-Scene ratios (So/Sc) of -18, -15 and -12 dB. Somewhat surprisingly, the incongruent sounds were more easily identified than the congruent ones, except at the very poor So/Sc ratio of -18, when there was no significant difference. It appears from these data that context, at least the naturalistic contexts used here, did not enable identification of expected sounds, but rather of unexpected sounds.

The authors interpreted the results as reflecting the tendency of the auditory system to detect change or novel stimuli, which is certainly adaptive from an evolutionary standpoint and not unique to hearing, but has been found in vision as well e.g., [41]. So we tend to group expected sounds into a background which do not compel attention, so it takes a greater saliency (in this case a louder stimulus) to stand out. In contrast, we tend to monitor an auditory scene for unexpected events, so as to be better able to react to them. Designers of auditory displays can utilize this tendency to either enhance the saliency of an auditory icon or to make it less noticeable, depending on the need. However, the fact that this finding appears to contrast sharply with the effects of context in speech stimuli indicates that it may be only applicable to naturalistic environments, which are based on probabilities of occurrence and does not hold for rule-based systems, such as grammar, in which there are stronger constraints on what sequences are possible. On the other hand, direct comparisons

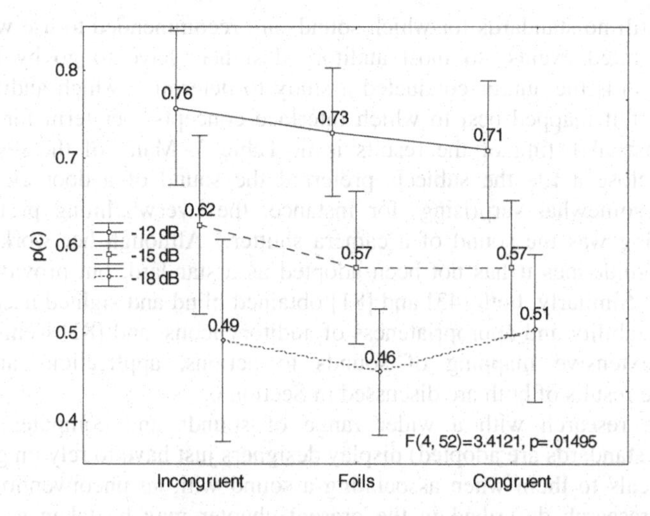

Fig. 3. Identifiability of environmental sounds in incongruent, congruent, or neutral contexts as a function of presentation level. From [40] with the author's permission.

between speech and environmental sound context effects should be treated with caution because of the differences in what specifically is being examined in each experimental paradigm. Since auditory displays tend to fall between the two extremes, i.e., not anything is possible but they are still not entirely deterministic, it is possible that listeners might switch between the modes of attending given the setting and task demands.

4 Semantic Compatibility of Sound and Function

When a sound is identified, as noted in [17] there are multiple source characteristics and numerous associations activated in the listener's mind pertaining to knowledge of the source: how big the objects involved in the source are, what they are made of, how they were interacting. In order to use them effectively in a display, the function in the display should relate in some way to the source properties already known by the user [42]. However, the number of possible mappings from physical sources to auditory icons is limited, and effective ones may not always be obvious or intuitive. For example, as noted in [81], the "open file" action could be association with a number of sounds, such as unzipping trousers, opening a drawer or a beer can or pulling the curtains, all of which partially relate to opening a file. And to denote reversal of an action (such as closing a file), is it sufficient to present the original sound played backwards?

In the SonicFinder interface, Gaver developed a complex mapping of sounds to functions, some of which were listed above in Table 1. While some are obvious and have been incorporated into modern interfaces (the sound of dropping something in a trash can to denote deleting it), others are less intuitive and on the "symbolic" end of the scale, e.g., a pouring sound for copying. As a result, these mappings would likely be harder for the listener to learn.

There are still no standards for which sounds are recommended to use with which computer-generated events, so most auditory designers have to go by their own intuitions. In [14] the author conducted a study to determine which auditory icons eight subjects felt mapped best to which interface concepts (her term for computer events). A partial listing of the results is in Table 3. Many of the results were expected (to close a file the subjects preferred the sound of a door closing), but several were somewhat surprising: for instance, the overwhelming preference to indicate copying was the sound of a camera shutter. Although her work provided several good guidelines it has not been adopted as a standard, but provides a good starting point. Similarly, both [43] and [81] obtained blind and sighted users' ratings of the recognizability and appropriateness of auditory icons, and [81] went further to develop an extensive mapping of sounds to actions, applications and screen navigation; the results of both are discussed in Section 6.

Until more research with a wider range of sounds and computer events is conducted (or standards are adopted) display designers just have to rely on guesswork and what appeals to them when associating a sound with an unconventional event, although the research described in the present chapter may be taken as an initial guideline. Since environmental sounds activate a range of semantic features, if one knew what those features are, it might point the way to a good mapping. The perceptual dimensions of 145 environmental sounds were measured in [44] using the semantic differential method, in which listeners rated sounds on twenty different scales based on binary pairs of qualities, such as "tense – relaxed", "light – heavy" or "compact – scattered". The ratings can then be subjected to factor analysis to determine the primary underlying perceptual dimensions. The authors found four main perceptual dimensions: harshness, complexity, appeal and size. So if a sound designer has an activity that might seem tense, or complex to the user, (s)he can use a sound that rated highly on those scales (the loading of each sound on each factor is supplied in the article).

5 Using Environmental Sounds to Represent Continuous Data

In the SonicFinder interface, the sound of something dropping in a basket denoted incoming mail, and the larger the incoming mail, the louder the sound. This is an example of how environmental sounds can represent not just events but a range of values for an event. However, to do this requires parameterizing the acoustics of the sounds to match the range of values of data displayed. This is a nontrivial problem. While certain salient acoustic variables such as loudness are quite easily manipulated, others (such as spectral centroid,) are not. Further, as was pointed out in [45], for auditory icons to be consistent with the nomic relations of the underlying sound, the parameterization should not just be affected in an arbitrary acoustic feature, but must reflect source properties of the sound. So, in the example cited above, the louder sound was an effective indicator of the size of an incoming mail message because larger things do tend to make louder sounds.

Unfortunately, the acoustical effects of variations in sound-producing events are seldom so neat. If one wanted to indicate an approaching or receding object using sound, one could also make the sound louder or softer, but that would only be a crude

Table 3. A partial listing of auditory interface concepts and users' preferred associated auditory icon adapted from [2]

Concept	X^2	α	Sounds	$X_c^{2_s}$	$\#_c$ [0-8]
Copying	136	<.001	camera shutter	96.86	2
Closing	113	<.001	closing car door	59.10	4
			zipping down	30.62	2
A text field	101	<.001	typing	76.82	6
A slider	92	<.001	whistle up	59.10	2
Check boxes	89	<.001	keystroke	59.10	2
Dragging	78	<.001	whistle up	43.70	0
			cars driving by	11.42	2
Opening	75	<.001	open door	19.86	3
			whistle up	19.86	3
			motorcycle drive	11.42	1
A push button	73	<.001	crack whip	19.86	0
			short pop	11.42	2
			keystroke	11.42	1
			water drop	11.42	1

approximation of the actual effect because there are also significant spectral changes when objects approach or recede [46] – high frequencies tend to be more greatly attenuated than low frequencies with distance. For sounds that are produced by the interactions of two objects, such as impact sounds, the acoustics are affected by the materials of the interacting objects, the shapes of the interacting objects and of course the type of interaction[7, 9, 47-49]. So to be able to adequately manipulate the sounds to approximate changes in source properties, fairly detailed physical models of the sounds are needed. Although standard equations exist in acoustic textbooks for a number of interactions, Gaver [15, 50] pointed out that the parameters of the models should reflect physical source attributes, which are realized in the acoustic domain, as shown in Table 4.

Since the goal of everyday listening is sound source recognition, Gaver developed a taxonomy of basic sound-producing events [33], which he described as impacts, liquid sounds and aerodynamic sounds and looked at physics of the events to develop models. Gaver implemented this in a model of a class of impact sounds (striking an object with a mallet) with parameters that could be constrained to patterns typical of various object configurations, material, and mallet hardness. He tested sounds generated using these on four listeners who confirmed that the impact sounds generated in this way were realistic [50]. Gaver also proposed models for liquid sounds, scraping sounds and machine sounds, although he said that none of the other models were as successful and he did not report any quantitative tests of them.

Gaver's approach has been very influential and has lead to numerous physical models for environmental sounds based on source properties, although more recent work involving similarity judgments [51] has indicated that listeners tend to regard liquids and aerodynamic sounds as a single class of sound-producing events, and consider harmonic sounds as a separate class of sounds. Most of these models have been for impact sounds, e.g. [52-55], but there have also been models of rolling sounds [56, 57], liquid sounds [58] and rubbing sounds [59]. For more information, the website for the Sounding Objects project (SOb), www.soundobject.org, has a large repository of papers and reports in this area.

The numerosity of these models has made a comprehensive evaluation difficult, but it is important to remember that there may be no "best" way to synthesize sounds parametrically. As long as the pertinent information is available and audible, listeners are remarkably good at attending to the relevant features for extracting that information [60]. In fact [61] showed that due to the redundancy of spectral-temporal information in environmental sounds even listeners who do not adopt the most efficient strategy for identifying a sound still manage to perform nearly as well as an ideal listener. The sensitivity of listeners to minute changes in environmental sounds has received very little attention: one study showed that thresholds for discriminating changes in specific spectral regions of a group of environmental sounds were relatively large, on the order of 7-10 dB, although there were large individual differences [62]. In addition, it is not necessarily the case that the most "realistic" sound is the best. Cartoonification is as useful for auditory icons as it is for visual icons in terms of memory storage and interface continuity [63]. There are even instances where "fake" sounds are judged to be better than the "real thing" [64]. So when assessing physical models, ease of implementation should be a consideration. Along those lines it should be noted that the website for Dick Hermes' Sound Perception class has downloadable Matlab code home.tm.tue.nl/dhermes/lectures/sd/ChV.html for generating several different type of sound-producing events, such as impacts, bouncing sounds, rolling sounds, dripping and machine sounds.

Table 4. Physical parameters of an impact event and their acoustic manifestations, either in the frequency to the time domain. Adapted from [50].

Frequency Domain	*Temporal Domain*
Restoring force (material feature)	Interaction type
Density (material feature)	Damping (material feature)
Size (configurational feature)	Internal Structure (material feature)
Shape (configurational feature)	Support (configurational feature)
Support (configurational feature)	

Alternatively, if a designer really wants to use actual sounds and have some variation in the source properties, the website for the Auditory Perception Laboratory www.auditorylab.org has numerous high-quality recordings of actual sound-producing events such as rolling, deformation, liquid, air and impact events, with a variety of materials and type of interacting objects. For example, the rolling sounds have a variety of accelerations so that a parameterizing could be simulated (although the memory requirements would be quite substantial.)

6 Auditory Icons for Special Populations

Most of the previous research has been conducted on subjects with normal hearing and cognitive abilities. However, given that a great number of Web applications are for special populations, it is useful to know what considerations need to be taken into account regarding use of environmental sounds for these populations.

6.1 Hearing-Imparied Users

An obvious group that must be considered are hearing impaired users, those using hearing aids (HA), or cochlear implants (CI), as well as those without any prosthetic hearing device at all, especially in the area of alerts or warning sounds. Although all of these groups have greatly diminished auditory sensitivity, the nature of the information they are able to perceive varies drastically. Hearing impaired persons most often have varying losses of sensitivity in the upper frequencies (typically > 2 kHz) although hearing at 1000 Hz and below may be near normal. This effect can be approximated by having a lowpass filter with a cutoff at somewhere between 2 and 4 kHz, and the sounds that were found in [25] to be identifiable with only low-frequency energy preserved (see Figure 2) will be best-suited for hearing-impaired people who are not using any amplification device.

For hearing aid users the situation is somewhat different. Although hearing aids adjust the gain on those frequencies the user have trouble hearing to compensate for these losses, many hearing aids users report extreme sensitivity to loudness (recruitment) particularly in those regions and as a result the dynamic range between what is audible and what is painful is drastically reduced, from 100 dB to 60 dB or less [65]. So when using environmental sounds as auditory icons, care should be taken that the overall level of the icons does not greatly exceed the level of the other sounds in the display and that sounds with a large amount of high frequency content are avoided. It should also be noted that a comparison of responses on an open – ended questionnaire of individuals with hearing loss with those of experienced hearing aid users revealed that the number of problems with environmental sound perception was comparable between the two groups [66].

Cochlear implants users pose quite different challenges for the auditory display designers. Cochlear implants are given to deafened persons, who have little or no residual hearing and cannot benefit from sound amplification provided by hearing aids. The processing in cochlear implants involves filtering using a bank of bandpass filters, extracting the envelope from each of the filter bands, and using each envelope modulate the amplitude of electrical pulses delivered by implant electrodes. The results can be simulated by using the envelopes to modulate the amplitude of a particular carrier signal (typically a noise band or a sine wave) specific to the frequency of each band. A schematic is shown in Figure 4 with a white noise carrier and a six-channel bandpass filterbank to produce a six-channel Event-Modulated Noise (EMN). The result is the temporal information of the original signal is largely preserved, while the spectral information is greatly reduced.

What this means is that CI users have access to temporal cues but much less to spectral cues. While this is adequate for speech perception purposes (at least in the quiet –listeners can achieve near-perfection speech recognition scores with as few as four channels [67]) the effect on perception of environmental sounds is more complex. Studies testing environmental sounds identification using CI simulations on normal-hearing listeners [25, 28] and in actual CI users [29, 68, 69] have both

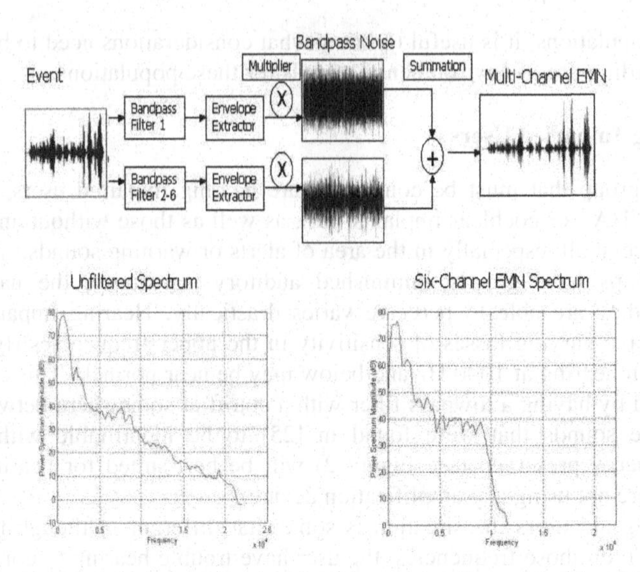

Fig. 4. A schematic of the processing involved in creating Event Modulated Noises (EMF) which simulate the effects of cochlear implants on environmental sounds

indicated that some environmental sounds are perceived quite accurately using CI processing, while others are almost unidentifiable.

The main factor determining this seems to be the amount of temporal versus spectral information in the target sound. Sounds which have a distinct envelope but whose spectra are similar to a broadband noise, such as a horse running, hammering or a ping-pong ball bouncing, are perceived quite accurately. In contrast, sounds with a steady-state envelope, such as a car horn, wind blowing or a flute, tended not to be recognized well at all. One mitigating factor is the number of channels used in the processing scheme, since the greater number of channels, the more spectral information is available to listeners. So sounds such as a baby crying or a bird chirping, which have harmonic spectra, but also a unique temporal patterning, can be quite well identified with a greater number (4-8) of channels. [28] also showed that the number of frequency channels required for 70% correct identification (i.e., 8 channels or less or 16 channels or more) can be predicted with relatively high accuracy of 83% based on the number of bursts and the standard deviation of centroid velocity in the original environmental sound.

It should be noted that overall performance on identifying environmental sounds actually may decrease with greater than 16 channel processing [28]. However, the designer of auditory displays for CI users has no knowledge of, or any way of knowing, the number of channels in the CI users' processing scheme, so the best strategy may just be assume the processing scheme that provided the best overall performance in [28] (8 channels) and use sounds that were well identified in that condition (the appendix of [28] has a handy listing of the sounds that were identified with at least 70% accuracy in each channel condition).

One issue that is common to all the hearing impaired populations is a decrease in the ability to separate sounds in a mixture of sounds [70-72], even among hearing impaired listeners fitted with hearing aids. This failure of the "cocktail party effect" [73] means that auditory displays designed for hearing impaired users cannot contain too many concurrent sounds as the users will likely not be able to hear them, which means that many of the context effects mentioned earlier will be difficult to utilize (i.e. it will be more difficult to make an incongruous sound stand out in hearing impaired users).

6.2 Cognitive Factors in Identifiability

Although peripheral processing is a major determinant of identifiability, there are also cognitive factors which can cause hearing impairment. One of these is the general cognitive slowing which occurs with aging [74], which has a strong effect on speech perception: [75] found that declines in speech perception in elderly listeners went above that expected from purely audiometric factors.

In terms of environmental sound perception, the study mentioned earlier which tested baseline identification of a large group of environmental sounds in young normal listeners (YNH) [1] also tested 50 elderly listeners with normal audiograms, adjusted for age (ENH). The mean identification accuracy for the ENH was significantly less than for the YNH, 0.89 vs. 0.95, even though the same tokens were presented in the clear to both groups. Fig. **5.** plots the mean performance on the different sounds for both. Other than the elderly listeners being overall worse at identifying, there are some notable differences on particular sounds: for example, elderly listeners were quite bad at identifying a basketball bouncing, which the young listeners were nearly perfect at. There were also some interesting discrepancies in which particular token was judged to be the "most typical" token for a certain sound: among the airplane tokens, YNH judged a recording of a jet to be the most typical of airplane sounds, whereas ENH thought a prop plane was.

As with the hearing impaired, normal-hearing elderly also have greater difficulty with isolating a sound in a mixture of sounds, which is true both for speech [76] and environmental sounds. The study testing target sounds in natural backgrounds described earlier [40] was replicated with ENH listeners. In addition to overall poorer identification of sounds in all contexts, the Incongruency Advantage described earlier for YNH only occurs in elderly listeners at about 6 dB greater So/Sc levels.

6.3 Visually-Impaired Users

One of the main groups that can benefit from improved auditory displays is visually-impaired users, who must get the vast majority of their information through the auditory modality. [43] showed that using auditory icons significantly speeded up blind users' performance on a word processing task. A large number of Web browsers for the visually-impaired have merely translated the browser into text using a screen reader which is read to the user. There are some applications that go further and try to represent some of the embedded images and functions aurally, such as WebbIE and eGuideDog. A fuller overview of previous commercial products is available in [81] and in the related chapter in this book. Unfortunately, few commercial applications have taken advantage of the work that has been on developing richer auditory interfaces for the blind.

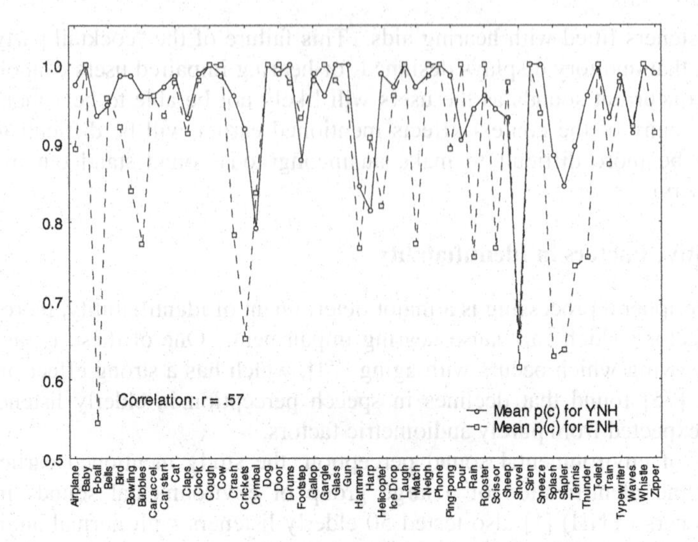

Fig. 5. Mean p(c) for young and elderly listeners for the sounds tested in [1]

One major problem with translating Web pages for the blind is converting the dense visual information which is presented in two dimensions to an auditory presentation which is time-bound. It is possible to map visual coordinates to binaural using Head-Related Transfer Functions (HRTFs) but as noted in the Introduction, spatial resolution for hearing is much poorer than for vision and in binaural presentations there is nearly always masking between components even if they are quite far apart in the display. Such a system was implemented using auditory icons in [77], but no evaluation data were supplied, nor were the types of auditory icons used described.

A more difficult problem is how to present auditory icons that need to be continually available onscreen because the display will very quickly become cluttered with continually sounding icons, causing the user to be unable to discriminate them and overwhelmed with too much information. Other issues are on-screen navigation, handling multiple screens in browsers and help menus. In [81] and in the related chapter in this book by G. Wersényi a detailed treatment of issues in translating GUIs for the blind is provided.

A notable early attempt at designing an auditory-only graphical interface was the Mercator Project [78]. The author carefully examined visual GUIs in term of the structure and function of the objects and translated those to auditory objects .in a hierarchical arrangement (similar to telephone menus) rather than a spatial one. To create auditory icons that resembled visual ones, she tried to find sounds that convey the affordances of the visual icons. A listing of the objects and their associated auditory icons is in Table 5. To parameterize the sounds, as mentioned earlier, filtering was used to "thin" (high pass filter) or muffle (low-pass filter) the sounds to represent highlighting or deselecting of an icon, and to represent properties such as the number of lines (for a text object) of the number of children for a container object. Finally, a number of actions of the system were also associated with auditory icons, which are also listed in Table 5. For example if a user wants to cancel and close a

popup window, the sequence would be:
"Rip" - the selection is successful.
"Whistle-down" - the pop-up disappear from the screen.
"Ca-chunk" "Reply" - the user is moved back to the reply push button.

According to the authors, reactions from the blind users trained in Mercator was positive, (and the use of environmental sounds was "overwhelmingly positive") although no controlled studies were done on the blind. A study conducted with sighted persons showed that although users learned the auditory interface quite quickly, there was little benefit from first becoming familiar with a standard GUI, suggesting that navigating the two system required different skills (not surprising, given the differences between the two).

Although, unlike some of the other system described here, Mercator was developed commercially by Sun, it seems to have gradually disappeared from their supported software and as of 2006 was completely discontinued. A recent attempt at providing a comprehensive auditory interface for the blind is described in [81] and in the related chapter in this book by G. Wersényi. The reader is referred to those references for the complete details, but the project first involved surveying blind users to find the most important functions for them in day-to-day computer use, including actions, applications and on-screen navigation, and then collecting a large number of sounds to represent those functions. The sounds used included auditory icons, emoticons, spearcons (time-compressed speech samples) and what the author terms "audtiory

Table 5. Auditory icons associated with graphical objects. Adapted from [78].

Interface Object	*Sound*
Editable text area	Typewriter, multiple keystrokes
Read-only text area	Printer printing out a line
Push button	Keypress (ca-chunk)
Toggle button	Pull chain light switch
Radio button	Pa pop sound
Check box	One pop sound
Window	Tapping on glass (two taps)
Container	Opening a door
Popup dialog	Spring compressed then extended
Application	Musical sound

Action	*Nonspeech Auditory Feedback*
Selection	Ripping papers
Switching between apps.	Paper shuffling
Navigation error	Ball rebounding against wall
Entering text mode	Rollin or rocking sound (drawer pulled out)
Moving edit cursor in text area	Click &itch based on position in text)
Popup appearing or disappearing	Whistle up or down
Application connecting to Mercator	Winding
Application disconnecting	Flushing
Selection	Ripping papers
Switching between applications	Paper shuffling

emoticons.“ Some of the uses were quite novel, such as using so the sound is a locking sound of a door to save a file, and as an extension, using the same sound for “save as” with an additional human “hm?” sound indicating that the securing process needs user interaction: a different file name to enter.

The sounds were then evaluated by blind and sighted users for identifiability, and “comfortability“ (how pleasing it is to listen to it). The project is ongoing and is in the first stages of implementation. So there is still a need for a functional auditory interface that can truly translate visual graphics into appropriate auditory icons.

For the designer who wishes to undertake such a project, there are some factors to keep in mind. Blind users can be assumed to have normal hearing and cognitive skills. In terms of sensitivity to environmental sounds, [79] reported no differences between blind and sighted persons in accuracy for identifying environmental sounds (although the overall performance in both groups was rather low, p(c) = 0.76-0.78. Comparing sighted with blind users' ratings of mapping from auditory icons to interface events, [80] found blind users gave significantly lower overall ratings of the appropriateness of the mappings, perhaps indicating that blind users have stronger associations of sounds to physical events, and so to abstract away from the established nomic relationship to a more metaphorical one is more difficult. This may be alleviated with training, but it does pose a challenge for designing an auditory interface that will be immediately usable for visually-impaired listeners.

7 Summary

This chapter has attempted to outline some of the factors auditory display designers need to be mindful of when using environmental sounds for auditory icons:

1) Identifiability. The source of the original sound must be easily and rapidly identifiable for the users. A large number of common environmental sounds can be readily recognized by users, but not all, and the identifiability suffers if there is any filtering or editing of the sounds.

2) Context. If one auditory icons is embedded in a number of other sounds, the context can either enhance identifiability or detract from it, depending on contextual congruency of the target icon sound and other background sounds. An incongruent sound can be more identifiable if listener is monitoring for expected sounds.

3) Mapping (i.e. choosing the right sound for the right icon). Some sounds more readily lend themselves to certain functions in a display (for instance the sound of a closing door when shutting down a program) but these are not standardized. The semantic attributes of a number of environmental sounds have been established and the designer can use these to find an appropriate sound for an icon.

4) Parameterization of the underlying sounds. To represent continuous data with an auditory icon, the underlying sound must be manipulable in terms of the sound source properties. There are several physical models of environmental sounds that can achieve this; choosing the right one will depend on ease of implementation, memory requirements and degree of fidelity to the original sound desired.

5) Target population. When designing displays for special populations, special considerations must be made if the target population is hearing impaired or has a cognitive impairment. For hearing impaired without hearing aids, sounds that depend

on high frequency information for identification should be avoided. For hearing aid users, the dynamic range of the sounds should be compressed to avoid causing pain to the listeners. Cochlear implant recipients will perform best with sounds that rely largely on temporal structuring. Elderly listeners generally perform less well on sound identification than young listeners, even elderly with normal hearing. However, for some sounds they perform similarly to young listeners. All of these groups have difficulty when there are numerous sounds in the display and they have to focus on a single sound. For these users, the display designer should take care than the auditory display does not have too many concurrent sounds.

Acknowledgments

This work was supported by a Merit Review Training Grant from the Department of Veterans Affairs Research Service, VA File # 06-12-00446, and the Rush University Medical Center.

References

1. Gygi, B.: Studying environmental sounds the Watson way. J. Acoust. Soc. Am. 115(5), 2574 (2004)
2. Mynatt, E.: Designing with auditory icons. In: Proceedings of the 2nd International Conference on Auditory Display (ICAD 1994), Santa Fe, NM, U.S, pp. 109–119 (1994)
3. Zhao, F., Stephens, S.D.G., Sim, S.W., Meredith, R.: The use of qualitative questionnaires in patients having and being considered for cochlear implants. Clin. Otolaryngol. 22, 254–259 (1997)
4. Warren, R.M.: Auditory perception and speech evolution. Annals of the New York Academy of Sciences, Origins and Evolution of Language 280, 708–717 (1976)
5. Gaver, W.W.: Auditory icons: using sound in computer interfaces. Hum. Comput. Interact. 2(2), 167–177 (1986)
6. Gaver, W.W.: The SonicFinder: An Interface that Uses Auditory Icons. Hum. Comput. Interact. 4(1), 67–94 (1989)
7. Kunkler-Peck, A.J., Turvey, M.T.: Hearing shape. J. Exp. Psychol. Human 26(1), 279–294 (2000)
8. Carello, C., Anderson, K.L., Kunkler-Peck, A.J.: Perception of object length by sound. Psychol. Sci. 9(3), 211–214 (1998)
9. Li, X., Logan, R.J., Pastore, R.E.: Perception of acoustic source characteristics: Walking sounds. J. Acoust. Soc. Am. 90(6), 3036–3049 (1991)
10. Keller, P., Stevens, C.: Meaning From Environmental Sounds: Types of Signal-Referent Relations and Their Effect on Recognizing Auditory Icons. J. Exp. Psychol.-Appl. 10(1), 3–12 (2004)
11. Absar, R., Guastavino, C.: Usability of Non-Speech Sounds in User Interfaces. In: Proceedings of the 14th International Conference on Auditory Display Paris, France, pp. 1–8 (2008)
12. Fernstroem, M., Brazil, E., Bannon, L.: HCI Design and Interactive Sonification for Fingers and Ears. IEEE MultiMedia 12(2), 36–44 (2005)

13. Watson, C.S., Kidd, G.R.: Factors in the design of effective auditory displays. In: Proceedings of the 2nd International Conference on Auditory Display (ICAD 1994), Santa Fe, NM, U.S, pp. 293–303 (1994)
14. Mynatt, E.: Designing with auditory icons. In: Proceedings of the 2nd International Conference on Auditory Display (ICAD 1994), pp. 109–119. International Community for Auditory Display, Santa Fe, NM, U.S (1994)
15. Gaver, W.W.: Using and creating auditory icons. In: Kramer, G. (ed.) SFI studies in the sciences of complexity, pp. 417–446. Addison Wesley, Longman (1992)
16. Ballas, J.: Delivery of information through sound. In: Kramer, G. (ed.) SFI studies in the sciences of complexity, pp. 79–94. Addison Wesley, Longman (1992)
17. Ballas, J.A.: What is that sound? Some implications for sound design. In: Design Sonore, Paris, pp. 1–12 (2002)
18. Lucas, P.A.: An evaluation of the communicative ability of auditory icons and earcons. In: Proceedings of the 2nd International Conference on Auditory Display (ICAD 1994), Santa Fe, NM, U.S, pp. 121–128 (1994)
19. Lass, N.J., Eastman, S.K., Parrish, W.C., Ralph, D.: Listeners' identification of environmental sounds. Percept. Motor. Skill. 55(1), 75–78 (1982)
20. Miller, J.D., Tanis, D.C.: Recognition memory for common sounds. Psychon. Sci. 23(4), 307–308 (1973)
21. Lawrence, D.M., Banks, W.P.: Accuracy of recognition memory for common sounds. Bull. Psychonom. Soc. 1(5A), 298–300 (1973)
22. Vanderveer, N.J.: Ecological acoustics: Human perception of environmental sounds. Dissertation Abstracts International 40(9-B), 4543 (1980)
23. Ballas, J.A.: Common factors in the identification of an assortment of brief everyday sounds. J. Exp. Psychol. Human 19(2), 250–267 (1993)
24. Fabiani, M., Kazmerski, V.A., Cycowicz, Y.M.: Naming norms for brief environmental sounds: Effects of age and dementia. Psychophys. 33(4), 462–475 (1996)
25. Gygi, B., Kidd, G.R., Watson, C.S.: Spectral-temporal factors in the identification of environmental sounds. J. Acoust. Soc. Am. 115(3), 1252–1265 (2004)
26. Marcell, M.M., Borella, D., Greene, M., Kerr, E., Rogers, S.: Confrontation naming of environmental sounds. J. Clin. Exp. Neuropsyc. 22(6), 830–864 (2000)
27. Myers, L.L., Letowski, T.R., Abouchacra, K.S., Kalb, J.T., Haas, E.C.: Detection and recognition of octave-band sound effects. J Am. Acad. Otolayrn 7, 346–357 (1996)
28. Shafiro, V.: Identification of environmental sounds with varying spectral resolution. Ear. Hear. 29(3), 401–420 (2008)
29. Shafiro, V., Gygi, B., Cheng, M.-Y., Mulvey, M., Holmes, B.: Perception of speech and environmental sounds in cochlear implant patients. J. Acoust. Soc. Am. 123(5), 3303 (2008)
30. Guillaume, A., Pellieux, L., Chastres, V., Blancard, C.: How long does it take to identify everyday sounds. In: ICAD 2004 -Tenth Meeting of the International Conference on Auditory Display, Sydney, Australia, pp. ICAD04-1–ICAD04-4 (2004)
31. Gygi, B.: From acoustics to perception: How to listen to meaningful sounds in a meaningful way. J. Acoust. Soc. Am. 113(4), 2326 (2003)
32. Warren, W.H., Verbrugge, R.R.: Auditory perception of breaking and bouncing events: A case study in ecological acoustics. J. Exp. Psychol. Human. 10(5), 704–712 (1984)
33. Gaver, W.W.: What in the world do we hear? An ecological approach to auditory event perception. Ecol. Psychol. 5(1), 1–29 (1993)
34. Mynatt, E.D.: Transforming graphical interfaces into auditory interfaces for blind users. Hum. Comput. Interact. 12(1), 7–45 (1997)

35. Bilger, R.C., Nuetzel, J.M., Rabinowitz, W.M., Rzeczkowski, C.: Standardization of a test of speech perception in noise. J. Speech. Hear. Res. 27, 32–48 (1984)
36. Hafter, E.R., Saberi, K.: A level of stimulus representation model for auditory detection and attention. J. Acoust. Soc. Am. 110(3), 1489 (2001)
37. Schlauch, R.S., Hafter, E.R.: Listening bandwidths and frequency uncertainty in pure-tone signal detection. J. Acoust. Soc. Am. 90, 1332–1339 (1991)
38. Watson, C.S., Foyle, D.C.: Central factors in the discrimination and identification of complex sounds. J. Acoust. Soc. Am. 78(1), 375–380 (1985)
39. Ballas, J.A., Mullins, T.: Effects of context on the identification of everyday sounds. Hum. Perform. 4(3), 199–219 (1991)
40. Gygi, B., Shafiro, V.: The Incongruency Advantage for Environmental Sounds Presented in Natural Auditory Scenes. Submitted (2008)
41. Gordon, R.D.: Attentional allocation during the perception of scenes. J. Exp. Psychol. Human. 30, 760–777 (2004)
42. Lucas, P.A.: An evaluation of the communicative ability of auditory icons and earcons. In: Proceedings of the 2nd International Conference on Auditory Display (ICAD 1994), pp. 121–128. International Community for Auditory Display, Santa Fe, NM, U.S (1994)
43. Petrie, H., Morley, S.: The use of non-speech sounds in non-visual interfaces to the MS Windows GUI for blind computer users. In: Proceedings of the 5th International Conference on Auditory Display (ICAD 1998), pp. 1–5. University of Glasgow, U.K (1998)
44. Kidd, G.R., Watson, C.S.: The perceptual dimensionality of environmental sounds. Noise Cont. Eng. J. 51(4), 216–231 (2003)
45. Gaver, W.W.: How do we hear in the world? Explorations in ecological acoustics. Ecol. Psychol. 5(4), 285–313 (1993)
46. William, W.G.: Auditory icons: using sound in computer interfaces. Hum. Comput. Interact. 2(2), 167–177 (1986)
47. Coleman, P.D.: An analysis of cues to auditory depth perception in free space. Psychological Bulletin 60(3), 302 (1963)
48. Freed, D.: Auditory correlates of perceived mallet hardness for a set of recorded percussive sound events. J. Acoust. Soc. Am. 87(1), 311–322 (1990)
49. Grey, J.M.: Multidimensional perceptual scaling of musical timbres. J. Acoust. Soc. Am. 61(5), 1270–1277 (1977)
50. Lakatos, S., McAdams, S., Caussé, R.: The representation of auditory source characteristics: Simple geometric form. Percept. Psychophys. 59(8), 1180–1190 (1997)
51. Gygi, B., Kidd, G.R., Watson, C.S.: Similarity and Categorization of Environmental Sounds. Percept. Psychophys. 69(6), 839–855 (2007)
52. Aramaki, M., Kronland-Martinet, R.: Analysis-synthesis of impact sounds by real-time dynamic filtering. IEEE T Speech Audi P 14(2), 1–9 (2006)
53. Avanzini, F.: Synthesis of Environmental Sounds in Interactive Multimodal Systems. In: Proceedings of the 13th International Conference on Auditory Display (ICAD 2007), Montreal, Canada, pp. 181–188 (2007)
54. Lee, J.-F., Shen, I.Y., Crouch, J., Aviles, W., Zeltzer, D., Durlach, N.: Using physically based models for collision-sound synthesis in virtual environments. J. Acoust. Soc. Am. 95(5), 2967 (1994)
55. Cook, P.R.: Physically inspired sonic modeling (PhISM): Synthesis of percussive sounds. Comput. Music J. 21, 38–49 (1997)

56. Rath, M.: An expressive real-time sound model of rolling. In: The 6th International Conference on Digital Audio Effects (DAFX 2003), pp. 165–168. University of London, Queen Mary (2003)
57. Stoelinga, C., Chaigne, A.: Time-Domain Modeling and Simulation of Rolling Objects. Acta Acust United Ac 93, 290–304 (2007)
58. van den Doel, K.: Physically-Based Models for Liquid Sounds. In: Tenth Meeting of the International Conference on Auditory Display, Sydney, Australia, pp. 1–8 (2004)
59. Avanzini, F., Serafin, S., Rocchesso, D.: Modeling Interactions Between Rubbed Dry Surfaces Using an Elasto-Plastic Friction Model. In: Proceedings of the COST-G6 Conf. Digital Audio Effects (DAFX 2002), Hamburg, pp. 111–116 (2002)
60. Lakatos, S., Cook, P.C., Scavone, G.P.: Selective attention to the parameters of a physically informed sonic model. J. Acoust. Soc. Am. 107(5,Pt1), L31-L36 (2000)
61. Lutfi, R.A., Liu, C.-J.: Individual differences in source identification from synthesized impact sounds. J. Acoust. Soc. Am. 122(2), 1017–1028 (2007)
62. Reed, R.K., Kidd, G.R.: Detection of Spectral Changes in Everyday Sounds. Indiana University, Unpublished data (2007)
63. Fernström, M., Brazil, E.: Human-Computer Interaction Design based on Interactive Sonification - Hearing Actions or Instruments/Agents. In: Proceedings of the 2004 International Workshop on Interactive Sonification, pp. 1–4. Bielefeld University, Germany (2004)
64. Heller, L.M.W.: When sound effects are better than the real thing. J. Acoust. Soc. Am. 111(5 pt.2), 2339 (2002)
65. Moore, B.C.J.: Cochlear hearing loss, pp. 47–88. Whurr Publishers, London (1998)
66. Badran, S., Osama, E.L.: Speech and environmental sound perception difficulties by patients with hearing loss requiring and using hearing aid. Indian J. Oto. 4(1), 13–16 (1998)
67. Shannon, R.V., Zeng, F.-G., Kamath, V., Wygonski, J., et al.: Speech recognition with primarily temporal cues. Science 270(5234), 303–304 (1995)
68. Reed, C.M., Delhorne, L.A.: Reception of Environmental Sounds Through Cochlear Implants. Ear. Hear. 26(1), 48–61 (2005)
69. Inverso, D.: Cochlear Implant-Mediated Perception of Nonlinguistic Sounds, Unpublished Thesis, Gallaudet University (2008)
70. Bronkhorst, A.W., Plomp, R.: Effect of multiple speechlike maskers on binaural speech recognition in normal and impaired hearing. J. Acoust. Soc. Am. 92(6), 3132 (1992)
71. Festen, J.M., Plomp, R.: Effects of fluctuating noise and interfering speech on the speech-reception threshold for impaired and normal hearing. J. Acoust. Soc. Am. 88(4), 1725 (1990)
72. Loizou, P.C., Hu, Y., Litovsky, R., Yu, G., Peters, R., Lake, J., Roland, P.: Speech recognition by bilateral cochlear implant users in a cocktail-party setting. J. Acoust. Soc. Am. 125(1), 372 (2009)
73. Cherry, C.: Some experiments on the recognition of speech with one and with two ears. J. Acoust. Soc. Am. 26, 975–979 (1953)
74. Salthouse, T.A.: The processing-speed theory of adult age differences in cognition. Psychol. Rev. 103(3), 403–428 (1996)
75. Divenyi, P.L., Stark, P.B., Haupt, K.: Decline of Speech Understanding and Auditory Thresholds in the Elderly. J. Acoust. Soc. Am. 118, 1089–1100 (2005)
76. Humes, L.E., Lee, J.H., Coughlin, M.P.: Auditory measures of selective and divided attention in young and older adults using single-talker competition. J. Acoust. Soc. Am. 120(5), 2926 (2006)

77. Mynatt, E.D.: Transforming graphical interfaces into auditory interfaces for blind users. Hum. Comput. Interact. 12(1), 7–45 (1997)
78. Roth, P., Petrucci, L., Pun, T., Assimacopoulos, A.: Auditory browser for blind and visually impaired users. In: CHI 1999 extended abstracts on Human factors in computing systems, pp. 1–2. ACM, Pittsburgh (1999)
79. Cobb, N.J., Lawrence, D.M., Nelson, N.D.: Report on blind subjects' tactile and auditory recognition for environmental stimuli. Percept Mot Skills 48(2), 363–366 (1979)
80. Petrie, H., Morley, S.: The use of non-speech sounds in non-visual interfaces to the MS Windows GUI for blind computer users. In: Proceedings of the 5th International Conference on Auditory Display (ICAD 1998), pp. 1–5. British Computer Society, University of Glasgow, U.K (1998)
81. Wersényi, G.: Evaluation of auditory representations for selected applications of a graphical user interface. In: Proceedings of the 15th International Conference on Auditory Display (ICAD 2009), Re: New – Digital Arts Forum, Copenhagen, Denmark, pp. 41–48 (2009)

Simulating the Soundscape through an Analysis/Resynthesis Methodology

Andrea Valle[1], Vincenzo Lombardo[1], and Mattia Schirosa[1,2]

[1] CIRMA, Università di Torino
via Sant'Ottavio, 20, 10124 Torino, Italy
[2] Music Technology Group, Universitat Pompeu Fabra
Roc Boronat, 138, 08018 Barcelona, Spain
andrea.valle@unito.it, vincenzo@di.unito.it, mattia.schirosa@iua.upf.edu
http://www.cirma.unito.it/
http://mtg.upf.edu/

Abstract. This paper presents a graph-based system for the dynamic generation of soundscapes and its implementation in an application that allows for an interactive, real-time exploration of the resulting soundscapes. The application can be used alone, as a pure sonic exploration device, but can also be integrated into a virtual reality engine. In this way, the soundcape can be acoustically integrated in the exploration of an architectonic/urbanistic landscape. The paper is organized as follows: after taking into account the literature on soundscape, we provide a formal definition of the concept; then, a model is introduced, and finally, we describe a software application together with a case-study[1].

1 Introduction

The term "soundscape" was firstly introduced (or at least, theoretically discussed) by R. Murray Schafer in his famous book *The tuning of the world* [6]. Murray Schafer has led the research of the World Forum For Acoustic Ecology, a group of researchers and composers who empirically investigated for the first time the "environment of sounds" in different locations both in America and in Europe. Murray Schafer and his associates studied for the first time the relation between sounds, environments and cultures. Hence on, the diffusion of the term has continuously increased, and currently the concept of soundscape plays a pivotal role at the crossing of many sound-related fields, ranging from multimedia [7] to psychoacoustics [8], from working environment studies [9] to urban planning [10], from game design [11] [12] to virtual reality [13], from data sonification [14] to ubiquitous computing [15] [16]: soundscape is a fundamental notion for acoustic design [17] [18], electroacoustic composition [19], auditory display studies [20].

Indeed, such a diffusion of the term is directly proportional to the fuzziness of its semantic spectrum. It is possible to individuate three main meanings of the term "soundscape", related to three different areas of research:

[1] Preliminary works to this paper includes [1], [2], and [3]. The GeoGraphy system has been initially discussed in [4], [5].

S. Ystad et al. (Eds.): CMMR/ICAD 2009, LNCS 5954, pp. 330–357, 2010.

- Ecology/anthropology [21]. Since Murray Schafer's pioneering work, this perspective aims at defining the relevance of sound for the different cultures and societies in relation to the specific environment they inhabit. A soundscape is here investigated through an accurate social and anthropological analysis, with two goals. On the one side, the researchers are interested in documenting and archiving sound materials related to a specific socio-cultural and historical context. On the other side, they aim at leading the design of future projects related to the environmental sound dimension.
- Music and sound design [22]. The musical domain is particularly relevant. All along the 20th century, ethnomusicological studies, bruitism, "musique d'ameublement" and "musique anecdotique" have pushed the research discussion on environmental sound dimension as acoustic scenery [23]. At the same time, musique concrète has prompted composers to think about sounds as sound objects. During the '60-'70s many composers started working with sound field recording. Sharing the *musique concrète* attitude towards sound, they have been strongly influenced by the soundscape studies. Not by chance, many of Murray Schafer's associates were composers. Thus, the concept is widely present in many contemporary musical forms, as the soundscape itself is regarded as a form of "natural" music composition (in general, cf. [24]). More, "soundscape composition" identifies a mainly electro-acoustic genre, starting from natural acoustic environmental sounds, sometimes juxtaposed to musical scores. Also, sound designers working for cinema and TV have contributed to the diffusion of the term, indicating with "soundscape" the idea of an acoustic scenario to be added/adapted to the moving image ([25], cf. [26]).
- Architecture/urban planning [27]. In recent years, electro-acoustic technology and architectural acoustics have allowed to think about the relation between sound and space in a new form, in order to make citizens aware of the sonic environment (the soundscape) they live in, so that they can actively contribute to its re-design. Many architectural projects have been developed descending from these assumptions [28]. The concept of "lutherie urbaine" was proposed as a combined design –of architecture and of materials– for the production of monumental elements located in public spaces and capable of acting like resonators for the surrounding sound environment [29].

It must be noted that such a complex and rich set of features related to soundscape is extremely relevant because it demonstrates that the problem of the relation between sound and space cannot be only solved in acoustic or psychoacoustic terms. An acoustic or psycho-acoustic approach considers the relation among sound, space and listener in terms of signal transfer [30]. Acoustic ecology, through a large body of studies dedicated to soundscape description and analysis [6], has pointed out that the soundscape perception implies the integration of low-level psychoacoustic cues with higher level perceptual cues from the environment, its cultural and anthropological rooting, its deep relations with human practices. The integration of soundscape in a landscape documentation/simulation is crucial in order to ensure a believable experience in human-computer interaction [31]. A consequence of the integration among different perceptual domains and among multilevel information is that the study of soundscape requires to include phenomenological and semiotic elements. In this sense, the study of soundscape can benefit from the research in "audio-vision", i.e. the study of the relation between audio and video in audiovisual products (film, video etc) [32]. More, soundscape studies have highlighted the relevance of different listening strategies in the perception of the sonic environments: from a phenomenological perspective ([33], [34]) it is possible to identify an "indexical" listening (when sounds are brought back to their source), a "symbolic" listening (which maps a sound to its culturally-specific meanings), an "iconic" listening (indicating the capabilites of creating new meanings from a certain sound material, in general see [35]).

2 For a Definition of Soundscape

As the semantic spectrum of the term "soundscape" is quite fuzzy, a modeling of soundscape requires first an explicit definition. With this aim, we need to introduce a number of concepts. A "sound object" is a cognitive and phenomenological unit of auditory perception [33]. It can be thought as an "auditory event" [36] and integrated in terms of ecological and cognitive plausibility in the auditory scene analysis approach [37]. Its nature of "object" is intended to emphasize its semiotic quality. This means that a sound object is always related to a specific listening practice, so it is not exclusively placed at the perceptual level but also related to a specific cultural context. In short, we can say that a sound object (*SO*) is a function of perceptual organization and cultural context. Assigning the cultural context a pivotal role, traditional soundscape studies have insisted on a tripartite typology of sounds in relation to their socio-cultural function [21]: keynote sounds, signal sounds, soundmarks. Keynote sounds are the sounds heard by a particular society continuously, or frequently enough, to form a background against which other sounds are perceived (e.g. the sound of the sea for a maritime community). Signals stands to keynotes sounds as a figure stands to a background: they emerges as isolated sounds against a keynote background (e.g. a fire alarm). Soundmarks are historically relevant signals (e.g. the ringing of the historical bell tower of a city). While this classification is intended as a guidance for the analysis of the soundscape in its cultural context, here we propose a pivotal role for the perceptual and indexical organization of the soundscape. So, we yield a different classification for the sound objects of a soundscape:

- atmospheres: in relation to sound, Böhme has proposed an aesthetics of atmospheres [38]. Every soundscape has indeed a specific "scenic atmosphere", which includes explicitly an emotional and cultural dimension. An atmosphere is an overall layer of sound, which cannot be analytically decomposed into single sound objects, as no particular sound object emerges. Atmosphere characterizes quiet states without relevant sound events. While keynote sounds are intended as background sounds (i.e. they are a layer of the soundscape), atmospheres identify the whole sound complex.

- events: an event is a single sound object of well-defined boundaries appearing as an isolated figure. In this sense, it is equivalent to a signal as defined in soundscape studies.
- sound subjects: a sound subject represents the behavior of a complex source in terms of sequencing relations between events. In other words, a sound subject is a description of a sound source in terms of a set of events and of a set of sequencing rules.

In other words, an atmosphere is a sound object which source cannot be identified, as the source coincides with the whole environment. Events and sound subjects are sound objects related to specific sources. In the case of an event, the behavior of the source is simple, and can be thought as the emission of a specific sound object. In the case of a sound subject, the behavior is complex and is specified as a set of generation/sequencing relations.

However, the previous classification of sound objects is not enough to exhaustively define a soundscape. A soundscape is not only a specific structure of sound objects arranged in time (otherwise, every piece of music could be defined a soundscape), but is related to a space, and the exploration of such a space reveals other aspects of the same soundscape. This exploration is performed by a listener, not to be intended as a generic psycho-acoustic subject but as a culturally-specific one: through the exploration of the space, the listener defines a transformation on the sound objects that depends on the mutual relation between her/himself and the objects. The transformation is both spatial –as it depends on features related to the listening space (e.g. reverberation)– and semiotic –as it depends on cultural aspects (e.g. specific listening strategies). By coupling the spatial and semiotic aspects, Wishart [39] has discussed in depth the symbolic construction of landscape in acousmatic listening conditions, by taking into account this overall sound dimension. For Wishart "the landscape of a sound-image" is "the imagined source of the perceived sounds" ([39]: p. 44). In this sense, for Wishart the landscape of the sounds heard at an orchestral concert is musician-playing-instruments, exactly as the landscape of the same concert heard over loudspeakers through recording. So, the first component of the reconstructed landscape is a semiotic construction based on the cultural coding of soundscape. Wishart proposes a semiotic/phenomenological description of natural soundscapes. For instance: moorlands reveal a lack of echo or reverberation, sense of great distance, indicated by sounds of very low amplitude with loss of high-frequency components; valleys display a lack of distance cues and possibly include some specific image echos; forests are typified by increasing reverberation as the distance of the source from the listener increases ([39]: p. 45). Such a characterization can be semantically described through three continuous parameters: dynamics (cf. in music: from *ppp* to *fff*), reverberation (expressed along the axis dry/wet), brightness (in timbral studies, typically along the bright/dull axis)[2]. These three categories provide three descriptive axes for a qualitative evaluation of the global acoustic behavior of the soundscape.

[2] It can be noted that a similar semantic description is strictly correlated to psychoacoustic cues, respectively intensity, direct-to-reverberant ratio, spectrum [31].

The second component of the reconstructed landscape is the distribution of sound objects over the space indexically related to the soundscape. Many scholars have noted that a soundscape can be decomposed as a group of several acoustic scenographies, which are then recomposed through the listener's exploring experience [28] [27] [29] [24]. As soundscapes are not uniform, the listener's experience is enhanced when s/he encounters aural transitions during her/his exploration of the environment [38]. When a listener is spatially exploring the soundscape, he can notice several perceptual differences in the sounds. In particular, the sound environment can be decomposed into internally homogeneous sub-parts. These sub-parts are here referred to as "sound zones". Sound zones can differ in dimension and number of elements, but they are characterized by typical sources, i.e. sound emissions that are often present in a region and absent (or only rarely heard) in the others. The soundscape will then results from the summation of all the sound zones that the listener will be able to explore. As an example, we can consider the following situation: in a classroom with acoustic insulation walls, closed doors and windows, a teacher is speaking in front of a very silent audience. The professor voice is loud and clear all over the classroom, without any relevant irregularity. By contrast, we can imagine the opposite situation: doors and windows are open, the thin walls are incapable of blocking any environmental sound, outside there are roadworks, a party is running in the hall just behind the door, a few students joke and laugh while the professor keeps explaining loudly. This second soundscape (represented in Figure 1) is completely different from the first one. Considering a wide classroom, it would be very simple to move around the space and run across several recognizable sub-soundscapes ("zones"). Someone near the door can notice that reception party sounds are louder than any other sound source coming from the classroom. As s/he moves to the desk, s/he can hear the teacher's voice. And so on. In the first insulated room case it is possible to identify a soundscape consisting of one only zone; in the second case four sound zones are clearly defined. Thus, even if their boundaries can be fuzzy, each zone can be considered as completely independent on another. This means that it is possible to describe the character of each zone. To sum up, a zone is a region of the space that can include all the different kinds of sound objects previously discussed (atmospheres, events and sound subjects) and is perceptually separated from another zone. That is, the listener can detect a perceptually different soundscape when crossing over zone edges. Empirically we can determine a threshold that represents a measure of the just noticeable difference (JND) over zones.

A soundscape results to be composed of several zones and sound edges.

No matter the definition of the discrimination function, the term "soundscape" indicates the summation of all the sound zones that a listener will be able to explore. In order to take into account the composite nature of the soundscape, it is thus necessary to model the exploration process performed by the listener, and the resulting aural transitions. From the previous discussion, we can provide the following definition:

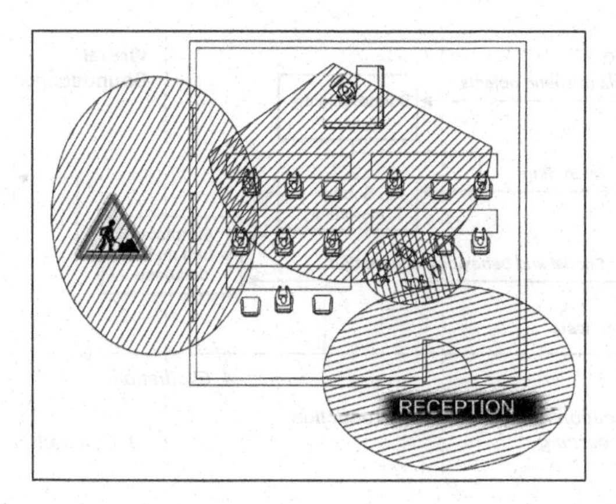

Fig. 1. The classroom example: four sound zones

A soundscape is a temporal and typological organization of sound objects, related to a certain geo-cultural context, in relation to which a listener can apply a spatial and semiotic transformation.

3 Towards a Formalized Model

Despite the profusion of usages, there are neither models nor applications aiming at a simulation of a soundscape starting from the analysis of an existing soundscape. Listen [40] works on the generation and control of interactive soundscapes, but does not include an explicit modeling of the soundscape itself. Tapestrea [22] is designed to generate "environmental audio" in real time, but does not define any relation between sound and space.

In devising a computational model to be implemented in an application, a first issue concerns the epistemological status of sound objects. As discussed above, they can be described as a function of perceptual and cultural factors. Cultural context results from encyclopedic knowledge on the environment and from interviews to different typologies of users (e.g. usual/occasional goers/workers, musicians/non-musicians, etc.). A related issue concerns the relation between the acoustic and the phenomenological levels. For each sound object we must identify a corresponding signal. The problem is particularly relevant when the sound objects play simultaneously. The more complex a soundscape, the more difficult to unambiguously identify a signal corresponding to a sound object. Hence on, a "sound material" will be an audio signal corresponding to a certain sound object. Thus, in the analysis of a complex soundscape, there are at least two difficult issues: first, the decomposition of the sound continuum into sound objects; second, the retrieval of the corresponding signal for each sound object. As an example, in the soundscape of a restaurant kitchen, the decomposition into sound objects can be quite hard, as some elements are easy recognizable (clashing cutlery, water in the basin, voices of cooks), but on the other side there is a diffuse texture made of a large amount of microsonic events that can

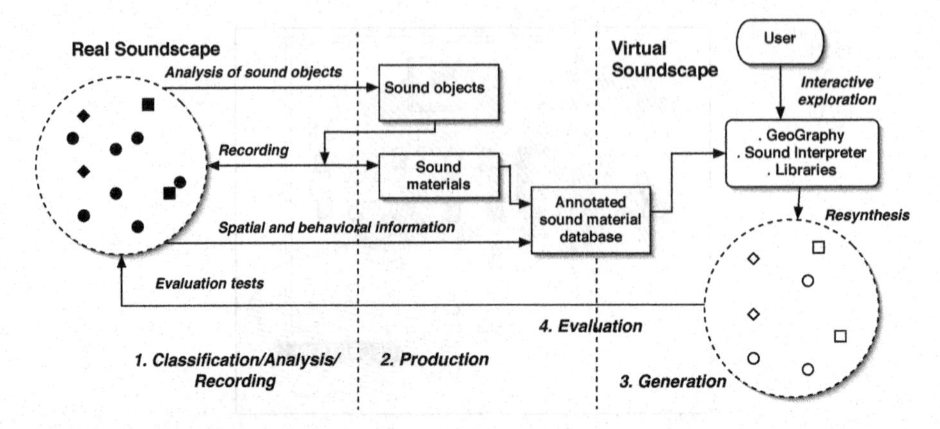

Fig. 2. Method. The generated, vitual, soundscape (bottom right) contains a reduced set of relevant sound objects from the real soundscape (top left).

be hardly brought back to their sources. Even after having identified the sound objects, it can be very difficult to extract isolated signals from the global soundscape. The first issue can be identified as "semiotic discretization", the second as "acoustic discretization". So, we define a semiotic discretization error (*Sde*) and an acoustic discretization error (*Ade*). In the perspective of the simulation of a soundscape (see Figure 2, more details later), we can then identify the following relation between the real soundscape (*RSC*) and the targeted, virtual one (*VSC*):

$$RSC = VSC + Sde + Ade$$

Although the characterization of *Sde* and *Ade* is a very interesting topic, it is outside the scope of this paper. Now, the virtual soundscape *VSC* is characterized by a set of sound objects *SO*:

$$SO = \{so_1, so_2, so_3, \ldots, so_n\}$$

Each sound object *so* is a tuple

$$so = \langle typ, mat, att, pos, ra \rangle$$

Where:

- *typ*: the sound object type (*A*, atmosphere, *E*, event, *SS*, sound subject)
- *mat*: the sound material related to the object and represented as an audio file
- *att*: a sublist representing variable attributes: iteration pattern (for sound subjects), size (e.g. for atmospheres), radiation pattern, etc.
- *pos*: the position of the object in the space, represented as a tuple of coordinates
- *ra*: the radiation area, that is how far from *pos* this sound object gets

A soundscape results from the interaction between a listener and a set of sound objects. A listener is given a certain position and orientation in the space. More, an acoustic transformation is associated to the listener: each sound object is affected by such a transformation in relation to its distance from the listener. Hence, the listener L is a triple

$$L = \langle pos, or, map \rangle$$

where the position *pos* is a tuple of coordinates (as above), the orientation *or* a set of Euler angles, and *map* is the mapping function of the distance/displacement between the sound object and the listener. A soundscape simulation of a real environment should include factors like sound propagation, radiation patterns, reflections, impulse responses. The mapping function encodes all the factors that compose the sound transformations. So, to summarize the virtual soundscape VSC is an audibility function Au defined over a set of sound objects and a listener:

$$VSC(t) = Au(SO, L, t)$$

The function Au takes a input the listener and the set of sound objects to return a set of sound sources in the space related to the listener. Given such a definition, it is possible to propose a model for the simulation of soundscapes. The model features four phases (Fig. 2): 1. classification, analysis and recording, 2. production, 3. generation, 4. evaluation.

4 Classification, Analysis and Recording

The first phase aims at gathering data from the real environment. It includes the classification of sound objects, their perceptual analysis and the recording of the related sound material (see the schema in 3). First (Figure 3, 0), general information on the space is collected, including cultural features (e.g. if it is a religious or a secular space), topographical organization (e.g. if it contains pedestrian areas), global acoustic properties (e.g. if it is reverberant or not and how much). This information is intended as guidance for the subsequent steps.

Then, we proceed at the identification of sound objects relying on qualitative parameters. Here is where the Semiotic discretization error arises. This error depends on two aspects: first, perceptual organization as defined by auditory scene analysis studies ([37]) depends on a set of many heuristics available for each listener to the auditory systems, and the prevalence of certain heuristics for a certain listening subject does not exclude the applicability of other heuristics for the same subject; second, heuristics depend on both primitive and culturally-based schemata, and the relevance of a heuristic with respect to other available ones can be strongly influenced by the competences of the differently involved listeners. The resulting (semiotic discretization) error is not quantifiable (as it does not involve physical entities), but it is nevertheless relevant to take it into account in order not to loose crucial qualitative features of the soundscape. The identification of sound objects consists in a two-step procedure. This limits the

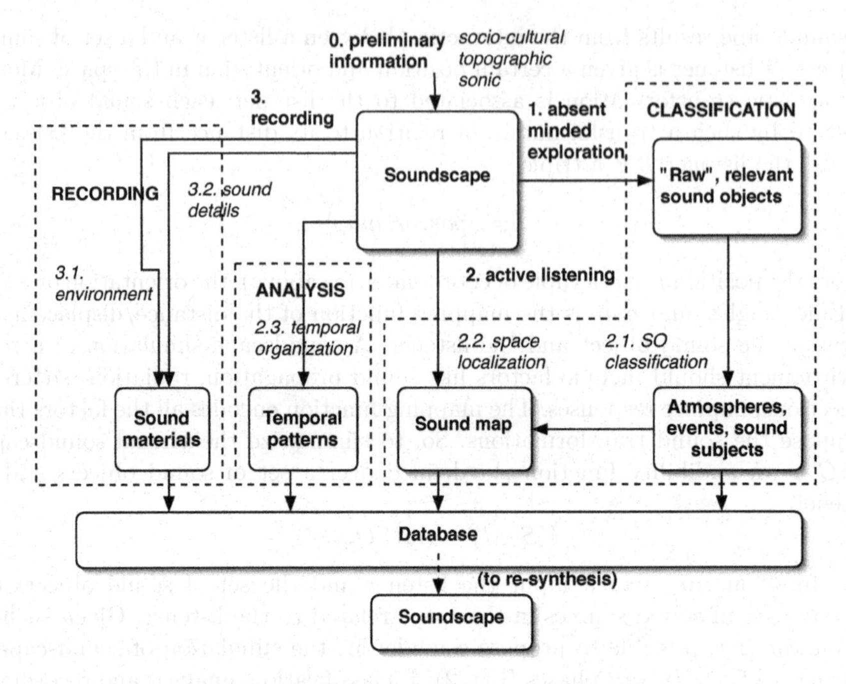

Fig. 3. Classification, Analysis and Recording

subjectivity and arbitrariness of identification and reduces the complexity of the whole procedure. The first step (Figure 3, 1) is an "absentminded" exploration of the soundscape: the analyst must be perceptually open, adhering to a passive listening strategy [26]. In this way s/he identifies the most relevant sound objects of the overall soundscape, i.e. the ones that are evident even to the least aware listeners. Moreover, the analyst carries out interviews with different kinds of listeners, dealing with their global experience of the soundscape at different levels (perceptual, emotional, cultural). In the second step (Figure 3, 2), an active listening strategy locates the sound objects in the space. The soundscape is investigated in depth, so that now even less prominent sound objects are detected and analyzed (Figure 3, 2.1). The step consists in an on-site exploration, so that the eye can complement and help the ear in the retrieval process while aiming at identifying areas with homogeneous sound objects. As an example, in case of a market, different areas can be identified in relation to different stands, pedestrian crossovers, loading/unloading areas, parking zones. It is thus possible to create a sound map partitioned into areas with sound objects assignment (Figure 3, 2.2, an example is shown in Fig. 10).

Then, we focus on the analysis of specific sequences of sound objects (Figure 3, 2.3). As an example, loading/unloading procedures in a market are characterized by specific sequences of sounds: it can be said that they show a specific syntax and temporal ordering. The analysis of the temporal behavior of the syntactical properties of sound objects is fundamental for the parameterization of the generative algorithm (see later).

Finally (Figure 3, 3), we record the raw audio material from the environment. The recording process tries to avoid information loss. The recording procedure features a double approach. On one side, large portions of soundscape are recorded via an omnidirectional microphone (Figure 3, 3.1): so, a large quantity of raw material is available for editing and processing. On the other side, high directivity microphones are used to capture a wide variety of emissions while minimizing undesired background (Figure 3, 3.2). Directional recordings are typically mono, as we devise to add the spatial information within the generative framework.

The data gathered in this phase are furtherly refined in the production phase.

5 Production

The production phase focuses on the creation of the soundscape database. The database is intended as a catalogue of the soundscape information and contains all the previously discussed information. The fundamental operation in this phase is the creation of sound materials. As discussed, a sound material is the audio signal associated with a sound object. The phase consists of two steps.

First, recordings are analyzed through an acousmatic listening, while the classification/recording phase relies on symbolic and indexical listening, trying to identify culturally relevant objects and locate them in the space. The production phase focuses on iconic listening strategy, since it considers the sounds as perceptual objects, regardless of their meaning or relation to the environment. After an accurate, acousmatic listening, the final set of sound objects is identified. In case different sound objects reveal analogous phenomenological features, they can be grouped into a single sound material. For example, the sounds of forks, spoons and knifes are indexically different; at an iconic listening, they can reveal a substantial phenomenological identity, so that they are realized through the same sound material. Indeed, this process –as discussed above– can introduce a form of Semiotic discretization error.

Then, sound materials are created: this operation involves some editing on audio samples (e.g. noise reduction, dynamics compression, normalization). The realization of sound materials introduces a second form of error, the Acoustic discretization error. This error depends on the difference between the target sound object and the physical signal. A typical case for a soundscape is a sound object retrieved inside a noisy, dense continuum. Even if the sound object might be very well isolated by auditory scene analysis, it could not be easily possible to obtain an equally isolated sound material from an environmental recording. In a case like this, the resulting sound material can be heard as a noisy representation of the sound object or in the worst case still revealing other parallel sounds. Noise reduction can indeed improve the quality of the sound material, but on the other side can remove audio information in such a measure that, again, the sound material is heard as a "distorted" version of the intended sound object.

The second step is a general reviewing phase of the previous information in order to create a final sound map. If some sound material is lacking, a new recording session is planned, targeted to that specific sound object. In this way a feedback loop is defined from production to classification.

6 Generation

The information retrieved from the annotation/analysis of the real soundscape is then used to generate a synthesized soundscape. The generation process involves two components: a formal model for the definition of sound object sequences and an interpreter to feed the synthesis algorithms. The formal model defines the sequencing of the sound objects. It is a generative model, i.e. able to create an infinite set of sequences of sound objects. This sequence represents a continuous variation from a finite set of sound objects. An algorithm merges the information coming from the sequencing model with the user's navigation data. In this way, a soundscape can be simulated and explored interactively. The generative component is based on graphs and extends the GeoGraphy system [4] [5]. The interpret (the second component) is responsible for the interpretation of the data generated by the model in a sonic context. Hence, it is named "Sound Interpreter". In the following subsections we first describe the two-layered GeoGraphy system, and then the Sound Interpreter.

6.1 The GeoGraphy System

Graphs have proven to be powerful structure to describe musical structures ([41]): they have been widely used to model sequencing relation over musical elements belonging to a finite set. It can be disputed whether non-hierarchical systems are apt for music organization, as hierarchical structures have proven to be useful in modeling e.g. tonal music [42]. It is common to consider soundscape in terms of multiple parallel and independent layers of sound [6]. A common feature of all these graph representations devised for music is that they generally do not model temporal information: the GeoGraphy model relies on time-stamped sequences of sound objects. The sequencing model is a direct graph (Figure 4, left), where each vertex represents a sound object and each edge represents a possible sequencing relation on a pair of sound objects. This graph is actually a multigraph, as it is possible to have more than one edge between two vertices; it can also include loops (see Figure 4 on vertex B). Each vertex is labeled with the sound object duration and each edge with the temporal distance between the onsets of the two sound objects connected by the edge itself. The graph defines all the possible sequencing relation between adjacent vertices. A sequence of sound objects is achieved through the insertion of dynamic elements, called "graph actants". A graph actant is initially associated with a vertex (that becomes the origin of a path); then, the actant navigates the graph by randomly following the directed edges. Each vertex emits a sound object at the passage of a graph actant. Multiple independent graph actants can navigate a graph

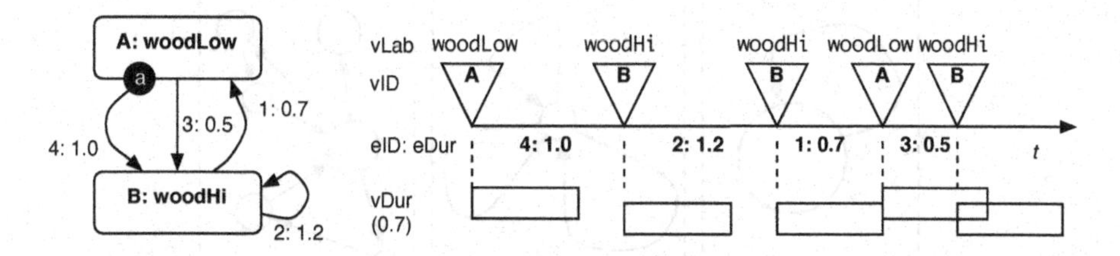

Fig. 4. A graph (left) with an actant "a" starting from edge 4. The resulting sequence is shown on the right. A and B: vertices; 1,2,3: edges. The duration of the vertices is 0.7.

structure at the same time, thus producing more than one sequence. In case a graph contains loops, sequences can also be infinite. As modeled by the graph, the sound object's duration and the delay of attack time are independent: as a consequence, it is possible that sound objects are superposed. This happens when the vertex label is longer than the chosen edge label.

A soundscape includes a set of sequences, which are superposed like tracks: in a soundscape there are as many sequences as graph actants. The generation process can be summarized as follows. Graph actants circulate on the graph: there are as many simultaneous sound object sequences as active graph actants. In the generation process, when an actant reaches a vertex, it passes to the level II the vertex identifier: the ID will be used in the map of graphs to determine if the vertex itself is heard by the Listener (see later).

An example is provided in Figure 4. The graph (left) is defined by two vertices and four edges. The duration of both vertices is set to 0.7 seconds. In Figure 4 (right), vertices are labeled with an identifier ("A", "B"). More, each vertex is given a string as an optional information ("woodLow", "woodHi"), to be used in sound synthesis (see later). A soundscape starts when an actant begins to navigate the graph, thus generating a sequence. Figure 4 (left) represents a sequence obtained by inserting a graph actant on vertex A. The actant activates vertex A ("woodLow"), then travels along edge 4 and after 1 second reaches vertex B ("woodHi"), activates it, chooses randomly the edge 2, re-activates vertex B after 1.2 seconds (edge B is a loop), then chooses edges A, and so on. While going from vertex A to vertex B by edge 3, vertex duration (0.7) is greater then edge duration (0.5) and sound objects overlap. The study of the temporal pattern of the many sound objects provides the information to create graphs capable of representing the pattern. Every graph represents a certain structure of sound objects and its behavior. Different topologies allow to describe structure of different degrees of complexity. This is apparent in relation to the three types of sound objects previously introduced. Atmospheres are long, continuous textural sounds: they can be represented by a single vertex with an edge loop, where the vertex duration (typically of many seconds) coincides with the edge duration (Figure 5, a). In this sense, atmospheres simply repeat themselves. Analogously, events can be represented by graphs made of a single vertex with many different looping edges, which durations are considerably larger than the duration of the vertex

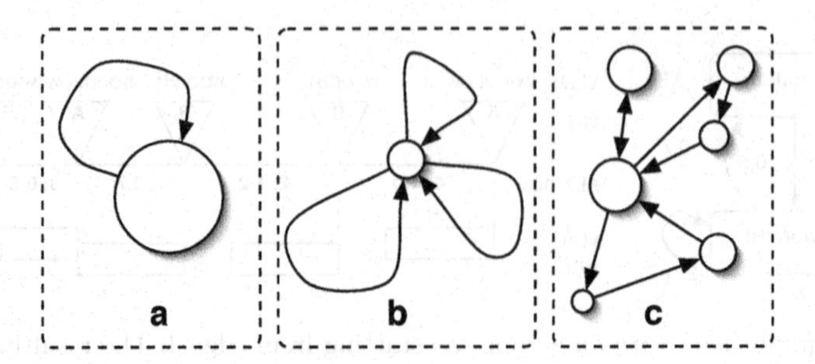

Fig. 5. Possible topologies for atmosphere, events and sound subjects (sizes of vertices and edge lengths roughly represents durations)

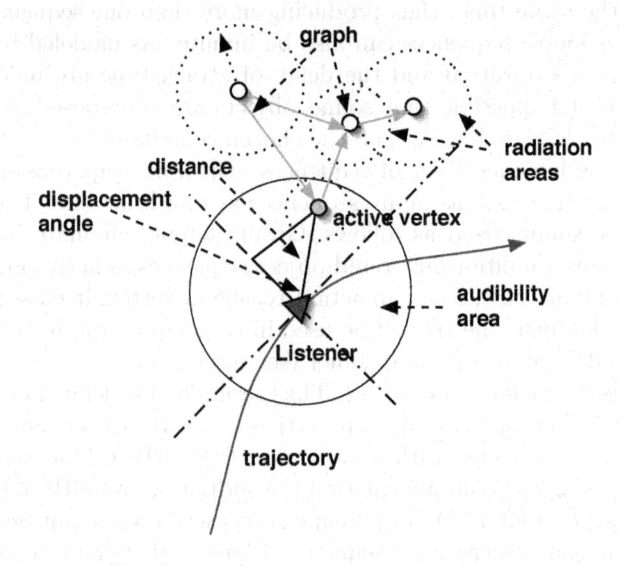

Fig. 6. Listener in the map of graphs. The audibility radius filters out active vertices falling outside.

(Figure 5, b). In this way, isolated, irregularly appearing events can be generated. Indeed, the graph formalism is mostly useful for sound subjects. A sound subject is a complex, irregular pattern involving many sound objects: it can be aptly described by a multigraph (Figure 5, c). The multigraph can generate different sequences from the same set of sound objects: in this sense, it represents a grammar of the sound subject's behavior. Vertices are given an explicit position: in this way, the original location of a sound source is represented. Each vertex is given a radiation area: the radius indicates the maximum distance at which the associated sound object can be heard. The space is named *map of graphs*. A map contains a finite number of graphs (n), which work independently, thus generating a sequences, where a is the total number of the graph actants that navigate in all the graphs. As there is at least one graph actant for each graph, there will be a minimum of n tracks ($a \geq n$), i.e. potential layers of the soundscape. Inside the map of graphs, a dynamic element, a "Listener" determines the actually heard soundscape. The Listener is identified by a position, an orientation and an audibility area (see Fig. 6). The position is expressed as a point in the map; the orientation as the value in radiant depending on the user's interaction control; the audibility area defines the perceptual boundaries of the Listener. The Listener can be thought as a function that filters and parameterizes the sequences of sound objects generated by the graph actants. Every time a vertex is activated by a graph actant, the algorithm calculates the position of the Listener. If the intersection between the Listener's audibility area and the vertex's energetic area is not void, then the Listener's orientation and distance from the vertex are calculated, and all the data (active vertex, position, distance and orientation of the Listener) are passed to the DSP module. To sum up, the two-layer system outputs a sequence of time-stamped vertex IDs with positional information added.

Actually, the model represents the space as a 2-dimensional extension, and assumes that the sound sources (represented by vertices) are static. Through the GeoGraphy model it is possible to generate a target complex soundscape. In Figure 7, a soundscape is represented by two disconnected subgraphs. As discussed, sound objects can overlap their audibility with other objects, depending on their radiation area. The radiation area is set according to dynamics annotations taken during the analysis phase. It is thus possible to regulate the radius of each element to interbreed a sound object with its neighbors. In Figure 7 a soundscape subcomponent is represented by the loop between vertices 4 and 5, respectively having radii I_4 and I_5 (the grouping is represented by the

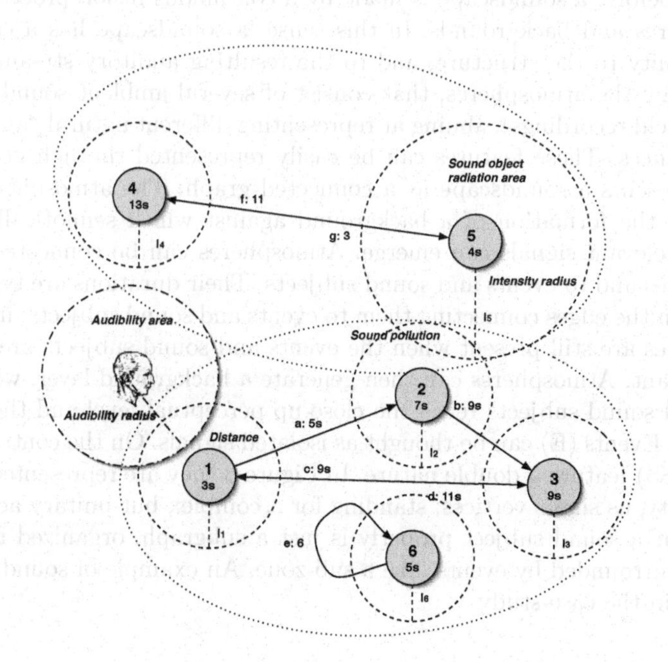

Fig. 7. Graphs and Listener

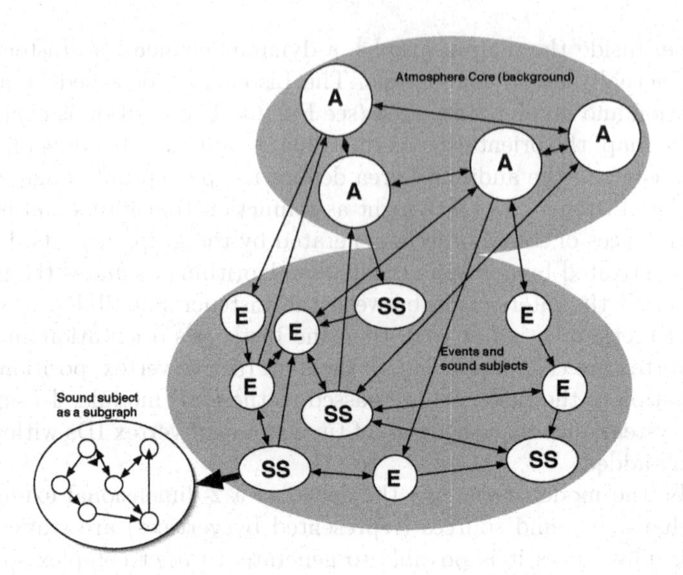

Fig. 8. A graph representing the relations among atmospheres, events and sound subjects in a soundscape

dotted ellipse). Similarly, the loop among vertices 1, 2, 3, 6 individuates another subcomponent, and the intersection between the elements 2 and 5 is the portion of space where the sounds of the two subcomponents can be hearable. As discussed before, a soundscape is made by a continuous fusion process between sound figures and backgrounds. In this sense, a soundscape has a core granting continuity to the structure, and to the resulting auditory stream: the core is formed by the atmospheres, that consist of several ambient sound materials (i.e. long field recordings), aiming at representing different natural "almost-quiet state" nuances. These features can be easily represented through graphs. Figure 8 represents a soundscape as a connected graph. The atmospheres (A, on top) allow the formation of a background against which semiotically and indexically relevant signals can emerge. Atmospheres can be connected to other atmospheres and to events and sound subjects. Their durations are typically set longer than the edges connecting them to events and sound subjects: in this way, atmospheres are still present when the events and sound subjects are activated by the actant. Atmospheres can then generate a background layer, while sound events and sound subjects reach the close-up perceptual level and then quickly disappear. Events (E) can be thought as isolated signals. On the contrary, sound subjects (SS) feature a double nature. In Figure 8 they are represented, for sake of simplicity, as single vertices, standing for a complex but unitary acoustic behavior. But a sound subject properly is just a subgraph, organized recursively in a core surrounded by events, like a sub-zone. An example of sound subject is discussed in the case-study.

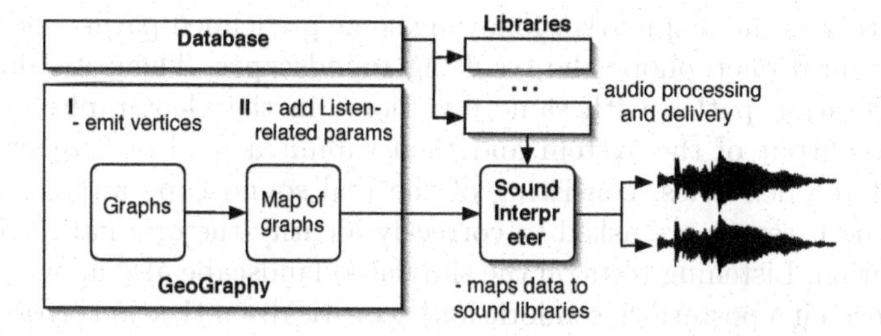

Fig. 9. An overview of the generation process. In this case the final delivery is stereo.

6.2 The Sound Interpreter

The graph-based model does not make any assumption about sound objects, whose generation is demanded to an external component. It defines a mechanism to generate sequences of *referred* sound objects (grouped in sequences). During the generation step, the data from the model are passed to the Sound Interpreter (Figure 9). The Sound Interpreter is responsible for audio retrieval from memory (e.g. samples), synthesis/processing (e.g. reverberation) and final delivery (e.g. stereo). As discussed, for each event the data include attributes of space and sources, and movement. The Interpreter defines the audio semantics of the data by relating them to transform functions. These transform functions are grouped into libraries containing all the necessary algorithms to generate the audio signal. They define a mapping schema associating the vertex IDs (resulting from graph sequencing, Figure 9, I) to sound materials in the database, and spatially-related data (resulting from Listener's movement in the Map of graphs, Figure 9, II) to audio DSP components, e.g. relating distance to reverberation or displacement to multi-channel delivery. By using different libraries, the system allows to define flexible mapping strategies. As an example, a library can define a realistic mapping in which the data from the model are used to control DSP algorithms aiming at a physical simulation of the acoustic properties of the space. Other libraries can include "fictional" rendering of the soundscape. In this way, the continuous nature of the space (populated by the same sound objects) is preserved, even if the global result can sound "alien". Alien mappings are useful to create artificial spaces (for artistic purposes, from music to sound design) and to test the degree of soundscape invariance over different space models.

7 Evaluation

The resulting simulation is evaluated through listening tests, taking into account both the sound materials and the transformations induced by space. As the competences about sound can vary dramatically from a user to another, the evaluation procedure takes into account four different typologies of listeners: occasional visitors, regular goers, non-sensitized listeners, sensitized listeners (musicians/sound designers). In order to evaluate the simulation, we use a comparative procedure. First, we record a set of explorations in the market. That is, a listener travels the real market following some predefined paths, and we record with a binaural microphone the resulting soundscapes. Then, we simulate the same exploration paths, with same durations, in the GeoGraphy system. We record the output of the system and then submit a pool of listeners different pairs of stereo deliveries, consisting of the real soundscape and the simulated version. The listeners are asked to correctly identify the original recording and the simulation. Listening tests on the simulated landscape also allow to partially evaluate, even if a posteriori, semiotic and acoustic discretization errors. Acoustic discretization error is related to the segmentation/extraction of sound materials from the recorded audio signal. An assessment of audio quality, both in terms of overall quality and in terms of single sound materials, is a good indicator of a low discretization error, meaning that single materials have not lost their prominent audio features, and their sum –instead of amplifying distortions introduced during audio editing– produces a smooth audio experience. Semiotic discretization results from the identification of a finite set of sound objects, which are considered as the only relevant sound objects of the soundscape. The reported richness and completeness of the soundscape (in terms of relevant sound objects for the listeners) are two possible indices of a low semiotic discretization error.

8 Case-Study: The Market of "Porta Palazzo"

The model has been tested on a simulation of the soundscape of Porta Palazzo, Turin's historical market (see [43]). The market is a typical case of a socio-cultural relevant soundscape. In particular, the market of Porta Palazzo has a long tradition (it has been established more than 150 years ago): it is the greatest outdoor market in Europe and represents the commercial expression of the cultural heritage of the city of Turin. During the last century, it has tenaciously retained its identity, characterized by the obstinate will of the workers of sharing its government's responsibility.

The analysis of the case-study initially focused on the socio-cultural dimension of the market. First of all, we gathered bibliographic information to better understand the cultural features of such a historical place. The municipality of Turin, that has provided the most up-to-date report on the subject [43] and a detailed map of the whole area, prepared for the urban renovation and environmental improvement of the market. From the report it has emerged that the market of Porta Palazzo is probably the part of Turin where the largest number of different social realities and cultures inhabit. People speaks include languages and dialects from all the regions of Italy, South America, Eastern Europe, North Africa. More, every day the market serves 20,000 persons (80,000 on Saturday), and 5,000 persons work there every day. Not by chance, its soundscape manifests an impressive acoustic richness, as there are many qualitatively different sound sources (both of linguistic and non-linguistic nature). In order to evaluate relevance and features of the soundscape for its usual listeners, we made short informal interviews to local workers, customers and worker representatives. These interviews helped us to easily identify the most pervasive sound objects.

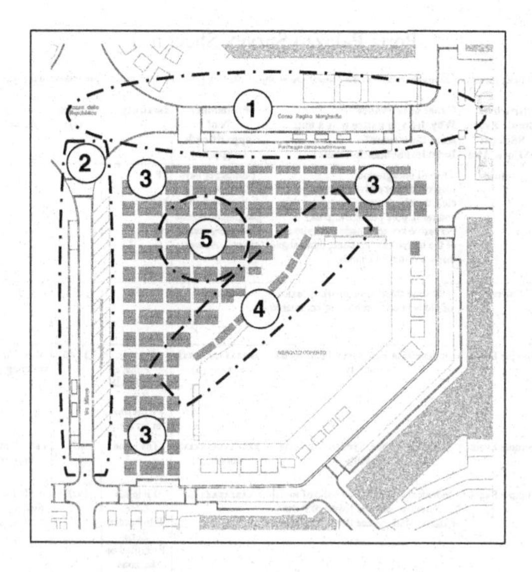

Fig. 10. Map of a portion of Porta Palazzo market: numbers and lines indicate sound zones identified during the annotation phase

As an example, the sound of plastic shopping-bags is a unique keynote sound represented as a mass of sound events. The shouts of the merchants are another multi-particle keynote in which the listener of the marketplace soundscape is immersed (so, the most intense, vibrant, repetitive, significant advertising messages have been recorded to be simulated). In some sense, their sum is the pervasive call of the market: the Porta Palazzo market voice.

We then investigated the topographic structure of the market. The map provided by the municipality (Figure 10) has been a fundamental resource in order to effectively simulate the market, as it is the source for the creation of the map of graphs. In other terms, by parametrizing the dimensions of the map of graphs on the topographic map, we are then able to place the sound sources over its surface. The market area is square-shaped, but the market stands are asymmetrically placed in it. In fact, stands are placed along two conjunct sides, while the opposite corner is substantially free and deputed to pedestrian passage. In sum, market stands are located in the north-west half of the square cut by the diagonal. Unlike the pedestrian south-east corner, the north and west sides are characterized by the presence of motor vehicles. Yet, the two sides present some differences, as the north one borders an ordinary street (typically with car traffic), while the west one is occupied by the market load area. Also, busses and trams pass through the same area, but not cars. The analysis revealed a specific keynote sound in certain border regions that invades all the space: the noise of motor vehicles and carriages. As already suggested during the interviews, in the opinion of the customers the arrival of the streetcar number 4 is the unique sound source that can be heard throughout the whole soundscape, acquiring specific

Porta Palazzo Survey Sheet n_1_

From "Name" until "Intensity" field, the informations are related to the general sound object. Starting from "Length" the info are related to the specific recording of sound material

Name	Geographical Reference	Structural Simple Event, SSEvent SSCore, Atm	Cause of relevance: Why did you notice it as a unit from the stream? Indexical/Semantic/Iconic	Probability N of apparition/h or app/day	Intensity	Length (s)	Record Distance	Generative Structure If external relationship with other sounds & note recording of sound material
Pure Market Zone Atmosphere	5 (A map with number reference is needed)	Atm	Overall background Icon of market activity. Sense of closed ambient (transit ways are thin and rare), pure pedestrian area, Pervasive keynote made from shouts of the merchants. Frenetic, disarranged inspected impact sounds.	1 times per days Each during 2 hours (when geographical zone crowed by market clients). At noon.	PP P M F FF	...	...	Continuously. Relationship impact events generative rate. Omnidirectional microphone used for recording
Balance emptied	All market stands with old style balance	SSEvent	Old style stand Soundmark Indexical of old balance Symbolic of commerce	xxx (h)	P	1	5 m	Sequencing relationship with Paper, Plastic shoot, Coins. Shoot gun mic.
Bus1	2	Simple Event	Indexical of bus, Symbolic of social mobility.	xxxxxxxxxxxxxxxxxxx (h)	F in his Zone. M In border Zone PP muffled in far zone	31,46	Very Far (40 meter approx)	Long sound with amplitude envelope of far bus passing by. Material Reffered to P, muffled in far zone. Delivery truck activity in background. Shoot gun mic.
Bus2 (Further sound material)	2	Simple Event	Indexical of bus, Symbolic of social mobility.	xxxxxxxxxxxxxxxxxxx (h)	The same	4,73	Far (20 meter approx)	Sound Material with P intensity. To be used in border Zone of Zone 2. Shoot gun.
Tram 1	2	Simple Event	Indexical of Tram, Symbolic of social mobility. Specific Iconic of Porta Palazzo Marketplace (from interview).	xxxxxxxx (h)	FF in his Zone. F In border Zone P muffled in far zone	19.09	Medium distance (15 meter approx)	Sound Material with FF intensity. To be used in Zone 2. Shotgun mic.

Fig. 11. Chart for sound object annotation

nuances in each zone (i.e. due to reverberation and to low frequency distance attenuation). Hearing this sound object makes one think immediately of Porta Palazzo market.

Following the devised methodology, we then proceeded to the identification and classification of sound objects. The procedure was based on two steps. In both cases, we devised a soundscape survey sheet helping for collecting information, even some that will probably not be used for implementation. The chart (see Fig. 11) allows to describe each sound object through different fields: a figurative label (name), the market area to which it belongs (Geographical Reference), degree of complexity (Structure), relevant feature of the sound object (Cause of relevance), repetition rate (Probability), intensity (expressed in musical form, e.g. *pp*, *ff* etc), duration (Length), an assessment of distance of the source (Record distance), the sequencing behavior (Generative structure). In order to define the intensity level for the sound object, we used a five-step scale, based on musical dynamics (p, pp, m, f, ff). Intensity evaluation was performed in relation to the overall soundscape, as we first assigned the *ff* events (the loudest) and the *pp* events (the faintest); then, we grouped all the sound objects showing intensity similarity; finally, we assigned each group an intensity. A rough approximation between qualitative categories and sound pressure is the following:

- *ff* (fortissimo): -3dB
- *f* (forte) = -12dB
- *m* (medio) = -21dB
- *p* (piano) = -30dB
- *pp* (pianissimo) = -39dB

In relation to the recording distance, we need to notice that, because of many logistic issues, it was difficult to estimate the recording distance (as an example, it may not be possible to ask workers to produce sound object when we were in need of a second evaluation).

The aforementioned interviews occurred while performing the first absent-minded exploration of the place. In this way, by comparing interviewer's perception of the soundscape and by practicing an informal listening, it has been easy to individuate and annotate the most common sound objects: the sound of plastic shoppers (acting like a Murray Schafer's keynote sound), the shouts of the merchants, the pervasive noises of vehicles. Then, a second, more analytical, step has been performed. Preliminary data were refined by deeply investigating the market place, that is, not only relying on auditory perception, but including visual/spatial clues. In this way, starting from the identification of the sources (e.g. a mechanical device), it has been possible to identify some sound objects within larger sound objects. The result of the two-step analysis is a complete sound object list. More, during the second step, each sound object was associated to a source and thus received a location in the space. At this stage, it became possible to individuate sound zones, in particular the analysis of the soundscape has led to five independent zones formed by distinctive elements. In Figure 10 all the zones were assigned an identifying index. Not surprisingly, sound zones respect the main topographic features of the market area, as discussed above. Zones 1 and 2 are characterized by the sounds of motor vehicles. Zone 1 is mainly characterized by a sound atmosphere made up of little delivery trucks, hand-carts and gathering of packing boxes from stands. It is the only street accessible by any vehicles as bus, trams, cars and motorbike. Instead, in the zone 2 there are two important sound features: the load area of big delivery trucks and the street dedicated to public transport, with rail system allowing streetcar passage. Both the zones present sounds related to bread, mint and spice hawkers. Zone 3 is a diffused area showing a mixup of sounds related to market and to street/parking areas. This feature has required to aptly adjust the radius of sound sources to describe its fuzzy sonic boundaries. In addition, some emissions related to the daily process of assembling/disassembling stands are present. Zone 4 is formed by different and rare stands; it presents a less prominent sound density because the passage area is bigger, so the sound of walking costumers, hand-cart distribution, empty box collecting process, are louder than other sound objects. More, many atypical stands are positioned in this zone, making its atmosphere unique. The motor sound is almost imperceptible, with the exception of some very loud source (as streetcar 4). Zone 5 presents only vegetable and fruit stands: transit ways are thin and rare, and only walking people can pass through. The shouts of the merchants reach the highest intensity and mask all the pollution sound coming from the other zones, while the many sound signals (activities and voices) make the soundscape particularly frenetic, a disarranged composition of sound objects making it a "pure" example of market soundscape. After such an analysis of sound objects in relation to space, we took into account iterated sequences of sound objects, to be modeled during the generation phase by sound subjects. In particular, five typical stand sounds have been analyzed and have revealed complex rhythmical patterns. As an example, "shopping" showed a particular sequence of events: plastic rustle, paper rustle, clinking coins, cash opening, clinking coins, cash closing. The stands of the anchovy sellers proved to be very different from all other stands: they included sounds of metal cans, of anchovies being beaten over wood plates, of olives thrown in oil, of noisy old scales. These analytic steps drove field recordings, with the aim of creating for each sound object the proper sound materials. All recordings were at mono, 44.100 Hz sampling rate, carried out through a DAT recorder. Many different recordings related to the same sound object were stored, so that the complexity of the original soundscape would not be lost. In order to minimize background noise, sound objects have been recorded with Super-cardioid highly directional microphones. Super-cardioids are especially suitable for picking up quiet signals in noisy or acoustically live environments, as they discriminate against sound not emanating from the main pick-up direction. More, some omnidirectional recordings were carried out for each sound zone, to be used both as material for atmosphere and as references for evaluating the simulation. The data collected during the analysis and recording phases were processed in order to be inserted into a database. The creation of sound materials proved to be a particularly complex task, as the recorded environment of the market was very noisy. Tests with noise reduction techniques were unsatisfactory, as a lot of signal was removed together with noise. As a consequence, the Acoustic discretization error was very high, because a relevant difference between sound objects (as identified by classification and analysis) and the resulting sound materials emerged. After some unsatisfactory recording tests, we substantially solved the problem by using, in the recording phase, very high directional microphones, thus ensuring a high "presence" of sound in relation to background. Acousmatic listening of the recorded sounds allowed to identify a large quantity of unexpected sound objects, thus reducing the Semiotic discretization error. In some cases, this has led to the realization of other recording sessions in the market. More, acousmatic listening has allowed to perform data reduction. That is, different sound objects, originally identified during the analysis phase mainly with indexical strategies, have proven a substantial identity in terms of sonic features, and can then be represented by the same sound material. The generation phase uses information into the database to generate sequences of sound objects. To do so, it must also include the definition of graph structures representing different sound objects and their temporal organization (events, atmospheres, sound subjects). Graphs have proven to be capable of expressing very different behaviors, thus allowing to model a variety of relations among sound objects. As an example, the butcher's knife beating the meat generates a single sound object repeated with a specific pattern, which can be expressed by a graph made of a single vertex with a looping edge. Payment procedures have revealed a chain of many different sound objects: there is a specific pattern of sound objects, involving the rustle of the wrapping paper and the shopper, the tinkling of coins, the different noises of the

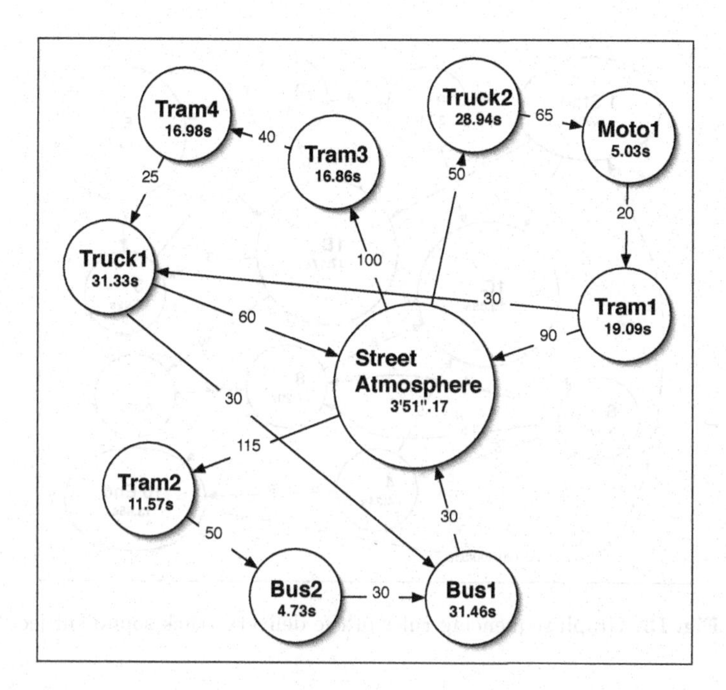

Fig. 12. Graph sequencing rules for a sub-part of zone 2

cash register marking the many phases of the action. In this case, a much more complex graph is needed.

Figure 12 shows the graph structure of a sub-part of zone 2. Edges define the sequencing rules between a possible atmosphere (labelled "XAtmo...", a four minute recording) of that specific sub-part, and nine indexical sound events. The atmosphere describes the almost quiet state of that area, generated by the continuous walking of costumers and the activity of some mint hawkers. The sound events describe activities by different vehicles. The number of repetitions of a sound object (i.e bus, tram, delivery truck, motorbike) is proportional to its statistical relevance: there are four tram objects, then two for bus and trucks, and only one for motorbike. No car was noticed here. The graph is cyclic, thus generating potentially infinite tracks. In this case, each possible path is designed to have the same duration of the atmosphere. So the time duration of edge connection $Edur_{x_y}$ between vertices are set according to the following rule:

$$Edur_{Atm_2} + Edur_{2_3} + ... + Edur_{x_{Atm}} = Vdur_{Atm}$$

In this way a long, looping background is continuously varied by the superposition of different other sound objects. By only using nine objects it has been possible to represent a complex soundscape. Figure 13 depicts the graph of a sound subject. The graph represents the behavior of a delivery truck. The delivery trucks arrive at that zone, unload the products, and leave back. By connecting three "core" objects 1*b*, 1*c*, 5 and nine sound events, it allows the

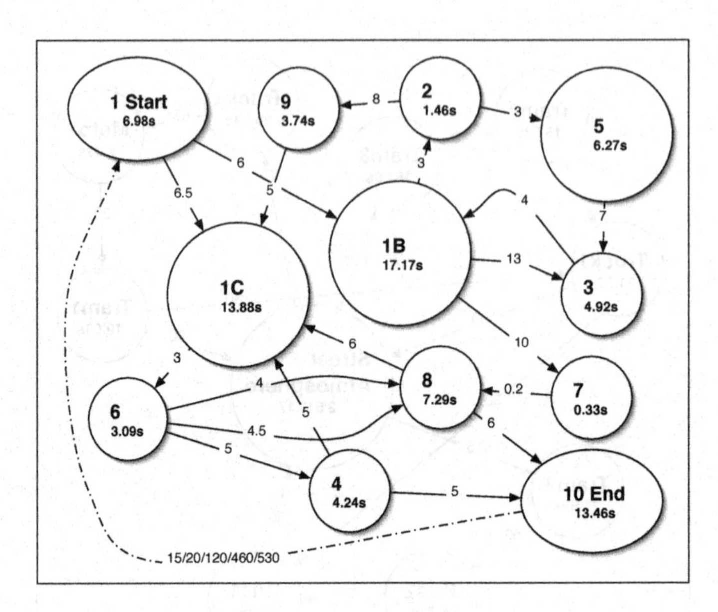

Fig. 13. Graph sequencing rules of the delivery truck sound subject

simulation of several instances of the truck. Here the sound subject reveals its sub-zone nature. The topological structure of the graph includes a start event (1) and an end event (10). The core objects are almost quiet recordings. As an example, 1*b* refers to a stationary truck with running motor while 1*c* refers to a truck making some accelerations. They are placed topologically in the center of the graph structure, providing a continuous background against which other smaller sound objects appear and disappear. As in the previous example, all the possible paths re-activate a core before its duration has finished. But there is an exception: after the end event the auditory stream stops, as the graph is acyclic. The graph can be made cyclic by the addition of edges connecting the end vertex 10 to the start vertex 1. These looping edges can have durations spanning over a large interval, from 15 to 530 seconds. After a path simulating the truck delivery has reached the end event, it is thus possible that the start event is emitted straight after that: the sound result will then be perceived as an activity of the same acousmatic subject. By contrast, when the path restarts a long time after the end vertex has finished, the result can be perceived as the arriving of a new truck in the soundscape.

As previously discussed, the Sound Interpreter –implemented in SuperCollider– has a twofold role: it maps vertices to sound material and spatially-related data to DSP parameters. In the implementation used for the simulation of Porta Palazzo market, while the sound materials were the ones inserted into the database, the mapping strategy to DSP algorithm for spatial cues rendering endorses a cartoonified approach. Rocchesso et al. [31] have proposed "cartoonification" techniques for sound design, i.e. simplified models for the creation of sounds related to physical processes and spatial cues. The cartoonification process starts from

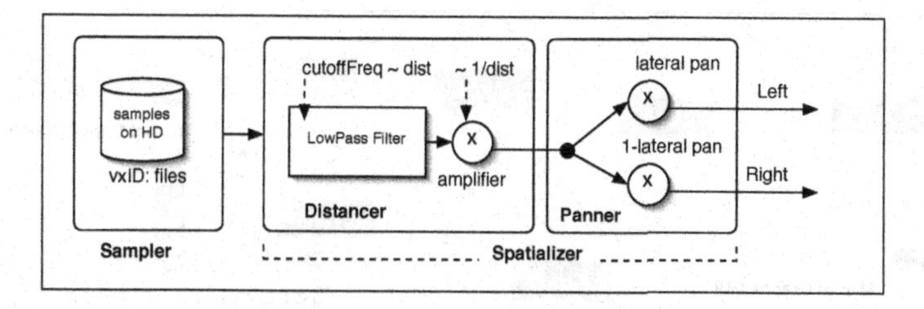

Fig. 14. Structure of a "Space Cartoonifier"

an analysis of the physical situation and simplifies it, retaining only the perceptually and culturally relevant features. Cartoonification is particularly relevant for our research as our approach is not intended as a physical modeling, but as a semiotic/phenomenologic reconstruction, aiming at the creation of "sound symbols" ([39], as discussed in section 2) of the whole landscape by providing global, semiotically recognizable, perceptual cues[3]. Hence, the implementation of the Sound Interpreter as a "Space Cartoonifier" (Fig. 14), including two modules, a Sampler and a Spatializer. The Sampler first uses vertex IDs to retrieve sound samples from disk. The Spatializer aims at cartoonifying monaural distance cues and dislocation cues [44]. In order to provide distance cues, in the Spatializer samples are then low-pass filtered and amplitude-scaled by the Distancer sub-module, so called because it maps distance to DSP parameters. The cutoff frequency of the lowpass filter is proportional to the distance, while the amplifier coefficient is inversely proportional to distance. So, on the one hand, a sample is heard –respectively– duller and quieter, when the listener moves away from the source. On the other hand, a sample is heard –respectively– brighter and louder, when the listener moves towards the source. Concerning dislocation, in the Panner sub-module, pan is used to feed a two channel equal power pan, where -1 places the sound source at the extreme left and $+1$ at the extreme right, with 0 indicating a perfectly frontal source.

In the occasion of the case study research, the system has been implemented in the audio programming language SuperCollider ([45], see Fig. 15), which features a high-level, object-oriented, interactive language together with a real-time, efficient audio server. The SuperCollider language summarizes aspects that are common to other general and audio-specific programming languages (e.g. respectively Smalltalk and Csound), but at the same time allows to generate programmatically complex GUIs. The application includes both GUIs and scripting capabilities (see Fig. 15). Graph structures are described textually (with a dot language formalism, [46]) and displayed graphically. Both the activation of vertices and the interactive exploration process can be visualized in real time. The Open Sound Control (OSC, [47]) interface, natively implemented in SuperCollider, allows for a seamless network integration with other applications. As a typical example, the GeoGraphy application can be connected to a virtual

[3] In fact, the map of graphs in itself can be considered as a cartoonification of the real space.

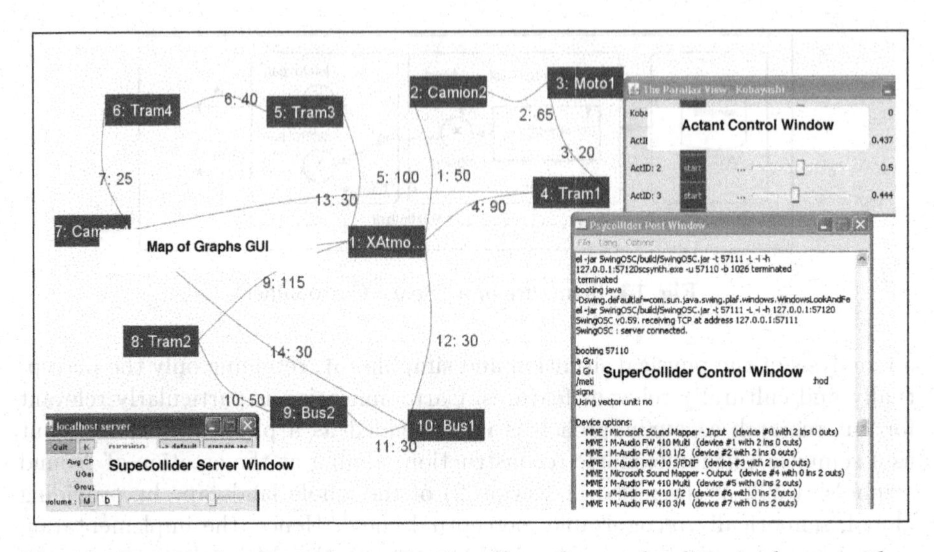

Fig. 15. A screenshot from the application. Here the graph editor is shown, with a graph width edge/vertex durations (background), the Actant Control Window and other SuperCollider-related GUIs for real-time control.

reality engine, in order to allow an audio-visual integration of an architectonic-urbanistic space. We carried out a preliminary evaluation, and we are planning to include both experts (sound designers, soundscape researchers) and occasional listeners. These preliminary listening tests have given promising results. Users reported about the exploration through the interface and the navigation of the market by referring a situation that seemed slightly different from the live experience. Reportedly, a prominent feature lies in the generative nature of the system: even if based on a discrete set of sound materials, the use of graph-based sequencing avoids the feeling of artificiality typical of sound sample looping, as the soundscape is in continuous transformation. For what concerns the actual interaction design, improvements are required. At the moment, navigation is made possible by using the keyboard. Users have reported difficulties in managing their orientation in the space. A smoother controlling interface is needed to ensure an easier exploration of the soundscape, so that users can entirely focus on soundscape.

9 Conclusions and Future Work

The notion of soundscape is increasingly relevant not only in contemporary culture, but also in the world of sound-related studies. Still, a rigorous definition of the concept is lacking. By providing such a formal definition, it is possible to propose a generative model for the simulation of (not only real) soundscapes. GeoGraphy provides a theoretical framework for the modeling of soundscape in terms of both temporal information describing sound time patterns (via the vertex/edge labeling) and spatial information encoding the site/observer relation (via vertex positioning). The proposed system is able to generate soundscapes from original sound materials but without relying on straightforward loops. This helps in preserving the typical "sound mood" of the original space: at the same time, the resulting soundscape is not fixed, but undergoes a continuous variation thanks to the graph dynamics, due to probabilistic connections. The Sound Interpreter allows to create different soundscapes from the same set of sound objects by defining specific mapping strategies to different libraries. GeoGraphy implementation in SuperCollider can operate interactively in real time and it can be integrated in other multimedia applications. A major drawback of the method is the manual, time consuming generation of multigraphs. We are planning to extend the system so to include the automatic generation of graphs starting from information stored in the database or from sound-related semantic repertoires (see [48]). The database itself can probably include not only sound materials created from direct recording but also samples from available sound libraries. An interesting perspective is to investigate user-generated, online databases such as Freesound[4]: in this case, the graph generation process can be governed by social tagging.

References

1. Schirosa, M., Valle, A., Lombardo, V.: Un modello per la generazione dinamica di paesaggi sonori. In: Rocchesso, D., Orvieto, A., Vidolin, A. (eds.) Atti del XVII Colloquio di Informatica Musicale, Venezia (15–17 Ottobre), pp. 35–41 (2008)
2. Valle, A., Lombardo, V., Schirosa, M.: A graph-based system for the dynamic generation of soundscapes. In: Aramaki, M., Kronland-Martinet, R., Ystad, S., Jensen, K. (eds.) Proceedings of the 15th International Conference on Auditory Display (ICAD 2009), Copenhagen, Denmark, May 18–21 (2009)
3. Valle, A., Lombardo, V., Schirosa, M.: A framework for soundscape analysis and re-synthesis. In: Gouyon, F., Barbosa, A., Serra, X. (eds.) Proceedings of the SMC 2009 - 6th Sound and Music Computing Conference, Porto, 23-25 July 2009, pp. 13–18 (2009)
4. Valle, A., Lombardo, V.: A two-level method to control granular synthesis. In: Bernardini, N., Giomi, F., Giosmin, N. (eds.) XIV CIM 2003. Computer Music: Past adn Future, Firenze, 8-10 May 2003, pp. 136–140 (2003)
5. Valle, A.: Geography: a real-time, graph-based composition environment. In: NIME 208: Proceedings, pp. 253–256 (2008)
6. Murray Schafer, R.: The Tuning of the World. In: Knopf, New York (1977)
7. Burtner, M.: Ecoacoustic and shamanic technologies for multimedia composition and performance. Organised Sound 10(1), 3–19 (2005)
8. Fontana, F., Rocchesso, D., Ottaviani, L.: A structural approach to distance rendering in personal auditory displays. In: Proceedings of the International Conference on Multimodal Interfaces (ICMI 2002), Pittsburgh, PA, USA, 14-16 October (2002)
9. McGregor, I., Crerar, A., Benyon, D., Macaulay, C.: Sounfields and soundscapes: Reifying auditory communities. In: Proceedings of the 2002 International Conference on Auditory Display, Kyoto, Japan, July 2-5 (2002)

[4] http://www.freesound.org/

10. Rubin, B.U.: Audible information design in the new york city subway system: A case study. In: Proceedings of the International Conference on Auditory Display 1998, Glasgow (1998)
11. Droumeva, M., Wakkary, R.: The role of participatory workshops in investigating narrative and sound ecologies in the design of an ambient intelligence audio display. In: Proceedings of the 12th International Conference on Auditory Display, London (2006)
12. Friberg, J., Gärdenfors, D.: Audio games: New perspectives on game audio. In: Proceedings of the 2004 ACM SIGCHI International Conference on Advances in computer entertainment technology, pp. 148–154. ACM Press, New York (2004)
13. Serafin, S.: Sound design to enhance presence in photorealistic virtual reality. In: Proceedings of the 2004 International Conference on Auditory Display, Sidney, Australia, July 6-9 (2004)
14. Hermann, T., Meinicke, P., Ritter, H.: Principal curve sonification. In: Proceedings of International Conference on Auditory Display (2000)
15. Butz, A., Jung, R.: Seamless user notification in ambient soundscapes. In: IUI 2005: Proceedings of the 10th international conference on Intelligent user interfaces. ACM Press, New York (2005)
16. Kilander, F., Lönnqvist, P.: A whisper in the woods - an ambient soundscape for peripheral awareness of remote processes. In: Proceedings of the 2002 International Conference on Auditory Display, Kyoto, Japan, July 2-5 (2002)
17. VV.AA.: The tech issue..to be continued. Soundscape 3(1) (July 2002)
18. VV.AA.: Acoustic design. Soundscape 5(1) (2004)
19. Westerkamp, H.: Linking soundscape composition and acoustic ecology. Organised Sound 7(1) (2002)
20. Mauney, B.S., Walker, B.N.: Designing systems for the creation and evaluation of dynamic peripheral soundscapes: a usability study. In: Proceedings of the Human Factors and Ergonomics Society 48th Annual Meeting 2004, New Orleans (2004)
21. Truax, B.: Acoustic Communication. Greenwood, Westport (1984)
22. Misra, A., Cook, P.R., Wang, G.: Musical Tapestry: Re-composing Natural Sounds. In: Proceedings of the International Computer Music Conference, ICMC (2006)
23. LaBelle, B.: Background noise: perspectives on sound art. Continuum, New York–London (2006)
24. Mayr, A. (ed.): Musica e suoni dell'ambiente. CLUEB, Bologna (2001)
25. Murch, W.: In the blink of an eye, 2nd edn. Silman-James Press, Los Angeles (2001)
26. Agostini, L.: Creare Paesaggi Sonori. Lulu.com (2007)
27. Augoyard, J.F., Torgue, H.: Repertorio degli effetti sonori. Lim, Lucca (2003)
28. Amphoux, P.: L'identité sonore des villes européennes, Guide méthodologique à l'usage des gestionnaires de la ville, des techniciens du son et des chercheurs en sciences sociales. publication IREC, EPF–Cresson, Lausanne–Grenoble (1993)
29. VV.AA.: Résumé de l'étude de conception et d'aménagement du paysage sonore du secteur de la Sucrerie - St. Cosmes. Technical report, Acirene–atelier de traitement culturel et esthétique de l'environnement sonore (2007)
30. Truax, B.: Models and strategies for acoustic design. In: Karlsson, H. (ed.) Hör upp! Stockholm, Hey Listen! - Papers presented at the conference on acoustic ecology, Stockholm (1998)
31. Rocchesso, D., Fontana, F. (eds.): The Sounding Object. Edizioni di Mondo Estremo, Firenze (2003)
32. Chion, M.: L'audiovision. Son et image au cinéma, Nathan, Paris (1990)

33. Schaeffer, P.: Traité des objets musicaux. Seuil, Paris (1966)
34. Chion, M.: Guide des objets sonores. Pierre Schaeffer et la recherche musicale. Buchet/Castel-INA, Paris (1983)
35. Lombardo, V., Valle, A.: Audio e multimedia, 3rd edn. Apogeo, Milano (2008)
36. Handel, S.: Listening. An Introduction to the Perception of Auditory Events. The MIT Press, Cambridge (1989)
37. Bregman, A.: Auditory Scene Analysis. The Perceptual Organization of Sound. The MIT Press, Cambridge (1990)
38. Böhme, G.: Atmosfere acustiche. Un contributo all'estetica ecologica. In: Ecologia della musica: Saggi sul paesaggio sonoro, Donzelli (2004)
39. Wishart, T.: The Language of Electroacoustic Music. In: The Language of Electroacoustic Music, pp. 41–60. MacMillan, London (1986)
40. Warusfel, O., Eckel, G.: LISTEN-Augmenting everyday environments through interactive soundscapes. Virtual Reality for Public Consumption, IEEE Virtual Reality 2004 Workshop, Chicago IL 27 (2004)
41. Roads, C.: The computer music tutorial. MIT Press, Cambridge (1996)
42. Lerdahl, F., Jackendoff, R.: A Generative Theory of Tonal Music. The MIT Press, Cambridge (1983)
43. Studio di Ingegneria ed Urbanistica Vittorio Cappato: 50 centesimi al kilo: La riqualificazione del mercato di Porta Palazzo dal progetto al cantiere. Technical report, Comune di Torino, Torino (2006)
44. Fontana, F., Rocchesso, D.: Synthesis of distance cues: modeling anda validation. In: The Sounding Object, pp. 205–220. Edizioni di Mondo Estremo, Firenze (2003)
45. Wilson, S., Cottle, D., Collins, N. (eds.): The SuperCollider Book. The MIT Press, Cambridge (2008)
46. Gansner, E., Koutsofios, E., North, S.: Drawing graphs with dot (2006)
47. Wright, M., Freed, A., Momeni, A.: Opensound control: State of the art 2003. In: Thibault, F. (ed.) Proceedings of the 2003 Conference on New Interfaces for Musical Expression, NIME 2003 (2003)
48. Cano, P., Fabig, L., Gouyon, F., Koppenberger, M., Loscos, A., Barbosa, A.: Semi-automatic ambiance generation. In: Proceedings of the International Conference of Digital Audio Effeccts (DAFx 2004), pp. 1–4 (2004)

Effect of Sound Source Stimuli on the Perception of Reverberation in Large Volumes

Ilja Frissen[1,2], Brian F.G. Katz[3], and Catherine Guastavino[1,2]

[1] Center for Interdisciplinary Research in Music, Media and Technology, Montreal, Canada
ilja.frissen@mcgill.ca
http://www.cirmmt.mcgill.ca/
[2] McGill University, School of Information Studies, Montreal, Quebec H3A 1X1, Canada
[3] LIMSI-CNRS, Orsay, France

Abstract. The aim of the presented research is to determine whether the perception of reverberation is dependent on the type of sound stimuli used. We quantified the discrimination thresholds for reverberations that are representative for large rooms such as concert halls (reverberation times around 1.8 s). For exponential decays, simulating an ideal simple room, thresholds are around 6% (Experiment 1). We found no difference in thresholds between a short noise burst and a male voice spoken word, suggesting that discrimination is not dependent on the type, or spectral content, of the sound source (Experiment 2). In two further experiments using a magnitude estimation paradigm we assessed the perceived amount of reverberation as a function of various types of stimuli. Whereas the discrimination of reverberant stimuli does not seem to be affected by the sound stimulus, the perceived amount of reverberation is affected. Vocal stimuli are perceived as being more reverberant than non-vocal stimuli. The results are discussed in light of current neuroscientific models of auditory processing of complex stimuli but also with respect to their consequences for the use of reverberation in auditory display.

1 Introduction

In our daily lives we are continually confronted with echoic and reverberant environments. Reverberation is the accumulated collection of reflected sound from the surfaces in a volume, typically characterised by the rate of decay. Not only are we continuously confronted with reverberant sound, it can be argued that humans evolved in reverberant spaces; the ancestral cave comes to mind. Questions arise as to how humans have adapted to such reverberant environments.

A thorough understanding of human sensitivity to reverberant stimuli is not only of theoretical value but has practical potential as well. It will aid the designers of rooms specifically intended for listening, such as concert halls. But reverberation could also be used as an auditory cue to convey the structure of auditory displays. An example of a potential application is the use of simulated rooms of different sizes to cluster similar information.

S. Ystad et al. (Eds.): CMMR/ICAD 2009, LNCS 5954, pp. 358–376, 2010.

This paper presents four experiments on the perception of reverberation in terms of discrimination thresholds and magnitude estimation, with a focus on a range of reverberation typically found in orchestral performance spaces. Specifically, we investigate the effect of properties of the sound source stimuli (type and length) on the perception of reverberation.

1.1 Acoustics and Simulation of Reverberation

Reverberation is a crucial, though not all encompassing, acoustical parameter for characterizing the sound quality of an auditorium [1]. In a simple single volume the energy dissipates exponentially (or linearly on a log scale), and the reverberation time (RT_{60}) is classically defined as the time required for the sound level to decay by 60 dB relative to the initial level.

In our everyday life the listening environment is often not a simple single room but a number of connected rooms or volumes with different acoustical properties. An office, for instance, with its door open into a hallway, or a concert hall, with the orchestra pit as the primary and the audience area as the secondary volume. Kahle [2] and Bradley and Wang [3] showed that that energy does not decay linearly over time in actual concert halls but rather exhibit double (or multiple) slope decays. Some halls are designed specifically to behave in this manner, as shown in such examples as [4,5].

Here we only consider the single volume. For the purpose of simulating a single volume room with a particular RT_{60} on needs to construct the impulse response (IR) of the room. It consists of the direct sound and early reflections from the walls. Early reflections are particularly important in creating a sense of space (e.g., [6]). The number of early reflections continuously increases over time and transitions into a more random signal with an exponential decay called subsequent reverberation (see Figure 1). A simple IR can be simulated by the application of an exponential decay to a normally distributed random number sequence. In terms of signal processing, the post-processing of an audio signal with room simulation corresponds to the convolution of an audio signal with the room impulse response.

1.2 Perceptual Research Involving Reverberation

A considerable amount of research has concerned itself with the perceptual consequences of reverberation which can be positive or negative depending on the context. Various perceptual faculties have been investigated, such as speech intelligibility [7,8], the segregation of multiple people talking [9], sound localization [10,11], and auditory distance perception [12]. An increased understanding of the effects of reverberation will have consequences for more applied problems such as the design of hearing aids (e.g., [13]). Perceptually, reverberation, or rather, the ratio of the direct sound to reverberation is the major cue for determining the distance of an auditory source [14,15].

There is little work available on the perceptual discrimination of reverberant stimuli (see [16]). Cox et al. [7] investigated how sensitive people are to

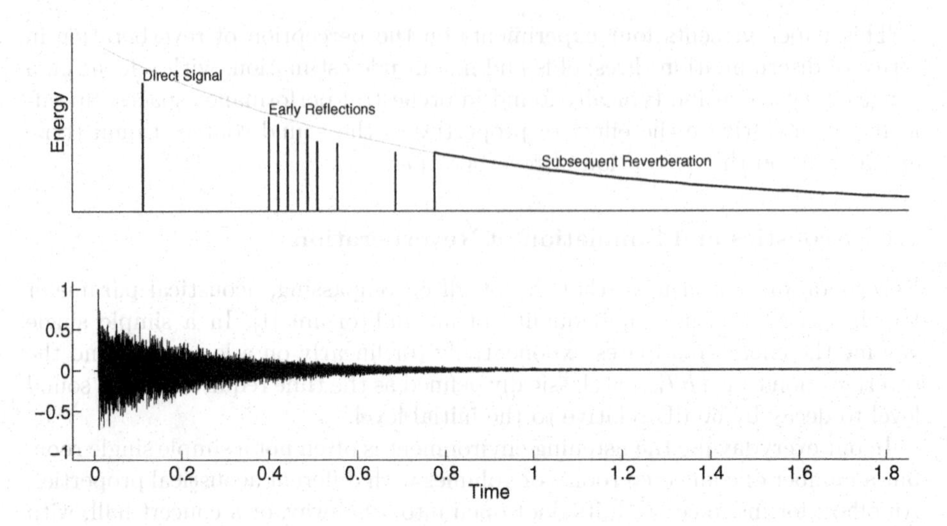

Fig. 1. Room impulse response. *Top panel.* The temporal component classifications of an IR. *Bottom panel.* A simulated IR with an RT_{60} of 1.8s.

changes in the various acoustical parameters that define the early sound field in auditoria, and whether these thresholds vary according to different music stimuli. Using realistic concert hall simulations they determined the discrimination thresholds for, among other, clarity and center time (or the barycentre of the squared IR). They did not, however, directly address thresholds for reverberation time. Seraphim [17] determined the discrimination thresholds for linear decay of narrowband noise bursts with RTs ranging from 170 ms up to 10 s, and found the best discrimination (3-5%) between RT of 0.8 s and $\tilde{4}$ s. He also stated that this performance depended neither on the frequency of the stimulus nor on its length. Niaounakis and Davies [18] investigated perceptual thresholds in relatively small rooms (RT $<$ 0.6 s) using very long (i.e., 21 s) musical excerpts. Interestingly, despite major differences in stimulus type, they found thresholds similar to Seraphim [17], around 6%.

Recently, perceptual aspects of coupled volumes have also received scientific attention [3,19,20]. For instance, Bradley and Wang [3] measured the perceived reverberation with a rating task (assign a value between 1 and 9) while manipulating the ratio of the two coupled volumes and the size of the aperture connection the volumes using numerical room simulations. Increasing the values of both factors led to corresponding increases in the perceived amount of reverberation. Both factors affected changes in perception. Perceived reverberation increased as the ratio and/or aperture area increased. These changes in turn correlated with objective measures of coupled volume reverberation. Frissen et al. [21] attempted to find a perceptual match between exponential decays and non-exponential decay profiles, using classical psychophysical methods. Regrettably, the tasks were found to be difficult for the participants and in several cases reliable measures were not obtained. The suggested reason was that participants were unable to discriminate the stimuli in a systematic fashion, making it difficult to construct consistent criterion by which to make their judgments.

1.3 The Present Study

A comparison of Seraphim's [17] and Niaounakis and Davies' [18] results suggests that the discrimination of reverberation is not strongly dependent on the semantics or even the spectral content of the sound source. Such comparisons are informative about the perceptual level of processing of reverberation. If the sound source stimulus is not relevant then this would have practical consequences if one wants to use reverberation in auditory information display systems. If, on the other hand, sound stimuli do have differential effects in the perceived amount of reverberation, as could, for instance, be inferred from the results of Cox et al. [7], such conclusions and consequences would have to be revised.

There are other reasons to expect an effect of the type of sound used. Listener's subjective preference for reverberation, for instance, has been shown to depend on the sound material. Preferences for RTs vary with the type of music. Kuhl [1] investigate RT preference for orchestral music and observed a preferred RT of 2.1s for Romantic music (Brahms), 1.5s for classical music (Mozart) and less than 1.5s for modern music (Stravinsky). These preference judgments were found to be stable over time and participants. In addition, for opera singing (using extracts of Puccini), Sakai et al. [22] observed a preferred RT of 0.8s.

From a neuroscientific point of view it is becoming increasingly clear that the central nervous system (CNS) treats the multitude of everyday sounds that impinge on our hearing in different ways and with different neural pathways depending on the nature of sounds. Belin et al. [23] found voice-selective areas in the auditory cortex, that showed greater neuronal activity when participants listened to vocal sounds than to non-vocal environmental sounds (including human non-vocal sounds). Zatorre et al. [24], proposed a model in which the CNS (predominantly) processes speech in the left hemisphere because of its superior temporal resolution in comparison to the right. The right hemisphere, on the other hand, is better suited for spectral analysis, making it more appropriate for music, for instance. The distinction between vocal and non-vocal sounds was also observed in behavioral studies using sorting tasks of everyday life sonic environments [25] and isolated sounds [26]. Converging evidence comes from the linguistic analysis of free-format verbal descriptions [27,28].

Given such differential processing of acoustic material it can be argued, a priori, that this could affect the appreciation of the amount of reverb associated with the different sounds. Because the studies thus far were not explicitly designed to test for the effects of the sound source stimuli on the perception of reverberation, we address it here in a series of experiments employing various psychophysical techniques. The first experiment is a partial replication of the Seraphim [17] experiment using white noise as a sound stimulus. In the following experiments we explore the speculation that the type of sound stimuli used does not affect the perception of reverberation. In the second experiment we use a standard psychophysical same/different task to determine the thresholds for the noise stimulus used in the first experiment and compare it to those for a vocal stimulus. The noise stimulus should not necessarily be favoured by either hemisphere, whereas the voice should receive particular consideration from the left hemisphere. In experiments 3 and 4 we changed the protocol to a magnitude estimation method which allows us to greatly extend the types and lengths of the sound stimuli as well as the range of RTs. We employ vocal (speech and singing) and non-vocal stimuli (piano and drums).

2 Experiment 1: Reference Experiment

The motivation for the first Experiment was to partially replicate the original experiment by Seraphim [17]. Thus, we determine the discrimination threshold (limen) of exponential decay reverberation profiles. In focusing interest on typical large orchestral performance rooms, we propose a study for variations around a reference RT of 1.8 s. We used the standard method of constant stimuli with a 2 interval forced-choice (2IFC) task.

2.1 Methods

Participants. Eight participants, including two of the authors, completed the experiment. All participants were tested for normal hearing using a standard audiometric test (over the range of 250-8000 Hz).

Stimuli. A series of nine impulse responses were generated in Matlab (the MathWorks). The synthesized IR is the result of the application of a simple exponential decay to a normally distributed random number sequence. The RT was varied from 1.48 to 2.12 s in equal steps of 80 ms. The experimental stimulus presented to subjects was the product of the convolution of the generated IRs with a 170 ms white noise burst. Sounds were presented over headphones (AKG K-271 Studio), played from a MacPro connected to an audio interface (Motu, mkII 828) at a sample rate of 44.1 kHz in an acoustically treated room.

Procedure. The experiment employed a standard method of constant stimuli paradigm, with a 2-interval, forced-choice task. The participant was presented with a sequential pair of stimuli. The task was to judge "which one had the most reverberation". The stimulus with an RT of 1.8 s served as the reference. Each of the nine comparisons was tested 12 times for a total of 106 trials. To reduce any inadvertent effects of response biases, the order of presentation of the reference and comparison was randomized such that in half of the trials the reference was presented first.

Presentation of the stimuli was controlled through a simple graphical user interface (GUI) developed and run in Matlab. The GUI featured only three buttons. One was a large "play" button that would play the stimulus pair with a random pause (0.5 -2 s) in between. The participants were free to listen to the stimulus pair as many times as desired (although this option was used only rarely). To enter their response, they clicked one of two buttons corresponding to the "first" and "second" sound in the stimulus pair.

Data Analysis. The proportion of trials in which each comparison was perceived to have more reverberation as the standard was calculated. To obtain psychometric functions, the data were fitted with cumulative Gaussians that were free to vary in position and slope using the software package psignifit (see http://bootstrapsoftware.org/psignifit/; [29]). The discrimination threshold (or, just-noticeable-difference, JND) was determined from the slope of the psychometric function. It was defined, as per convention, as the difference between the RTs that correspond to the 75% and 50% points of the cumulative Gaussian. Thus, the steeper the psychometric function, the more sensitive the corresponding discrimination, and therefore the smaller the JND.

2.2 Results

The results are summarized in Figure 2, with the left panel showing the individual psychometric functions. What can be seen is that six out of the eight participants showed very similar performance. Participants 3 and 8, on the other hand, found the task much more demanding as evidenced by their relatively flat psychometric functions. To illustrate the difference, the best and worst individual performances are highlighted in white circles and grey squares, respectively. The individual JNDs, and their average(s), are shown in the right panel. The

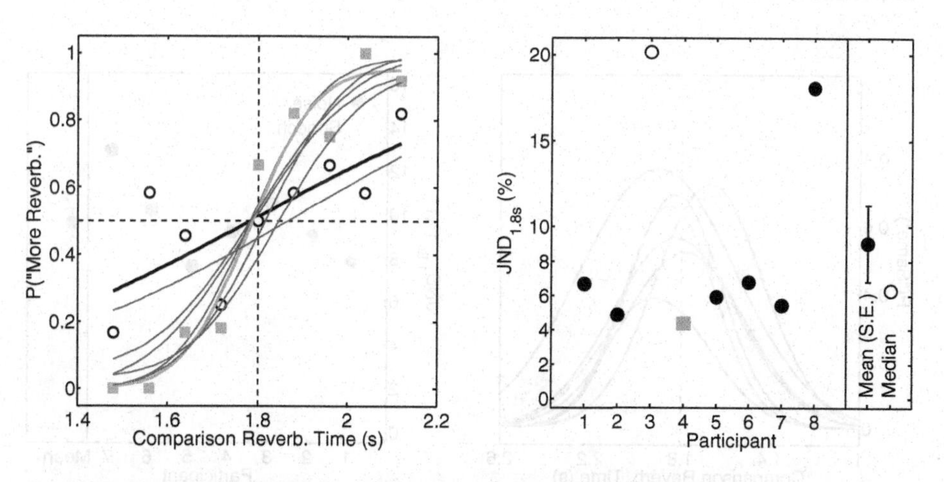

Fig. 2. Experiment 1. *Left panel.* Individual psychometric functions. Highlighted in different shades are the curves and data points for the best (participant 4; grey) and the worst (participant 3; white circles) performance. *Right panel.* Individual thresholds (JNDs) with same symbols and their averages. The errorbars correspond to the standard error of the mean.

mean JND was around 9%. However, because this value is heavily skewed by the two "outliers", we chose to utilize the median, which is approximately 6%.

2.3 Discussion

The discrimination thresholds were in the same range as those found in [17,18]. We were therefore able to successfully replicate and confirm parts of these earlier studies using a different methodology.

3 Experiment 2: Effect of Sound Source Type on Discrimination

The following three Experiments are concerned with the effect of the type of sound sources on the perception of reverberation. Here we determine the discrimination thresholds for two types of sound sources, the same noise as in Experiment 1 and speech.

3.1 Methods

Participants. Seven participants, including the first author, completed the experiment. Four of them also did Experiment 1.

Stimuli and Procedure. Sound stimuli were the 170 ms noise burst from Experiment 1 and a 600 ms recording of the French word "poussez", which was

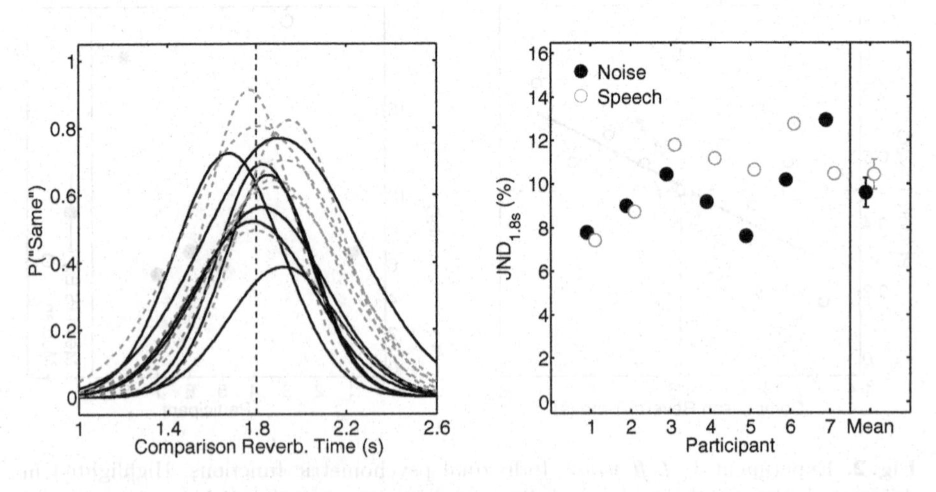

Fig. 3. Experiment 2. *Left panel.* All individuals psychometric functions for noise (black lines) and speech (grey dotted lines). *Right panel.* Individual thresholds (JNDs) and their averages. The errorbars correspond to the standard error of the mean.

extracted from a male speaker recording made in an anechoic chamber. As in Experiment 1, the IRs had a simple linear exponential decay. The speech stimuli and the noise stimuli were calibrated in level to have the same root-mean-square value.

The reference RT was 1.8 s and the tested RTs for comparison ranged from 1.48 to 2.12 s in seven equal steps of 80 ms. Each comparison for both sounds was presented nine times, for a total of 126 completely randomized trials. The task was to judge whether the pair of stimuli had the same or different amount of reverberation.

Data Analysis. From the raw data we calculated the proportion of "same" responses for each comparison and fitted to a Gaussian function. The standard deviation of the Gaussian corresponds to the JND.

3.2 Results

The obtained curves are shown in the left panel of Figure 2. The right panel shows the individual JNDs and their mean. There was no statistically significant difference between the JND for the Noise (9.6%) and the Speech (10.4%) stimuli, $t(6) = 1.14$, $p > 0.29$.

3.3 Discussion

There is no effect of the type of stimulus on the ability to discriminate reverberation. This is consistent with our original speculation based on the comparison of Seraphim's (1958) and Niaounakis and Daviess (2003) results.

It, however, does not mean that the absolute amount of perceived reverberation is the same for the different types of stimuli. For instance, it could still be that speech stimuli are perceived as having a longer reverberation time than for other stimuli with the same actual reverberation time. This is separate from the discrimination test performed in this experiment, since judgments of perceived reverberation were made between different reverberation times with the same stimulus. If there was a difference in perceived reverberation as a function of stimulus, such a change would not necessarily affect a discrimination task when the underlying perceptual scale changes proportionally as well.

The generality of this statement is obviously constrained to these very simple stimuli. Experiments 3 and 4 were designed to further investigate the effect of source signal using a broader range of stimuli.

The thresholds obtained in this experiment are noticeably higher than those found in Experiment 1. The difference between is most likely attributed to the nature of the task, whereas in Experiment 1 people were forced to directly compare the amount of reverberation, in Experiment 2, they "merely" had to decide whether the stimuli were the same or different. It seems that they used a less conservative criterion in performing the latter task.

4 Experiment 3: Effect of Source Type and Length on Magnitude Estimation

This experiment proposes another psychophysical method, magnitude estimation, championed by S.S Stevens [30] applied to reverberation perception. Participants make direct numerical estimations of a particular sensory magnitude produced by various stimuli. Magnitude estimation is highly efficient and permits rapid acquisition of substantial amounts of data. It is therefore one of the most frequently used psychophysical scaling methods [31]. It was selected for this experiment in order to extend our investigation to a wider range of source signals, varying in terms sound source and duration.

4.1 Methods

Participants. A total of twelve people participated in the experiment.

Stimuli. As previously described, stimuli were created by convolving a sound stimulus with a range of stimulated room IRs.

Three types of sound sources were used, all extracted from anechoic recordings: a drum riff, female singing, and male speaking. The drums featured only beats on a snare drum. The female singing was a recording of Händel's "Lascia Ch'io Pianga"[1]. The speech was a sentence in French spoken by a native male speaker. For the drums and speech stimuli, three different excerpts (short, medium, and long) were extracted from the original files. Due to the fact that the singing was very slow, it was impossible to create a short excerpt that was of comparable length to the drums and speech without noticeable artefacts. Therefore, only two excerpts were created for the singing. An overview of the sound sources' duration can be found table 1.

Table 1. Length (ms) of the sound stimuli used in Experiment 3

Length	**Short**	**Medium**	**Long**
Drums	396	800	2850
Singing		2328	6420
Speech	569	1293	3800

A series of six IRs was created, with linear exponential decays varying from 0.5 to 3.6 s in equal steps of 0.62s. The three sound stimuli for the various lengths, each combined with 6 IRs, produced a total of 48 test stimuli. These stimuli were presented over headphones (AKG K-271 Studio), played from a MacPro connected to an audio interface (Motu, mkII 828) at a sample rate of 44.1 kHz.

[1] Taken from the website of Prof. Angelo Farina; http://www.angelofarina.it/Public/Anecoic/

Procedure. Each participant completed two consecutive sessions with a small break in between. Per session all 48 stimuli were presented twice, for a total of 96 trials per session. Each session was of variable length as the participants set their own pace. On average, a session lasted about 15 minutes. The instructions to the participant were an adaptation of those suggested by S.S. Stevens (in [31], p. 239):

> "You will be presented with a series of different kinds of sound clips in random order. Your task is to tell how much reverberation there was in those clips by assigning numbers to them. Call the first clip a number that seems appropriate to you. Then assign successive numbers in such a way that they reflect your subjective impression. There is no limit to the range of numbers that you may use, except that they should be whole numbers. Try to make each number match the amount of reverberation as you perceive it. You can repeat the excerpt by entering "0" in the rating box. Otherwise, after entering a rating, press enter to advance to the next excerpt. There are no right or wrong answers-your honest opinion is what counts. Try to use the same strategy throughout the experiment, be consistent!"

Data Analysis. Because participants were free to choose any (integer) value they pleased it is necessary to normalize the individual results before any group mean analysis can be performed. For this we apply an individually determined gain that rescales each participant's results such that their overall mean becomes (arbitrarily) 100. The rescaling is implemented as follows:

$$\hat{X}_{ijk} = \frac{X_{ijk}}{\bar{X}} * 100 \qquad (1)$$

Where $\hat{X}_{ijk}$ represents a participant's rescaled data, the indices represent the different experiment parameters (i.e., sound source, length, and RT). Finally, X_{ijk} stands for the original data, and $\bar{X}$ for the participants overall mean. The 100 is an arbitrarily chosen value, which becomes the new common overall mean.

S.S. Stevens [30] argued in favor of the geometric mean as a representative measure of the group's performance. Here we also calculated the simple arithmetic mean for purposes of statistical analysis and to obtain an unbiased estimate of spread.

4.2 Results

The arithmetic and geometric mean were virtually identical (correlation, $\rho = 0.9996$). We therefore used the more common mean. The main results are shown in Figure 4 below . The three panels correspond to the three sound sources used. Plotted are the scaled estimated magnitudes as a function of RT. The parameter is sound stimulus length (see legend).

A number of observations can be made. First there was a clear monotonic increase in the estimate of reverberation as a function of RT. This means that

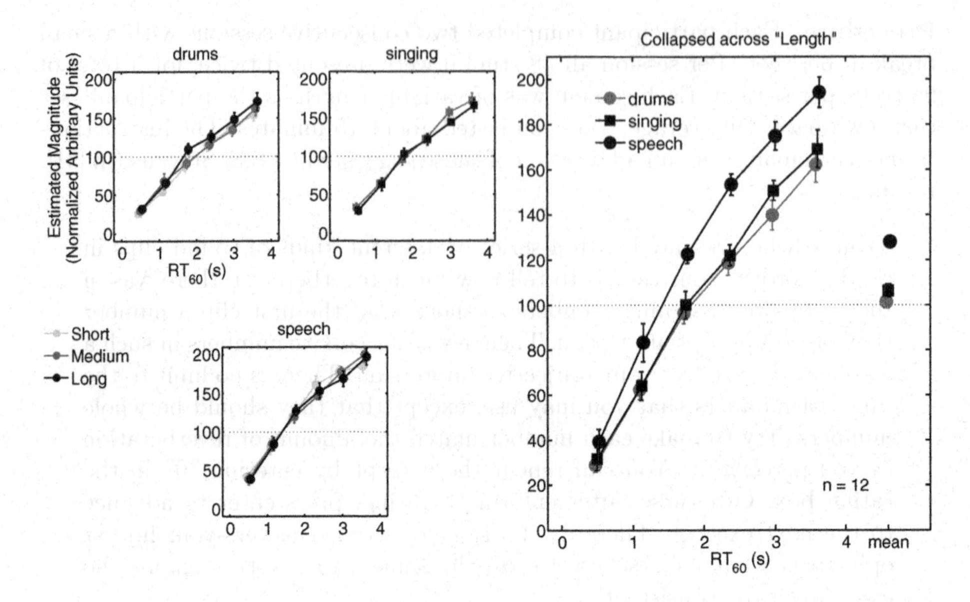

Fig. 4. Experiment 3. The three panels on the left show the normalized mean estimates for 12 participants per stimulus type and for each length of the source stimuli. The panel on the right shows the mean values collapsed across source length. Vocal and non-vocal stimuli are grouped in black and grey, respectively On the far right of the panel are plotted the overall means across RT. All errorbars correspond to the standard error of the mean.

people were generally able to reliably distinguish between the different levels of RT. Second, there seems to be no difference in estimated magnitude between the various sound-source lengths used. Finally, a comparison across panels suggests that there is no difference between the different types of stimuli. To further explore this observation the estimates for the different sound source length were collapsed. We next plotted the three resulting curves together. This can be seen in the figure to the left. Plotted this way there does seem to be an effect of type of sound stimuli in that for relatively long RTs (i.e., RT > 1.8 s) the reverberation of speech tends to be overestimated relative to either drums or singing.

To support these visual appreciations of the data a number of repeated measure ANOVAs were conducted. We first conducted ANOVAs for each sound source separately with Length and RT as independent variables.

Drums. There was a significant effect of RT, $F(5,55) = 87.63$, $p = .000$, but surprisingly also of Length, $F(2,22) = 8.42$, $p = .002$. The interaction between the two factors was not significant ($F < 1$). Individual, Bonferroni corrected t-tests, showed that there was only a significant difference between the short and long drum stimuli, $t(11) = 5.14$, $p < 0.05$. *Singing.* There was no effect of Length ($F < 1$), only an effect of RT, $F(5,55) = 79.37$, $p = .000$. *Speech.* There was no effect of Length ($F < 1$), and again only an effect of RT, $F(5,55) = 86.89$, $p = .000$.

Despite the one significant effect of Length for the Drums stimuli, we collapsed the data across Length in order to compare across the different types. A 3 (Type) x 6 (RT) ANOVA confirmed the visual appreciation of the results. Both Type, $F(2,22) = 20.35$, $p = .000$, and RT, $F(5,55) = 118.92$, $p = .000$, were significant, but not their interaction, $F(10,110) = 1.89$, $p = .115$.

4.3 Discussion

We have used a magnitude estimation paradigm to gauge people's perception of reverberant stimuli. Participants were free to assign a number to the amount of perceived reverberation. From these responses we were able to construct the underlying psychological scales [30] and compare them between the various conditions.

We found that the length of the source stimulus had no substantial effect on the amount of reverberation perceived. The ratio of the lengths of "long" and "short" reverberant test stimuli (with the same RT) was on average 1.8 and could be as high as 4.0. So, despite these large differences in overall stimulus length, participants judged the amount of reverberation to be the same. We can therefore conclude that people were not simply basing their judgments on overall length of the stimuli. This is consistent with Seraphim's [17] observation that the discrimination threshold of reverberation was independent from the length of the signal.

The pertinent result, however, came from the comparison between the different types of sound stimuli used. It was found that speech stimuli lead to higher estimates of reverberation than either singing or drums. This was especially the case for relatively long RTs (i.e., > 1.8s). Thus the sound stimulus does have an effect. The proposed cause for this difference could be the central nervous system's differential treatment of various different types of sounds (see 1.3. of the Introduction).

There are however two methodological issues with this experiment that could render this conclusion premature. First, the lengths of the various sound sources were not well matched, which invalidates the categorical nature of the groups in which they were put in the statistical analysis. This could have lead to spurious significant effects (or the masking thereof). Second, the sound stimuli were not equalized in terms of overall RMS level. It could be that presentation level has specific effects on the perceptual appreciation of reverberant stimuli. Even though both are unknowns, we cannot exclude the possibility that having unequal lengths and intensities compromised the comparison of the JNDs for the two conditions. Experiment 4 was performed to control for these two factors.

5 Experiment 4: Magnitude Estimation Revisited

The final Experiment was similar to Experiment 3, with a few changes. We now control for the lengths of the sounds sources which are now closely matched. We also calibrated the various sound sources in level to have the same RMS value.

In addition we reduced the number of lengths to 2, but increased the range of stimuli to four: two vocal stimuli (singing and speech) and two "instrumental", or non-vocal, stimuli (drums and piano).

5.1 Methods

Participants. A total of thirteen people participated in the experiment. The results of one participant (#9) were discarded because s/he had apparently misunderstood the instructions and was consequently not performing the task at hand.

Stimuli. Test stimuli were created by convolving various sound stimuli with a range of simulated room impulse responses (IRs). There were four types of sound stimuli, a female singing, piano, drums, and a male speaking. For each source a new short and long excerpts were selected. The drums were taken from the same recordings as in Experiment 3. Speech was now taken from a recording in English of somebody counting. The singing and piano stimuli were extracted from a recording (taken from the same website mentioned in 4.1.2) of Mozart's "Cosi' fan tutte". Piano was recorded from the electrical output of an electric piano whereas the singing was recorded from an outdoor performance on a 40 cm thick layer of grass. An overview of the sound sources' duration can be found in table 2.

Table 2. Length (ms) of the sound stimuli used in Experiment 4

	Short	Long
Singing	250	2260
Piano	250	2110
Drums	280	2300
Speech	250	2290
Mean	258	2240

A series of five IRs with exponential decays were synthesized as previously described. The RT was varied from 0.5 to 3.6 s in equal steps of 0.775 s. The 4 sound stimuli with 2 lengths, combined with 5 IRs, results in a total of 40 test stimuli. These stimuli were presented over headphones (AKG K-271 Studio), played from a MacPro connected to an audio interface (Motu, mkII 828) at a sample rate of 44.1 kHz.

Procedure and Data Analysis. Each participant completed two consecutive sessions with a small break in between. Per session all 40 stimuli were presented twice, for a total of 80 trials per session. The instructions to the participant were the same as in Experiment 3. Before the actual data collection started, the participant listened to all the 40 stimuli once, in random order, to have a clear impression of the range of reverberation used. The data analysis was identical to that of Experiment 3.

5.2 Results

Figure 5 shows the results of Experiment 4. A similar pattern of results as in Experiment 3 was observed. We again observed no apparent effect of the length of the stimulus. There is a clear grouping of non-vocal (in black) and vocal (grey) stimuli. Moreover, it seems that the non-vocal and vocal stimuli dissociate as early as at ≈1.2 s, only at the very smallest value of RT are judgments virtually identical.

Omnibus ANOVA. The data were submitted to a 4 (Sound Stimulus) x 2 (Length) x 5 (RT) overall repeated measures ANOVA. All effects, except for the highest order interaction and the interaction between Sound Stimulus and Length, were significant; Length ($F(1,11) = 13.5$, $p = 0.004$), Sound Stimulus ($F(3,33) = 11.3$, $p = 0.001$), RT ($F(4,44) = 246.6$, $p < 0.001$), Length and RT ($F(4,44) = 3.52$, $p = 0.014$), and Sound Stimulus and RT ($F(12,132) = 3.33$, $p = 0.005$).

Effect of Sound Stimulus Length. To explore the significant effects involving the factor Length we performed separate repeated measures ANOVAs per sound stimulus. Obviously, the main effect of RT was significant in all ($F(4,44) \geq 99.9$, $p < 0.001$).

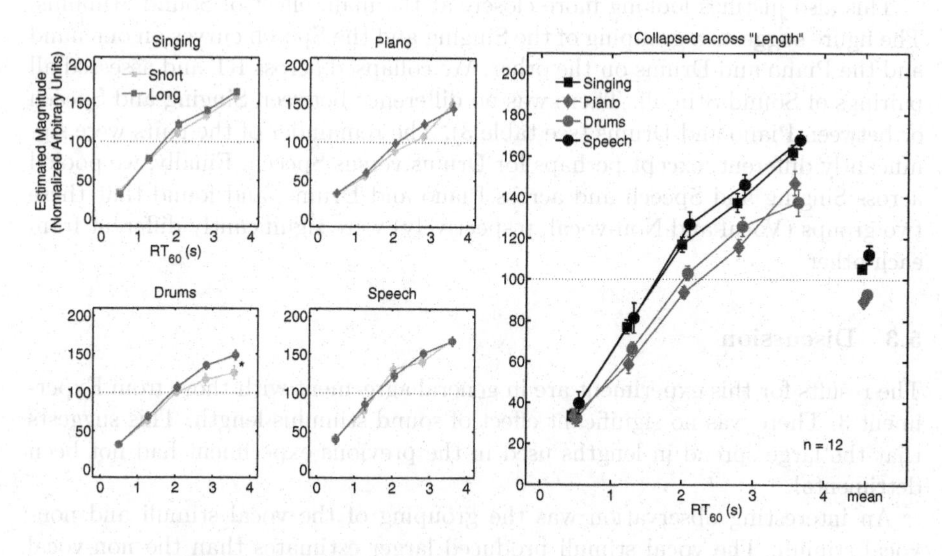

Fig. 5. Experiment 4. Same conventions as in Figure 4.

Singing. Neither the main effect of Length was not significant ($F(1,11) = 2.20$, $p = 0.166$), nor its interaction with RT ($F(4,44) = 1.11$, $p = 0.362$). *Speech.* Neither Length ($F < 1$), nor the interaction ($F(4,44) = 1.65$, $p = 0.178$) was significant. *Piano.* Both the effect of Length ($F(1,11) = 11.66$, $p = 0.006$), and the interaction ($F(4,44) = 3.21$, $p = 0.034$) were significant. *Drums.* There was a significant effect for Length ($F(1,11) = 6.02$, $p = 0.032$) and a near significant one for the interaction ($F(4,44) = 2.59$, $p = 0.050$).

Given the (near) significant interactions for the Piano and Drums stimuli, separate paired t-tests were conducted between the individual pairs of short and long stimuli at each RT. After Bonferroni correction only one pair showed a significant difference, namely the pair of the Drums stimuli at the longest RT (3.6s, marked by an asterisk in the figure above), $t(11) = 3.25$, $p = 0.04$.

Thus for all practical purposes there is no effect of sound source length on the perceived amount of reverberation. We therefore ignore the contributions of Length in the further analysis.

Effect of Sound Stimulus Type. There are three significant effects left, the two main effects of Sound Stimulus and RT, but both are qualified by their interactions, which we address now.

Inspection of the curves shows they are monotonously diverging except at one point. At the highest level of RT the Piano and Drums lines cross, which suggest a source for the interaction. However, when we exclude the values for the largest RT and only use the first four we still get a significant interaction, $F(9,99) = 3.43$, $p = 0.004$. Thus the interaction must be due to the divergence in the curves.

This also justifies looking more closely at the main effect of Sound Stimulus. The figure suggests a grouping of the Singing and the Speech curves on one hand and the Piano and Drums on the other. We collapsed across RT and assessed all pairings of Sound Stimuli. There was no difference between Singing and Speech or between Piano and Drums (see table 3). The remainder of the pairs were significantly different, except perhaps for Drums versus Speech. Finally, we pooled across Singing and Speech and across Piano and Drums, and found that these two groups (Vocal and Non-vocal, respectively) were significantly different from each other.

5.3 Discussion

The results for this experiment are in general agreement with those from Experiment 3. There was no significant effect of sound stimulus length. This suggests that the large spread in lengths used in the previous experiment had not been detrimental.

An interesting observation was the grouping of the vocal stimuli and non-vocal stimuli. The vocal stimuli produced larger estimates than the non-vocal ones. In the light of the lateralization model [24] the grouping of the vocal stimuli seems obvious. The challenge, however, comes from explaining the grouping of

Table 3. Results of t-tests comparing the various pairing of sound stimuli. The last column presents the p-values after Bonferroni correction, where applicable. The degrees of freedom for all test was 11.

Pair	**t**	**p**	**p corr**
Singing vs. Speech	1.23	0.245	n/a
Piano vs. Drums	1.24	0.242	n/a
Singing vs. Piano	6.70	0.000	0.000
Singing vs. Drums	6.01	0.000	0.000
Piano vs. Speech	4.05	0.002	0.012
Drums vs. Speech	3.10	0.010	0.060
Vocal vs Non-vocal	5.50	0.000	n/a

the piano and drums stimuli. Whereas the piano stimuli, as music, clearly require the right brain's spectral analytical abilities, the same cannot reasonably be said for percussion.

6 General Discussion

6.1 Summary of the Findings

This study presents an investigation of the human perception of reverberation which would be commonly found in large orchestral performance rooms, through a series of experiments. It was found that RT discrimination is relatively sensitive with thresholds around 6% (Experiment 1), which is comparable to previous studies [17,18].

When looking explicitly at the effects of the type sound stimulus on the perception of reverberant stimuli, *discrimination* ability is not affected (Experiment 2). In contrast, the *perceived amount* of reverberation is dependent on the sound stimulus. Generally speaking vocal stimuli lead to larger estimates in comparison to the other stimuli types. This could be relevant in considering previous findings of studies on preferred reverberation. Operatic music listeners have been found to prefer relatively short RTs [22] relative to instrumental musical [1]. One can speculate that a preference for a short RT is consistent with "compensating" for the heightened perceived amount of reverberation in vocal stimuli (Exp 4). In addition, the observed distinction between vocal and non-vocal stimuli is generally in keeping with neuroscientific and behavioral studies presented earlier, demonstrating that vocal sounds are processed differently.

6.2 Theoretical Implications

In the introduction we discussed the lateralization model proposed by Zatorre et al. [24]. The results of Experiments 3 and 4 are not necessarily in line with the expectations from this model. If the lateralization in the brain contributed to the current task the expectation was a grouping of the psychological scales in terms of the temporal or spectral nature of the stimuli. The two vocal stimuli were grouped, which could suggest similar engagement of the left hemisphere. The grouping of the two non-vocal stimuli, however, is less clear. Whereas the piano stimuli, being harmonic and melodic, would engage the right hemisphere, the drum stimuli, being percussive and not melodic, would likely not.

Although the difference in psychological scales can still be caused by differential treatment of the sounds by the CNS, the present results suggest that this is not (completely) based on the lateralization model [24]. This does not invalidate the lateralization model and this study was not designed to test it. The results of this study however do suggest that the perceptual processes in play are not entirely based on the simple lateralization of temporal and spectral processes.

There seems to be a dissociation of the effects of sound stimulus between processes for discrimination and perceived magnitude of reverberation. However, the differential effects on the perceived amount of reverberation found in Experiments 3 and 4 would suggest that the discrimination of reverberation can be modulated by the type of sound stimulus used. This is not consistent with the lack of an effect observed in Experiment 2. However, the effect of sound stimulus is larger for longer RTs and the RTs used in Experiment 2 were in the midrange (i.e., around 1.8s) of those used in the magnitude estimation experiments. Thus, it remains possible that discrimination for larger RT stimuli is sensitive to the source stimulus type.

6.3 Practical Implications

The fact that the sound stimulus has differential effects on the amount of perceived reverberation could have practical consequences. It has long been recognized that a room's acoustic should be tailored to the activity taking place in the room (e.g., [32]). Thus depending on whether the room is intended for speech (a lecture room) or certain kinds of music affects design choices.

The results of this new study can be pertinent to the application of reverberation to auditory display. As suggested in the introduction, reverberation can be used as a parameter tag to help perceptually group acoustic items. However, the observation that the type of sound stimulus affects the amount of perceived reverberation puts some limits or constraints on its usability. As long as one uses like stimuli there should be little concern and one can create a generous number of different groups that listeners can still reliably distinguish from each other. Based on current results, complications could arise if one used a larger range of stimuli, such as for instance speech and music, resulting in unreliable comparisons between different stimuli type reverberation. In this case, pairing speech and music that have the same physical RT could very well lead to miscategorizing them into different groups, which runs counter the very purpose of the application of the reverberation as an independent perceptual parameter.

Acknowledgements. We thank Aaron Rosenblum for helping with the data collection, and Prof. Angelo Farina for making available some of the anechoic recordings used in Experiments 3 and 4. This work was supported by NSERC, FQRSC and CFI grants to C. Guastavino. Testing took place at the McGill Multimodal Interaction Laboratory and at the Centre for Interdisciplinary Research on Music Media and Technology, Montreal.

References

1. Kuhl, W.K.: Ueber versuche zur ermittlung der guenstigsten nachhallzeit grosser Musikstudios. Acustica 4, 618–634 (1954)
2. Kahle, E.: Validation d'un modèle objectif de la perception de la qualité acoustique dans une ensemble de salles de concerts et d'opéras. Unpublished Ph.D. dissertation, Université du Maine, Le Mans (1995)
3. Bradley, D.T., Wang, L.M.: The effects of simple coupled volume geometry on the objective and subjective results from exponential decay. J. Acoust. Soc. Am. 118, 1480–1490 (2005)
4. Katz, B.F.G., Kahle, E.: Design of the new Opera House of the Suzhou Science & Arts Cultural Center. In: Proceedings of the 9th Western Pacific Acoustics Conference, WESPAC IX, Seoul, Korea, 26-28 June (2006)
5. Kahle, E., Johnson, R., Katz, B.F.G.: The new konzertsaal of the KKL Center, Lucerne, Switzerland. II Preliminary acoustical measurements. Acta Acust. 85, S2 (1999)
6. Barron, M., Marshall, A.H.: Spatial impression due to early lateral reflections in concert halls: the derivation of a physical measure. J. Sound Vibr. 77, 211–232 (1981)
7. Cox, T.J., Davies, W.J., Lam, Y.W.: The sensitivity of listeners to early sound field changes in auditoria. Acustica 79, 27–41 (1993)
8. Lavandier, M., Culling, J.F.: Speech segregation in rooms: monaural, binaural, and interacting effects of reverberation on target and interferer. J. Acoust. Soc. Am. 123, 2237–2248 (2008)
9. Culling, J.F., Hodder, K.I., Toh, C.Y.: Effects of reverberation on perceptual segregation of competing voices. J. Acoust. Soc. Am. 114, 2871–2876 (2003)
10. Hartmann, W.M.: Localization of sounds in rooms. J. Acoust. Soc. Am. 74, 1380–1391 (1983)
11. Shinn-Cunningham, B.G., Kopco, N., Martin, T.J.: Localization of nearby sources in a classroom: binaural room impulse responses. J. Acoust. Soc. Am. 117, 3100–3115 (2005)
12. Mershon, D.H., Ballenger, W.L., Little, A.D., McMurtry, P.L., Buchanan, J.L.: Effects of room reflectance and background noise on perceived auditory distance. Perception 18, 403–416 (1989)
13. Kates, J.M.: Room reverberation effects in hearing aid feedback cancellation. J. Acoust. Soc. Am. 109, 367–378 (2001)
14. Bronkhorst, A.W., Houtgast, T.: Auditory distance perception in rooms. Nature 397, 517–520 (1999)
15. Mershon, D.H., Bowers, J.N.: Absolute and relative cues for the auditory perception of egocentric distance. Perception 8, 311–322 (1979)
16. Katz, B.F.G.: International round robin on room acoustical impulse response analysis software 2004. Acoust. Res. Lett. Online 5, 158–164 (2004)
17. Seraphim, H.P.: Untersuchungen über die unterschiedsschwelle exponentiellen abklingens von raushbandimpulsen. Acustica 8, 280–284 (1958)

18. Niaounakis, T.I., Davies, W.J.: Perception of reverberation time in small listening rooms. J. Audio Eng. Soc. 50, 343–350 (2002)
19. Picard, D.: Audibility of non-exponential reverberation decays, Unpublished thesis, Rensselaer Polytechnical Institute, Troy, NY (2003)
20. Sum, K.S., Pan, J.: Subjective evaluation of reverberations times of sound fields with non-exponential decays. Acta Acust. United Ac. 92, 583–592 (2006)
21. Frissen, I., Katz, B.F.G., Guastavino, C.: Perception of reverberation in large single and coupled volumes. In: Proceedings of the 15th International Conference on Auditory Display, Copenhagen, Denmark, pp. 125–129 (2009)
22. Sakai, H., Ando, Y., Setoguchi, H.: Individual subjective preference of listeners to vocal music sources in relation to the subsequent reverberation time of sound fields. J. Sound Vibr. 232, 157–169 (2000)
23. Belin, P., Zatorre, R., Lafaille, P., Ahad, P., Pike, B.: Voice-selective areas in human auditory cortex. Nature 403, 309–312 (2000)
24. Zatorre, R.J., Belin, P., Penhune, V.B.: Structure and function of auditory cortex: music and speech. Trends Cog. Sci. 6, 37–46 (2002)
25. Guastavino, C.: Categorization of environmental sounds. Can. J. Exp. Psychol. 60, 54–63 (2007)
26. Giordano, B.L., McDonnell, J., McAdams, S.: Hearing living symbols and nonliving icons: category specificities in the cognitive processing of environmental sounds. (Submitted to Brain Cogn)
27. Guastavino, C.: The ideal urban soundscape: Investigating the sound quality of French cities. Acta Acust. United Ac. 92, 945–951 (2006)
28. Guastavino, C., Cheminée, P.: Une approche psycholinguistique de la perception des basses fréquences: Conceptualisations en langue, représentations cognitives et validité écologique. Psychol. Française. 48, 91–101 (2003)
29. Wichmann, F.A., Hill, N.J.: The psychometric function: I. Fitting, sampling, and goodness of fit. Percept. Psychophys. 63, 1293–1313 (2001)
30. Stevens, S.S.: Issues in psychophysical measurement. Psychol. Rev. 78, 426–450 (1971)
31. Gescheider, G.A.: Psychophysics, the Fundamentals. Erlbaum, New York (1997)
32. Meyer, J.: Acoustics and the Performance of Music. Verlag Das Musikinstrumenter, Frankfurt/Main, Germany (1978)

Towards Timbre Modeling of Sounds Inside Accelerating Cars

Jean-François Sciabica[1,2], Marie-Céline Bezat[1], Vincent Roussarie[1], Richard Kronland-Martinet[2], and Sølvi Ystad[2]

[1] PSA Peugeot-Citroën, 2 route de Gisy, 78943 Vélizy-Villacoublay, France
{jean-francois.sciabica,marieceline.bezat}@mpsa.com, vincent.roussarie@mpsa.com
[2] LMA, Centre National de la Recherche Scientifique, 31 chemin Joseph-Aiguier, 13402 Marseille cedex 20
{kronland,ystad}@lma.cnrs-mrs.fr

Abstract. Quality investigations and design of interior car sounds constitute an important challenge for the car industry. Such sounds are complex and time-varying, inducing considerable timbre variations depending on the driving conditions. An interior car sound is indeed a mixture between several sound sources, with two main contributions, i.e. the engine noise on the one hand and the aerodynamic and tire-road noise on the other. Masking phenomena occur between these two components and should be considered when studying perceptive attributes of interior car sounds in order to identify relevant signal parameters. By combining sensory analysis and signal analysis associated with an auditory model, a relation between a reduced number of signal parameters and perceptive attributes can be found. This approach has enabled us to propose timbre descriptors based on the tristimulus criterion that reflect the dynamic behavior of a sound inside an accelerating car.

Keywords: Interior car noise, auditory representation, timbre modeling.

1 Introduction

Since global noise level in and outside cars has considerably decreased in the past decades, interior car sounds are no longer considered as a discomfort, but rather as a sound design question. Interior car sounds indeed contribute to the identity and the perceived quality of the car and the perceived dynamism can, in the case of an accelerating car, contribute to the evocation of power and sportivity.

Sounds perceived in car passenger compartments are the result of three well known acoustic sources: the engine source – called the harmonic part-, the tire-road source and aerodynamic source –called the noise part.

In order to study perceptive attributes of interior car sounds, we would like to develop a perception-based synthesis model that can be used as a tool for the investigation of the perceptive relevance of the signal parameters. The identification of these parameters will make it possible to directly control the timbre and the dynamics of such sounds. Dynamics is indeed an important contribution to the

S. Ystad et al. (Eds.): CMMR/ICAD 2009, LNCS 5954, pp. 377–391, 2010.

perception of interior car sounds because of their non-stationarity that clearly evolves during a car acceleration. The first step towards the construction of our synthesis model consists in determining the components of the harmonic part which are actually audible in the presence of the noise part. A considerable amount of information contained in the harmonic part can be reduced by considering masking phenomena. In addition to the masking due to the noise part, masking phenomena between harmonics should also be considered due to the small frequency range between engine harmonics. Finally, masking phenomena also depend on the dynamics, since the noise part increases with the speed of the car.

In the literature, the investigation of car sound quality has most often been studied through investigations of perceptive aspects and evocations. Bisping [1] described two perceptive factors forming a four quadrant scheme of sound quality with one axis defined by pleasant-unpleasant" ant the other axis defined by "powerful-weak". Kubo [2] proposed a car sound map with two axes "not sporty…sporty" and "simple...luxurious". A semantic differential method and a factor analysis method were used to create their sound map. With semantic differential techniques, the meaning of engine sounds has been widely investigated and a quantity of adjectives was used to describe this kind of sounds. Chouard [3] proposed a list of 232 pairs of adjectives to illustrate the sound of an engine noise.

However, some studies aimed at finding the link between perceptive attributes and signal parameters. Hansen [4] analyzed the impact of tonal components on the sound quality. With an algorithm based on DIN 45681, he identified and extracted engine harmonics that are considered as prominent in interior car noise and showed their relation to the perception of tonal phenomena like whistling.

In a previous study [5], we proposed a procedure to predict audibility threshold of harmonics produced by the engine noise. Our algorithm is a linear model based on the knowledge of noise level in the critical bands centered on the harmonic frequency. We reproduced an audibility threshold experiment [6] with stimuli and conditions adapted to automotive context.

Indeed, the psychoacoustic approach seems to be the right way to obtain a robust description of interior car sounds. Bezat [7] also proposed an original analysis/synthesis approach to investigate the link between the quality of car door closure noise and signal parameters. The perceptive aspects of such sounds were first evaluated with a sensory panel. This method consists in finding acoustic descriptors to qualify the sound. Then, the panel can rate the sound for each descriptor and create a sensory profile. Secondly, a time/ERB description was chosen to analyze the signal. Combining these two ways to analyze signals, she proposed a synthesis model controlled by perceptive criteria.

Hence, in order to apply this approach to interior car sounds, we analyzed our signals with an auditory model in order to visualize harmonic components which are audible in the presence of the noise part.

In this paper we describe the specificities of interior car sounds and the different source contributions. We then describe the signal using a time-frequency analysis technique and a synthesis tool. Further, the perceptive analysis of the signal effectuated with a sensory panel is presented, and finally the auditory model is applied to the signal to point out perceptually relevant signal components. In order to characterize timbre changes during car acceleration, we developed a method inspired by the so-called tristimulus criterion in order to describe the energy transfer observed in the auditory representation.

2 The Interior Car Sound

Sounds perceived in car passenger compartments are the result of three well-known acoustic sources: the engine source – called the harmonic part-, the tire-road source and aerodynamic source –called the noise part. In this section a description of these sources as well as their interaction is given.

2.1 From the Piston Rotation to the Spectrum of the Harmonic Part

Good knowledge of motor mechanics is important for the comprehension of automotive sound. The harmonic part of an interior car sound is the result of the quasi-periodic movement of the piston caused by the rotation of the crankshaft. These harmonics are called engine harmonics (with the prefix H) and the number of an engine harmonic is called engine order.

If we consider the case of a 4 cylinder motor, pistons fire successively every half motor rotation so a given piston fires every second motor rotation. The period between two firing pistons and between successive firing pistons is not constant and these variations have to be taken into account. Nevertheless, we consider that the fundamental component called H1 corresponds to a motor rotation even if this phenomenon doesn't physically generate the harmonics. In this way we combine the two phenomena which generate their own fundamental and their own partials. If we call *Nrpm* the number of motor rotations per minute, H1 is given by the following relation:

$$H1 = \frac{Nrpm}{60}. \tag{1}$$

The first phenomenon, the most energetic, corresponds to the periodicity between two firing pistons (half a motor rotation) and generates harmonics called "odd harmonics" with it own fundamental frequency called H2 which is twice H1 (H2 = 2xH1) and with its own partials (H4 = 2xH2 = 4xH1, H6 = 3xH2=6xH1, H8 = 4xH2=8xH1). These harmonics are very often the most energetic in the spectrum. If the periodicity between firing pistons were identical, engine sounds would only contain these odd harmonics.

Due to the variations in periodicity between firing pistons, we have to consider a second phenomenon, the periodicity of one given firing piston which corresponds to 2 motor rotations and generates harmonics called "semi-harmonics ". The corresponding fundamental frequency is called H0.5 which is half H1 (H0.5 = 0.5xH1).

These harmonics are less energetic, but their contribution to the perceived sound cannot be neglected. They also respect the same labeling and numbering (H1=2xH0.5, H1.5=3xH0.5, etc.)

2.2 Aerodynamic Noise and Tire Road Noise

The noise part of an interior car sound is composed of aerodynamic noise and tire road noise. Aerodynamic noise is a broadband noise whose global sound level increases with speed. It mainly has a low frequency energy distribution (below 400 Hz), but its perceptual contribution is also important in the high frequency domain. Indeed, aerodynamic noise mainly masks high engine orders (~>12), but its impact can also be observed at low engine orders. Tire-road noise depends on three main parameters: car speed, tire texture and road texture. The contact between tire and road generates low frequency noise.

2.3 Impact of Driving Situation on the Interior Car Sound

The spectrum of an interior car sound also depends on the driving situation. Depending on the speed of the car, the balance between the noise part and the harmonic part changes. At slow motion, the harmonic part is predominant, while at fast motion, the noise part becomes more and more audible.

Moreover, the distribution of the harmonic amplitudes can vary rapidly in time. Indeed, engine sound transfer from the motor to the car interior depends on frequency. Some harmonics which were clearly audible at a given rotation speed can be less energetic at an other one. The balance between harmonics will also be modified in time, changing the timbre perception. In this case, masking phenomena between harmonics must be considered.

During an acceleration, the engine speed varies from 800 rpm to 6 000 rpm, which means that H2, the fundamental frequency of odd harmonics spans from 27 Hz to 200 Hz. Similarly, the frequency gap between semi-harmonics varies form 6 to 50 Hz creating good conditions for roughness phenomena. The density of semi harmonics in an ERB band is indeed important. For example, engine harmonics H3.5, H4, H4.5 and 5 are separated by 25 Hz at 3000 rpm. They belong to the same ERB band and interact to create sound modulations. The resolution of harmonics by the auditory system will therefore depend on the engine speed and the corresponding frequency spans of the harmonics.

3 Analysis Tools

3.1 Time/ Frequency Analysis

In order to study the frequency evolution of the signal obtained inside an accelerating car, we performed a time-frequency analysis (short Time Fourier with a window length of 0.3 seconds and a window step of 10 ms).

The signal was recorded in an accelerating car with 100% accelerator load (i.e. full acceleration during the whole acceleration period) and a dummy head in the passenger position. The chosen frequency range enables the visualization of harmonics up to the 12th engine order.

The spectrogram on figure 1 shows that a lot of harmonics are physically present in the signal. We can notice that many semi-harmonics contribute to the signal (H4.5,

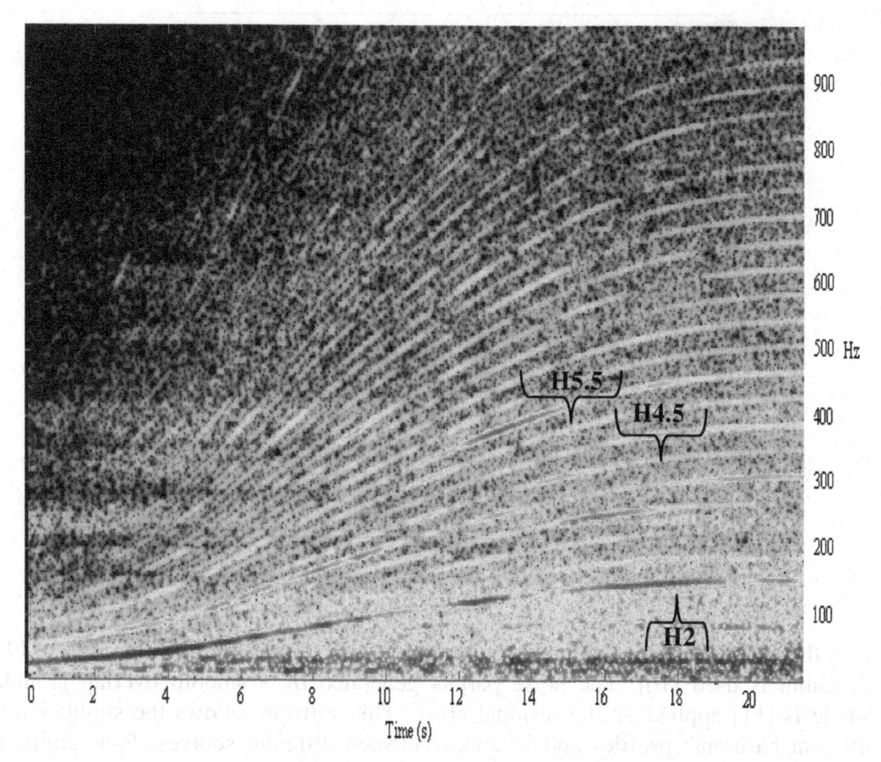

Fig. 1. Time-frequency representation of a signal recorded in an accelerating car

H5, and H5.5). We also notice that some harmonics emerge during the acceleration. However, this kind of representation does not tell us which harmonic is responsible for possible timbre variations of the interior car sound.

3.2 Analysis by Synthesis

In order to test some perceptive signal properties, a synthesis system call Hartis (Harmonic Real Time Synthesis) was developed at the PSA Research Center [8]. This tool is used to create signals for listening tests and to study the perceptual impact of specific signal parameters.

The first step of this sound design process is the separation of the original sound in a harmonic part and a noise part. For this purpose, we use the Additive software [9] to extract acoustic characteristics of the engine partials (frequency, amplitude and phase) according to the number of motor rotations per minute (RPM). Additive also extracts a residual noise which corresponds to the original sound minus the detected harmonic components. This residual noise contains aerodynamic noise and tire-road noise.

The synthesis software Hartis is further used. The files obtained with the Additive Software (residual noise and engine partials) are used as input. This software can replay the original sound and allows real time control by acting on basic parameters of the signal like engine partials and noise level. To resynthetize the harmonic part

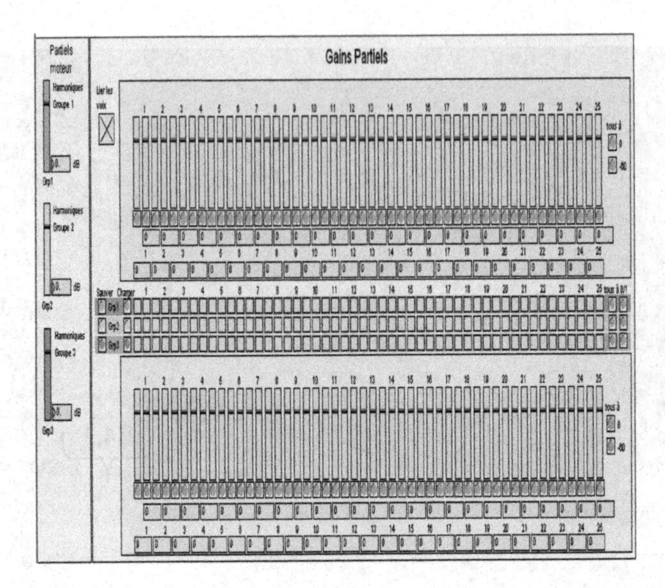

Fig. 2. Hartis Interface: Software for the real-time synthesis of interior car sound

from the characteristic of engine partials, an additive synthesis with the inverse FFT algorithm is used [10]. The noise part is generated by a smooth overlap granular synthesis [11] applied on the residual noise. This software allows the simulation of different harmonic profiles and balances between different sources. Four grains of 160 ms are played together to create the sound. By modifying the amplitude of 50 harmonics, we can create a large variety of sounds in order to test some relevant parameters of the harmonic part while the noise part remains unmodified.

4 Perceptive Analysis

The perceptive analysis has been effectuated in 2 steps. First a sensory analysis has been conducted to identify descriptors that correspond to onomatopoeia that characterize the sounds. Second, an auditory model has been applied to the sounds to identify signal parameters responsible for the perceived phenomena revealed by the sensory panel in the first experiment.

4.1 Sensory Analysis

We used a sensory panel as analytic noise evaluation tool. The panel was made of 9 naïve and normal hearing subjects, called judges, who were paid for their participation. After some training sessions, the panel was asked to define a minimal number of descriptors to qualify the sound they heared (2 x 2 hours). Then the panel proposed a scaling for each descriptor (4 x 2 hours). The panel gave an evaluation (2 x 2 hours) of 12 stimuli of 2 seconds recorded in standard and sporty cars with 100% accelerator load in 3rd gear. The beginning of the presented stimuli corresponds to the moment when the motor rotates at 3500 rpm. Sounds were recorded in cars from different constructors. The total procedure was divided in 8 sessions of 2 hours. Through this method, we obtain a sensory profile for each sound. The success condition of a sensory panel is sound discriminability, the judges' repetitivity in time and their consensus. This is the reason why such a long procedure is necessary to transform naive subjects into efficient judges in order to obtain their consensus on all the descriptors.

The sensory panel identified three main descriptors [8]. First, what they called an "ON" which characterizes engine noise booming. It mainly depends on the audibility of odd harmonics (H2 –H4 – H6). The second descriptor, called "REU" is used to describe the roughness of the sound. This roughness is strongly related to the interaction between semi-harmonics (for example the interaction between high engine order harmonics H9-H12). A third descriptor is identified by the Panel as an "AN". It differs from the "ON" by the relative importance of intensity level of odd harmonics. This phenomenon is less common than the ON Phenomenon, but we take it into account because we believe that the transition between "ON" and "AN" contributes to the perception of dynamism.

4.2 Application of Auditory Model

To point out signal parameters responsible for the descriptors obtained from the sensory panel and other perceptually relevant aspects, we applied an auditory model to the interior car sound. This approach also reveals masking phenomena which can give us information about the audibility threshold in the noise part masking between engine harmonics.

The auditory model simulates different stages of our auditory system. The transfer of sounds through the outer and middle ear can be modeled using a single FIR filter [12]. It is also constructed to simulate the effects of head-phone presentation. The output of the filter can be considered as symbolizing the sound reaching the cochlea.

The cochlea can be described as a bank of bandpassfilters called auditory filters which center frequencies span from 50 to 15000 Hz and which bandwidth increases with increasing center frequency. We chose a 4th order linear gammatone filter, which is well adapted to our automotive preoccupations and which parameters are the center frequency and the damping coefficients [13]. The derivation of those parameters leads to the design of a non-linearly spaced auditory filterbank. The frequencies of the auditory filterbank are linearly spaced on the so called ERB frequency scale. The step size determines the density of the filters on the ERB Scale. In this study we chose an important number of filters by ERB in order to improve the visualization at the output of our model. The center frequencies of each filter were distributed according to the harmonic frequencies. Hereby we had the possibility to follow the perceptive impact of each harmonic. Frequency centers are calculated with the knowledge of the RPM in time. We used a 20 ms interval between two updates of the center frequency filters. We draw the output of our auditory model by indexing the output of each gammatone filter by the engine number. That's why we have all the information of a given harmonic on the same line. This representation is helpful to describe the interaction between harmonics.

The next stage is the modeling of the inner cells: This stage was well described by Meddis [14]. The unidirectional excitation of inner hair cells is modeled by a half-wave rectification (HWR) followed by a low pass filter (LPF). We add compression with a power law in order to accentuate the variation [15].

The output of our model can be considered as the cochleagram defined by Slaney et al. [16] and can be interpreted as the cochlea movement.

4.3 Results and Discussion

4.3.1 Qualitative Results on Auditory Representation

The representation on Figure 3 confirms the important role played by the harmonics around H2-H4 and H6 to the perception of the "ON" phenomenon revealed by the sensory panel. This description first shows that high order harmonics do not have a great impact on the perception of the global timbre of an interior car sound. They are perceptually mixed with aerodynamic noise. Secondly, we believe that the booming phenomenon we can hear in the sound can be identified thanks to such a representation. Perceptively, the booming corresponds to a resonance generated by one of the even harmonics and can be identified on the cochleagram by an important increase in energy of this spectral component (For example between time t=16 s and t = 20 s on the engine harmonic 2). The representation, figure 3, facilitates the understanding of the transitions between these different events.

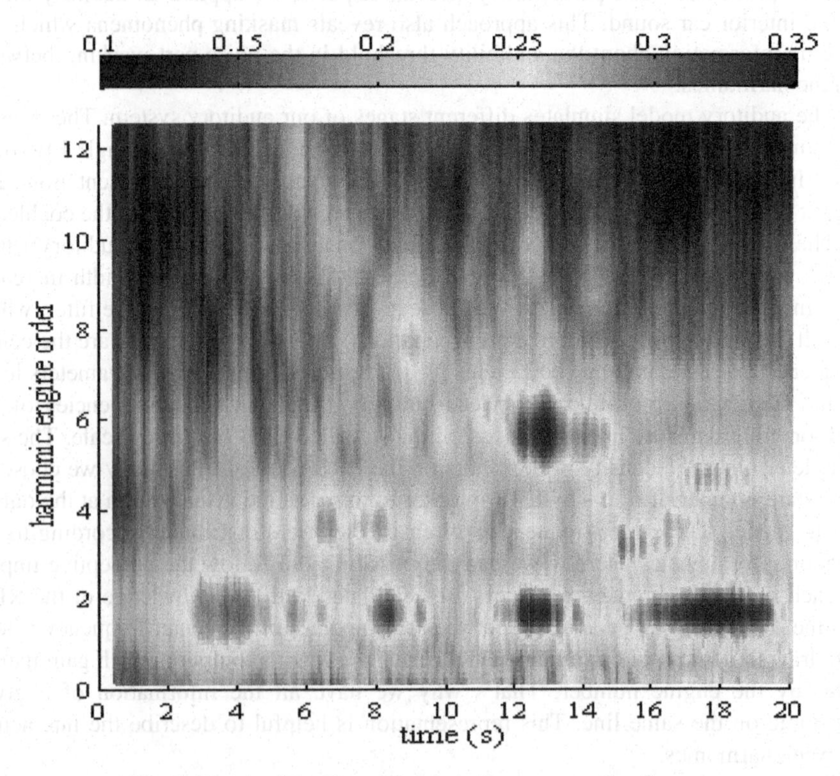

Fig. 3. Output of Auditory Model on the same accelerating car

Moreover, we can identify perceived formants on our representation. Actually, a combination of harmonics that seems to contribute to the perception can be observed. For example, we can see on Figure 3 that at the instant t=12s, we have two formants: a first formant resulting from a combination between the harmonics H1.5, H2 and H2.5 and a second formant linked to the harmonics H5, H5.5, H6 and H6.5. The presence of this second formant clearly changes the sound. The next question we have to study is how these formants interact to quantify timbre variations. Formant trajectories can also inform us about timbre evolution and give a score of timbre change. We can suppose that there is a relationship between these formant distributions and the descriptors used by the sensory panel. In the sound we present in Figure 3, we can hear a timbre change when we have two formants on our representation (for example at 12 seconds). The sound is different when we only have one formant. It sounds like an "ON" phenomenon while with two formants the sound is more ambiguous between an "ON" phenomenon and a "AN" phenomenon.

By considering the harmonic distribution through a few formants, we can reduce the information and hereby concentrate our investigation on fewer and perceptively relevant parameters to characterize perceptive aspect of the interior car sound. Further investigations are needed to interpret the cochleogram. In particular, masking phenomena that occur between the harmonic part and the noise part should be identified through the analysis of such representations.

4.3.2 Characterization of Timbre Change by Tristimulus

Auditory representation was also applied on certain interior car noises recorded in accelerating cars in order to characterize the timbre change phenomena during acceleration. Our goal is to characterize the timbre change by a relevant indicator whose fluctuation is clearly correlated with timbre evolution. We propose an extension of the tristimulus criterion as defined by Pollard [17]. Tristimulus is a three-coordinate representation for representing the timbre of sounds taking into account the loudness for three separate parts of the spectrum. It has shown to be appropriate to describe timbre of sounds which partial distribution evolves with time. Initially, the first component of the tristimulus was defined as the loudness of the fundamental frequency, the second as the loudness of the 2nd to 4th harmonics and the third as the loudness of the rest of the partials.

We have slightly modified the definition of each tristimulus component to adapt the tristimulus criterion to the specificities of our interior car sounds. Hence, the first component, called X computes the relative weight of the harmonics from H0.5 to H3. It includes H2 witch is perceived as the fundamental frequency of the interior car sound. The second component (Y) measures the relative weight of the harmonics from H3.5 to H6.5 in order to evaluate the impact of the 2nd and the 3rd pair harmonics. The definition of Pollard also includes the 4th pair harmonic but we don't take it in account because it doesn't seem relevant from the auditory representation. The third component (Z) determines the relative influence of all the remaining harmonics. As we formerly noticed, this distribution of harmonic bands was chosen in order to match the observation of the auditory representation.

As X+Y+Z = 1, it is adequate to use two of the coordinates to draw the tristimulus diagram as shown in figure 4. When the values of Y and Z are small, the fundamental

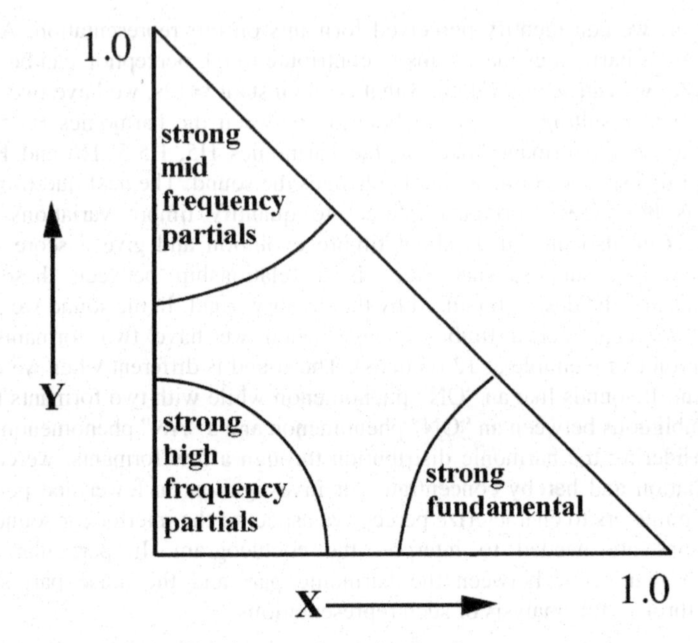

Fig. 4. Acoustic tristimulus diagram

frequency is perceptively dominant while a high y-value reflects strong mid-frequency partials and a high X-value strong high-frequency partials.

We slightly modified the tristimulus diagrams applied to our car sounds. The axes are unchanged, but their upper limits are decreased (corresponding to a zoom) to facilitate the visualization.

We applied this extension of tristimulus on both the auditory model and the spectral representation. By comparing the two representations, we showed that the auditory model gives clearer representations of timbre as it evolves in time. We compared the two kinds of representation in order to show that auditory representations are more relevant to explain the perception of interior car sounds. In both cases, the evolution of the second band (Y) as a function of the 1st band (X) is plotted.

Voice recordings: To test whether the tristimulus criterion could reveal "ON-AN" transitions, we first applied the auditory representation on voice recordings. For this purpose, we asked native, French speakers to imitate an accelerating car by singing French "ON" and progressively reach French "AN". In this way we also obtained a natural morphing between these two voice timbres.

By listening to these recordings, we can clearly notice that we have an "ON" at the beginning of the sound and a "AN" at the end. The transition's length is approximately 2 seconds. Our goal is to get a clear transition between these two panel descriptors of car sounds that we previously introduced, and to observe the transition between these voice timbres both on spectral and on auditory representations. We can notice that we have an energy transfer between two well identified bands. The first band is made of the fundamental frequency of the sound and the 2nd and 3rd harmonics. The second

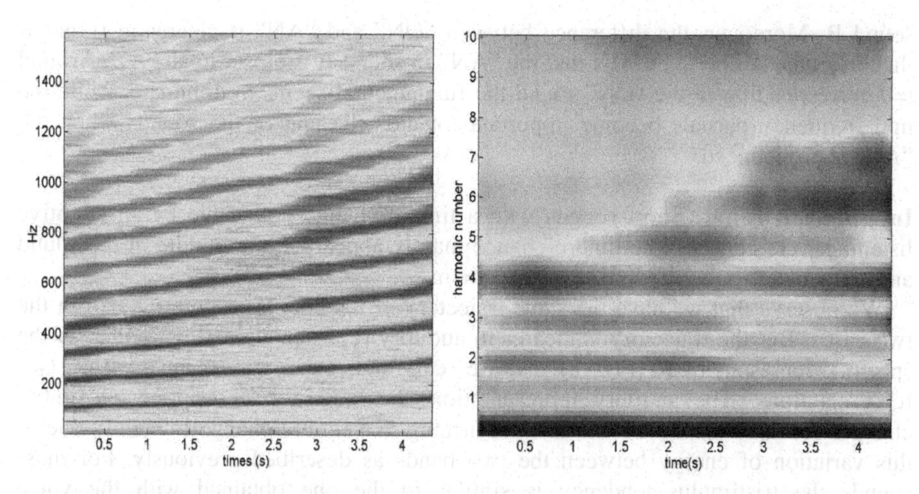

Fig. 5. From left to right: spectral representation and auditory representation applied on a voice recording imitating an accelerating car

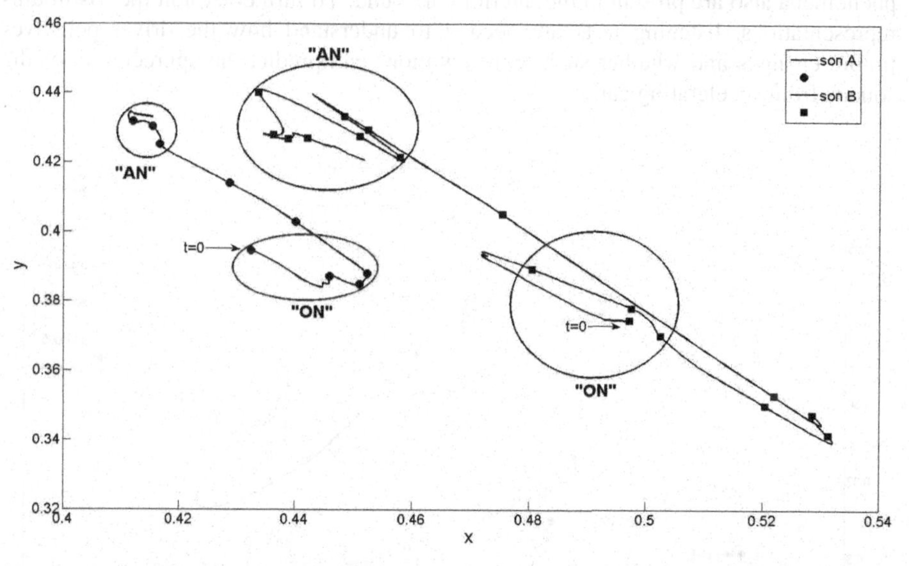

Fig. 6. X vs. Y of the tristimulus applied on two voice recordings. There is a mark each 0.5 second. The initial time is indicated by an arrow on the figure.

band is made of the harmonics from the 4th to the 8th band. Figure 5 confirms that timbre change is due to the different groups of harmonics involved and that the energy transfer between the fist two bands seems to be responsible for the change in timbre.

From the tristimulus plot we can see that both plots have similar tendencies although the scales are different. The length of sound A is 4 seconds whereas the length of sound B is 8 seconds. We can hear that the fundamental frequency is more present in sound B than in sound A and the transition between ON and AN is faster in sound B. Moreover, the difference between "ON" and "AN" is clearer in sound B than in sound A due to the fact that the "AN" in sound B is closer to an "A" . Around t=0 corresponding to the "ON" sound the fundamental seems to dominate, while the mid frequency partials become important towards the end of the sounds when the "AN" is pronounced.

Interior car sound: Short sounds with a timbre change noticeable by an attentive listening were chosen. The timbre change mainly appears at the middle of the sound and contributes to the perception of dynamism.

We observe that we have the same trajectory of the couple of point (X, Y) in the two cases. But the trajectory is clearer in auditory representation (Fig 7) than in the spectral representation (Fig 8) as we can notice if we compare the two representations. The auditory representations facilitate the description of timbre changes revealed by the tristimulus, confirming that timbre change is mainly due to this variation of energy between the two bands as described previously. For most sounds the tristimulus tendency is similar to the one obtained with the voice recordings, figure 6, with a fundamental frequency dominance at the beginning of the sound and mid frequencies taking over towards the end, suggesting that the "ON-AN" phenomena also are present in the interior car sound. To further exploit the tristimulus representations, listening tests are needed to understand how the driver perceives timbre changes and whether such representations can predict the appreciation of the sounds from accelerating cars.

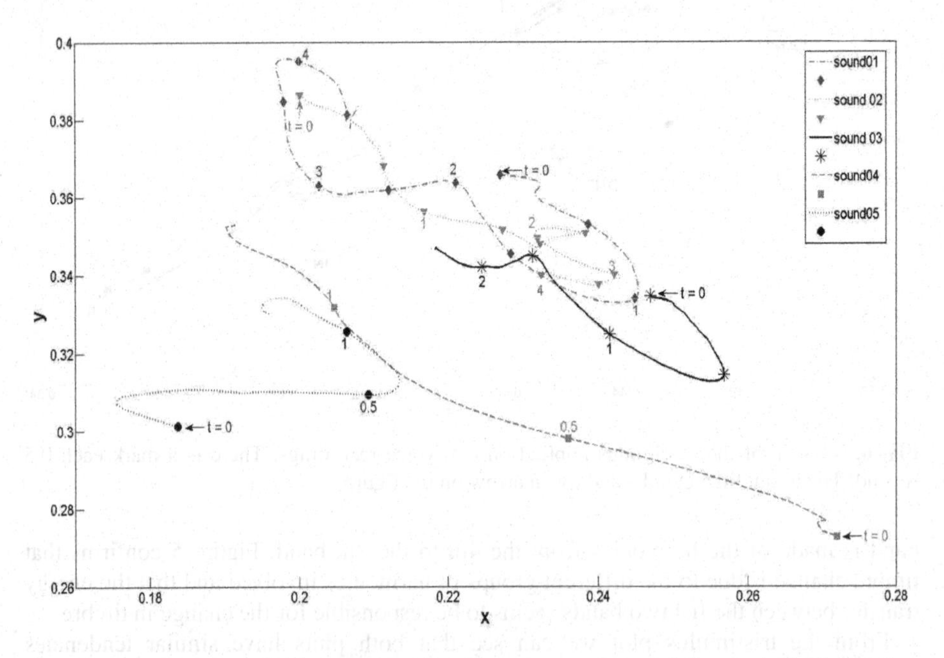

Fig. 7. X vs. Y for the auditory representation. There is a mark each 0.5 second. The initial time is indicated by an arrow on the figure.

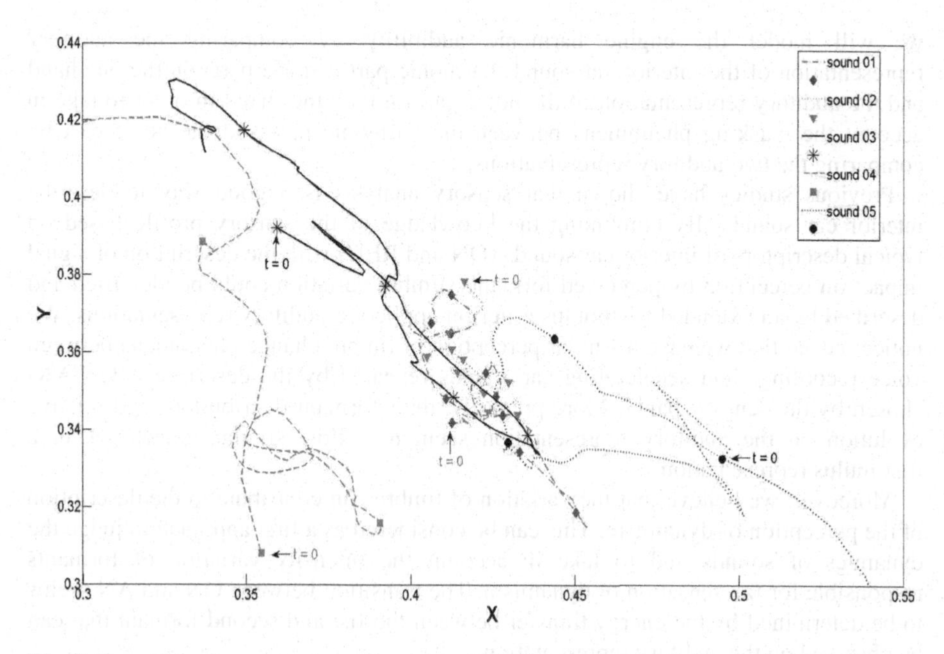

Fig. 8. X vs. Y for the spectral representation. There is a mark each 0.5 second. The initial time is indicated by an arrow on the figure.

We can notice a global trend in the evolution of the tristimulus trajectory. The sounds evolve when we observe a reduction of fundamental frequency (H2) and an increase of mid frequency partials (H4-H6). The sounds that we chose are real and recorded in different cars and driving situations that's why the trajectories differ.

Nevertheless, we can make the hypothesis that the slope of the trajectory could explain the perception of dynamism during the transition between ON and AN. Indeed, even if the transition is not as clear as in a voice imitation of an accelerating car (which does not contain the noise contribution), we can observe similarities between the two kinds of trajectories. It's difficult to conclude about the position in the XY plane, since the "ON" area and the "AN" area are indeed not clearly identified in this plane. Anyway, we presume that the trajectory between the beginning ("ON") and the end ("AN") of the interior car sound is more important than the XY coordinates of the beginning and end of the sound.

5 Conclusion and Perspectives

Time-frequency representations are inadequate to describe the perceived interior car sounds because of the presence of multiple masking phenomena between the harmonic part and the noise part. Auditory models are more efficient to highlight the perceived impact of signal parameters. We show here our first auditory representation of interior car sounds and describe our first attempt to build a complete model of interior car sound timbre. The first step is the reduction of signal parameters by determining the audibility of engine harmonics in the interior car noise. In the future, we will model the engine harmonic audibility by comparing the auditory representation of the interior car sound (harmonic part + noise part) on the one hand and the auditory representation of the noise part on the other hand in order to take in account the masking phenomena between the harmonic part and the noise part by comparing the two auditory representations.

Previous studies have shown that sensory analysis is a good way to describe interior car sounds. By combining the knowledge of the sensory profile based on typical descriptors of interior car sounds (ON and REU) with the description of signal impact on perception by perceived formants, timbre variation could be identified and described by an extended tristimulus criterion applied to auditory representations. We notice indeed that we have a similar perception of timbre change phenomena between voice recordings and accelerating car sounds revealed by the descriptors (ON/AN) chosen by the sensory panel. More precisely, their formant distributions and relative evolution on the auditory representation seem to follow similar trajectories in a tristimulus representation.

Moreover, we believe that the variation of timbre can contribute to the description of the perception of dynamism. This can be considered as a first approach to judge the dynamics of sounds and to take in account the intensity variation of formants responsible for the sensation of dynamism. The transition between ON and AN seems to be determined by the energy transfer between the fist and second formant that can be observed on the auditory representation.

In future experiments, we will evaluate the perception of dynamism as a function of formant distribution by listening tests. We will also evaluate the importance of the transition length between "ON" and "AN" during the acceleration in order to test its influence on the perception of dynamism.

References

1. Bisping, R.: Car Interior Sound Quality: Experimental Analysis by Synthesis. Acta Acustica united with Acustica 83, 813–818 (1997)
2. Chouard, N., Hempel, T.: A semantic differential design especially developed for the evaluation of interior car sounds. J. Acoust. Soc. Am. 105(2) (1999)
3. Kubo, N., Mellert, V., Weber, R., Meschke, J.: Categorization of engine sound. In: Proceedings of the Internoise, Prague (2004)
4. Hansen, H., Weber, R., Letens, U.: Quantifying tonal phenomena in interior car sound. In: Proceedings of the forum acusticum, Budapest (2005)
5. Richard, F., Costes, F., Sciabica, J.-F., Roussarie, V.: Vehicle Acoustic specifications using masking models. In: Proceedings of the Internoise, Istanbul (2007)
6. Hawkins, J.H., Stevens, S.S.: The masking of pure tones and of speech by white noise. J. Acoust. Soc. Am. 22, 6–13 (1950)
7. Bezat, M.C.: Perception des bruits d'impact: Application au bruit de fermeture de porte automobile. Thèse de Doctorat en acoustique, Université de Provence (2007)
8. Pollard, H., Jansson, B.: A tristimulus method for the specification of musical timbre. Acta Acustica united with Acustica 51, 162–171 (1982)
9. Roussarie, V., Richard, F.: Sound design in car passenger compartment: Process and tool for the control of engine sound character. Journées du Design Sonore (2004p)

10. Rodet, X.: Musical Sound Signal Analysis/Synthesis: Sinusoidal Residual and Elementary Waveform Models, TFTS 1997 (IEEE Time-Frequency and Time-Scale Workshop 1997), Coventry, Grande Bretagne (1997)
11. Rodet, X., Depalle, P.: Spectral envelopes and inverse FFT synthesis. In: Proceedings of the 93rd Audio Engineering Society Convention (1992)
12. Zadorin, L.: Granular synthesis: an introduction. Queensland University of technology (1997)
13. Glasberg, B.R., Moore, B.C.J.: Development and Evaluation of a Model for Predicting the Audibility of Time-Varying Sounds in the Presence of Background Sounds. J. Audio Eng. Soc. 53, 906–918 (2005)
14. Hohmann, V.: Frequency analysis and synthesis using a Gammatone filterbank. Acta Acustica United with Acustica 88, 433–442 (2002)
15. Hewitt, M.J., Meddis, R.: Implementation details of a computation model of the inner hair-cell/auditory-nerve synapse. J. Acoustic. Soc. Am. 87(4), 1813–1816 (1990)
16. Gnansia, D.: Modèle auditif en temps reel. Mémoire de stage, Rapport de stage du Master ATIAM, Université Pierre et Marie Curie (2005)
17. Slaney, M., Naar, D., Lyon, R.F.: Auditory Model Inversion for Sound Separation. In: Proc. IEEE-ICASSP, Adelaide (1994)

Spatialized Synthesis of Noisy Environmental Sounds

Charles Verron[1,4], Mitsuko Aramaki[2,3], Richard Kronland-Martinet[4], and Grégory Pallone[1]

[1] Orange Labs, OPERA/TPS,
Avenue Pierre Marzin, 22307 Lannion, France
{charles.verron,gregory.pallone}@orange-ftgroup.com
[2] CNRS - Institut de Neurosciences Cognitives de la Méditerranée,
31, chemin Joseph Aiguier 13402 Marseille Cedex 20, France
[3] Aix-Marseille - Université,
58, Bd Charles Livon 13284 Marseille Cedex 07, France
aramaki@incm.cnrs-mrs.fr
[4] CNRS - Laboratoire de Mécanique et d'Acoustique,
31, chemin Joseph Aiguier 13402 Marseille Cedex 20, France
kronland@lma.cnrs-mrs.fr

Abstract. In this paper, an overview of the stochastic modeling for analysis/synthesis of noisy sounds is presented. In particular, we focused on the time-frequency domain synthesis based on the inverse fast Fourier transform (IFFT) algorithm from which we proposed the design of a spatialized synthesizer. The originality of this synthesizer remains in its one-stage architecture that efficiently combines the synthesis with 3D audio techniques at the same level of sound generation. This architecture also allowed including a control of the source width rendering to reproduce naturally diffused environments. The proposed approach led to perceptually realistic 3D immersive auditory scenes. Applications of this synthesizer are here presented in the case of noisy environmental sounds such as air swishing, sea wave or wind sound.We finally discuss the limitations but also the possibilities offered by the synthesizer to achieve sound transformations based on the analysis of recorded sounds.

1 Introduction

The creation of auditory environments is still a challenge for Virtual Reality applications such as video games, animation movies or audiovisual infrastructures. The sensation of immersion can be significantly improved by taking into account an auditory counterpart to the visual information. In this context, synthesis models can constitute efficient tools to generate realistic environmental sounds. A comprehensive review of environmental sound synthesis can be found in [1,2] based on the classification of everyday sounds proposed by W. Gaver [3,4]. The author defined three main categories from the physics of the sound-producing events: vibrating solids (impacts, deformations...), aerodynamic sounds (wind, fire...) and liquid sounds (sea wave, drop...). Physically-based synthesis models were proposed for these categories. For instance, one can cite [5,6,7] for contact sounds, [8,9] for aerodynamic sounds and [10] for liquid sounds.

In this paper, we focus on the synthesis of noisy environmental sounds that represent a huge variety of everyday sounds such as sea wave, wind, fire or air swishing

S. Ystad et al. (Eds.): CMMR/ICAD 2009, LNCS 5954, pp. 392–407, 2010.

("whoosh") sounds. Note that they cover most of the categories of everyday sounds defined by W. Gaver. Several authors proposed relevant models describing the physical phenomena to simulate the resulting sound. For example, [8,9] studied the vortex sounds produced by aerodynamic phenomena such as wind and combustion. First they pre-compute sound textures by making use of computational fluid dynamics. Then they use the sound textures for realtime rendering of aerodynamic sounds. The synthesis is driven by high-level parameters such as the fluid velocity. The model described in [9] handles the rubbling combustion noise and the authors can synthesize a complete fire by adding pre-recorded crackling sounds. They couple their sound modeling with graphics rendering to achieve a complete audiovisual simulation of aerodynamic phenomena.

Usually, especially for noisy sounds, the physics beyond environmental phenomena becomes rapidly complex (they refer to stochastic processes) and the physical approach may not answer to the real time constraints imposed by Virtual Reality applications. In particular, these applications require constraints of interactivity so that the surrounding sounds should be continuously updated according to the users' locations and actions in the virtual world. Thus, we considered signal-based synthesis models that aim at generating signals from their time-frequency representations, independently from their physical correlates. These models provided a wide palette of timbres and were extensively used for analysis, transformation and synthesis of musical sounds. Environmental sounds are also efficiently modeled with this approach. For instance, a wavelet approach was presented in [11] for analysis and synthesis of noisy environmental sounds. The authors proposed a method with four stages: analysis, parameterization, synthesis and validation. During the analysis, a wavelet decomposition of the sound was computed. The parameterization consisted in finding relevant manipulations of the wavelet coefficients so as to produce new sounds. The synthesis reconstructed the sounds from the manipulated wavelet coefficients. Finally the validation consisted in perceptual evaluations of the model quality. Several models were presented for sounds such as rain, car engine or footsteps... In addition, an "had-hoc" synthesis approach, that concentrates on the perception of environmental sounds by the listener rather than on analysis/synthesis has been presented in [12]. The author used time-varying filtered noise (subtractive synthesis) for creating and controlling sea waves and wind sounds. The goal was not to obtain audiorealistic sounds but to reproduce the main acoustic invariants, so that the sounds were easily recognizable and conveyed informations about the source. The synthetic sounds were used as auditory icons in an auditory display to monitor the progress of various operations. Finally, several analysis/synthesis techniques dedicated to stochastic signals were developed in the context of the additive signal model in which the sound is defined as a sum of deterministic and stochastic contributions. The deterministic part is composed of sinusoids whose instantaneous amplitude and frequency vary slowly in time while the stochastic part is modeled by a time-varying colored noise. We will describe the main synthesis techniques in the following Section.

The notion of source position and spatial extension is of great importance for the creation of immersive auditory environments. For example, the generation of sea or windy environments involves the synthesis of extended sea wave or wind sounds around a fixed 3D position. It may also be of interest to control the spatial extension of an initial point-like sound source according to the distance source-listener: for example, the perceived width of a fire in a chimney increases when we move closer to the source and decreases when we move away (note that the timbre of the fire sound also varies). Thus, to take into account this aspect of sound rendering, we proposed the design of a spatialized synthesizer that efficiently combines synthesis and spatialization techniques at the same level of sound generation.

The paper is organized as follows: we first present the modeling of stochastic signals developed for additive signal models. Then, we describe the implementation of the spatialized synthesizer with a specific interest in noisy sound synthesis. Finally, we discuss the limitations and the possibilities offered by the synthesizer to generate realistic 3D environments.

2 Analysis/Synthesis of Stochastic Signals

Sound synthesis can be implemented either in the time (using oscillator banks) or in the frequency domain [13]. We here focus on synthesis processes based on the short-time Fourier transform (STFT) since the proposed synthesizer (section 3) is based on this technique. The pioneer works were conducted by McAulay and Quatieri for speech synthesis [14]. The speech signal was approximated by a sum of short-time sinusoids which amplitudes, frequencies and phases were determined from the STFT. In [15], Serra and Smith brought improvements to the McAulay and Quatieri's model by developing the Spectral Modeling Synthesis (SMS). This method provided a complete analysis/transformation/synthesis scheme by taking into account both deterministic and stochastic parts of the signal. In this model, the stochastic residual $s(t)$ was defined by:

$$s(t) = \int_0^t h(t,\tau)x(\tau)d\tau \qquad (1)$$

where $x(t)$ is a white input noise and $h(t,\tau)$ the impulse response of a "time-varying" filter. This stochastic part is usually assumed to represent a minor part of the original signal. However, for noisy environmental sounds, it would be predominant compared to the deterministic contribution.

In [16], Hanna and Desainte-Catherine presented an analysis/synthesis scheme to model noisy signals as a sum of sinusoids (CNSS). The frequency and phase of the sinusoids were randomly chosen at each frame (following a uniform distribution). A spectral density (number of sinusoids per frequency band) was additionally estimated in the analysis stage.

In [17,18], the authors proposed the Bandwidth-Enhanced Additive Model that allowed synthesizing sinusoidal signals with noisy components by representing the noisy components as part of each partial. Their model follows ridges in a time-frequency analysis to extract partials having both sinusoidal and noise characteristics. They also proposed to use time-frequency reassignment methods to enhance the analysis accuracy [19].

2.1 Characterization of the Time-Varying Spectral Envelope

The stochastic signal is fully defined by its power spectral density (PSD), i.e., the expected signal power with respect to the frequency. This means that the instantaneous phases can be ignored for the resynthesis. In the SMS model, the analysis of the stochastic part consisted in measuring at each time step the average energy in a set of contiguous frequency bands covering the whole frequency range. The obtained amplitude curve corresponded to a piecewise linear function and is commonly called the "time-varying spectral envelope". To take into account human hearing system, Goodwin [20] proposed to define this spectral envelope on the ERB (Equivalent Rectangular Bandwidth) scale defined by:

$$ERB(f) = 21.4 \log_{10} \left(4.37 \frac{f}{1000} + 1\right) \quad (2)$$

where f is expressed in Hz. The amplitude was assumed to be constant in each ERB subband. The spectral modeling based on ERB scale allowed efficiently coding the residual part (excluding transients) without loss of sound quality.

This approach is close to the analysis/synthesis scheme implemented in the channel vocoder developed by Dudley in 1939 [21,22,23]. Formerly used for speech coding, the channel vocoder reconstructs an original signal based only on its short time power spectrum (by contrast with the so-called "phase vocoder" that keeps the phase information). The whole process is illustrated on Figure 1. At the analysis stage, a short time power spectrum of the input signal $s[n]$ is measured with a bank of M contiguous bandpass filters ($H_1[z]$,...,$H_M[z]$). Then, a time-varying envelope $e_m[n]$ is estimated on each subband signal $s_m[n]$ by using an envelope follower, and can be defined by:

$$e_m[n] = \sqrt{\frac{1}{I} \sum_{i=0}^{I-1} (v[i] s_m[n+i])^2} \quad (3)$$

where $v[n]$ is an analysis window of size I (a rectangular window for instance). At the synthesis stage, a pulse train (that simulates the glottal excitation) or a noise (that simulates the transient parts of the speech) noted $b[n]$ feeds the same filterbank ($H_1[z]$,..., $H_M[z]$) and is weighted by the subband spectral envelope $E[n] = (e_1[n], \ldots, e_M[n])$ so that the output $\hat{s}[n]$ is close to the original signal $s[n]$.

Note that for efficiency, the short time Fourier transform is typically used to implement the analysis/synthesis filterbank [23]. Furthermore, only a discrete version of the envelopes is usually computed:

$$E^r = (e_1^r, \ldots, e_M^r) = (e_1[rR], \ldots, e_M[rR]) \quad (4)$$

where r is the index of the frame and R the analysis hop size. This representation allows saving a significant amount of data compared to the original signal. When using an analysis hop size of 512 samples (e.g., a 1024-tap analysis window with an overlap factor of 50%) and 32 subbands, the compression ratio is $512/32 = 16$. Note that this ratio can be increased for relatively stationary sounds since longer analysis windows can be used without degrading the quality.

2.2 Time-Frequency Domain Synthesis

For a given time-varying spectral envelope, the stochastic signal can be synthesized in the time-frequency domain using the inverse fast Fourier transform (IFFT) algorithm

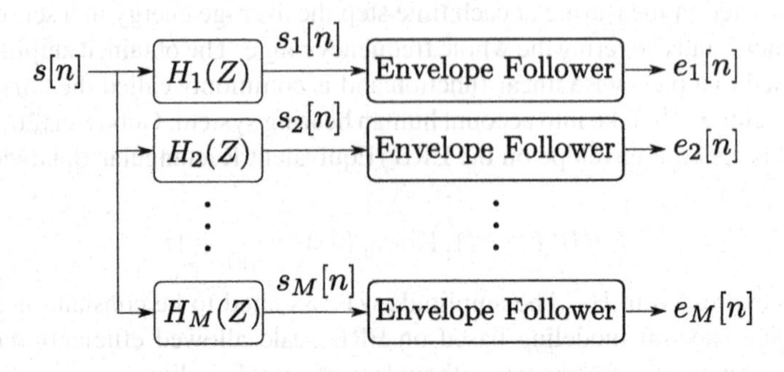

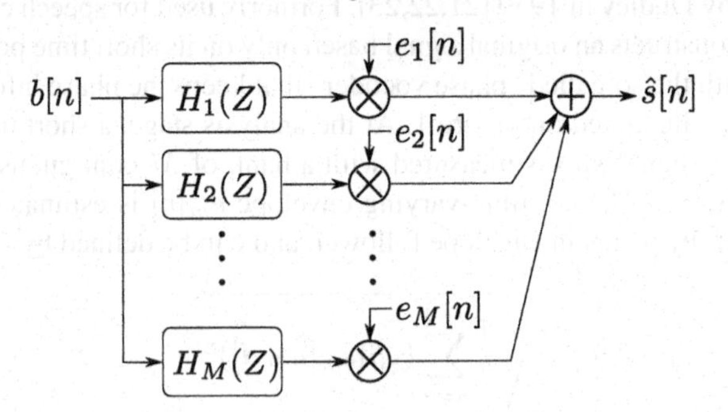

Fig. 1. Channel vocoder. *Analysis stage:* the original sound $s[n]$ is passed through a bank of M bandpass filters $H_m[z]$ and an envelope $e_m[n]$ is estimated in each subband, resulting in time-varying spectral envelope $E[n] = (e_1[n], \ldots, e_M[n])$. *Synthesis stage:* the input signal $b[n]$ is passed through the same filterbank $H_m[z]$ and weighted by the estimated set of spectral envelopes $E[n]$.

developed by Rodet and Depalle [24] and commonly called "IFFT synthesis". From a theoretical point of view, an approximation to the STFT is computed from the synthesis parameters (i.e., the expected spectral envelope for noise), then the inverse STFT is processed. IFFT synthesis is an implementation with a frame by frame pattern. Short-time spectra are created and the IFFT is performed. The resulting short-time signals are weighted by the synthesis window and overlap-added (OLA) to reconstruct the temporal signal[1].

These successive steps illustrated in Figure 2 are now detailed in the case of stochastic signal synthesis. Let N be the block size of the synthesis window $w[n]$ and $S^l[k]$ the N-point short-time spectrum (STS) to be reconstructed from a given spectral envelope E^l at frame l. This envelope is first resampled for $k = 0, \ldots, N/2$ since only positive frequencies (i.e., $k = 0, \ldots, N/2$) need to be considered for a real-valued signal (the

[1] If the initial spectral envelope is estimated from the analysis of a natural sound, the analysis window, the synthesis window and hop size can be different. The spectral envelope is interpolated at the synthesis frame rate.

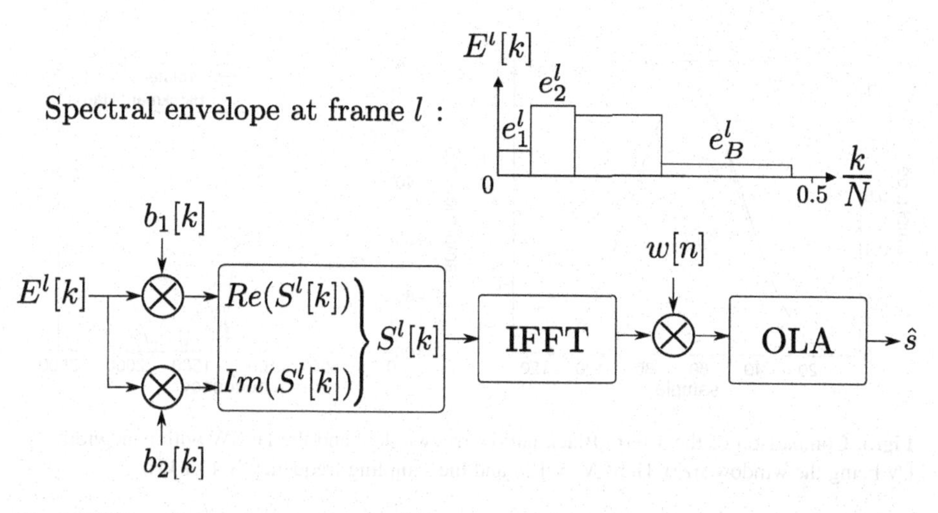

Fig. 2. IFFT synthesis of a noisy signal $\hat{s}[n]$ from a given time-varying spectral envelope $E^l[k]$. At each frame l, the real and imaginary part of the short-time spectrum $S^l[k]$ is reconstructed by multiplying $E^l[k]$ by two random sequences $b_1[k]$ and $b_2[k]$. The resulting frames are inverse fast Fourier transformed (IFFT), weighted by the synthesis window $w[n]$ and finally overlap-added (OLA) to obtain the signal $\hat{s}[n]$.

corresponding spectrum is conjugate-symmetric). Then, the obtained envelope $E^l[k]$ is multiplied by two Gaussian random sequences $b_1[k]$ and $b_2[k]$ to get the real and imaginary parts of the STS $S^l[k]$:

$$\begin{cases} \Re\{S^l[k]\} = b_1[k]E^l[k] \\ \Im\{S^l[k]\} = b_2[k]E^l[k] \end{cases} \tag{5}$$

Additionally, $b_1[k]$ and $b_2[k]$ should satisfy [15,20]:

$$b_1[k]^2 + b_2[k]^2 = 1 \quad \text{for} \quad k = 0, \ldots, N/2 \tag{6}$$

so that the magnitude of $S^l[k]$ fits the desired spectral envelope $E^l[k]$. However, if we consider the discrete Fourier transform $G[k]$ of a zero-mean N-point sequence of Gaussian white noise with variance σ^2 , it is shown in [25] that the magnitude of $G[k]$ follows a Rayleigh distribution and the phase a uniform distribution. The real and imaginary parts of $G[k]$ are independent Gaussian sequences with variance $\sigma^2 N/2$. Informal listening tests confirmed that letting $b_1[k]$ and $b_2[k]$ be two independent Gaussian sequences, i.e., not satisfying Equation (6), leads to good perceptive results.

Then, the short-time spectrum $S^l[k]$ is inverse fast Fourier transformed to obtain the short-time signal $s^l[n]$:

$$s^l[n] = \frac{1}{N} \sum_{k=0}^{N-1} S^l[k] e^{j2\pi n \frac{k}{N}} \tag{7}$$

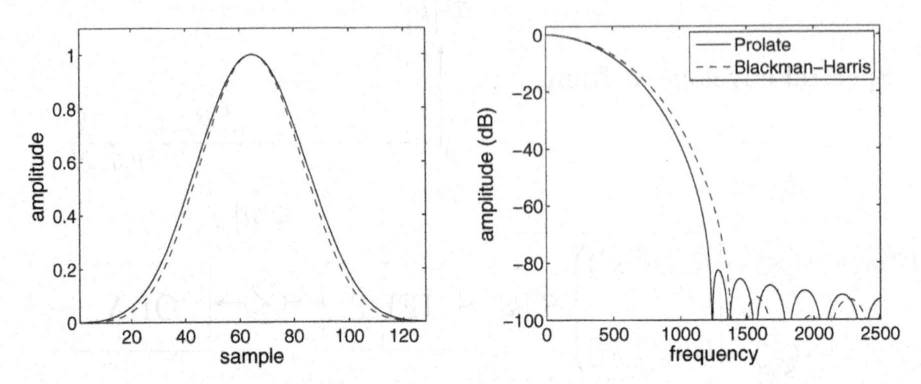

Fig. 3. Comparison of the 4-term Blackman-Harris window and the DPSW with bandwidth $\frac{3.5}{N}$ (N being the window size). Here $N = 128$ and the sampling frequency is 44.1 kHz.

and after weighted by the synthesis window $w[n]$. Finally, the overlap-add is processed on the weighted short-time signals to get the whole reconstructed signal $s[n]$:

$$s[n] = \sum_{l=-\infty}^{\infty} w[n-lL]s^l[n-lL] \tag{8}$$

where L is the synthesis hop size.

For the purpose of combining the stochastic synthesis with sinusoidal synthesis, the multiplication by the synthesis window is performed as a convolution in the frequency domain [26]. In that case, to reduce the computational cost of the convolution, the synthesis window is usually defined as a 4-term Blackman-Harris window (side-lobe level at -92 dB) whose spectrum is truncated to its main-lobe and sampled on 9 frequency bins. For our concern, we use a digital prolate spheroidal window (DPSW) since this window obtains the optimal energy concentration in its main-lobe [27]. The DPSW family constitutes a particular case of the "discrete prolate spheroidal sequences" developed by Slepian [28]. We compute the DPSW with bandwidth $\frac{3.5}{N}$ (N being the window size), truncate its spectrum to its main-lobe (the side-lobe is at -82 dB) and sample it on 7 frequency bins. Comparison between the Blackman-Harris window and this DPSW is illustrated on Figure 3. For an accurate synthesis, [29] showed that the following condition:

$$\sum_{l=-\infty}^{\infty} w[n-lL]^2 = 1 \qquad \forall n \in Z \tag{9}$$

should be satisfied to avoid modulations in the resulting signals and to make sure to obtain a flat power spectrum in the case of white noise. The latter condition (Equation (9)) was satisfied by the DPSW window provided a small synthesis hop size, typically corresponding to an overlap of 75% between two successive frames. To increase the efficiency of the method, Rodet and Depalle proposed to reduce the overlap to 50% and to smooth discontinuities between frames with an additional window, that is different from the synthesis window [24]. In practice, Bartlett, Hann or Bartlett-Hann windows are well adapted to this use.

3 The Spatialized Synthesizer

Based on this IFFT synthesis, we proposed the design of a spatialized synthesizer to simulate sound sources in a 3D space. By contrast with traditional implementations that consist in synthesizing a monophonic source before spatialization, the synthesizer has the advantage to efficiently combine synthesis and spatialization modules at the same level of sound generation in a unified architecture. In the following sections, we summarize the implementation of the synthesizer by briefly describing the source positioning module and how it was included in a one-stage architecture. Then, we present a method for the spatial extension rendering that was included in the synthesizer. This latter effect will be of great importance to simulate naturally diffused noisy sources such as sea wave or wind sounds. We refer the reader to [30,31] for more details on the implementation of the synthesizer.

3.1 Source Positioning

Several approaches exist for positioning sound sources in virtual environments. For instance, High Order Ambisonics (HOA) and Wave Field Synthesis aim at reconstructing the sound field in a relatively extended area with a multichannel loudspeaker setup. Discrete panning (time or amplitude panning) reconstructs main aspects of the sound field at the "sweet spot". In the case of binaural synthesis, the sound field at the entrance to the ear canals is reconstructed by filtering the monophonic sound with Head Related Impulse Responses. The binaural synthesis is mainly for headphone reproduction but can also be extended to loudspeaker setup commonly referred to as "Transaural" setup. A general implementation strategy applicable to all the techniques cited above can be found in [32].

For the synthesizer, we proposed an architecture relying on "amplitude-based" positioning, for which the spatial filterbank is reduced to a vector of position-dependent gains. Consequently the synthesizer was compatible with several 3D positioning methods such as Ambisonics, HOA, amplitude panning and some multichannel implementations of binaural synthesis.

3.2 One-Stage Architecture

The one-stage implementation was made possible while the synthesis and the spatial encoding can be performed in the frequency domain. Thus, it handled all positioning methods that use only gains in the spatial encoding module. WFS was excluded since it required delays in the spatial encoding module, expensive to compute in the frequency domain. The integration was effectuated following three main stages:

1. *time-frequency domain synthesis:* based on the IFFT synthesis described in Section 2.2, the real and imaginary parts of the STS are computed at each frame from a given spectral envelope associated to each source.
2. *3D positioning:* the spatial encoding is processed by directly applying spatial gains to the STS of each source. The encoded STS are summed channel by channel in the mixing stage. The spatial decoding is performed by matrixing and/or filtering the multichannel signal. Note that the decoding stage is common for all sources and do not depend on individual source position.

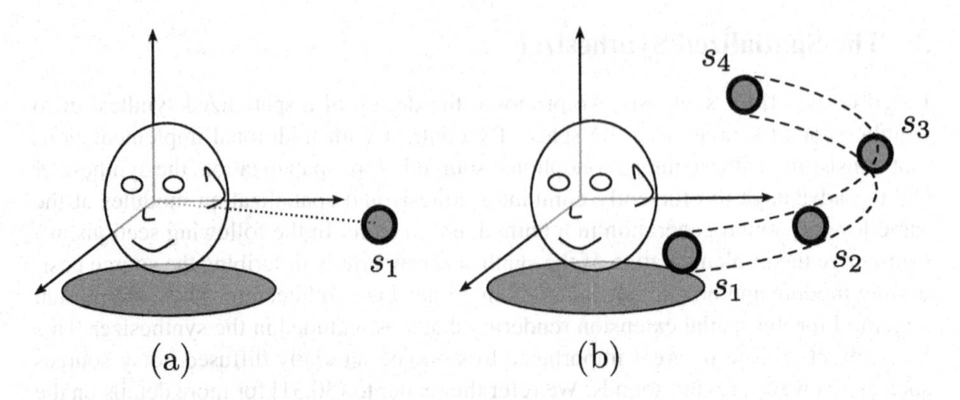

Fig. 4. (a) A single point-like source produces a narrow auditory event. (b) Several decorrelated copies of the source located at evenly spaced positions around the listener produce a wide auditory event. The perceived source width can be adjusted by changing the relative contributions (i.e., gains) of the decorrelated sources.

3. *reconstruction of the time-domain signal:* the decoded STS are inverse fast Fourier transformed and the successive short-time signals are overlap-added to get the synthetic signal for each channel of the loudspeaker setup.

The proposed architecture reduced the computational cost since it requires only one IFFT per frame for each loudspeaker channel, independently of the number of sound sources. This is particularly attractive for applications over headphones using binaural synthesis because only two IFFT are computed per frame, while the scene can contain hundreds of sources.

3.3 Source Width Extension

The extension effect cannot be reproduced by simply feeding a multichannel loudspeaker setup with duplicated copies of a monophonic signal: in that case, a relatively sharp phantom image is created. The creation of a wide spatial image necessitates feeding the loudspeaker setup with decorrelated versions of the monophonic signal (see Figure 4). Each decorrelated copy is called "secondary source". In this context, various filtering techniques were proposed to create decorrelated versions of an original signal [33,34,35,36] that were further used to create wide source images [36,37,38,39]. Nevertheless, filtering techniques may alter the transients and the timbre of the original sound.

The proposed architecture presents the advantage to overcome this drawback by effectuating the decorrelation at the synthesis stage, i.e., without filtering process. In particular, different versions of the STS $S^l[k]$ (noted $S^l_i[k]$) can be created from a same original spectral envelope $E^l[k]$ with different noise sequences $b_i[k]$ (see Figure 2). The resulting signals $\hat{s}_i[n]$ correspond to different versions of the same original sound, and they are statistically uncorrelated. This way an unlimited number of decorrelated secondary sources can be created from different random sequences. Based on this

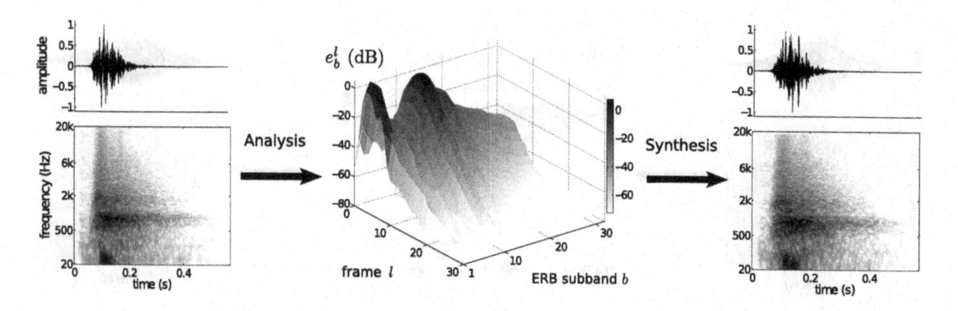

Fig. 5. Analysis/Synthesis of an air swishing ("whoosh") sound, analysis with 32 subbands using and a 1024-tap window with 75% overlap. The amount of data is reduced by a factor 8 compared to the original sound. The synthesis uses a 1024-tap window with 75% overlap.

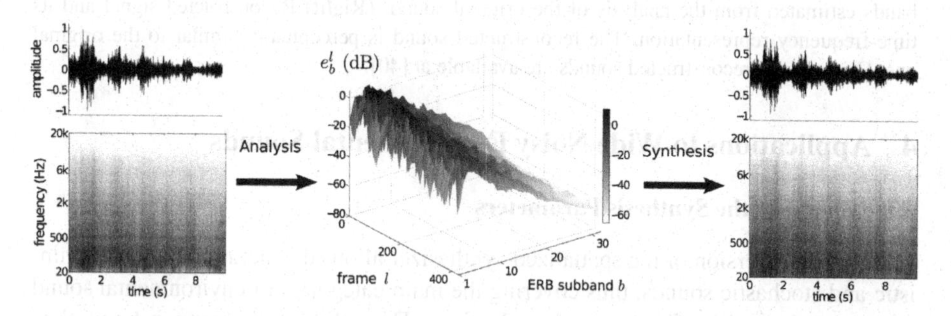

Fig. 6. Analysis/Synthesis of a wind sound, analysis with 32 subbands using and a 4096-tap window with 75% overlap. The amount of data is reduced by a factor 32 compared to the original sound. The synthesis uses a 1024-tap window with 75% overlap.

technique, we included a spatial extension effect in the synthesizer by using a maximum of eight virtual secondary sources evenly spaced on a circle surrounding the listener. The control of source width acted on the relative contributions of the eight sources via a set of extension gains.

Furthermore, it was of interest to control the interchannel correlation. For that purpose, a correlation C was introduced between the signals $s_1[n]$ and $s_2[n]$ that can be accurately controlled by creating $S_2^l[k]$ with:

$$\begin{cases} \Re\{S_2^l[k]\} = C \times \Re\{S_1^l[k]\} + \sqrt{(1-C^2)} \times b_1[k]E^l[k] \\ \Im\{S_2^l[k]\} = C \times \Im\{S_1^l[k]\} + \sqrt{(1-C^2)} \times b_2[k]E^l[k] \end{cases} \tag{10}$$

where $b_1[k]$ and $b_2[k]$ are two independent Gaussian noise sequences. For instance, this control allowed, for stereo (2-channel) applications, going progressively from a sharp spatial image of the sound source towards a completely diffused source between the two loudspeakers.

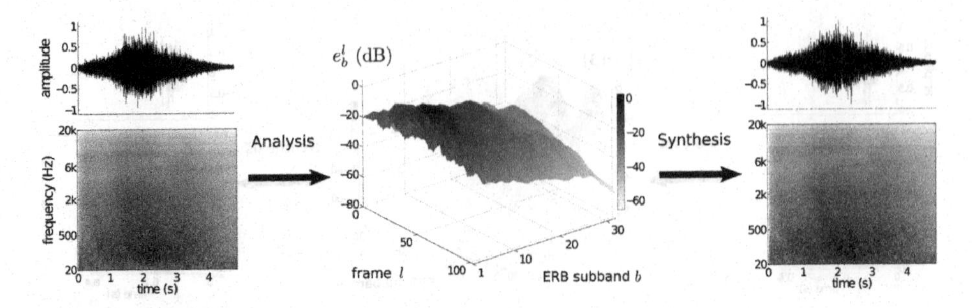

Fig. 7. Analysis/Synthesis of a wave sound, analysis with 32 subbands using and a 8192-tap window with 75% overlap. The amount of data is reduced by a factor 64 compared to the original sound. The synthesis uses a 1024-tap window with 75% overlap. **(Left)** Original signal and its time-frequency representation. **(Middle)** Time-varying spectral envelope defined in 32 ERB subbands estimated from the analysis of the original sound. **(Right)** Reconstructed signal and its time-frequency representation. The reconstructed sound is perceptually similar to the original one. Original and reconstructed sounds are available at [40].

4 Applications to Wide Noisy Environmental Sounds

4.1 Choice of the Synthesis Parameters

The complete version of the spatialized synthesizer allowed generating both deterministic and stochastic sounds, thus covering the main categories of environmental sound categories defined by Gaver (see Introduction). Thus, the synthesizer can be used to simulate various 3D auditory environments. As mentioned previously, we here focus on the generation of specific class of noisy environmental sounds. For the synthesis, we propose to find a set of parameters satisfying time and frequency resolution constraints to generate most of these types of sounds.

We examined several prototypes of noisy sounds among the main environmental sound categories, i.e., wind, sea waves and air swishing sounds. Their time-frequency representations were investigated. We observed that all sounds usually do not have very sharp transients and that 32 subbands evenly spaced on the ERB scale were sufficient to reproduce their salient spectral properties. Based on these observations, we experimented different synthesis window sizes and concluded that a 1024-tap digital prolate window led to a good compromise for an accurate reproduction of the sound source. Regarding the time resolution, this window was short enough to reproduce signals with relatively fast variations (i.e., 21 milliseconds at 48kHz sampling frequency). Regarding the frequency resolution, this window was sufficiently long for synthesizing the required 32 ERB subbands.

Figures 5, 6 and 7 illustrate the analysis/synthesis process of air swishing, wind and wave sounds by using the 1024-tap synthesis window. Both temporal and spectral properties of the original sounds were accurately reproduced with the chosen synthesis window. Sound examples are available at [40].

Finally, the spatialized synthesizer allowed to easily extend sound sources around the listener based on the decorrelation technique described in Section 3.3. The perceptual evaluation of the extension effect was effectuated by conducting a formal listening test [31]. The test was performed on several items representative of the main categories of environmental sounds, in particular, sea wave and wind sounds using two reproduction systems (headphones and a standard 5.0 loudspeaker setup). Results showed that the extension control implemented in the synthesizer accurately modified the perceived width of these noisy sound sources. Sounds are available at [41].

4.2 Limitation of the IFFT Synthesis

A limitation of the IFFT synthesis remains the inherent trade-off between time and frequency resolutions set by the length of the synthesis window. In particular, narrowband noise synthesis requires long windows (for example, more than 1024 taps; longer the window, narrower the bandwidth) with which short transient signals cannot be synthesized. Freed proposed a variant version of the IFFT synthesis algorithm [24] to reproduce noisy signals with narrow frequency bands [42,43]. His algorithm synthesized short-time sinusoids whose phase was randomly distributed between 0 and 2π at each frame. The resulting signal was perceived as a narrow-band noise. This approach is equivalent to the stochastic synthesis method described in section 2.2 and considering a spectral envelope that is non-zero only for one frequency bin (the non-zero bin corresponding to the desired center frequency). The resulting noisy signal has the narrowest bandwidth that can be generated by IFFT synthesis. The multiplication by the synthesis window being equivalent to a convolution in the frequency domain, this narrowest bandwidth is equal to the bandwidth of the synthesis window. Choosing a long synthesis window guarantees a narrow bandwidth, but a bad resolution in the time domain. On the contrary, choosing a short synthesis window guarantees a good time resolution, but a wide bandwidth in the frequency domain. Consequently, IFFT synthesis may not accurately generate noisy sounds presenting both short transients and narrowband components.

The fire sound is a good example to illustrate the limitation of the IFFT synthesis since it is constituted of both crackling (noisy transients impacts), hissing (narrowband noises) and combustion. A way to overcome this trade-off consisted in using different windows according to the different contributions of the sound. In particular, we chose 128 taps

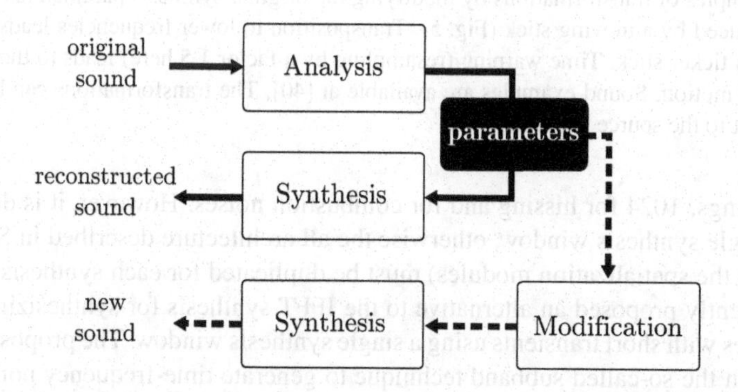

Fig. 8. Analysis/transformation/synthesis framework: the synthesis parameters extracted from the analysis of an original sound are used to resynthesize the original one or to create new sounds by signal transformations (e.g., by pitch-shifting, time-stretching, equalization or morphing)

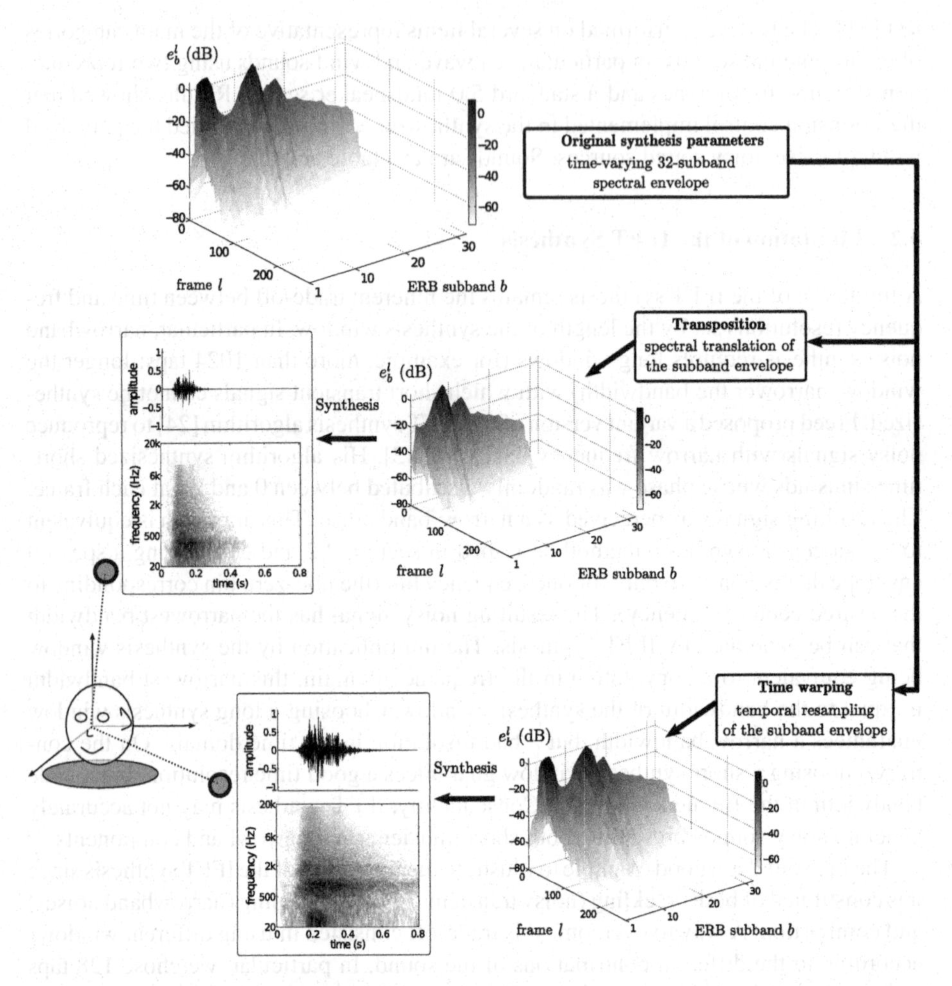

Fig. 9. Examples of transformations by modifying the original synthesis parameters of a whoosh sound produced by a moving stick (Fig. 5). Transposition to lower frequencies leads to the perception of a ticker stick. Time warping (resampling by a factor 1.5 here) leads to the perception of a slower motion. Sound examples are available at [40]. The transformations can be achieved with respect to the source position.

for cracklings, 1024 for hissing and for combustion noises. However, it is desirable to have a single synthesis window, otherwise the all architecture described in Section 3.2 (including the spatialization modules) must be duplicated for each synthesis window.

We recently proposed an alternative to the IFFT synthesis for synthesizing narrow-band noises with short transients using a single synthesis window. The proposed method is based on the so-called subband technique to generate time-frequency noise with an auto-correlation function so that the resulting output PSD fits any arbitrary spectral envelope at the expense of extra computations [44]. In particular, the bandwidth of the generated noise can be narrower than the one of the synthesis window. We are currently investigating the possibilities offered by this method.

4.3 Sound Transformations

Compared to classical wavetable synthesis, the spatialized synthesizer allows parametric transformations of the generated signals. In the context of analysis/synthesis, the analysis leads to the determination of the synthesis parameters for reconstructing the original sound. Modifying synthesis parameters allows the creation of new sounds (see Figure 8).

In [45] the authors used the analysis/transformation/synthesis framework for generating complex scenes from a small set of environmental recordings. In particular, they decompose the original sounds into deterministic and stochastic parts. They apply different parametric transformations to these components to generate new sound sequences avoiding unnatural repetitions.

In the synthesizer, classical signal transformations such as pitch-shifting or time-stretching can be easily applied on the spectral envelopes. Stretching or warping the envelopes in time by simple interpolation techniques results in a temporal stretching/warping of the reconstruded sound without pitch alterations. Also interpolating and shifting the spectral envelopes in the frequency domain produces transpositions without temporal alterations. These transformation processes are illustrated on Figure 9. Sound morphing can also be realized to go from one set of spectral envelopes to another. The simplest morphing is a simple linear interpolation between the two sets of envelope. More elaborate solutions can be found in [46] for morphing between spectral envelopes of speech signals. Subband equalization of the reconstructed signal is also possible by weighting the spectral envelope with a set of coefficients before the synthesis process. It provides an efficient and intuitive control of the spectral shape of the reconstructed sound.

In addition, since synthesis and spatialization are processed at the same level of sound generation, sound transformations can be achieved on timbre parameters with respect to the source position, as shown in Figure 9.

5 Conclusion

In this paper, we focused on noisy environmental sounds. We proposed an overview of the existing stochastic models generally developed in the context of additive signal model. We described an efficient frequency-domain synthesis technique based on IFFT algorithm. The use of stochastic modeling in the frequency domain allowed us to propose a one-stage architecture where spatialization techniques operate directly at the synthesis stage. Listening tests have shown that our technique produced realistic extended environmental sound sources such as sea waves and wind.

References

1. Cook, P.R.: Real Sound Synthesis for Interactive Applications. A. K Peters Ltd. (2002)
2. Rocchesso, D., Fontana, F.: The Sounding Object (2003), http://www.soundobject.org/
3. Gaver, W.W.: What in the world do we hear? an ecological approach to auditory event perception. Ecological Psychology 5(1), 1–29 (1993)

4. Gaver, W.W.: How do we hear in the world? explorations in ecological acoustics. Ecological Psychology 5(4), 285–313 (1993)
5. van den Doel, K., Kry, P.G., Pai, D.K.: Foleyautomatic: physically-based sound effects for interactive simulation and animation. In: Proceedings of the 28th annual conference on Computer graphics and interactive techniques, pp. 537–544 (2001)
6. O'Brien, J.F., Shen, C., Gatchalian, C.M.: Synthesizing sounds from rigid-body simulations. In: Proceedings of the 2002 ACM SIGGRAPH/Eurographics symposium on Computer animation, pp. 175–181 (2002)
7. Raghuvanshi, N., Lin, M.C.: Interactive sound synthesis for large scale environments. In: Proceedings of the 2006 symposium on Interactive 3D graphics and games, pp. 101–108 (2006)
8. Dobashi, Y., Yamamoto, T., Nishita, T.: Real-time rendering of aerodynamic sound using sound textures based on computational fluid dynamics. ACM Transactions on Graphics Proc. SIGGRAPH 2003 22(3), 732–740 (2003)
9. Dobashi, Y., Yamamoto, T., Nishita, T.: Synthesizing sound from turbulent field using sound textures for interactive fluid simulation. EUROGRAPHICS 23(3), 539–546 (2004)
10. van den Doel, K.: Physically-based models for liquid sounds. In: Proceedings of ICAD 04-Tenth Meeting of the International Conference on Auditory Display (2004)
11. Miner, N.E., Caudell, T.P.: Using wavelets to synthesize stochastic-based sounds for immersive virtual environments. In: Proceedings of of ICAD 1997-The fourth International Conference on Auditory Display (1997)
12. Conversy, S.: Ad-hoc Synthesis of auditory icons. In: Proceedings of of ICAD 1998-The fifth International Conference on Auditory Display (1998)
13. Goodwin, M.: Adaptive Signal Models: Theory, Algorithms and Audio Applications. PhD thesis, University of California, Berkeley (1997)
14. McAulay, R.J., Quatieri, T.F.: Speech analysis/synthesis system based on a sinusoidal representation. IEEE Transactions on Acoustics, Speech and Signal Processing 34(4) (1986)
15. Serra, X., Smith, J.O.: Spectral modeling synthesis: A sound analysis/synthesis system based on a deterministic plus stochastic decomposition. Computer Music Journal 14(4), 12–24 (1990)
16. Hanna, P., Desainte-Catherine, M.: A statistical and spectral model for representing noisy sounds with short-time sinusoids. EURASIP Journal on Applied Signal Processing 5(12), 1794–1806 (2005)
17. Fitz, K., Haken, L.: Bandwidth enhanced sinusoidal modeling in lemur. In: Proceedings of the International Computer Music Conference (1995)
18. Fitz, K., Haken, L., Christensen, P.: Transient preservation under transformation in an additive sound model. In: Proceedings of the International Computer Music Conference (2000)
19. Fitz, K., Haken, L.: On the use of time-frequency reassignment in additive sound modeling. JAES 50(11), 879–893 (2002)
20. Goodwin, M.: Residual modeling in music analysis-synthesis. In: Proceedings of the IEEE International Conference on Acoustics, Speech, and Signal Processing (1996)
21. Dudley, H.: The vocoder. Bell Labs Record 17, 122–126 (1939)
22. Gold, B., Rader, C.M.: The channel vocoder. IEEE Transactions on Audio and Electroacoustics 15(4), 148–161 (1967)
23. Smith, J.O.: Spectral Audio Signal Processing (October 2008), Draft (online book), `http://ccrma.stanford.edu/jos/sasp/`
24. Rodet, X., Depalle, P.: Spectral envelopes and inverse fft synthesis. In: Proceedings of the 93rd AES Convention (1992)
25. Hartmann, W.: Signal, Sound and Sensation. In: American Institute of Physics (2004)
26. Amatriain, X., Bonada, J., Loscos, A., Serra, X.: Spectral Processing. In: DAFX: Digital Audio Effects, John Wiley & Sons Publishers, Chichester (2002)

27. Verma, T., Bilbao, S., Meng, T.H.Y.: The digital prolate spheroidal window. In: Proceedings of the IEEE International Conference on Acoustics, Speech, and Signal Processing, ICASSP (1996)
28. Slepian, D.: Prolate spheroidal wave functions, Fourier analysis, and uncertainty. V- The discrete case. Bell System Technical Journal 57, 1371–1430 (1978)
29. Hanna, P., Desainte-Catherine, M.: Adapting the overlap-add method to the synthesis of noise. In: Proceedings of the COST-G6 Conference on Digital Audio Effects, DAFX 2002 (2002)
30. Verron, C., Aramaki, M., Kronland-Martinet, R., Pallone, G.: Spatialized additive synthesis of environmental sounds. In: Proceedings of the 125th AES Convention (2008)
31. Verron, C., Aramaki, M., Kronland-Martinet, R., Pallone, G.: A 3d immersive synthesizer for environmental sounds. IEEE Transactions on Audio, Speech, and Language Processing (to accepted)
32. Jot, J.M., Larcher, V., Pernaux, J.M.: A comparative study of 3-d audio encoding and rendering techniques. In: Proc. 16th Int. Conf. AES (1999)
33. Schroeder, M.R.: An artificial stereophonic effect obtained from a single audio signal. JAES 6(2) (1958)
34. Orban, R.: A rational technique for synthesizing pseudo-stereo from monophonic sources. JAES 18(2) (1970)
35. Gerzon, M.A.: Signal processing for simulating realistic stereo images. In: AES Convention 93 (1992)
36. Kendall, G.: The decorrelation of audio signals and its impact on spatial imagery. Computer Music Journal 19(4), 71–87 (1995)
37. Sibbald, A.: Method of synthesizing an audio signal. United State Patent No. US 6498857 B1 (december (2002)
38. Potard, G., Burnett, I.: Decorrelation techniques for the rendering of apparent sound source width in 3d audio displays. In: Proc. Int. Conf. on Digital Audio Effects, DAFX 2004 (2004)
39. Jot, J.M., Walsh, M., Philp, A.: Binaural simulation of complex acoustic scene for interactive audio. In: Proceedings of the 121th AES Convention (2006)
40. `http://www.lma.cnrs-mrs.fr/ kronland/spatsynthIcad09/index.html`
41. `http://www.lma.cnrs-mrs.fr/ kronland/spatsynthIEEE/ index.html`
42. Freed, A.: Real-time inverse transform additive synthesis for additive and pitch synchronous noise and sound spatialization. In: Proceedings of the 104th AES Convention (1998)
43. Freed, A.: Spectral line broadening with transform domain additive synthesis. In: Proceedings of the International Computer Music Conference (1999)
44. Marelli, D., Aramaki, M., Kronland-Martinet, R., Verron, C.: Time-frequency synthesis of noisy sounds with narrow spectral components. IEEE Transactions on Audio, Speech, and Language Processing (to accepted)
45. Misra, A., Cook, P.R., Wang, G.: A new paradigm for sound design. In: Proc. Int. Conf. on Digital Audio Effects, DAFX 2006 (2006)
46. Rodet, X., Schwarz, D.: Spectral Envelopes and Additive + Residual Analysis/Synthesis. In: Analysis, Synthesis, and Perception of Musical Sounds: Sound of Music, pp. 175–227. Springer, Heidelberg (2007)

Imagine the Sounds: An Intuitive Control of an Impact Sound Synthesizer

Mitsuko Aramaki[1], Charles Gondre[2], Richard Kronland-Martinet[2], Thierry Voinier[2], and Sølvi Ystad[2]

[1] CNRS - Institut de Neurosciences Cognitives de la Méditerranée, 31 Chemin Joseph Aiguier, 13402 Marseille Cedex, France
aramaki@incm.cnrs-mrs.fr
[2] CNRS - Laboratoire de Mécanique et d'Acoustique, 31 Chemin Joseph Aiguier, 13402 Marseille Cedex, France
{gondre,kronland,voinier,ystad}@lma.cnrs-mrs.fr

Abstract. In this paper we present a synthesizer developed for musical and Virtual Reality purposes that offers an intuitive control of impact sounds. A three layer control strategy is proposed for this purpose, where the top layer gives access to a control of the sound source through verbal descriptions, the middle layer to a control of perceptually relevant sound descriptors, while the bottom layer is directly linked to the parameters of the additive synthesis model. The mapping strategies between the parameters of the different layers are described. The synthesizer has been implemented using Max/MSP, offering the possibility to manipulate intrinsic characteristics of sounds in real-time through the control of few parameters.

1 Introduction

The aim of the current study is to propose an intuitive control of an additive synthesis model simulating impact sounds [1]. This is of importance within several domains, like sound design and virtual reality, where sounds are created from high-level verbal descriptions of the sound source and are to be coherent with a visual scene [2]. In this context, the challenge consists in being able to synthesize sounds that we have in mind. Efficient synthesis models that enable perfect resynthesis of natural sounds were developed in different contexts. In spite of the high quality of such models, the control, and the so-called mapping strategy, is an important aspect that has to be taken into account when constructing a synthesizer. To propose an intuitive control of sounds, it is in the first place necessary to understand the perceptual relevance of the sound attributes and then to find out how they can be combined to propose a high-level evocative control of the synthesizer. The sound attributes can be of different types and can either be directly linked to the physical behavior of the source [3], to the signal parameters [4] or to timbre descriptors obtained from perceptual tests [5][6][7]. In this particular study, perceptually relevant descriptors together with physical parameters linked to wave propagation phenomena such as dispersion and dissipation are considered.

S. Ystad et al. (Eds.): CMMR/ICAD 2009, LNCS 5954, pp. 408–421, 2010.

Based on these findings, we propose a complete mapping strategy that links three control layers: top layer (verbal description of the imagined sound source), middle layer (descriptors related to the characteristics of the signal) and bottom layer (parameters related to the synthesis model). The top layer offers the most intuitive way for a non-expert user to create impact sounds by specifying the perceived properties of the impacted object (like the material category, size and shape) and of the nature of the action (force, hardness, excitation point). The middle layer is composed of sound descriptors that characterize impact sounds from a perceptual point of view. The bottom layer directly depends on the parameters of the synthesis process. Finally, the mapping between the top and middle layers is based on results from previous studies on the perception of physical characteristics of the sound source (i.e., perception of material, object and action). The mapping between middle and bottom layers is defined based on results from synthesis experiments.

The paper is organized as follows: we first describe the theoretical model of impact sounds based on physical considerations and the real-time implementation of the synthesizer. Then, we define sound descriptors that are known to be relevant from a perceptual point of view in the case of impact sounds. The three-layer control strategy based on these descriptors is presented and the mappings between the different layers are detailed. We finally present some additional uses allowing analysis of natural impact sounds or real-time control in a musical context.

2 Signal Model of Impact Sounds

From a physical point of view, impact sounds are typically generated by an object under free oscillations that has been excited by an impact, or by the collision between solid objects. For simple cases, the vibratory response of such vibrating system (viewed as a mass-spring-damper system) can be described by a linear PDE:

$$\frac{\partial^2 x}{\partial t^2} = \frac{E}{\rho} L x \tag{1}$$

where x represents the displacement, E the Young modulus and ρ the mass density of the material. L represents the differential operator describing the local deformation and corresponds to the Laplacian operator for strings (in 1D) or membranes (in 2D) and to the Bi-Laplacian for bars (in 1D) or thin plates (in 2D). To take into account loss mechanisms, the Young modulus generally is defined as complex valued [8] so that the solution $d(t)$ of the movement equation can be expressed by a sum of eigen modes $d_k(t)$, each of them decreasing exponentially:

$$d(t) = \sum_{k=1}^{K} d_k(t) = \sum_{k=1}^{K} A_k e^{2i\pi f_k t} e^{-\alpha_k t} \tag{2}$$

where A_k is the amplitude, f_k the eigen frequency, α_k the damping coefficient of the kth mode, and K the number of components. The damping coefficient

α_k, generally frequency-dependent, is linked to the mechanical characteristics of the material and particularly to the internal friction coefficient [9]. The eigen frequencies f_k are deduced from the eigen values of the operator L with respect to the boundary conditions. Note that for multidimensional structures, the modal density increases with frequency so that the modes may overlap in the high frequency domain.

Consequently, we consider that from a signal point of view, an impact sound is accurately modeled by an additive model that consists in decomposing the signal $s(t)$ into deterministic $d(t)$ and stochastic $b(t)$ contributions :

$$s(t) = d(t) + b(t) \tag{3}$$

where $d(t)$ is defined in (2) and $b(t)$ is an exponentially damped noise defined by :

$$b(t) = \sum_{n=1}^{N} b_n(t) e^{-\alpha_n t} \tag{4}$$

where N is the number of frequency subbands. To take into account perceptual considerations, the subbands are defined on the Bark scale corresponding to the critical bands of human hearing [10]. We assume that the damping coefficient α_n is constant in each Bark band so that the damping of the noise signal is defined by 24 values.

3 Implementation of the Synthesizer

The real-time implementation of the theoretical model (defined in (3) and (4)) was made with MaxMSP [11]. The whole architecture is shown in Figure 1. The input signal consists of a stochastic contribution providing the noisy broadband

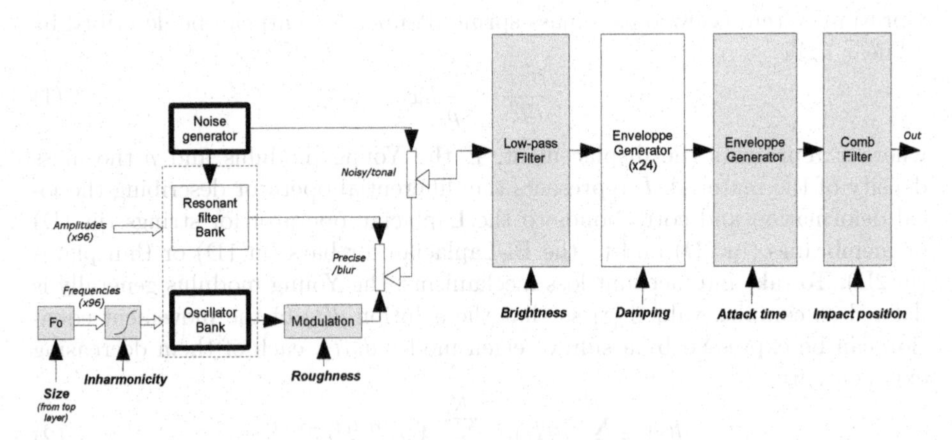

Fig. 1. Implementation of the impact sound synthesizer based on the additive signal model defined in Equation 3. The boxes represented in grey correspond to the modules added for the mapping toward higher levels (Section 5.2).

spectrum and a tonal contribution simulating the modes (boxes in bold). The stochastic contribution is produced by a noise generator which statistics can be defined by the user (we here used a gaussian noise). The tonal contribution is obtained by combining a sum of 96 sinusoids (oscillators) and 96 narrow-band filtered noises (obtained by filtering the output of the noise generator with resonant filters). The respective output level of sinusoids and filtered noises can be adjusted by a fader (*precise/blur* control), enabling the creation of interesting sound effects such as 'fuzzy' pitches. The output levels of stochastic and tonal contributions may also be adjusted by a fader (*tonal/noisy* control).

Then, the signal is damped by an envelope generator providing exponentially decaying envelopes in the form of $e^{-\alpha t}$. These envelopes differed with respect to frequency to take into account the frequency-dependency of the damping. Based on equation (4), we considered 24 envelopes, i.e., one per Bark band, characterized by the damping coefficient α_n. In each frequency subband, the same envelope is applied on both stochastic and deterministic parts of the signal to increase the merging between them. Nevertheless, for a pure deterministic signal, a damping coefficient can be defined for each partial of the tonal contribution. At signal level, the sound generation necessitated the manipulation of hundreds of parameters and consequently, was only intended for experts. Thus, the large number of signal parameters necessitates the design of a control strategy. This strategy (generally called mapping) is of great importance for the expressive capabilities of the instrument, and it inevitably influences the way it can be used in a musical context [12]. For that reason, different mapping strategies can be proposed with respect to the context of use.

4 Perceptually Relevant Sound Descriptors

In this paper, we aim at proposing a mapping providing an intuitive control of the synthesizer from verbal descriptions of the sound source to the acoustic parameters of the signal. For that, we focused on previous psychoacoustic studies that investigated the links between the perception of the physical properties of the source and the acoustic attributes of the resulting sound. They revealed important sound features that uncover some characteristics of the object itself and the action. In particular, the perceived object size is found to be strongly correlated with the pitch of the generated sounds while the perceived shape of objects is correlated with the distribution of spectral components [13][14][15][16][17][18][19]. The perception of material seems to be mainly correlated with the damping of spectral components [9][14][3][20][21] and seems in addition to be a robust acoustic descriptor to identify macro-categories (i.e., wood-plexiglass and steel-glass categories) [22]. Regarding perception of excitation, [23] has shown that the perceived hardness of a mallet striking a metallic object is predictable from the characteristics of the attack time (a measure for the energy rise at sound onset). In addition, the perceived force of the impact is related to the brightness of the sound commonly associated with the spectral centroid, i.e, a measure for the center of gravity of the spectrum.

The attack time and spectral centroid were also identified as relevant descriptors in studies investigating the timbre perception for other types of sounds (e.g., sounds from musical instruments [5][6]). These studies revealed that timbre is a complex feature that requires a multidimensional representation characterized by several timbre descriptors. The most commonly used descriptors in the literature, in addition to attack time and spectral centroid, are spectral bandwidth, spectral flux and roughness. The spectral bandwidth is a measure for the spectrum spread. The spectral flux is a spectro-temporal descriptor that quantifies the time evolution of the spectrum. Its definition is given in [7]. The roughness is closely linked to the presence of several frequency components within the limits of a critical band and is closely linked to the notion of consonance/dissonance [24][25].

The control proposed in the synthesizer will be based on results from these psycho-acoustical studies since they give important cues about the intuitive aspect of the mapping strategy.

5 Control Strategy of the Synthesizer

We propose a control strategy based on three hierarchical layers allowing us to route and dispatch the control parameters from an evocative level to the signal level (see Figure 2). The top layer represents the control parameters that the user manipulates in an intuitive manner. Those control parameters are based on verbal descriptions of the physical source that characterize the object (nature of material, size and shape) and the excitation (impact force, hardness and position). The middle layer is based on sound descriptors that are known to be relevant from a perceptual point of view as described in Section 4. Finally,

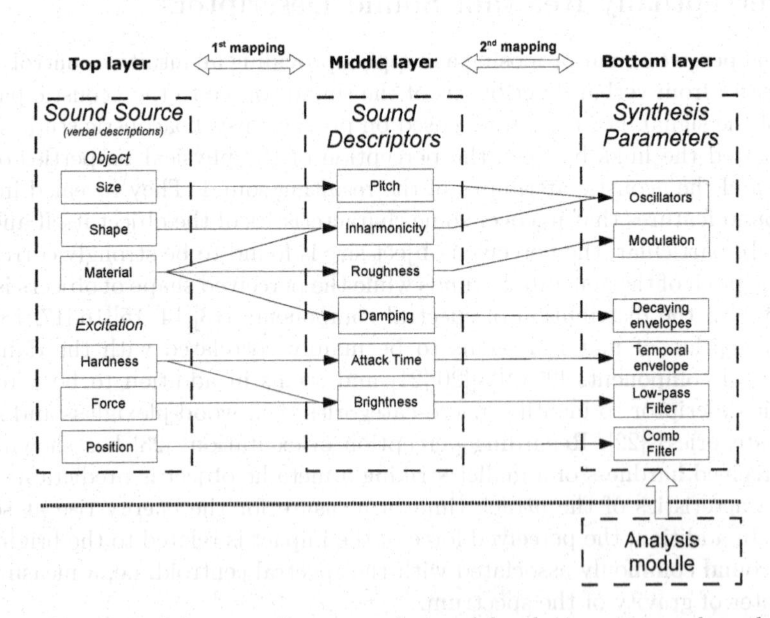

Fig. 2. Overview of the control strategy designed for the impact sound synthesizer

the bottom layer is composed of synthesis parameters as described in Section 3. Note that by default, the user only has access to the top layer. Nevertheless, we give the possibility for an expert user to directly access the middle or bottom layers. Such features are in some cases useful for sound design and musical experimentation to study the perceptual influence of specific parameters.

Between these three layers, two different mappings are to be implemented (represented as black arrows in Figure 2). As the parameters that allow intuitive controls are not independent and might be linked to several signal characteristics at a time, the mappings are far from being straight-forward. We describe these two mapping strategies in the following sections.

5.1 First Mapping: From Verbal Descriptions of Sound Source to Sound Descriptors

The first mapping links verbal descriptions characterizing the sound source to the perceptually relevant sound descriptors.

Object (material, size and shape). The characteristics of the object are defined by its perceived material, shape and size. As described in Section 3, previous studies have shown that the perception of material is related to the damping but also to additional cues mostly linked to the spectral content of sounds. In particular, the roughness has shown to be important to distinguish metal from glass and wood [26]. Consequently, the control of the perceived material involves the control of Damping but also of spectral sound descriptors such as Inharmonicity or Roughness.

The perception of the size of the object is mainly correlated with the pitch. Indeed, based on the physics, the pitch is related to the dimension of the object: actually, a big object is generally vibrating at lower eigenfrequencies than a small one. For quasi-harmonic sounds, we assume the pitch to be related to the frequency of the first spectral component. By contrast, complex sounds (i.e., numerous and overlapping modes), may elicit both spectral and virtual pitches [27]. Spectral pitches correspond to existing spectral peaks contained in the sound, whereas virtual pitches are deduced by the auditory system from upper partials of the spectrum. The virtual pitches may not correspond to any existing peak owing to the presence of a dominant frequency region situated around 700 Hz for which the ear is particularly pitch-sensitive. Thus, the pitch of complex sounds is still an open issue. In the present case, the perceived size of the object is directly linked to the fundamental frequency of the sound.

Furthermore, for impacted objects presenting a cavity (e.g., empty bottle), physical considerations (Helmholtz resonance) led to the prediction of a resonant frequency value with respect to the air volume inside the cavity [28]. In practice, the size of the cavity is directly mapped to the bottom layer and is simulated by adjusting the gain of a second-order peak filter which center frequency is set to the fundamental frequency, and which quality factor is fixed to $\frac{\sqrt{2}}{2}$.

Finally, the shape of the impacted object determines the spectral content of the generated impact sound from a physical point of view. As described in

Section 2, the frequencies of the spectral components correspond to the so-called eigenfrequencies that are characteristic of the modes of the vibrating object. Consequently, the perceived shape of the object is linked to the control of the Inharmonicity together with the pitch.

Excitation (impact force, hardness and position). The control of the excitation is based on the hardness of the mallet, the force of the impact as well as the excitation point. The excitation characterizes the nature of the interaction between the excitator and the vibrating object. From a physical point of view, this interaction can be described by a contact model such as the model proposed by [29] based on the contact theory of Hertz. The author found that the contact time τ can be defined by:

$$\tau = \pi \sqrt{\frac{\delta H}{E} \frac{LS*}{S}} \tag{5}$$

where δ is the density, H the height and $S*$ the section of the impacting object. E is the modulus of elasticity, S the section and L the length of the impacted object. This expression allows us to notably deduce that the harder the mallet (i.e., higher the modulus of elasticity), the shorter the contact time (the $S*/S$ ratio also acts on the contact time). In addition, the frequency range sollicited by the impact is governed by the contact time. From the theoretical spectrum $S_C(\omega)$ of the contact expressed by [30]:

$$S_C(\omega) \propto \frac{\sqrt{2}}{\left|\omega_0 \left(1 - (\omega/\omega_0)^2\right)\right|} \sqrt{1 - \cos\left(\frac{\pi\omega}{\omega_0}\right)} \tag{6}$$

where $\omega_0 = 1/\tau$, we can conclude that the shorter the contact time (the limit case would be the dirac), the larger the spectrum spread and the brighter the resulting sound. Based on these considerations, we linked the control of the hardness to the Attack time and the Brightness.

Concerning the force of the impact, its maximum amplitude is determined by the velocity of the impacting object. [31] showed that the contact time is weakly influenced by the force. Thus, the force is linked to the Brighness so that the heavier the force, the brighter the sound.

The excitation point, which strongly influences the amplitudes of the components by causing envelope modulations in the spectrum, is also taken into account. In practice, the impact point position is directly mapped to the bottom layer and is simulated by shaping the spectrum with a feedforward comb filter, defined by the transfer function :

$$H_\beta(z) = 1 - z^{-\lfloor \beta P \rfloor} \tag{7}$$

where P is the period in samples, $\beta \in (0, 1)$ denotes a normalized position, and $\lfloor . \rfloor$ the floor function [32].

5.2 Second Mapping: From Sound Descriptors to Signal Parameters

The second mapping (connection between middle and bottom layers) is intended to act upon the signal parameters according to the variations of the sound descriptors.

Inharmonicity. As already mentioned in Section 5.1, the distribution of the spectral components is an important parameter, as it may change one's perception of the size, shape and material of the impacted object. Its control is an intricate task since many strategies are possible. Based on the physical considerations described in Section 2, the inharmonicity induced by dispersion phenomena produces changes in the distribution of spectral components, and has shown to be an efficient parameter to control different shapes. Thus, we propose a control of the inharmonicity that allows the user to alter the spectral relationship between all the 96 initial harmonic components of the tonal contribution using three parameters a, b and c of the inharmonicity law defined by:

$$\widetilde{f_k} = a f_k \left(1 + b \left(\frac{f_k}{f_0}\right)^2\right)^c \tag{8}$$

where $\widetilde{f_k}$ is the modified frequency, f_k the frequency of the kth harmonic, and f_0 the fundamental frequency.

Thus the inharmonicity control changes the frequency ratio f_k/f_0 of each spectral component and provides an efficient way to get different types of inharmonicity profiles. Setting $a \geq 1$ and $b > 0$ leads to spectral dilations (i.e., frequencies will be deviated to higher values) providing a way to get stiff string or bell-like inharmonicity profiles, while setting $a < 1$ and $b < 0$ leads to spectral contractions (deviation to lower values) such as membrane or plate inharmonicity profiles. For example, a piano string inharmonicity is obtained for $a = 1$, $c = 0.5$ and b between 5.10^{-5} and 70.10^{-5} in the lower half of the instrument compass [33]. Large values of parameter c allows to strongly increase the frequency deviation. Some pre-defined presets offer a direct access to typical inharmonity profiles. Besides the proposed inharmonicity law, the possibility is given to the user to freely design desired behaviors by defining an arbitrary frequency ratio, independent for each component.

Roughness. Roughness is strongly linked to the presence of several spectral components within a bark band. Thus the control of roughness involves the generation of additional spectral components associated to the original ones. Based on this concept of presence of several components within a critical band, several methods have been proposed for the estimation of roughness for stationary tonal sounds [24][34]. A roughness estimation is obtained from the frequencies and amplitudes of the components. It is more difficult to evaluate the roughness of noisy and/or rapidly time-varying sounds. A computation model based on the auditory system has to be used. Several models have been developed [35][36], and for our investigations we used a model [37] that leads to a 'time-frequency representation' of the roughness. This representation reveals, for a given sound, the critical bands that contain roughness, and how the roughness varies with respect to time. These investigations show that roughness is not equally distributed on the whole sound spectrum. For many impact sounds roughness exists in some frequency regions or 'roughness formants'.

This observation governed the roughness control implementation. For that, we implemented a way to increase the roughness independently for each bark band by means of amplitude and frequency modulations. Both methods are applied on each component at the oscillator bank level (Figure 1):

– Amplitude modulation :

$$d_k(t) = [1 + I\cos(2\pi f_m t)] \times A_k \cos(2\pi f_k t) \tag{9}$$

$$\begin{aligned} d_k(t) = A_k \cos(2\pi f_k t) + \frac{A_k I}{2} \cos((2\pi f_k + 2\pi f_m)t) \\ + \frac{A_k I}{2} \cos((2\pi f_k - 2\pi f_m)t) \end{aligned} \tag{10}$$

where $I \in [0, 1]$ is the modulation index, f_m the modulating frequency, and A_k and f_k the kth partial's amplitude and frequency respectively. Thus, for each partial, the amplitude modulation creates two additional components on both sides of the original partial, that consequently increases locally the roughness.

– Frequency modulation :

$$d_k(t) = A_k \cos(2\pi f_k t + I\cos(2\pi f_m t)) \tag{11}$$

$$d_k(t) = A_k \sum_{n=-\infty}^{\infty} J_n(I) cos((2\pi f_k + n2\pi f_m)t) \tag{12}$$

where $n \in \mathbb{N}$, and J_n is the Bessel function of order n. Thus, for each partial, the frequency modulation creates an infinite number of additional components whose amplitudes are given by the partial's amplitude and the value of the Bessel function of order n for the given modulation index. In practice, the modulation index I is confined between 0 and 1, so that only a limited number of those additional components will be perceived.

In both the amplitude and frequency modulations, the user only defines the modulating frequency and the modulation indices. The modulating frequency is defined as a percentage of bark bandwidth. Based on [38], the percentage was fixed at 30%, while the modulation indices are controlled through 24 frequency bands, corresponding to the bark scale.

Note that the control of roughness can be considered as a local control of inharmonicity. Indeed, both controls modify the modal density (by creating additional components or by dilating the original spectrum) but the control of roughness has the advantage of being controlled locally for each component.

Brightness. The brightness, that is linked to the impact force, is controlled by acting on the amount of energy in the signal. In practice, the signal is filtered with a second order low pass filter of cut-off frequency f_c. Thus brightness perception decreases by progressively removing the highest frequencies of the broadband spectrum.

Damping. The material perception is closely linked to the damping of the spectral components. Since the damping is frequency dependent (high frequency components being more rapidly damped than low-frequency components), it necessitates a fine control of its frequency dependent behavior. Based on equation (4), the damping is controlled independently in each Bark band by acting on 24 values. To provide a more meaningful indication of the dynamic profile, the damping coefficient values were converted to duration values, i.e., the time necessary for the signal amplitude to be attenuated by 60dB. In addition, we defined a damping law expressed as an exponential function:

$$\alpha(\omega) = e^{a_{\mathrm{g}} + a_{\mathrm{r}}\omega} \tag{13}$$

so that the control of damping was reduced to two parameters: a_{g} is defined as a global damping and a_{r} is defined as a frequency-relative damping. The choice of an exponential function enables us to efficiently simulate various damping profiles characteristic of different materials by acting on few control parameters. For instance, it is accepted that in case of wooden bars, the damping coefficients increase with frequency following an empirical parabolic law which parameters depend on the wood species [39]. The calibration of the Damping was effectuated based on behavioral results from our previous study investigating the perception of sounds from different material categories based on a categorization task [26]: sounds from 3 impacted materials (i.e., Glass, Metal and Wood) were analyzed and synthesized, and continuous transitions between these different materials were further synthesized by a morphing technique. Sounds from these continua were then presented randomly to participants who were asked to categorize them as Glass, Metal or Wood. The perceptual limits between different categories were defined based on participants' responses and a set of unambiguous 'typical' sounds were determined. The acoustic analysis of these typical sounds determined the variation range of Damping parameter values (i.e., the global damping a_{g} and the relative damping a_{r}) for each category. Thus, the control of these two parameters provided an easy way to get different damping profiles directly from the label of the perceived material (Wood, Metal or Glass).

Attack time. The Attack time, which characterizes the excitation, is applied by multiplying the signal with a temporal envelope defined as a dB-linear fade-in function. The fade-in duration is set-up as the attack time duration.

5.3 Further Functionalities

Extracting synthesis parameters from natural sounds. An analysis module providing the extraction of the signal parameters (i.e., amplitudes, frequencies, damping coefficients, PSD of the noisy contribution) from natural percussive sounds was implemented in Matlab [40] (see also [4]). This module provided a set of signal parameters for a given impact sound and was linked to the synthesis engine at the bottom level. From these settings, the controls offered at different layers allowed the user to manipulate characteristics of the resynthesized sound and to modify its intrinsic timbre attributes. Then, the modified sound could be stored in a wave file. Note that if the initial spectrum was non-harmonic, the control of inharmonicity was still valid: in that case, f_k corresponded to the frequency of the component of rank k.

MIDI controls. The synthesizer can be also used in a musical context. In order to enhance the playing expressivity, parameters that are accessible from the graphical interface (e.g., presets, attack time, size, material, impact position...) can be controlled by using the MIDI protocol. In practice, parameters are mapped to any MIDI channel, and can be controlled using either "control change" or "note on" messages. For instance, if an electronic drum set is used to control the synthesizer, MIDI velocity provided by the drum pad can be mapped to the impact force and the pitch value can be mapped to the size of the object. This functionality enables the creation of singular or useful mappings when using MIDI sensors.

In addition, to control the high number of parameters (96 frequency-amplitude pairs), a tuning control based on standard western tonal definitions was implemented, which enables the definition of chords composed of four notes [1]. Each note is defined by a fundamental frequency and is then associated with 24 harmonics, so that the 96 frequencies are defined 'automatically' by only four note pitches. In this chord configuration, the controls of sound descriptors related to spectral manipulation is effectuated on the 24 spectral components associated with each note and replicated on all the notes of the chord. Such a feature is thus useful to provide an intuitive control to musicians, as it is to facilitate the complex task of structuring rich spectra.

6 Conclusion and Perspectives

In this study, we have developed an intuitive control of a synthesizer dedicated to impact sounds based on a three level mapping strategy: a top layer (verbal descriptions of the source), a middle layer (sound descriptors) and a bottom layer (signal parameters). The top layer is defined by the characteristics of the sound source (object and excitation). At the middle layer, the sound descriptors were partly chosen on the basis of perceptual considerations, partly on the basis of the physical behavior of wave propagation. The bottom layer corresponded to the parameters of the additive signal model. This mapping strategy offers various possibilities to intuitively create realistic sounds and sound effects based on few control parameters. Further functionalities were also added such as an analysis module allowing the extraction of synthesis parameters directly from natural sounds or a control via the MIDI protocol. The mapping design is still in progress and some improvements are considered. In particular, although the sound descriptors chosen for the control are perceptually relevant, the link between top and middle layers is far from being evident, since several middle layer parameters interact and cannot be manipulated independently. Additional tests will therefore be needed to choose the optimal parameter combinations that allow for an accurate control of sounds coherent with timbre variations.

Acknowledgment

This work was supported by the French National Research Agency (ANR, JC05-41996, "senSons").

References

1. Aramaki, M., Kronland-Martinet, R., Voinier, T., Ystad, S.: A percussive sound synthesizer based on physical and perceptual attributes. Computer Music Journal 30(2), 32–41 (2006)
2. van den Doel, K., Kry, P.G., Pai, D.K.: FoleyAutomatic: physically-based sound effects for interactive simulation and animation. In: Proceedings SIGGRAPH 2001, pp. 537–544 (2001)
3. McAdams, S., Chaigne, A., Roussarie, V.: The psychomechanics of simulated sound sources: material properties of impacted bars. Journal of the Acoustical Society of America 115(3), 1306–1320 (2004)
4. Kronland-Martinet, R., Guillemain, P., Ystad, S.: Modelling of Natural Sounds Using Time-Frequency and Wavelet Representations. Organised Sound 2(3), 179–191 (1997)
5. McAdams, S., Winsberg, S., Donnadieu, S., De Soete, G., Krimphoff, J.: Perceptual scaling of synthesized musical timbres: common dimensions, specicities, and latent subject classes. Psychological Research 58, 177–192 (1995)
6. Grey, J.M.: Multidimensional perceptual scaling of musical timbres. Journal of the Acoustical Society of America 61(5), 1270–1277 (1977)
7. McAdams, S.: Perspectives on the contribution of timbre to musical structure. Computer Music Journal 23(3), 85–102 (1999)
8. Valette, C., Cuesta, C.: Mécanique de la corde vibrante. Hermès, Lyon (1993)
9. Wildes, R.P., Richards, W.A.: Recovering material properties from sound. In: Richard, W.A. (ed.) ch. 25, pp. 356–363. MIT Press, Cambridge (1988)
10. Moore, B.C.J.: Introduction to the Psychology of Hearing, 2nd edn. Academic Press, New York (1982)
11. MaxMSP, Cycling 1974, http://www.cycling74.com/downloads/max5
12. Gobin, P., Kronland-Martinet, R., Lagesse, G.A., Voinier, T., Ystad, S.: Designing Musical Interfaces with Composition in Mind. In: Wiil, U.K. (ed.) CMMR 2003. LNCS, vol. 2771, pp. 225–246. Springer, Heidelberg (2004)
13. Carello, C., Anderson, K.L., KunklerPeck, A.J.: Perception of object length by sound. Psychological Science 9(3), 211–214 (1998)
14. Tucker, S., Brown, G.J.: Investigating the Perception of te Size, Shape, and Material of Damped and Free Vibrating Plates. Technical Report CS-02-10, University of Sheffield, Departement of Computer Science (2002)
15. van den Doel, K., Pai, D.K.: The Sounds of Physical Shapes. Presence 7(4), 382–395 (1998)

16. Kunkler-Peck, A.J., Turvey, M.T.: Hearing shape. Journal of Experimental Psychology: Human Perception and Performance 26(1), 279–294 (2000)
17. Avanzini, F., Rocchesso, D.: Controlling material properties in physical models of sounding objects. In: Proceedings of the International Computer Music Conference 2001, pp. 91–94 (2001)
18. Rocchesso, D., Fontana, F.: The Sounding Object (2003)
19. Lakatos, S., McAdams, S., Caussé, R.: The representation of auditory source characteristics: simple geometric form. Perception & Psychophysics 59, 1180–1190 (1997)
20. Klatzky, R.L., Pai, D.K., Krotkov, E.P.: Perception of material from contact sounds. Presence 9(4), 399–410 (2000)
21. Gaver, W.W.: How do we hear in the world? Explorations of ecological acoustics. Ecological Psychology 5(4), 285–313 (1993)
22. Giordano, B.L., McAdams, S.: Material identification of real impact sounds: Effects of size variation in steel, wood, and plexiglass plates. Journal of the Acoustical Society of America 119(2), 1171–1181 (2006)
23. Freed, D.J.: Auditory correlates of perceived mallet hardness for a set of recorded percussive events. Journal of the Acoustical Society of America 87(1), 311–322 (1990)
24. Sethares, W.A.: Local consonance and the relationship between timbre and scale. Journal of the Acoustical Society of America 93(3), 1218–1228 (1993)
25. Vassilakis, P.N.: Selected Reports in Ethnomusicology (Perspectives in Systematic Musicology). Auditory roughness as a means of musical expression, Department of Ethnomusicology, University of California, vol. 12, pp. 119–144 (2005)
26. Aramaki, M., Besson, M., Kronland-Martinet, R., Ystad, S.: Timbre perception of sounds from impacted materials: behavioral, electrophysiological and acoustic approaches. In: Ystad, S., Kronland-Martinet, R., Jensen, K. (eds.) CMMR 2008. LNCS, vol. 5493, pp. 1–17. Springer, Heidelberg (2009)
27. Terhardt, E., Stoll, G., Seewann, M.: Pitch of Complex Signals According to Virtual-Pitch Theory: Tests, Examples, and Predictions. Journal of Acoustical Society of America 71, 671–678 (1982)
28. Cook, P.R.: Real Sound Synthesis for Interactive Applications. A. K. Peters Ltd., Natick (2002)
29. Graff, K.F.: Wave motion in elastic solids. Ohio State University Press (1975)
30. Broch, J.T.: Mechanical vibration and shock measurements. In: Brel, Kjaer (eds.) (1984)
31. Sansalone, M.J., Streett, W.B.: Impact-Echo: Nondestructive Testing of Concrete and Masonry. Bullbrier Press (1997)
32. Jaffe, D.A., Smith, J.O.: Extensions of the Karplus-Strong plucked string algorithm. Computer Music Journal 7(2), 56–69 (1983)
33. Fletcher, H.: Normal Vibration Frequencies of a Stiff Piano String. Journal of the Acoustical Society of America 36(1), 203–209 (1964)
34. Vassilakis, P.N.: SRA: A web-based research tool for spectral and roughness analysis of sound signals. In: Proceedings of the 4th Sound and Music Computing (SMC) Conference, pp. 319–325 (2007)
35. Daniel, P., Weber, D.: Psychoacoustical roughness implementation of an optimized model. Acustica 83, 113–123 (1997)
36. Pressnitzer, D.: Perception de rugosité psychoacoustique: D'un attribut élémentaire de l'audition à l'écoute musicale. PhD thesis, Université Paris 6 (1998)

37. Leman, M.: Visualization and calculation of the roughness of acoustical musical signals using the synchronisation index model (sim). In: Proceedings of the COST-G6 Conference on Digital Audio Effects, DAFX 2000 (2000)
38. Plomp, R., Levelt, W.J.M.: Tonal Consonance and Critical Bandwidth. Journal of the Acoustical Society of America 38(4), 548–560 (1965)
39. Aramaki, M., Baillères, H., Brancheriau, L., Kronland-Martinet, R., Ystad, S.: Sound quality assessment of wood for xylophone bars. Journal of the Acoustical Society of America 121(4), 2407–2420 (2007)
40. Aramaki, M., Kronland-Martinet, R.: Analysis-Synthesis of Impact Sounds by Real-Time Dynamic Filtering. IEEE Transactions on Audio, Speech, and Language Processing 14(2), 695–705 (2006)

Algorithms for an Automatic Transcription of Live Music Performances into Symbolic Format

Stefano Baldan, Luca A. Ludovico, and Davide A. Mauro

Laboratorio di Informatica Musicale (LIM)
Dipartimento di Informatica e Comunicazione (DICo)
Università degli Studi di Milano
Via Comelico 39/41, I-20135 Milan, Italy
{ludovico,mauro}@dico.unimi.it

Abstract. This paper addresses the problem of the real-time automatic transcription of a live music performance into a symbolic format. The source data are given by any music instrument or other device able to communicate through a performance protocol. During a performance, music events are parsed and their parameters are evaluated thanks to rhythm and pitch detection algorithms. The final step is the creation of a well-formed XML document, validated against the new international standard known as IEEE 1599. This work will shortly describe both the software environment and the XML format, but the main analysis will involve the real-time recognition of music events. Finally, a case study will be presented: PureMX, a set of Pure Data externals, able to perform the automatic transcription of MIDI events.

1 Introduction

The study of automatic transcription tools is an interesting matter both in research and in commercial applications.

In this paper we will focus on the design and implementation of *ad hoc* automatic transcription algorithms for real-time instrumental performances. This research presents a number of application fields, ranging from the encoding of unique improvisations in symbolic form to speeding up the score writing process, like a musical dictation.

There are some problems to face even at this early stage. First, which kind of music devices should be supported? Which music features should be extracted and translated? Where and when the required computation should be carried out? Finally, which kind of encoding should represent the results of the process? Of course each question can have multiple answers, but our purpose here is just demonstrating that an efficient and effective solution can be implemented. Then our results can be applied with little or no effort to more general cases.

In short, the process described in this paper starts from a live performance, where MIDI-capable music instruments and devices are used. The resulting data stream is parsed by a real-time environment provided by Pure Data. Through this application, a number of algorithms to achieve an automatic transcription

S. Ystad et al. (Eds.): CMMR/ICAD 2009, LNCS 5954, pp. 422–437, 2010.

are implemented. The final format to represent music events is the new standard known as IEEE 1599-2008. Our choices will be justified in the next sections, and a brief description of the applications and adopted standards will be given.

We want to point out that in the following we will present the algorithms related to a specific implementation but they are suitable in a more general case. For instance any performance language or algorithm able to provide a series of pitches and durations could feed the system we describe in the following, as well as the choice of the IEEE 1599 standard is arbitrary.

2 A Short Overview of IEEE 1599

Even if the encoding in XML format represents only the final step of the process, it is important to describe this aspect immediately as all the algorithms will be affected by this choice.

The music code we adopt, namely IEEE 1599, is not a mere container for symbolic descriptions of music events such as notes, chords, rests, etc. Thanks to its multi-layer structure, illustrated in detail in [1], IEEE 1599 allows to describe many different aspects of music within a unique document. In particular, contents are placed within 6 layers:

- *General* - music-related metadata, i.e. catalogue information about the piece;
- *Logic* - the logical description of score symbols (see below);
- *Structural* - identification of music objects and their mutual relationships;
- *Notational* - graphical representations of the score;
- *Performance* - computer-based descriptions and executions of music encoded in performance languages;
- *Audio* - digital or digitized recordings of the piece.

The *Logic* layer has a central role in an IEEE 1599 document. In detail, it contains i) the main time-space construct aimed at the localization and synchronization of music events, known as *Spine* sub-layer; ii) the symbolic description of the score in terms of pitches, durations, etc., known as *Logically Organized Symbols* (*LOS*) sub-layer; and iii) information about a generic graphical implementation of symbolic contents.

The *Logic* layer is the only level directly involved in the process of live performance transcription, since it contains the music symbols written in Common Western Notation (CWN). Specifically, music events have to be listed, identified and sorted in a common data structure called *Spine.* Spine translates the typically 2-dimensional layout of a score in a 1-dimensional sorted list of music events, uniquely identified by an ID. Each symbol of spine presents a space and time distance from the previous one, expressed in relative way. In this work, only temporization of events is involved and not their placement on a graphical score; as a consequence, only time-related aspects of *Spine* sub-layer will be discussed.

After providing a list of IDs in *Spine*, music events can be defined in standard notation within the *LOS* sub-layer. Here pitches are described by note names and octaves, and rythmical values are expressed in fractional form. For example, a 𝅗𝅥 corresponds to the XML line:

```
<duration num="1" den="2" />
```

and a ♩· to the XML line

```
<duration num="3" den="8" />.
```

As regards durations, please note that a reduction to lowest terms of the fraction is not required by IEEE 1599 specifications, even if desirable. Similarly, dotted notations is supported in IEEE 1599 in order to try to obtain 1 as numerator, like in the latter example where the duration ♩· has been encoded as ♪⌒♪⌒♪; alternatively, the following XML lines could be employed:

```
<duration num="1" den="4" />
<augmentation_dots number="1" />.
```

The *Logic* layer - which defines music events from a logical point of view - takes a key role for all the other layers, as they refer to spine identifiers in order to bind heterogeneous descriptions to the same music events. In order to obtain a valid XML file, only spine is strictly required, so that even scores not belonging to CWN are supported by IEEE 1599.

In our context, the first advantage coming from IEEE 1599 consists in the possibility to encode contextually additional information: provided that a live performance can be correctly transcribed, within a unique IEEE 1599 document not only the logic score (notes, rests, etc.), but also the corresponding computer-based performance layer, the resulting digital audio, and even related structural information can be encoded. This aspect highlights the heterogeneity of media types and different kinds of description supported by IEEE 1599. Its multi-layer structure allows to organize such a variety as a broad and comprehensive picture of a unique music piece.

Besides, an IEEE 1599 document can host, for each layer, multiple descriptions of the same piece. For example, the file containing the "logic" score of a piece - namely a sequence of music symbols flowing like in the composer's mind - can present n different graphical instances, related to n score versions, in the *Notational* layer. Similarly, the *Audio* layer can host m sub-sections, corresponding to as many tracks (e.g. historical performances, live unplugged executions, transcriptions, variations, piano reductions, and so on).

As a consequence, IEEE 1599 in our opinion fits very well the purposes of this work. Nevertheless, the algorithms we will describe in the following sections can produce transcriptions in any other music code: binary (e.g. NIFF) as well as plain-text (e.g. DARMS, LilyPond), general markup (e.g. MML) as well as XML (e.g. MusicXML).

3 Pure Data as a Platform for Live Performances

In this section we will shortly introduce Pure Data, the open-source counterpart of MAX/MSP system. Both of them were developed by the same author, Miller Puckette; his contribution to the project is presented in [2].

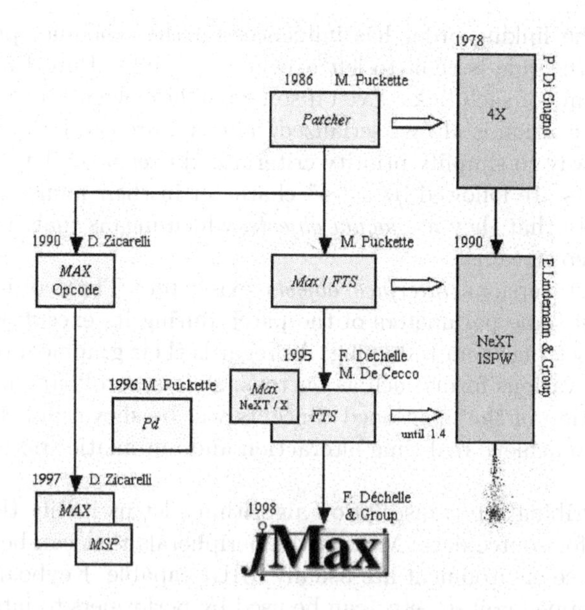

Fig. 1. The evolution of MAX and Pure Data

Pure Data is an integrated platform designed for multimedia, and specifically for musical applications. This graphical real-time environment can be successfully used by programmers, live performers, "traditional" musicians, and composers.

As illustrated in Figure 1, both the environments had a long evolution since their author started the development process in the eighties. Some of the key concepts have not changed over time, such as the overall flexibility and modularity of the system. Pure Data functions can be improved by the use of *abstractions*, i.e. sub-patches recalled by the user under other patches, and *externals*, i.e. newly created object programmed in C via the Pure Data framework and its API. Pure Data was written to be multi-platform and portable; versions exist for Win32, IRIX, GNU/Linux, BSD, and Mac OS X. Source code is available too.

The program interface is primarily constituted by two kinds of window: *PureData* and *patch/canvas*. The former gives access to the settings of the program and to the visualization of system messages, allowing the control of the correct workflow. The latter is the place where the user creates and interacts with the application by placing objects and linking them together.

Patches present two different states: *edit mode* and *run mode*. In *edit mode* the user can add *objects*, modify them and link them through *cords*. In *run mode* the patch follows its workflow and the user can interact with it in real-time.

Objects appear like "black boxes" that accept input through their *inlets* or as *arguments* (placed near their name) and return output data through their *outlets*. Programs are built disposing these entities on a canvas (the *patch*) and creating a data flow by linking them together through *cords*. Data are typed; as a consequence not all the possible links are available.

Choosing the linking order has influences on the scheduler priority. Unlike MAX, where the rule is right-to-left execution of links, Pure Data is ruled by the creation time of such links. Even if some patches suggest a certain degree of parallelism, execution is always serialized. This feature can be viewed as a limit but also as a way to simplify priority criteria and execution flows.

Some objects are followed by a "~" character in their name. This symbol is used to indicate that they are *signal objects*, which means that they can handle audio and video streams.

In the latest versions *interface objects* exist, too. These objects allow the user to control some parameters of the patch during its execution without the annoyance of setting them by typing. As regards their graphical representation, they can have various forms such as buttons, sliders, scrollbars, menus, etc.

An application of the mentioned concepts will be shown in Section 6, where the inferface to achieve real-time interaction and automatic trascription will be descrived.

Before describing the transcription algorithms, let us justify the adoption of MIDI format for source data. Most of the peripherals that can be attached in a live performance environment are usually MIDI capable. Keyboards, synthesizers, MIDI-equipped guitars, etc., can be used by performers to interact with the system. Pure Data can handle MIDI format though its primitives, thus allowing a simple but effective implementation of our work. However, the algorithms introduced in the next section make use of basic information that is available in a large number of formats. It would be virtually possible to adopt any other input format for the transcription. For example, Pure Data has primitives for OSC (OpenSound Control), thus the support for that format could be easily implemented.

Please note that, even if MIDI has a lower expressivity than IEEE 1599, it is widely used both for score encoding and for numeric communication among sound devices. Thanks to its extensive employment in standard live performance environments, it has been adopted as the base for music events transcription into IEEE 1599 format.

4 From MIDI to IEEE 1599

The simplicity, extensibility and power of Pure Data make it an ideal choice to develop musical applications based on the IEEE 1599 format, thus improving - as a side effect - the diffusion of this new standard. In order to demonstrate the effectiveness of our approach, we have developed PureMX, a library of Pure Data externals designed to convert a MIDI stream (played live by either performers or sequencers, or a combination of both) into a well-formed and valid IEEE 1599 document. By now, the program focuses just on the construction of the *Logic* layer of IEEE 1599, which mainly contains a sorted list of music events (*Spine* sub-layer) and their symbolic representations (*LOS* sub-layer). See Section 2 for further details.

The PureMX library is written in ANSI C, making exclusive use of the standard libraries and the Pure Data API, in order to be highly portable on a wide variety of platforms. It is also extremely modular, taking full advantage of the Pure Data object paradigm and simplifying the integration of new features not yet implemented in the library itself.

Once loaded in the system, PureMX objects can be used inside a Pure Data patch, in combination with the native primitives of the platform, other libraries of externals or Pure Data abstractions (sub-patches).

One of the most challenging aspects encountered in the developing process of PureMX lies in the conceptual and structural difference between the formats involved in the conversion. First of all, while MIDI is just a sequence of chronologically ordered events, IEEE 1599 represents musical information in a hierarchical and multilayered fashion. So, it is useful to organize the MIDI input stream inside a data structure which mirrors the nature of the IEEE 1599 format. In second instance MIDI, as a performance format, stands at a lower level of abstraction than the *Logic* layer of IEEE 1599. For instance, fundamental entities of symbolic score description in IEEE 1599 (such as clef and tonality), that are very rich in semantic content, are not explicitly present in the input stream. In fact, MIDI was designed to convey semantically poorer concepts such as the mechanical actions made by a performer on a musical instrument.

Going down the hierarchy of musical information layers, we can consider the lower levels a "practical instance" of the abstract concepts contained in the higher ones, so information is just translated into a new form but not completely lost. In our case we can consider the events of a performance format as a practical realization of the concepts which should be written in the IEEE 1599 *Logic* layer, so most of the information we need is "hidden" but still present in MIDI. It is possible to apply musical information retrieval algorithms on the input stream and obtain those missing elements, as they are implicitly contained in the relations among events and in the general context of the whole stream. For complete reference on MIDI see [3].

5 Algorithms for Pitch and Tempo Extraction

5.1 Event Segmentation

In MIDI - like in most performance formats - music events are represented by the succession of two message types: a *noteon* which activates a note and a *noteoff* which deactivates it. Please note that music events can be interleaved, e.g. a *noteon* could follow another *noteon* message before the *noteoff* of the former. Both message types contain a *pitch* parameter that identifies which note has been activated/deactivated, and a *velocity* parameter that indicates the intensity of the action on the instrument. The duration of a note can be easily calculated by counting the time between a *noteon* and a *noteoff* message sharing the same pitch. In the same way, the duration of a rest can be calculated by counting the time between a *noteoff* and the next *noteon* (in this case *pitch* does not matter).

In MIDI, durations are calculated in Midi Time Clock (MTC) units. Most sequencers use 24 MTC per quarter, and their absolute duration depends on the Beat Per Minute (BPM) value of the sequence. On the other side, in the *Spine*

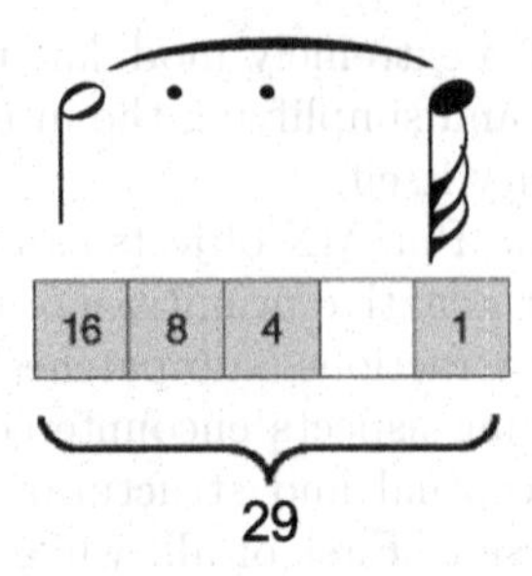

Fig. 2. Segmentation of a VTU duration into parts representable by CWN symbols

sub-layer of IEEE 1599 durations are calculated and stored in Virtual Time Units (VTUs). The only difference is the following: while MTC granularity is fixed, VTU granularity can change for each IEEE 1599 document. For example, if the shortest note of a given music piece is the quarter note, then in IEEE 1599 we can associate a single VTU to this rhythmic value.

In order to achieve a conversion from VTU-based temporal representation to Common Western Notation (CWN), it is necessary to identify and segment music events that are not representable by a single CWN symbol. A rhythmic value whose numerator is not power of 2 should be encoded by using two or more symbols tied together. For instance, the first note in Figure 2 presents a duration that can be expressed as 7/8. The segmentation algorithm implemented in PureMX exploits the similarity between CWN symbols and the binary system, providing an extremely fast and effective way to split complex durations into simple ones. We choose granularity in order to assign the time unit ($2^0 = 1$ VTUs) to the smallest CWN symbol. Going from short to long values, the following symbol will have exactly twice the duration of the first ($2^1 = 2$ VTUs), the third will have twice the duration of the second ($2^2 = 4$ VTUs) and so on. As a consequence, we can represent CWN rhythmic values through variable-length binary strings where the highest order bit is set to 1 and all the others are 0s.

As a further step, also augmentation dots should be considered. Each augmentation dot increases the duration of a note by half its value, so a dotted value can be represented as a sum of the value itself and of the immediately smaller one. In binary terms, notes dotted n times can be encoded through strings with the highest $n+1$ bits set to 1 and the others set to 0.

Unfortunately if we want to represent either events tied together or inside a tuplet we cannot assign the time unit to a basic CWN symbol. To deal with these situations we must assign a time unit so that each note in the score (both inside and outside the irregular groups) could be represented by a whole amount of VTUs. In other words, the fractional representation of each note segmented by the algorithm (e.g. 1/4 for the quarter, 1/8 for the eighth, 1/12 for the eighth under triplet etc.) multiplied by the amount of VTUs in a whole note (which will be called W from now on) must always be an integer.

Please note that W represents an input value to the algorithm, so the user has to choose it a priori on the basis of the supposed performance to be transcribed. A practical rule consists in taking into account all the "granularities" of rhytmical values to be considered. The term granularity implies the fact that some values can be expressed in terms of powers of 2, whereas other values could belong to irregular groups. In the latter case, the fractional value of the note is not related to the CWN symbol that encodes it, but to its real duration inside a tuplet. For instance, in a quintuplet of sixteenth notes (whose total duration is the one of a quarter note), the real duration corresponds to $1/5 \cdot 1/4 = 1/20$. Finally the value of W can be computed as the lcm (least common multple) of all those denominators.

Taking these new considerations into account, we can extend the set of duration values representable by a single CWN symbol: not only numbers in the form $\{1, 3, 7\} * 2^n, n \geq 0$ (i.e. simple, dotted and double dotted notes), but also each of the formers multiplied by one or more of the prime factors of W. If we proceed in the inverse order, dividing a valid duration by all the prime factors in common between itself and W (their GCD) will result in another valid duration, belonging to the base form defined before. Exploiting this property, the similarity between CWN notation and binary system broken by the insertion of tuplets and ties can be restored: it is sufficient to divide the binary string to be segmented by the GCD between itself and W, operate the actual segmentation following the principles explained above, then multiply again each segmented value by the GCD to obtain the final event durations.

The segmentation algorithm takes in input an arbitrary binary string which represents a duration in VTUs. If the string encodes a duration longer than $7/4$ (whole note with two augmentation dots), events lasting $7/4$ are repeatedly written and their duration is subtracted from the original string until the remaining part becomes shorter than this amount. The string obtained in this way is then divided by the GCD between itself and W, to remove from its factorization the terms corresponding to the tuplet ratios. It is then copied and right bit-shifted until it becomes a sequence of bits all set to `1` (e.g. `1`, `11` or `111`); finally, it undergoes a left bit-shift process by the same amount and it is multiplied again by the GCD found before. In this way, the algorithm finds the largest value representable by a single note symbol, possibly dotted. Such an event is written into the data structure and its duration is subtracted from the original string. The process is repeated until the remaining part of the original string is made of 0s.

For example, let a note be representable by the fraction $13/32$. This is the case of a dotted quarter ($3/8$) tied to a thirty-second note ($1/32$). If the latter value is the smallest one in the piece, the corresponding binary string would be `1101`. 13 is prime so the GCD with no matter which number will be 1. After verifying that $13/32 \leqslant 7/4$, the proposed algorithm calculates the GCD which, 13 being prime, will be 1. After that, it carries out 2 right bit-shifts, thus producing the string `11` (made of a pure succession of 1s). Then the process is inverted, and the string becomes `1100`. This bit configuration now corresponds to a single rhythmic value, namely a dotted quarter note. After subtracting the new string from the original one, we obtain the remaining part of the original duration:

1101-1100 = 0001, i.e. the value to add in order to obtain the whole duration. As a matter of fact, the 0001 string corresponds to a thirty-second note.

Even if metric and accent issues are ignored, the mentioned algorithm is guaranteed to find always a correct solution to the segmentation problem.

5.2 Clef Guessing

After facing the problem of complex rhythmical values, we need to define the clef in order to represent pitches by disposing symbols on a staff. In CWN the clef is used to associate a well-defined pitch to each line and space of the staff, thus creating an unambiguous correspondence between the vertical position of a symbol and the name of the note represented.

The problem of finding a clef that fits well a given musical sequence is quite easy to solve, as all the information we need is coded inside the *pitch* parameter of MIDI *noteon* and *noteoff* events. The *pitch* parameter is an integer between 0 and 127, and each number represents a different semitone, like a different key on a keyboard. In MIDI, pitch 60 corresponds to the Middle C, having a frequency of 261.63 Hz approximately. This pitch is usually referred as C4. Consequently, MIDI pitch 61 is assigned to C♯4 (277.18 Hz), MIDI pitch 62 to D4 (293.67 Hz) and so on.

The clef guessing algorithm designed for PureMX is based on the computation of a mean among the various pitches inside a measure, in order to find the "average pitch" of that measure. For the sake of simplicity, the mean is arithmetic: each symbol has the same weight, no matter what its duration is. Please note that this algorithm has to compute results in real time. The whole pitch range is divided into intervals, and a clef is associated to each of them. The clef that fits best the melodic sequence is the one that minimizes the use of additional cuts, as shown in Figure 3.

The choice of using just two of the seven available clefs avoids interval overlapping; moreover, in current notation the other five clefs are rarely used, e.g. in particular contexts (such as in vocal music) or for certain musical instruments (such as the alto clef for viola).

The average pitch calculation presents a problematic issue: as the concept of pitch makes no sense for rests, they should not be included in the mean; but in this case empty measures would not have an average pitch value and, consequently, they would not have a clef. The same clef of the previous measure could be assigned, but the problem remains if the first measure is empty too.

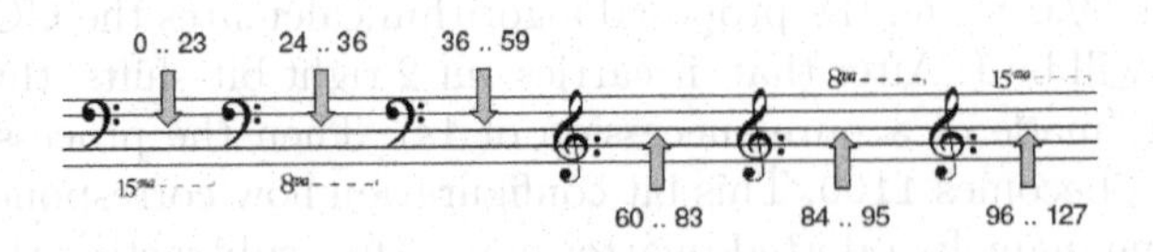

Fig. 3. Average pitch intervals for the PureMX implementation of the clef-guessing algorithm

The adopted solution consists in assigning a pitch value to rests, in particular the same pitch of the previous event, if any, otherwise the pitch of the following one.

The intervals proposed in Figure 3 are the ones used in the PureMX implementation of the algorithm, however there are many other alternatives: for example, creating a specific interval set for vocal music based on *tessitura* instead of average pitch; calculating this parameter by a weighted mean; taking note durations into account; performing this calculation on the whole part/voice instead of measure by measure.

5.3 Key Finding and Pitch Spelling

The *pitch* parameter is also useful in finding the tonal context of a given sequence of notes. This is a fundamental aspect to make a good CWN transcription because in the equal temperament, universally adopted by current western music, each semitone may correspond to 2 or 3 *enharmonically equivalent* pitches. For instance, MIDI pitch 60 can be written in a score as C, B♯ or D♭♭; and pitch 68 only as G♯ or A♭. Knowing the tonal context allows us to choose the correct note name, and therefore the right position on the staff, for each pitch value in the sequence.

In the least 20 years, many people have proposed studies and methods to solve the key finding and/or the pitch spelling problem, reaching good results. The common element shared by every approach is the analysis of musical events not as isolated entities but as part of a context, which affects their interpretation and at the same time is affected by their presence. Differences lie in the choice of parameters and rules to identify and quantify relations among those events. All key finding and pitch spelling algorithms contain heuristic information, namely prior knowledge about the problem they have to solve, mainly based on the rules of tonal harmony. For this reason, all those approaches do not work well (sometimes they fail at all) when applied to music belonging to different cultural areas or historical periods.

In order to give the PureMX library key finding and pitch spelling capabilities, many different solutions have been examined. With simplicity in mind, we wanted an algorithm reasonably efficient, easy to implement and able to solve both problems at the same time. The Krumhansl-Schmuckler algorithm (and its further improvement by David Temperley) [4] has all these features. It is based on a Bayesian approach: each note gives a certain amount of "points" to each possible tonal centre, and the one which gains the higher score is chosen as the most probable for that sequence of notes.

In the PureMX implementation of the algorithm, twelve possible tonal centres are defined, one for each semitone in an octave. In case of enharmonic equivalence, the tonality with less accidentals is preferred, following the principle of notational economy. Scores to tonal centres are then assigned following two probability distributions, one for major keys and the other for minor keys. Such distributions were experimentally deduced by Krumhansl and Kessler at the beginning of the 80s and then improved by Temperley in 1990. The experiment consisted in asking listeners to rate how well "probe tones" fitted into

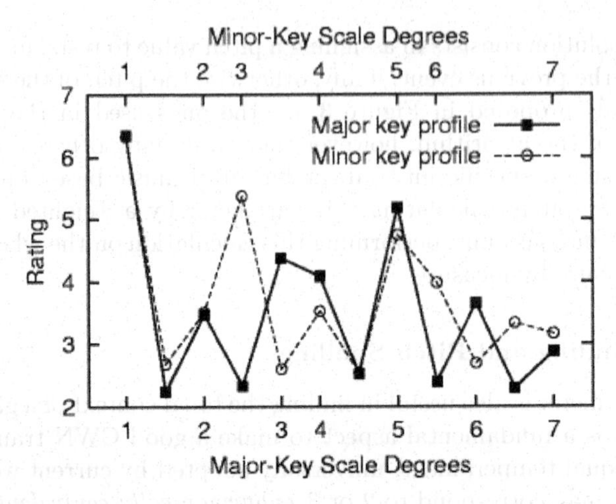

Fig. 4. Probability distributions for score assignment to tonal centres based on pitch values (Krumhansl-Kessler version). Picture taken from [5].

various musical contexts (cadences in major and minor) [5]. Results are shown in Figure 4.

The probability distributions provides a measure about the fitness of a particular pitch inside the scale of each tonality. For example, a pitch value of 60 (possibly corresponding to the notes C, B♯ or D♭♭) will give a high score to C major as its tonic, or to F major as its dominant, or to A minor as its median; but it will give a poor evaluation to B major because it is not a natural degree in that scale. Score assignment is also weighted on note durations, so that more importance is given to longer notes. It is worth to underline that small rhythmic values are often used for less relevant notes, such as chromatic passages and embellishments.

Once the tonality is determined, pitch spelling for notes on the scale is immediate and unambiguous. For notes outside the scale, note names representing stronger degrees are preferred because they give a better harmonic representation of the key. In particular, major and minor degrees are preferred over diminished and augmented ones.

The original version of the Krumhansl-Schmuckler algorithm uses an overlapping window technique to control its reactivity: small windows and few overlapping notes will produce many tonality changes, while longer and more overlapped windows will produce more uniform results. The PureMX approach is quite different as it uses an exponential mean to obtain the same effects: we define a multiplying factor α in the closed range $[0, 1]$ and a "history" array H made of n elements, with n equal to the number of the considered tonal centres. Let $H(x)$, with $x \in \mathbb{N}_0$, be the status of array H at step x. $H(0)$ is set to 0. Let $C(x)$ be an n-elements array where the score assigned to the current measure are stored. Then we calculate the tonality of a given measure by the equation:

$$H(x) = C(x) \cdot \alpha + H(x-1) \cdot (1-\alpha)$$

The maximum of the n elements of $H(x)$ is used to determine the tonality of the current measure, and the array is stored as the new history. The α factor is a measure of the algorithm reactivity, as greater values give more weight to the current measure evaluation whereas smaller values give more weight to history. As the history is iteratively multiplied by a factor smaller than 1, the contribution of earlier measures to the history becomes less important as time passes, until it becomes irrelevant for very distant measures.

Even if this version of the algorithm already provides satisfactory results for well-defined tonal contexts, further improvements could be applied. For example, it would be possible to implement a pitch spelling algorithm (the one proposed by [6] is simple yet efficient) and use its results to help the key finding algorithm, or could be employed other criteria (voice leading, accents pattern) in addition to the fitness functions in order to obtain a more accurate tonality guess. Anyway, this would add complexity to the algorithm, in contrast with our main goals.

5.4 From Absolute to Relative Time Representation

We have already described the concept of VTU in Subsection 5.1, and we have defined it as the temporal representation of events in an IEEE 1599 document. VTU values are stored in the *Spine* (see Section 2) and describe the chronological order of the events in a musical piece. We can see VTUs as ticks of a clock, where the absolute duration (in seconds) of a single tick depends on the chosen granularity and on the BPM value of the piece.These values are intentionally unexpressed at *Logic* level as the virtual temporization of score symbols can correspond to different instances at *Performance* or *Audio* layer. Anyway, a different concept of time is also present in the *LOS* sub-layer. Each note and rest is described as a CWN symbol with its relative duration, so we need to transform VTU durations in fractions of measure in order to obtain the logic description of music events.

Let us recall that the conversion from VTU-based timings to CWN fractional durations has to be achieved in real time, so a fast but effective algorithm has to be designed and implemented. Nevertheless, extemporary improvisation is often made of irregular rhythms, including nested tuplets, so the problem of rhythm transcription is not trivial and cannot be solved by using mere quantization.

In the following, the approach employed in PureMX is described. The inputs of the algorithm are:

1. The meter of the piece in fractional form $\frac{s}{t}$ (e.g. $\frac{4}{4}, \frac{6}{8}$, etc.). Please note that s is the number of subdivisions in a measure, whereas $1/t$ is the rhytmic value corresponding to a subdivision;
2. VTUs per measure (let it be v), or alternatively VTUs per subdivision (let it be w); v and w are related by the formula $v = s \cdot w$;
3. The duration of the music event to parse, expressed in VTUs and referred in the following as x. This amount is available as soon as a *noteoff* event follows the corresponding *noteon* in case of notes, or a *noteon* message follows a remote *noteoff* in case of rests.

The process proposed here will return the CWN symbol to write in the *LOS* sub-layer, even when located inside a tuplet. The duration will be represented through a numerator n and a denominator d, where n is not necessarily equal to 1 whereas d has to be a power of 2. In short, the following algorithm computes n and d starting from s, t, w (or v) and x.

1. Let g be the greatest common divisor (GCD) between x and $(w \cdot t)$;
2. Calculate $a = x/g$ and $b = (w \cdot t)/g$. The fraction a/b represents the reduction to lowest terms of the original fraction $x/(w \cdot t)$;
3. Evaluate the obtained denominator b. If $b = 2^n$, with $n \in \mathbb{N}_0$, namely it is a power of 2, then proceed to branch (a). Otherwise, the value to be parsed belongs to a tuplet. In this case, let d be the floor rounding of b to the closer power of 2, which will allow to write an existing rhythmic value in CWN. Jump to branch (b);
   (a) Set $n = a$ and $d = b$. Now the event duration can be encoded in the *LOS* sub-layer. If a more compact notation is required, the previously described segmentation algorithm can be employed in order to express the fraction as a sum of rhythmic values, possibily dotted or double-dotted. Finally, jump to Step 4;
   (b) In order to dimension the tuplet, namely to fix the number i of values the measure should be divided into, calculate $i = s \cdot w/x$. In natural language, this means: "Divide the measure in i equal parts and represent the current note as n/d under the resulting tuplet". In IEEE 1599, the tuplet encoding is in the form "Put i/d in the place of s/t" (e.g. "Put 6/8, i.e. 6 ♪, in the place of 2/4, i.e. 2 ♩"), and all the required values have been calculated. According to standard notation rules, necessarily $i \in \mathbb{N}$, but in general this is not guaranteed by the algorithm (refer to the last example of this subsection). In this case, a segmentation algorithm has to be used in order to split the original x into the sum $x_1 + x_2 + ... + x_n$, where each element makes the corresponding i an integer. It is possible to demonstrate that this operation can always be performed in a finite number of steps;
4. Write the n and d values in the corresponding attributes of the IEEE 1599 element `<duration>`. If the event is dotted, or it belongs to a tuplet, compile also those parts of the document.

Now we will provide a number of examples to clarify the applicability of the algorithm. Let us consider Figure 5, where time signature is $\frac{3}{4}$, thus $s = 3$ and $t = 4$. Let 30 be the number of VTUs per quarter, i.e. $w = 30$ and $v = 30 \cdot 3 = 90$. Finally,

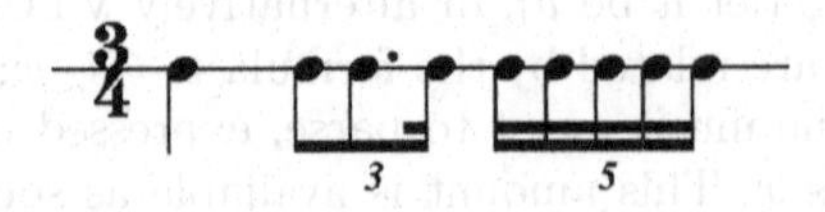

Fig. 5. A measure containing both standard durations and tuplets

let us apply the algorithm when $x = 30$, which intuitively corresponds to the first note in Figure 5. From Step 1, g is the GCD between $x = 30$ and $w \cdot t = 30 \cdot 4 = 120$, so $g = 30$. Step 2 provides the reduction to lowest terms by using g, so the numerator $a = 1$ and the denominator $b = 4$. The evaluation process at Step 3 confirms that $a = 1$ and $b = 2^n$, with $n = 2$. As a consequence, the obtained results corresponding to a ♩ are ready to be written in the XML description of the event.

When $x = 10$, g is the GCD between $x = 10$ and $w \cdot t = 30 \cdot 4 = 120$, so $g = 10$. Step 2 states that $a = 1$ and $b = 12$. Step 3 determines that b is not a power of 2, so branch (b) is entered. Through the algorithm, $b = 12$ is rounded to the value 8, and this value is assigned to d. This means that the symbolic value to be written under tuplet is ♪. Now the tuplet has to be determined. From $i = s \cdot w/x = 3 \cdot 30/10$ follows $i = 9$. Finally, the algorithm states how to compute the tuplet: "Put i/d in the place of s/t", namely "Put 9/8, i.e. 9 ♪, in the place of 3/4, i.e. 3 ♩". Even if already correct, the result can be improved by dividing the two numerators by their GCD, thus obtaining "Put 3/8, i.e. 3 ♪, in the place of 1/4, i.e. 1 ♩", which corresponds to the graphical representation of the second note in Figure 5.

When $x = 15$, $g = 15$, $a = 1$ and $b = 120/15 = 8$. Since the obtained b is a power of 2, branch (a) is entered. As a consequence, the third value in Figure 5 is recognized as ♪, even if its graphical representation was a dotted eighth note inside a triplet. From a mathematical point of view, this is correct: in fact the latter representation implies taking 1.5 parts of the subdivision of a quarter by 3, i.e. $3/2 \cdot 1/3 \cdot 1/4 = 1/8$, corresponding to an eighth note.

When $x = 6$, intuitively we are dividing the VTU duration of the subdivision by 5. This is the case of the last 5 notes in Figure 5. By applying the algorithm, $g = 6$, $a = 1$ and $b = 120/6 = 20$. Since the obtained b is not a power of 2, branch (b) is entered. The rhythmic value to use in the tuplet is 1/16, as 16 is the floor rounding of 20 to the nearest power of 2. From $i = s \cdot w/x = 3 \cdot 30/6$ follows $i = 15$. Finally, the algorithm says: "Put i/d in the place of s/t", namely "Put 15/16 in the place of 3/4", or alternatively "Put 5/16 in the place of 1/4".

Finally let us explain the process when applied to a more complex case. Let the meter be $\frac{7}{8}$, thus $s = 7$ and $t = 8$. Besides, let 9 be the number of VTUs per ♪, i.e. $w = 9$ and $v = 9 \cdot 7 = 63$. Please note that no problem is due to the complex time signature: e.g., when $x = 36$ a 𝅗𝅥 symbol is recognized; and when $x = 3$ the algorithm recognizes a sixteenth note in a tuplet made of 21 values of the same kind (in this case, the tuplet puts 21/16 in the place of 7/8, namely a ♪· in the place of a ♪). Rather, a problem arises when $x = 4$. In fact, $g = 4$, $a = 1$ and $b = 72/4 = 18$. Since the obtained b is not a power of 2, once again branch (b) is entered and the value to use in the tuplet results to be 1/16. From $i = s \cdot w/x = 7 \cdot 9/4$ follows that $i \notin \mathbb{N}$, and the original x has to be split into a sum of integer addends. For example, a simple way to solve the problem is considering $x = x_1 + x_2 = 3 + 1$, which corresponds to representing a unique music event as two tied symbols of different duration. Now the algorithm is able to work on such values.

6 PureMX Interface

As a library of Pure Data externals, PureMX shares the user interface of its host program. Figure 5 shows an example patch in which all of the PureMX externals are used together with some native Pure Data primitives.

The first external of the set is *mxsheet*, which records, segments and stores events of the input stream inside an organized and hierarchical data structure. Its leftmost inlet accepts some reserved messages that control the recording process (start, stop) and the pitch number of incoming events. The next two inlets accept velocity values and channel numbers, respectively. The other inlets are used to indicate BPM value, time signature (numerator and denominator), VTU per quarter and whether the recording should start automatically as soon as the first event is received or not. By now, these values are set manually through numeric inputs, but further development could involve the creation of beat tracking or metric structure recognition objects.

The *notein* object attached to the first three inlets is a native Pure Data primitive, and it is used to translate MIDI events into Pure Data messages. This means that, even if PureMX has been conceived with MIDI in mind, it is theoretically possible to use the library with any other performance format that has the same data organization of MIDI. For example, attaching an OSC translator instead of the *notein* object, we could make use of OSC messages instead of MIDI messages.

In terms of outlets, the first two simply send out *bangs* (sort of "wake up" messages) in time with the metronome, while the third sends out a custom message (*mxfeed*) once the recording stops. This last message contains, among other things, the pointer to the first memory cell of the recorded data structure, and it is used by other PureMX externals to read the recorded information or write new elements.

The other objects of the library are *mxcsig*, which implements the clef guessing feature; *mxksig*, which implements the key finding and pitch spelling features; and finally *mxspiner*, *mxlosser* and *mxbuild*, which respectively write the *Spine*, the *LOS* and the whole IEEE 1599 document to text files. It is not mandatory

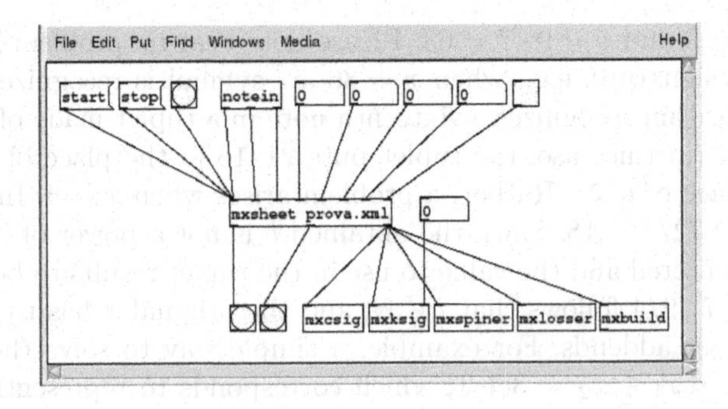

Fig. 6. Example patch showing PureMX externals, together with some Pure Data primitives

to use all of these objects: there are cases in which, for example, the IEEE 1599 document is used just for synchronization so the *LOS* element is not needed, or it makes no sense to guess tonality as the musical piece recorded does not belong to the tonal repertoire.

7 Conclusions

In this paper we have presented a working application that can fulfil the needs of musicians who want a transcription of their live performance. This can be useful for recording of live electronics, for improvisation, that are unique and always varying, for music archives, for analysis of the recorded material and so on. We have faced the problem adopting IEEE 1599 and Pure Data, namely open and extensible environments that allow the tailoring of the application to the user needs. In the paper we have also proposed the use of some algorithms, both designed *ad hoc* and adapted from literature, that can solve the various problems encountered in the definition of the problem domain and in its practical resolution.

References

[1] Ludovico, L.A.: Key Concepts of the IEEE 1599 Standard. In: Proceedings of the IEEE CS Conference The Use of Symbols To Represent Music And Multimedia Objects. IEEE CS, Lugano (2008)

[2] Puckette, M.S.: Pure Data: another integrated computer music environment. In: Proceedings of Second Intercollege Computer Music Concerts, Tachikawa, Japan, pp. 37–41 (1996)

[3] Authors, V.: The Complete MIDI 1.0 Detailed Specification. Document version 96.1. The MIDI Manufacturers Association (1996)

[4] Temperley, D.: What's Key for Key? The Krumhansl-Schmuckler Key-Finding Algorithm Reconsidered. Music Perception 17(1), 65–100 (1999)

[5] Krumhansl, C.L.: Cognitive Foundations of Musical Pitch. Oxford University Press, Oxford (1990)

[6] Cambouropoulos, E.: Pitch Spelling: a Computational Model. Music Perception 20(4), 411–430 (2003)

Raising Awareness about Complete Automation of Live-Electronics: A Historical Perspective

Javier Alejandro Garavaglia

SJCAMD – London Metropolitan University
41 Commercial Road, E1 1LA, London, UK
j.garavaglia@londonmet.ac.uk

Abstract. The article raises attention to the advantages and disadvantages of complete automation in DSP processes (including the triggering of events) during the performance of interactive music. By proposing a historic summary divided in three main periods according to the technologies, methods and processes available during each of them, (including examples of key works and composers), it shows how the usage of automation in live-electronics was dependent on the development of new technologies, specially digital. Further, it explains how full automation works in two works of mine, describing the features and techniques involved. Considering those examples, the advantages and disadvantages resulting from the introduction of complete automation of DSP in live performance are finally discussed. Even though automation is not a new technique in the field, I am keen to dedicate special attention to completely automated events -including their triggering- given the impact that automation can have on performances.

Keywords: Automation, Interactivity, Live-Electronics, DSP, Halaphone, 4X, MAX/MSP, Spatialisateur, ISPW, AUDIACSystem, UPIC, Synlab, SMPTE.

1 Introduction

What does the term *live-electronics* mean? The "ElectroAcoustic Resource Site project" (EARS) defines it as:

> 'A term dating from the analogue age of electroacoustic music that describes performance involving electronic instruments which can be performed in real-time. The term is more commonly expressed today as music involving interactive instruments.' [1]

This generally implies the interaction, which transforms and/or processes live or recorded sound in real time. Interactivity is considered nowadays as inseparable from computers or even a network of computers but this has not been always the case. In the last forty years, several and different types of interaction have been identified as *live–electronics*; hence many types of musical pieces can be included in this category. Through my research and practice in the area, I find it very helpful to analyse the history of live-electronics *considering the historical development of the technologies available in each period.* Those technologies have marked each period with their own particular sonorities.

S. Ystad et al. (Eds.): CMMR/ICAD 2009, LNCS 5954, pp. 438–465, 2010.

Following this perspective we may identify three main periods, which are not meant however to be inclusive or exclusive; their purpose is simply to give a general view of how interactive music has been changing in the past fifty years, with even contrasting aesthetics emerging in each of them. Further, they allow us to see how automation of real-time processes has evolved since the 1960s. The case that *interactivity* has not always been produced by computers but rather by several (and different) other devices, contradicts to some degree the definition at the EARS website, where it is described exclusively as a computer interaction.[1] In the early times, the implementation of interactive synthesis processes was operated by analogue equipment, and, in most of the cases, manually. Automation in all these cases was therefore mostly impossible. These facts are relevant to the type of sound, the different concert situations and the impact that music had in its own time, with regard to both performance-related arrangements in concert halls and perception by the audience. Different technologies have also created particular (and up to some extend, I dare to say, classical) sonorities, which are typical of the period in which they were created and/or implemented. The following section discusses this more in depth, with examples of technologies, compositions and the role automation played in each period.

2 Brief Summary of the Historical Periods

The periods mentioned above run respectively from ca. 1960 to ca. 1970 in the case of the first one, the second, from 1970 to around 1990 and the third from ca. 1990 to the present times. The first period includes only analogue equipment; the second sees the transition between pure-analogue interaction and the slow but unavoidable incorporation of digital techniques, including the first generations of personal computers; finally, the third period includes mainly computers, alongside with other digital devices. We will observe through these periods how the usage of automation evolved, as digital devices were slowly introduced and further developed, whilst analogue devices were slowly abandoned and replaced by digital ones.

Among the several devices which marked the introduction of digital techniques and technologies in the second period, the following are worth mentioning: Dean Wallraff's *DMX-1000* (late 1970s), the *4X*, developed at IRCAM in the early 1980s and the *UPIC*,[2] devised by Iannis Xenakis and developed at the Centre d'Etudes de Mathématique et Automatique Musicales (CEMAMu) in Paris and completed in 1977 [4]. These devices allowed for the creation and/or implementation of new and/or

[1] 'Interactivity refers broadly to human-computer musical interaction, or human-human musical interaction that is mediated through a computer, or possibly a series of networked computers that are also interacting with each other. Often, an interactive performance, composition or improvisation will involve the creation or programming of software that will respond to pre-determined aspects of a live performance. The software will then determine other aspects of the music, either generating synthesised sound, or modifying in some way all or some of the *live* sound.' Direct quote from [2].

[2] Unité Polyagogique Informatique du CEMAMu [3].

existing sound synthesis procedures. Granular synthesis[3] is a good example of this: Barry Truax's piece *Riverrun*, composed in 1986, was solely based on granular synthesis (possibly the first piece doing so to such an extent) using the *DMX-1000* controlled by the *PODX* system, an interactive compositional software environment developed at Simon Fraser University (Canada) around this time by Truax himself [6].

The third period consolidates the developments of the former two, adding more digital technologies (which became cheaper and smaller every year since then) and procedures to the palette of real-time sound treatment and/or synthesis.

2.1 Analogue Live-Electronics: ca. 1960-1970

During this time, scarcely any digital equipment was available.[4] Even though some electrical devices such as the Theremin (which could be truly considered a real-time sound processing device) were already in use several years before 1960, it was not until the middle of the 1960s, that Karlheinz Stockhausen, (Germany, 1928-2007) began to experiment, compose (and even write detailed scores for the electronic part in his pieces), that we can truly refer to live-electronics as previously defined. The 1960s saw not only a rapid development of analogue equipment (arguably the most significant may be the Moog synthesizer, introduced in 1964), but also plenty of experimentation with a rich palette of sound synthesis methods. During this period, Stockhausen's pieces including live-electronics used invariably filters and different types of modulation, mostly ring modulation (RM).

Some key works by Stockhausen in the 1960s are:

1964: *Mikrophonie I* (Tam-tam, 2x microphones & 2x filters)
1964: *Mixtur* (5 orchestral groups, 4x sine generators & 4x modulators)
1965: *Mikrophonie II* (Choir [12x singers], Hammond & 4x modulators)
1970: *Mantra* (2x pianos, RM, sine generators)

From those compositions, it is arguably *Mixtur*, due to the time it was composed,[5] the most challenging of the four. The work exists mainly in two versions:[6] the first, for five orchestral groups and a later version from 1967, for a smaller ensemble (maximal 37 players) instead of the five orchestral groups. In it, the usage of live interaction

[3] The concept of granular synthesis was first introduced by Gabor in 1947 and later by Xenakis (1971), but at those times, the concept was still more theory rather than a real possibility [5].

[4] Even though computers did exist since some twenty years before, it was not until the mid-70s, that the microchip technology (which was successfully implemented already in 1958) with the *microprocessor* was implemented to create the *microcomputer* (or PC, for *personal computer)*. Most of the development in real-time processing was not possible before microprocessors were introduced in computers and related devices [7, 8].

[5] Even though composed in July/August 1964 and therefore the first of these pieces to be fully composed, Stockhausen situated in his work catalogue *Mikrophonie I* before it, only because experimentation on the Tam-Tams for *Mikrophonie I* began as early as 1961; it also premiered in December 1964, while *Mixtur* had its premiere on November 9th, 1965 (Hamburg).

[6] There is a third revision, from 2003. Astonishingly, the revision is not about the electronics, but rewriting the score, to eliminate the indeterminacy (aleatoric) that the piece had in its two first versions. Therefore, the instrumentation is completely fixed, where no choices need to be taken by the musicians in the orchestra.

between instruments and electronics opened back in 1964 a new world of new sonorities, setting the ground for all interactive music yet to come. In *Mixtur*, the interaction is set between four sine generators, which ring-modulate four of the instrumental groups, while the fifth group is only amplified. The five orchestral groups (which, as defined by Stockhausen, can have in the first version, any number of players within them) are divided in: wood instruments, brass, strings (*arco*), strings (*pizzicato*) and percussion. From them, the first four are used for the ring modulation with the sine generators; the percussion instruments however, are the ones only amplified. For the smaller 1967 version, Stockhausen required 22x microphones for the main 4x groups, 2x contact microphones for the double basses plus 6x more for the percussion, 7x mixing desks on the stage plus a bigger one in the concert hall, 7x loudspeakers in the front and 2 x 2 loudspeakers at the back. These latter receive the signals from the modulators (he required the signal to be rather soft and to be evenly distributed from right to left, to cover the whole stereo image); the former seven speakers receive the amplified direct signal from the instruments' microphones. The modulations to be heard at the back were carefully planned and written down in the score, for which the composer calculated precisely the spectrum results (differential tones) produced by the ring modulation for each note of each particular instrument. Moreover, the players in charge of the four sine generators have their parts written down: they have to play mostly glissandi at different speeds, which must reach certain frequencies at exact points for the modulations to sound as planned. *Mantra*, composed six years later for two pianos and electronics, still worked with the same principle of ring modulation, added to short wave sounds for one of the pianos. These latter however, do not interact with the pianos; they must be played similarly to tape music (actually, for some performances, a recording of short waves is played back).

Another development worth mentioning during this period is the *Groove machine* by Max Matthews, which dates from the late 1960s; it primarily consisted of controlling analogue equipment via algorithms. However, it was not commercially available, as it existed only in laboratory environment [4].

Between the 1960s and mostly, in the 1970s (and beyond), there have been several DIY ("do it yourself") synthesisers, such as the *Elektor Formant*, which would make a long list to mention in this article, and therefore, it should suffice here only to acknowledge their existence.

It should be rather clear by now, that automation (not to mention completely automated pieces) was not actually possible during this period, and that the results from analogue equipment for these and other pieces depended almost exclusively on manual steering of the devices during the performance. This fact slowly changed in the next period, with the introduction of digital technologies and the first hybrid (digital/analogue) devices, although reasonable and general automation was not possible until the late second half of the 1980s and, in fact, not successful before the second half of the 1990s.

2.2 Analogue and Digital Live-Electronics: ca. 1970-1990

During this rather long period, the path from pure analogue equipment to digital devices occurs. A crucial event in this period is the creation in 1974 (May 14th) of the *Institut de Recherche er Coodination Acoustique-Musique* (IRCAM) in Paris. It was a project initiated by Pierre Boulez (France, 1925) back in 1972, which became since its creation one of the most important and prestigious centres for electroacoustic music in the world.

As mentioned above, the first *Digital Signal Processors* (DSP) also appeared (such as the *DMX-1000* or the *UPIC*) during this period, mostly by the end of the 1970s. Even though these and other systems -such as the *4X* at IRCAM- were in a position to generate different sound synthesis procedures, these were generally not performed in real time (mostly due to the slow reaction of computer chips at that time). This fact made them more suitable for the production of overwhelmingly tape pieces in studio. The piece *Riverrun* by Barry Truax is an excellent example (see above, section 2). Other important devices developed during this period were: the *Emulator,* with its versions *I* (1981), *II* (1984) and the *E-III* (1988),[7] by *E-mu Systems*; the *Fairlight* synthesisers/samplers;[8] the *Yamaha DX7* synthesisers (1983, using Chowning FM) and following models; the *Synclavier* with its different versions *Synclavier Digital Synth* (1977),[9] the *Synclavier II*[10] (1980), the *PS* model series and, at the end of the 1980s, the *Synclavier 9600 Tapeless studio*, the last model to come from the New England Digital company before it went out of business in 1991. Even though none of these devices were neither created nor used for real-time interaction, they were important, because, being digital (or hybrids), they marked a tendency towards a digital predominance in the market. From about 1975 onwards, there were some systems developed, which controlled analogue synthesisers via computers in real-time; some were programmed using the *BASIC* programming language. An example is the *PDP-11* at SYCOM (University of South Florida, USA), which needed paper tape for storage and input via teletype [4]. Between 1975 and 1978, the firm *Hofschneider* (Berlin) in collaboration with the Folkwang Hochschule Essen (Germany) produced the *SYNLAB*, a modular analogue synthesizer based on the design of the *ARP 2500*,[11] which could be controlled later also via computers, therefore becoming a hybrid device; the computers used across the years were: a *Tektronix 4051* Mini-Computer; then an *Apple II Plus* and finally an *APS 64/40*, built

[7] The *E-III* included a sequencer. With the *Emulator I,* breakthrough to digital sampling was achieved.

[8] *Fairlight* was founded in 1975. The first product was the *Fairlight Qasar M8*, based on the *Qasar M8* by Tony Furse (Creative Strategies). The famous *CMI* -Computer Musical Instrument- was one of the first integrated synthesiser, sampler and sequencer. The *CMI Series I* appeared on the market in 1979 and was a digital sampler and synthesiser (additive synthesis). The *CMI Series II*, from 1982, added optional MIDI. Series III appeared in 1985, enhancing (among other improvements) the number of channels, sampling rate and bit resolution form 8 to 16 bits, but basically the same chip architecture as the former model. Several later models were based on the *CMI III*, such as *Waveform Supervisor* (1988) and the *MFX1* (1990). This development continued in the 1990s, with the *MFX3.48* appearing in 2000 [9].

[9] 'Originally developed as the "Dartmouth Digital Synthesizer" by Dartmouth College professor Jon Appleton, in association with NED founders Cameron W. Jones and Sydney A. Alonso, - and subsequently under the marketing guidance of Brad Naples the Synclavier was one of the first synthesizers to completely integrate digital technology.' Direct quote from [10].

[10] The *Synclavier II* had a MIDI option, a sampling option and a stereo FM option.

[11] The *ARP 2500*, a monophonic, analogue, modular synthesiser with a set of matrix switches to connect both modules and patch cords, appeared in 1970 through the mid-70s [11].

by the company *MicroControl* (Essen). There were only two models of the *SYNLAB* built, a small one in the Technische Universität Berlin and the biggest, at the Folkwang Hochschule Essen, which counted in total 80 modules [12]. The modules in the *SYNLAB* included envelope following, tone generators, noise and random generators, sample and hold units, FM and AM, apart from diverse filters. While studying at ICEM (Folkwang Hochschule Essen), I composed in 1992-93 an acousmatic piece called *pizz.* entirely using this device.

Whilst all these developments took place, Stockhausen concentrated by the end of the 1970s in his major project *LICHT* and rather abandoned the type of live-electronics he initiated in the 1960s in favour of the usage of synthesisers, tapes, live mixing and transmitters. Other composers, such as Luigi Nono (Italy, 1924-1990), took the inverse path, composing pieces for tape in the 1960s and early 1970s and turning to live-electronics from the end of the 1970s. Nono, together with other composers such as Boulez and Iannis Xenakis (Greece 1922-2001) began to experiment with interaction, particularly spatialisation. Boulez distanced himself from tape music accompanying instruments, because (as he profusely stated in several interviews)[12] he was not satisfied with the results, proving for him to lack the flexibility he demanded. After having experimented back in 1958 with tape music and instruments with *Poésie pour pouvoir* (orchestra/tape), he began work on a composition in 1971, which since then (with or without electronics) has seen at least five different versions: *'...explosante, fixe...'*. In its original conception[13] it was a schematic, aleatoric work in one page (without even the indication of the instruments involved) in seven musical fragments, which were labelled *Originel* and *Transitoires II-VII*, with an original seven-note cell (the basic musical material for the entire composition) and single line in *Originel* and introducing more instruments and musical materials in the following *Transitoires*. In 1973/4, Boulez presented a revised version for solo flute, seven instruments[14] and electronics, which was performed in New York in 1974. The electronics were in charge of spatialising the sound of the instruments using an analogue device created by Hans-Peter Haller and Peter Lawo at the Experimental Studio of the Heinrich Strobel Foundation of the SWR in Freiburg, Germany in the early 1970s: the *Halaphone*. Even though Boulez abandoned this version of the piece soon after -as he was not satisfied with how the *Halaphone* worked in those early times-, after several versions of the piece,[15] the version that eventually satisfied his intentions, was composed at IRCAM in the early 1990s, thus, in the next period proposed in this article. During this second period of hybrid analogue-and-digital electronics, however, Boulez composed one of his finest pieces, *Répons,* which was performed in 1981 and 1982 in Donaueschingen, Paris and London. The electronics were the result of a collaborative work between IRCAM and the Experimental Studio (Freiburg), by still using the *Halaphone* as a sound distributor, this time attached to a small computer. Vogt states, that with the overwhelming reception of *Répons* in those years, both IRCAM and the Experimental

[12] Such as Josef Häusler's interview with Boulez about *Répons* in 1985 [14].

[13] The basic note is E flat (S in German, also the first letter in the surname of Igor Stravinsky), as he began the project as short musical tribute to Stravinsky after his death in 1971.

[14] The seven instruments are: clarinet, trumpet, harp, vibraphone, violin, viola and cello.

[15] Among them, one version is for vibraphone and electronics (1986) and another, purely instrumental, called *'Mémoriale ("...explosante-fixe..." originel)'* (1985) for flute and octet.

Studio finally legitimated themselves [13]. By this time, the *Halaphone* was ten years old and better developed, so it indeed was in a position to overcome the challenges posed by Boulez's inventiveness, which was not the case back in 1974 when '*...explosante, fixe...*' was premiered. *Répons* consists of six instruments (two pianos -No 2 also synthesiser-, harp, vibraphone, cimbalom xylophone/glockenspiel) and a big ensemble and it grew in the three versions between 1980 and 1984 from 17 to 45 minutes of duration. The title refers to the responsorial style typical of the Gregorian chant, in which there is an alternation between soloist and choir. Thus, the six soloists are positioned at the periphery of the concert hall whereas the instrumental ensemble is placed in the centre, meaning that the audience and the six loudspeakers surround the ensemble while the six soloists surround the audience. The live-electronics apply only to the six soloists. Boulez revised the piece further in the 1980s with the newest developments at IRCAM available at the time and the help of Andrew Gerzso: the last version of *Répons* was then performed using the *4X* and the *Matrix 32*.[16] The software developed for the spatialisation was called the *Spatialisateur* [15]. The *4X* was a computer firstly designed by Giuseppe Di Giugno in 1980 with its final version, released in 1984 by the French firm *SOGITEC* [15]. It had eight processors boards, which could be independently programmed to transform, analyse or synthesise sound. The *Spatialisateur* was a computer programme, which, like the *Halaphone,* was able to distribute and move signals across several speakers (six, in the case of *Répons*) at variable speeds (both 'accelerandi' and delays) in different densities and registers, modifying the original sounds. The variable speeds produced specific spatial effects, which created a virtual space [16]. The *Spatialisateur* on the *4X* worked using an envelope follower, which read incoming audio input; the results of the envelope were then applied to generate a signal, which changed its frequency with the variations in amplitude of the waveform envelope. Through the so-called flip-flop units (FFUs), outputs to different speakers were allowed or neglected, as FFUs can only be on one at each time. The frequency of the signal determined how long a given FFUs would remain on, for example, with a higher frequency, the time will be shorter, and so on. A very short overlap among the speakers was required, in order to compensate the time that the amplitude takes from its maximum level back to zero [15].

The *4X* remained rather unchallenged, despite its (for nowadays perception) lack of power to produce more complex synthesis processes in real-time. In the mid-1980s, Robert Rowe began to develop the *4xy* programming to control the *4X*. Later, Miller Puckette began with an early version of *MAX* (without graphical interface). The *Macintosh* version of *MAX* was developed further to enable control on the *4X* with a graphical-interface. [17]

At the same time Boulez and Gerzso were working with the *Spatialisateur*, Nono kept working with the *Halaphone* until his death in 1990. In his pieces, Nono worked together with Hans-Peter Haller for the implementation of the live processes, mainly at the Heinrich Strobel Research Institute in Freiburg (Germany). All pieces composed by Nono in the 1980s using live-electronics concentrate in the electronic transformation of the instrumental sounds, in some way, quasi liquefying them in the space by the usage of spatialisation and delay lines. Nono's main characteristics of his

[16] This unit was in charge of the distribution of sound from the microphones or the *4X* and to the loudspeakers.

mature style after completing his string quartet *Fragmente-Stille, An Diotima* in 1980, were the profuse usage of microtonalism, added to a tendency in the instrumentation for extreme low or high frequencies (such as bass flute, piccolo, tuba, etc) and for a favouritism for fragmented sections; these sections are often separated by pauses of different lengths, creating a transition that operates on the limits of perception.[17] An example of this style is the piece *Omaggio a György Kurtág* (alto, flute, clarinet, bass tuba & live electronics - 1983-1986), divided in fourteen sections of different lengths, each separated by long pauses, the longest lasting one minute for a total duration of the piece of ca. eighteen minutes. The most compelling piece from that period is however *Prometeo Tragedia dell'ascolto* given the big scale of all forces involved in its performance. The characteristics explained above are richly amplified, also by the use of big spaces (the premiere took place at the church of San Lorenzo in Venice, Italy in September 1984).

Before finishing with this period, it is worth also dedicating a few words to the *UPIC* system, conceived by Xenakis in 1977, which combines synthesis methods with a graphical user interface for compositional purposes. The performance device however was an electromagnetic pen to draw different sound parameters on a large, high-resolution graphics tablet [18]. In 1978, Xenakis composed a piece with the *UPIC* called *Mycenae-Alpha* for mono tape, to be projected onto either two or four sound sources around the audience. It was part of a bigger work by the composer, *Polytope of Mycenae,* a spectacle of light, movement and music, which took place at the Acropolis of Mycenae. The real interaction at this early stage was that the system reacted in real-time to the graphics drawn on the tablet. However, for any other interaction, the computer chip was too slow. Despite this, the next period, from 1990, saw the *UPIC* develop and extent its potential to allow for real-time interaction.

Summarizing, the most common interactive processes used in this period were *delays, reverberation, transpositions or frequency shifting*[18] and *spatialisation*, which can be found in the following key works:

Pierre Boulez:[19]

1973-74: *"...explosante-fixe..."* (version for solo flute, clarinet, trumpet, harp, vibraphone, violin, viola, cello and electronics)

1980: *Répons* (two pianos, harp, vibraphone, xylophone, cimbalom, ensemble and live-electronics). Revised and expanded in 1982 and further, in 1984.

Luigi Nono:[20]

[17] Indications in his scores such as ***'pppppp'*** are rather frequent in works from this period.

[18] They are not to be mistaken with pitch shifting in a phase vocoder, which uses real-time FFT (or DCT). Pitch shifting works with the transposition of the fundamental and the partials, keeping the ratio between them, while frequency shifting does not maintain this relationship, creating a different spectrum by raising all harmonics by the same amount of Hz. See Boulez and Gerzso for more details [15].

[19] Even though Boulez's *Dialogue de l'ombre double* for clarinet, piano resonance and tape (1984-5) works with similar spatialisation principles as *"...explosante-fixe..."* and *Répons*, the electronic part in 'Dialogue' is a prerecorded tape.

[20] Nono's piece *Con Luigi Dallapiccola* for '6 esecutori di percussione e live electronics', from 1979, is atypical for this period, as the technologies involved are stereo sound and three ring modulators with three sine generators, which modulate the percussion instruments in a similar way as Stockhausen did for *Mixtur* and *Mantra* in the former period.

1980-83: *Das atmende Klarsein* (bass flute, small choir, live-electronics)
1983: *guai ai gelidi mostri* (ensemble, live-electronics)
1983-86: *Omaggio a György Kurtág* (alto, flute, clarinet, bass tuba, live- electronics)
1981-1985: *Prometeo. Tragedia dell'ascolto* (solisti vocali e strumentali, coro misto, 4 gruppi strumentali e live electronics)
1985: *a Pierre, dell'azzurro silenzio, inquitum* (flute, clarinet, live-electronics)
1987: *Post-Prae-ludium n. 1 per Donau* (tuba, live-electronics)

In spite of the relatively big progress that can be observed from the analogue to the digital era of sound electronics during these two decades, the rather slow computer chips did not allow for much automation in real-time. All pieces listed above needed manual steering for most of the processes to occur. Only during the period described in the next section (2.3) it is possible to identify automation implemented as an effective tool for real-time interaction. The basis for this development was nevertheless set during this period, mostly with the research at IRCAM, which included the first versions of the *MAX* software.

2.3 Digital Live-Electronics: ca. 1990 – Present

The development of the digital techniques and devices might have begun in the 1980s, but it was not until the 1990s, that almost every kind of real-time processing could be achieved. The types of processes increased, whereas some new were discovered or, even though existent, they could only be properly applied for interaction purposes during this period. Synthesis processes depending on Fast Fourier Transform (FFT) such as *convolution, granulation, pitch shifting,* and *time stretching* became at last not only feasible, but certainly also common practice.

At the start of the 1990s, computers alone were not in the position to cope with the power needed for real-time processing. Hence, additional hardware was needed, such as the *ISPW (*IRCAM Signal Processing Workstation*)*, the *MARS*[21] system, the *Kyma* or the *AUDIACSystem.*

The *ISPW* development began in 1989 at IRCAM as a replacement to the *4X*, which by that time, has served unchallenged over ten years. One of the problems that the *ISPW* solved was that of transportability, as the *4X* was too big to be moved. The introduction of the *ISPW* in 1990, designed by Eric Lindemann, allowed for more transportability and also sinking costs. The system ran the *MAX* software (which included by now DSP objects, programmed by Miller Puckette) on the *NeXT* computer and the separate DSP unit was in charge of all real-time processes [17].

Similar systems were the *MARS* workstation (designed by Di Giugno in Italy, developed almost simultaneously in time as the *ISPW* [17]) and the *AUDIACsystem.* The *MARS* workstation was a

> 'programmable specialized digital machine for real time audio applications which has been entirely developed by the Italian Bontempi-Farfisa research institute IRIS. MARS has been conceived as an integrated environment in which a graphical user interface, an embedded real-time operating system and two IRIS digital audio processors are linked

[21] Musical Audio Research Station.

together to create a flexible and an interactive workstation for audio research, musical production and computer music pedagogy.' [19]

At the beginning, it ran on an *Atari* computer, but by 1997 was ported to *Windows*-PC platforms. The *AUDIACsystem* was also a hardware-and-software environment for real-time interaction. It was a collaborative research project by the ICEM (Institut für Computermusik und Elektronische Medien) and the company Micro-Control GmbH & Co KG, which took place at the Folkwang Hochschule-Essen (Germany). The people responsible for its entire design were: Dr. Helmut Zander, Dipl. Ing. Gerhard Kümmel, Prof. Dirk Reith, Markus Lepper and Thomas Neuhaus. The project began in 1987 and involved not only the hardware architecture, but also the software exclusively created for this particular environment called *APOS*. The hardware architecture of the *AUDIACSystem* used the principle of the specialised subsystems, not only to generate organised forms for musical production, but also to incorporate the generation and transformation of sounds in real-time. This implied at that time a huge demand in relation to its computing potential, which could only be solved with the subsystems mentioned [20]. The DSP unit's size was about one cubic meter, which allowed for a rather comfortable transportation and could be set anywhere.

The *Kyma* system, which still exists with its latest version, the *Kyma X* and its sound unit, the *Pacarana* (a DSP module in charge of the entire sound processing) was a development which began already in 1984; that year Kurt Hebel and Lippold Haken (University of Illinois, USA) designed the *Platypus*, a digital signal processor for real-time audio, whilst Carla Scaletti wrote the software, which she called *Kyma*. The next steps were taken in 1987, when *Apple* gave a grant to Scaletti to develop a graphical interface and, in 1989 Hebel replaced the *Platypus* by the *Capybara*, which stayed in the market for more than ten years. In recent years, the *Capybara* was finally replaced by the *Pacarana*. The software ran almost since the beginning on both *Windows* and *Apple* platforms [21].

This period is perhaps the richest in different possibilities of generating interactive DSP. The reason for this is that computer technology after the second half of the 1990s, with the introduction of the first *Intel Pentium* and *Motorola/IBM PowerPC* RISC processors, increased enormously not only in the speed of their clocks from MHz to GHz, but also in memory size and the slow but constant introduction of faster and broader internal busses. Another event was the introduction in 2001 of the *Unix* BSD based *MacOS X* operating system by *Apple*,[22] which enabled more compatibility with other systems based on *Unix*. Added to these advances in technology, equipment became not only smaller, but also much more affordable, so that not only studios were in a position to use these technologies, but almost every musician owing a computer or a laptop could too.

[22] Apple bought in 1995 *NeXT* computers, as Steve Jobs came back to Apple as CEO; the idea behind this move was to ensure that Apple had the licences on *NeXTStep*, the *Unix* BSD operating system on all *NeXT* computers (NeXT was a company also founded by Jobs). At that time, *NeXTStep* was supported by Motorola's CISC (68030 and 68040) and all *Intel* processors, but not by the *PowerPC* RISC platform of IBM/Motorola. It took Apple more than five years to adapt the *NeXTStep*, and it was only in March 2001, when the *MacOs X* Version 10.0 appeared on the market for desktop computers (a server version was available since 1999) [22, 23].

With the appearance in 1997 of the G3 processors on *Apple* desktop computers and laptops, the additional hardware began to be less indispensable and since then, most live-electronics could be played directly from a laptop or computer (with or without the inclusion of an audio interface). The most common platforms nowadays are *MAX/MSP*,[23] *PD*,[24] *Supercollider*,[25] and the *Kyma X*.

Apart from these, some systems from the former period developed further, while others simply disappeared or were replaced mostly by software programmes; the *UPIC* is among those that were further developed. By 1991 a new 64-oscillator synthesis real-time engine and the coupling to a *Windows* PC, permitted a much more sophisticated graphical interaction [18].

A system which was absorbed by another one was the *Spatialisateur*. The IRCAM software development programmed a piece of software, which was incorporated to the *MAX* environment as a typical *MAX* object (*spat~*) for both *MacOS X* and *Windows XP* platforms. *The Spatialisateur Project* began in 1991 as a collaborative project between IRCAM and Espaces Nouveaux, collecting and developing research carried out in IRCAM's room-acoustics laboratory. It integrates 3D stereo reproduction modes for headphones (binaural), 2/4 loudspeakers (transaural), Vector-Based Amplitude Panning and Ambisonics [27].

A different development can be found in the work by Luciano Berio (Italy, 1924-2003) and the center Tempo Reale.[26] With only three compositions using live-electronics, Berio was not as prolific as he was composing pure instrumental music. However, what makes his work in the field of particular interest is the fact, that all three pieces were composed using the facilities and research at Tempo Reale. Among those features that most interested Berio in the field of live-electronics were those of physical movement of sound, such as trajectories followed by sound events through space; this resulted in new, unconventional acoustic spaces. Even though this was already a characteristic of Berio's instrumental music since the 1960s (for example, the inclusion of a third violins group at the back of the orchestra in his *Sinfonia* [1968]) or the two harps close to the audience in *Formazioni* [1987]), it is in his interactive music where we can find this concept fully developed. Added to space, there are other considerations such as transpositions, continuous modulation of harmonic and dynamic

[23] *MAX* (without the *MSP* part, which means 'Max Signal Processing' was developed also at IRCAM (Paris-France) from 1980 by Miller Puckette. It works with a set of objects than can be connected with each other. The first development for the *Macintosh* included only MIDI. From 1997 David Zicarelli used the *PD* audio part developed the year before by Miller Puckette and released *MSP*, as the DSP part for *MAX* [24]. In 2008, Zicarelli's company, Cycling74, joined the *MAX/MSP* package with *Jitter* (real-time video objects) in the latest version, *MAX 5*.

[24] By 1995, as the *ISPW* hardware was being let aside due to cost reasons; Miller Puckette (creator of *MAX*) began by then to develop PD (Pure Data), an improved version of the latest *MAX/ISPW* and mainly *MAX/FTS* [25].

[25] A free environment and programming language by James McCartney, which appeared in 1996 for real-time audio synthesis and algorithmic composition. In 2002 it was released under the terms of the GNU General Public License. The latest version so far (3.3), was released in April 2009 [26].

[26] TEMPO REALE, Center for Music Production, Research and Education was founded by Luciano Berio in 1987. Since then, the centre's main activity has been the production of Berio's works.

levels, and the profuse usage of sound layers, which were at the core of Berio's compositional intentions [28]. The three pieces by Berio developed at Tempo Reale since 1987 are significant because they all share the facts mentioned above. Berio's interactive pieces are: *Ofanìm* (1988–1997) for female voice, two children's choirs, two instrumental groups; *Outis*, (azione musicale, 1996); and *Altra voce* (1999) for mezzo-soprano and flute. All the electronic parts were totally revised in 1999 [28]. The main challenge of the first two is the big instrumental groups involved, which turn the electronics rather complicated, as besides the system itself, several microphones, mixers, and other equipment are required.

The system developed at Tempo Reale is based on a number of different characteristics, which all serve the primary goal of achieving the maximum degree of automation possible, without having to excessively constrain the freedom during the performance. However, the three pieces mentioned use in most cases the common system of working with cues, with a cue manager at the core of it, as described by Giomi, Meacci and Schwoon (p. 37):

> 'The Cue Manager is based on a structured sequence of events in which each block represents a precise point in the score (cue) along with all of the electronics events associated with it. During a live performance, the list is scrolled manually by means of a computer keyboard or through a MIDI peripheral. (In *Ofanim*, for example, the keyboard player in the orchestra controls the Cue Manager.)' [28]

This is indeed a rather high degree of automation, but triggering the electronic events is made nonetheless manually.

Since Berio's death, the research and development of new tools has been continued, and among the latest we find *MEEG* and several *MAX/MSP* externals (for *MAX 5*). *MMEG* (or *Max Electronic Event Generator*) is a system conceived for data management integration into *MAX/MSP* for live-electronics purposes; it offers an interface supporting the modification of parameters in real time.[27]

In the recent past, other approaches such as game consoles and also *Nintendo's Wii* remote-control interfaces have been used for the performance of electroacoustic music. In 2004, James Paul Sain (USA, 1959) composed a piece for flute and *MAX/MSP*,[28] in which a game console controls the live-electronics.

The amount of interactive compositions that could be mentioned in this period exceeds the frame of this article. Therefore, the pieces and composers listed below (enriched by the examples already mentioned in this section) intend only to be a guidance and example of core works. Because my piece *Gegensätze (gegenseitig)* for alto flute, quadraphonic tape and live-electronics (1994) was the first work entirely composed to be performed in a concert situation using the *AUDIACSystem*, I would like to add it to this list.[29] [20].

---

[27] 'Max/MSP communicates with a relational database implemented in MySQL using a PHP interface. The entry and editing of data can be performed from any web browser through an html form.' [29].

[28] *ball peen hammer*, for flute and computer (2004) uses, apart from *MAX/MSP*, the *STEIM's JunXion* software, the *Logitech* Dual Action USB game controller, and the Logitech Attack 3 USB joystick [30].

[29] The piece was produced after a rather long period of research and development of the *AUDIACsystem* at ICEM – Folkwang Hochschule Essen (Germany).

Pierre Boulez
1991-93: *…explosante-fixe…* for MIDI flute, two flutes, ensemble and live-electronics.[30]
1997: *Anthémes 2* for violin and live-electronics.[31]
Cort Lippe (USA, 1953)
1992: *Music for Clarinet and ISPW*
1993: *Music for Sextet and ISPW*
1994: *Music for Flute and ISPW*
1995: *Music for Contrabass and Computer*
1998: *Music for Hi-Hat and computer* (*MAX/MSP*)

As it can be observed from this list, Cort Lippe was very prolific composing several pieces for one or more instruments and the *ISPW* in the first half of the 1990s. In *Music for Clarinet and ISPW*, he used a 'score-follower' (an idea also pursued by Boulez for pieces such as '*…explosante, fixe…*'), programmed in *MAX*, so that the electronics could follow the gestures and notes played by the clarinet.

A great portion of the development since 1990 has seen so far the practice of ideas on hold in the former decades as much as the search for entirely new ideas and possibilities. On the one hand, we can observe this fact in, for example, objects for the *MAX* software, such as *spat~*, (which is basically a nifty and increased development of the basic idea that gave birth to the *Halaphone*). On the other hand, several others, such as the *FFTease* package by Eric Lyon and Christopher Penrose [31], work with FFT, which allow for sound synthesis processes based on spectral analysis, something that was impossible to achieve until this period. With this degree of evolution in technology, automation became not only possible, but also a real alternative for the performance of interactive music. A way of achieving automation was *score following*, already mentioned afore. Barry Vercoe and Roger Dannenberg were the first to present this idea almost simultaneously (and independently) during the International Computer Music Conference 1984 [32]. Both authors defined it as:

> '[T]he process of tracking live players as they play through a pre-determined score, usually for the purpose of providing an automatic computer accompaniment to a live player.' [32]

By 1987, IRCAM was using it too. The main way of using *score following* is by pitch detection. Problems surged though, as some instruments such as woodwinds needed and acoustic-based pitch detection, complicating the process. Puckette and Lippe (p. 182) explain how the main algorithms worked in the early 1990s:

> 'The procedure for score following is rather simple: enter a score into a computer in some form, and then play the score as real-time input to the computer (either via a MIDI interface or a microphone and acoustic analysis), comparing the computer's stored score to the musician's playing on a note by note basis. If the comparison being made between the two is successful, the computer advances in the database (the stored score) in parallel

[30] Electronics programmed at IRCAM by Andrew Gerzso. This is the fifth and so far final version of the piece, consisting of *Transitoire VII – Interstitiel 1, Transitoire VI – Interstitiel 2* and *Originel*, with a total duration of ca. 37 minutes.

[31] Electronics (also programmed at IRCAM by A. Gerzso) include pitch shift, spatialisation in six channels, harmonizer and score following.

with the player, triggering electronic events at precise points in the performance score. In theory, one or many parameters can be followed' [32].

The IRCAM website points out two purposes of score following:

'The interest of score following is usually twofold: (1) technically for realtime alignment of audio signals to symbolic music score, and (2) musically for triggering and managing electronic music scores that are interpreted conform to musician's play.' [33]

The method described by Puckette and Lippe above (the same Lippe used in his piece for clarinet and *ISPW*) was running on *MAX*, either on the *Macintosh* or the *ISPW* [34]. Even though pitch detection algorithms or devices have improved in the last years compared to those in the early 1990s, they are sometimes still not fully reliable, and score following remains today an option, but not the only one. Further examples are discussed in section 3.2.

Other ways of achieving automation is the usage of computer-controlled timelines (for example, on *MAX/MSP*) and the usage of time-code, such as SMPTE. Section 3 deepens on the latter.

As for automation (partial or complete), the main subject in the following sections, this was the period in which they could be finally fully achieved. This was possible only due to the big step achieved by computer companies in the development of their products, for example, speeding their chips and the capacities of their systems, while at the same time procuring smaller sizes for hardware components, all facts, which are still on going.

3 General Considerations for Programming/Composing Pieces Including Live-Electronics

In the present times, most of the systems that appeared at the beginning of the 1990s are out of use, so that the main environments for the performance of real-time electronics are *MAX/MSP, PD* and *Supercollider.* In spite of the *Kyma,* which still uses a separate unit for the DSP processes, the trend since the end of the 1990s is that separate audio-interfaces are used only for the purposes of a better overall output quality and multi-tracking, but the synthesis processes are performed inside the main computer by mostly the three software environments mentioned above and not on external devices.

Since 1998 I have been actively composing pieces for instruments and live-electronics using *MAX/MSP*; in most of them, I use complete automation of the real-time processes. This decision is explained later in section 4. The following three main considerations for the composition of interactive pieces are based on my own experience, as a consequence of both teaching about and composing for this particular combination.

3.1 Programming

Programming live-electronics is a constitutional part of the compositional process. How much concordance between those processes and the musical (notated or improvised) part can vary rather substantially depending on the composer and also on the compositional/improvisation goals set for the pieces. In order to achieve these goals, the composer has achieved a high degree of control over both musical and programming aspects of the entire piece. *Automation* (either complete or partial) can be a comfortable and effective option, as it is demonstrated in section 4.

3.2 Choice of Equipment and Devices

For the choice of equipment and devices required for an interactive composition, aspects such as the type of venue where a particular piece can be performed must be thoroughly considered. At least since the 1990s, only a minority of pieces have been composed for a special and determined space. A vast majority of the concerts including pieces with live-electronics take place generally in academic circles; moreover, the pieces performed in a single concert could vary enormously regarding their instrumentation, type and number of devices involved, microphones, etc. Thus, the planning of each composition can be summarised under: *'keep it as simple as possible'*. This 'precept' implies:

(a) *A careful planning of the main platform/equipment*
As a general rule, the simplest way of setting up the equipment is generally also the most effective. Platforms that only work in fixed places present a big disadvantage, as they do not generally allow the performance of those specific pieces outside the context in which they were created. In this regard, using *MAX/MSP* (*Apple* or *Windows*), *PD* (multiplatform) or even *Supercollider* allows for rather unproblematic performances of interactive pieces in almost any concert hall. Using these environments, it is possible to transport the algorithms (patches) to any hardware (generally computers with or without audio interfaces), regardless of the hardware on which they were originally created. Automation is possible with all these platforms. This also impacts on the necessity or not of requiring the composer's compulsory attendance during rehearsals and performances, because the required files can be easily transported via a network or saved onto storage devices and installed in other computers without any difficulty.

(b) *Controllers and switches used by the performer on the stage to trigger the electronics vs. automation*
A very careful planning is required for this issue, as there is a direct relationship between the number of the extra activities required by some live-electronics processes to be initiated or triggered by the player and the increasing amount of added technical difficulties, which in many cases can be detrimental to the concentration of the performer and also of the audience (something that can have a big impact on the perception of the piece). Sometimes performers must wear different devices in their clothing, or activate several pedals or switches, which in some cases, must be triggered very fast and even with special movements, all what may contradict the intention of the music dramatic, apart from -in some cases, unnecessarily- additional difficulties for the performer. In other cases, the switches are activated by a second person (usually the composer), via controllers, a mouse, etc. This latter case has a positive effect on the concentration of both performers and audience. However, if the person in charge of that particular activity is not the composer, it will be in my

experience very difficult to find someone who will spend the necessary time rehearsing and learning how to activate the programmed live-electronics in an impeccable timing (essential for the synchronisation of the live interaction). If not the composer, this person should be a trained musician, capable of reading music and follow accurately the score, apart from knowing how to activate every single step of the live-electronics. This said, some pieces allow however for more freedom than others regarding the processes involved and the way they are musically used in a piece, so that the role of this second person can substantially vary and in some cases be of rather lesser importance than in others.

Hans Tutschku (1966, Germany), in his piece *Zellen-Linien* for piano and live-electronics (2007) makes usage of a MIDI pedal to trigger all events (32 events in total). The score of the piece[32] presents extremely carefully detailed indications about how to set up the software, which include a well prepared sound-check for the two existing versions (eight or six channels). Even though programmed completely on *MAX/MSP*, the patch itself is a stand-alone application (*Zellen-Linien.app*).[33] This is indeed a nifty solution, which provides for a comfortable transportation and installation of the software, as even the sound files required for the performance are contained in the download package. However, the set up requires loading the sound files manually. Each event is in itself rather fully automated, but their triggering is not: they are activated each time by the pianist via a MIDI pedal in a similar manner as Tempo Reale's *Cue Manager* described in section 2.3. It may be, that one of the reasons that this piece is not completely automated is that the sound files loaded to the memory (buffers) need to be tuned before the performance according to the actual tuning of the piano in the concert hall (although, I suppose there might be ways of getting around this problem). The system selected here is however very elegant: the score has two rows, the first (divided in fours systems, due to the extreme registers used in the piece) for the piano, and the second one for the activation of the pedal, written in actual beats. Figure 1 shows the first four bars of this piece, with indications of when the MIDI pedal needs activation for the events.

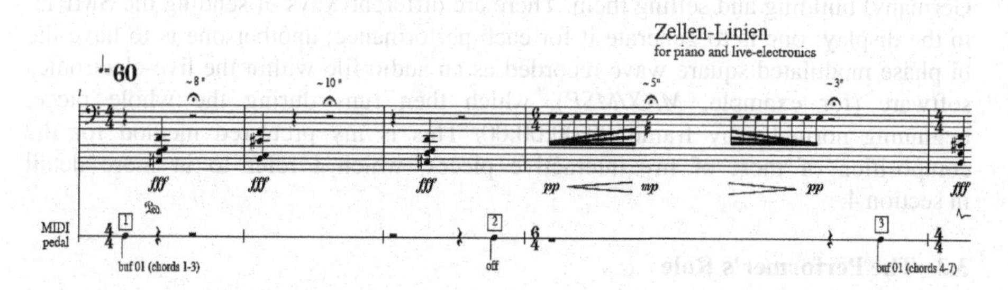

Fig. 1. Excerpt from the first four bars of the piece *Zellen-Linien* (2007) by Hans Tutschku[34]

[32] The score can be found at: http://www.tutschku.com/download/Zellen-Linien-score.pdf
[33] The application can be downloaded in three different versions (*Apple PPC, Apple Intel and Windows*) at: http://www.tutschku.com/content/works-Zellen-Linien.en.php
[34] Copyright with the composer. Printed with permission.

The player is not required a heavy load of extra work by switching the MIDI pedal. Nonetheless, as stated before, it is an extra activity, which in the end may impact on the performer's concentration and overall musical results, depending on many factors such as personality, experience, technology available, and so on. However, if specific processes are required at a specific times, in more complicated situations than the one just described in Tutschku's piece, manual steering of the system will not always produce the intended results. In those cases, *automation* is a very valid alternative for all those cases because its effects are twofold: it can set the performer as free as possible from any extra-musical activity *and* there is no need to have the composer or somebody versed in music to trigger the live-electronics (with the beneficial effect of allowing more frequent performances of pieces too). In this way the performer can concentrate solely on the interaction with the electronics and on the music itself, without the need of having to activate anything else.

As it was mentioned in section 2, one way of achieving this is using different types of *score-following* algorithms, which actually follow the pitches (or other parameters) played live; this is normally the case of pieces having a fully written score and therefore, it is not frequent in improvised pieces. Added to the former examples of score following are Noel Zahler's *Concerto for clarinet, chamber orchestra and interactive computer* (2003) composed under the same principle, even though using completely different algorithms than the ones applied by Lippe. In this case, the computer uses a -by that time- new score-following algorithm, which was created by Zahler, Ozgur Izmirli and Rob Seward; the algorithm was especially programmed for this composition using *MAX/MSP* [35]. In the past years, several composers have been using IRCAM's software *FTM* on *MAX/MSP*, which is basically a real-time object system and a set of optimized services to be used within *MAX/MSP* externals. These include score-following as well. A good example of this is *Etudes for listeners* by Miroslav Spasov (Macedonia, 1964), for piano and live-electronics.

Another way of achieving synchronisation is the usage of time-code (TC). TC such as SMPTE shown on a display on the stage makes a rather simple but extremely effective solution in the case of fully automated live-electronics. SMPTE displays are not easy to find though, but there are nonetheless several companies (most of them in Germany) building and selling them. There are different ways of sending the SMPTE to the display: one is to generate it for each performance; another one is to have the bi-phase modulated square wave recorded as an audio file within the live-electronics software (for example, *MAX/MSP*), which then runs during the whole piece, beginning normally by frame 00:00:00:00. This is my preferred method for the composition of most of my interactive pieces, which I refer to in more detail in section 4.

3.3 The Performer's Role

It is essential for performers to have a clear idea about their own role throughout a piece of music. Especially in the case of interactive pieces, this is primordial. A confused interpreter or a player who does not know exactly where the main aspects of a piece of music lie is not in a position of playing the piece in an adequate manner; as a consequence, the entire sound universe of that particular piece (including its dramaturgy) may not be projected to the audience as intended. The interaction with tape (i.e. the combination of a fixed medium with a live performer) has already been rather confusing for many interpreters since its introduction, mostly for those not acquainted with electroacoustic music. In the case of the interactivity with real-time DSP processes, the situation can be yet more challenging, mostly due to the variety of software, hardware and other devices that each particular piece may use. The fact that some processes cannot be practiced until shortly before the concert or rehearsal, demands a fair amount of concentration and energy from performers, for them to be up to the challenge. Also the acquaintance with typical electroacoustic sounds (for example, granular synthesis) plays an essential role.

In order to make sure, that the performer is supplied with all the information required, composers must provide for a fully edited score (or similar, such as for example, precise graphics -or both-), which should include the adequate level of guidance, giving a clear picture of what to do at each particular moment and how to interact sonically with the live-electronics. Concert organisers should also provide for good monitor systems, so that performers can listen to and provide for a good balance of their sound with the results coming out of the loudspeakers during both rehearsals and concert.

Interactive music is essentially music to be played live in concert. *The very core of interactivity lies in the live performance.* Thus, composers and concert halls must provide for the best possible framework. As it was mentioned above, the inclusion of pedals, switchers and similar devices do not allow in several cases for a complete concentration on pure musical aspects, which sometimes can be detrimental to the final artistic result of the pieces. At this point is where the inclusion of different degrees of automation can prove vital for the success or not of a piece of music.[35] Automation (and particularly complete automation) is the main subject of the next section, with examples taken from two of my most recent own compositional production.

4 Examples of Automation in Two of My Own Compositions

If automation at some degree is the choice to achieve a performance in which some of the topics already mentioned have been considered, the composer/programmer normally has two basic decisions to make: firstly, the degree of how much of the live-electronics' part should use automation added to how the performer can follow and understand it as simply as possible; secondly, the composer must weight the musical and technical advantages and disadvantages that automation implies for the intention (dramaturgic content) of the piece and its performance.

In order to give a practical description of the former statements, it will be attempted now to show them in two pieces of my own authorship: *Intersections (memories)*, for clarinet and live-electronics in 5.1 surround sound [*MAX/MSP*] (2007) and *farb–laut E-VIOLET* for viola and *MAX/MSP* in 5.1 surround sound (2008).

In all my interactive pieces, programming the electronics means that both music and electronics are essential parts of one-final product, the composition itself. This

[35] Success is meant here not only among different audiences, but also among musicians willing to play the piece several times.

implies a detailed and exact programming of each section of the piece, so that for each of them, both the desired musical and the real-time processes concur in the *desired aesthetical and dramaturgical effect.*

4.1 Description of the Works

Intersections (memories). This piece for clarinet and live-electronics with a 5.1 surround spatialisation was composed based on a hidden story regarding real facts of my personal life. It is not intended however, that the audience should have any previous knowledge about the programmatic issue. The electronics are a substantial part of how the story is musically told. Hence, complete automation was in my view the only way to achieve the required absolute accuracy between player and computer.

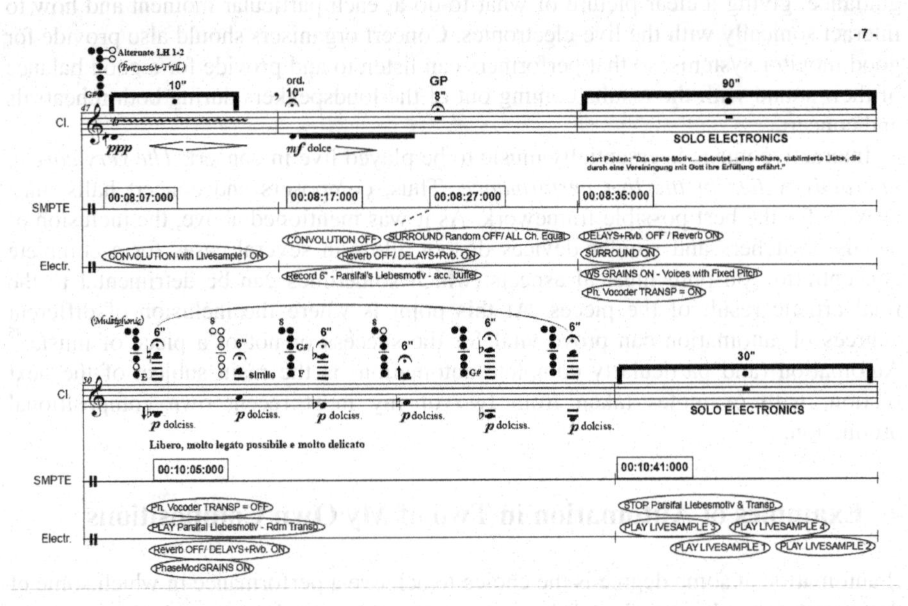

Fig. 2. *Intersections (memories)* for clarinet and live-electronics: excerpt of the final part (page 7).[36] The row in the middle shows different clue-times, which the performer should follow on the provided SMPTE display. The lower row shows the DSP processes.

The diverse DSP functions in the piece include: random and circular surround spatialisation, granular synthesis (phase modulated grains and time-stretching granulation), different types of reverberation, convolution,[37] pitch shifting and live recording of particular notes with specific durations throughout the piece. These durations are shown in the long notes (of variable duration) played throughout the piece, which are all the eighteen notes with their original duration of the first

---

[36] Copyright with the author. Printed with permission.

[37] Convolution: the method used for this piece is the re-synthesis of the spectral multiplication of the results of the FFT analysis of two different sounds (in this case, live clarinet against live recorded clarinet sounds).

Leitmotiv from Wagner's *Parsifal* (*Liebesmotiv*). As these notes are played mostly isolated, the audience has no idea of their real purpose when they first appear. However, the electronics record and store them in a cumulative buffer, so that when all notes have been played and recorded, the Leitmotiv is played back (granulated) in its original form, having a very precise dramaturgical function. In order to record all eighteen samples, so that they can be cumulated and afterwards played as a continuous melody, absolute accuracy was required; this would not have been possible without the programming of fully automated live-electronics.

The score presents three different lines, one of the instrument, the second for the SMPTE time and the third to describe the DSP process activated/deactivated at that particular stage (Fig. 2). The spatialisation of this piece (as it can be seen in Fig. 2, between SMPTE times 00:08:27 and 00:08:35) is also fully automated. The configuration for this latter is a system developed by myself, in which the 5.1 surround sound can move either in circle (clock-and-anticlockwise) or randomly, with the time of the movement of sound between loudspeakers measured in milliseconds. Time was dynamically programmed, so that for a given period of time (for example, 20 seconds), the sound must change constantly at a variable rate during that period. The range of this dynamical time set is variable, and can be, for example, from 5000 ms to 200 ms. If the changes between speakers occur at rates lower than 200 ms, the result will be a spatial granulation. This is because of the usage of envelopes for each localisation of sound in each speaker: the lower the rate, the more the results can be perceived as being 'granulated in the space'.

All processes in the piece respond to a particular dramaturgical reason, acting together with -and being excited by- the music composed for the live instrument (with no pre- recorded materials involved). Figure 2, above, shows also how synchronisation between the live part and the processes is written in the score, so that it can be achieved by reading a SMPTE display on the stage during the performance.[38]

farb-laut E–VIOLET. This piece, which was specially commissioned for the festival *farb-laut* in Berlin (Germany) in November 2008 is, at the time of writing this article, my last piece for viola and live-electronics (*MAX/MSP*).

Regarding the treatment of DSP, it is similar to the former piece: the live-electronics are fully automated for just the same reasons. However, automation has here a further implication: the piece was conceived, like all the others for viola and live-electronics I composed since 1998,[39] *to be played by myself.* Whilst *Intersections (memories)* needed complete automation for purely technical reasons (synchronisation), *farb-laut E-Violet*, on the other hand required automation to allow me to play the piece without further considerations.

As in *Intersections*, samples recorded during the performance interact throughout the piece in many ways, without the need of any pre-recorded materials. DSP

[38] Due to space reasons, I only show an excerpt of the score but not of the electronics. A similar example of a *MAX/MSP* patch is shown for the next piece, in 4.1.2. (this time, without the score, which is organised also in a rather similar manner).

[39] *Color Code (1998); NINTH (music for Viola & computer) (2002); Ableitungen des Konzepts der Wiederholung (for Ala) (2004).*

functions besides live recording include different types of granulation, dynamic delays,[40] reverberation, *COMB* filters, ring modulated *COMB* filters, pitch recognition and 5.1 surround sound.

Figure 3 shows the main part of the *MAX/MSP* patch of this piece. Here we can see different aspects of how complete automation works: the button at the left upper corner is the only activation needed. From then on, the SMPTE time-code will be displayed on the *MAX/MSP* patch on the right upper corner. If needed, the SMPTE can also be sent to an external display.

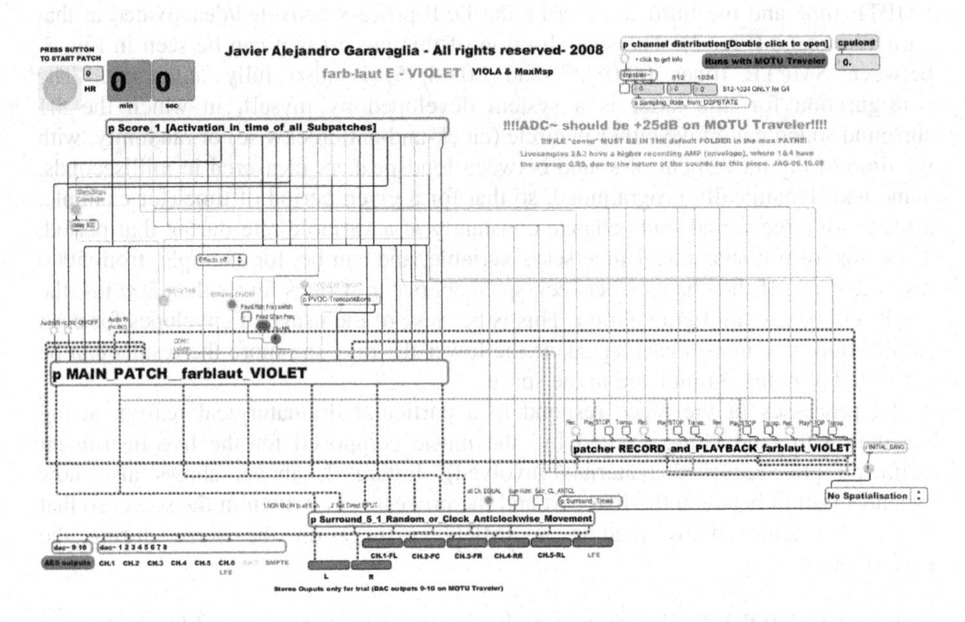

Fig. 3. *farb-laut E-Violet* for viola & live-electronics: screenshot of the main *MAX/MSP* patch

The first sub-patch (‘*p Score_1_[Activation_in_time_of_all_Subpatches]*’) is an electronic score, which automatically triggers the required DSP functions at precise moments. The sub-patch ‘*p MAIN_PATCH_farblaut_VIOLET*’ contains all processes (displayed in detail in Fig. 4). There is also a sub-patch for recording and playback of the samples during the performance, as well as a sub-patch for the spatialisation, similar to that explained in section 4.1.1.

Besides a laptop, an audio interface, a microphone and a surround sound system, this is all what is needed for the performance of both *farb-laut E-Violet* and *Intersections (memories)*. A second person on the mixing desk is normally desirable for balancing the level of the sound in the concert hall, but not to be in charge of any of the processes directly linked with the electronics, which ran automatically and without any human intervention.

---

[40] These work with a given random time and location for each delay.

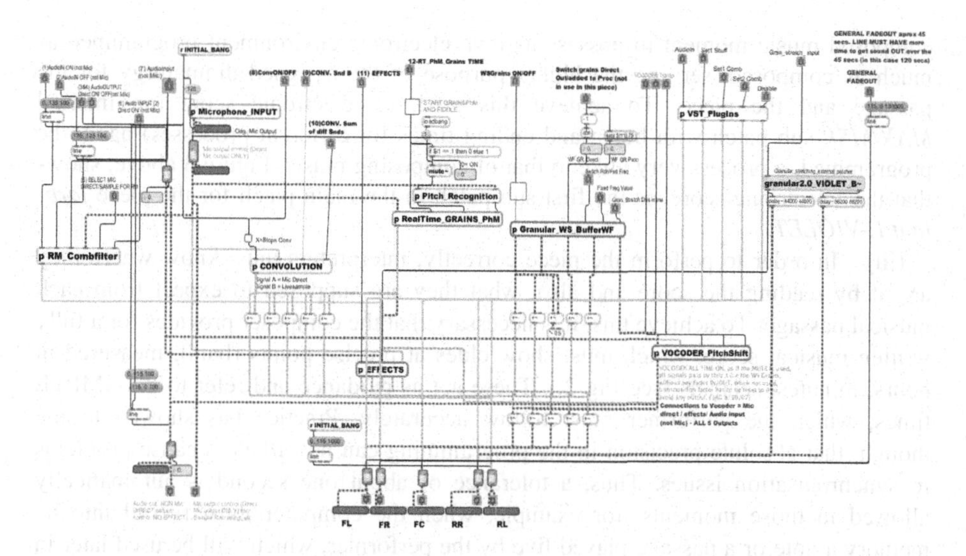

Fig. 4. *farb-laut E-Violet* for viola & live-electronics: screenshot of the sub-patch containing all processes included in the piece

4.2 Description of the Main Technical Characteristics in the Former Examples

Both pieces described in section 4.1 share the following features:

- *Minimal technical requirements*: both need one computer, a microphone and an audio interface with at least one INPUT and a maximum of 8 Outputs (for example, *MOTU* 828 mkII, *MOTU Traveler* or similar).

- *A MAX/MSP patch*: a main patch consists of multiple sub-patches, each of them containing algorithms with different DSP such as: granulation, convolution, phase vocoding (pitch shift), reverberation, dynamic delays, ring modulation, sample and hold, filtering, live recording, spatialisation, etc.

- *Solo passages*: in both pieces, the complete automation allows for the live-electronics to generate 'solo' electronic moments at particular points, in which no interaction between the soloist and the computer occurs (the performer has in fact a break during those sections). The sounds however, have their origin in some past interaction (normally sounds recorded at a previous moment during the performance).

- *Performance related issues*:

(i) The *MAX/MSP* patch is completely automated. The activation of the electronics takes place by pressing a button on the patch (*MAX*: *bang* function through the object *button*, see Fig. 3) at the very beginning of the work with a delay of about 10 or 15 seconds between pressing the button and the actual beginning of the patch/piece, to allow the player time to take position to play. All DSP events have a fixed duration, which require only mixing activity on the console and no manipulation of the electronics' part itself at all.

(ii) Because everything in the *MAX/MSP* patch is completely programmed in forehand, the evolution of every DSP event has to be accurately programmed and timed, so that the performer can accurately interact with them. This allows for each composed music moment to possess its own electronic environment programmed as much as composed, serving the higher purpose of the assigned dramaturgy for the passage and the piece. To achieve this goal, an electronic score (an internal *MAX/MSP* sub-patch with begin and ending times for different processes) has to be programmed, a process very alike to that of composing music. Figure 3 above, shows that the 'electronic score' is the first sub-patch in the main patch for the piece *farb-laut E-VIOLET*.

(iii) In order to perform the piece correctly, interpreters must know where they are at by reading the score and also, what they are supposed to expect from each musical passage. To achieve this, it is necessary that the composer provides for a fully written musical score, which must show clues at precise points (bars), measured in hours, minutes, seconds (see Fig 2.). These act as guidance and refer to the SMPTE times, which the performer must follow accurately. Practice has showed to me though, that absolute precision in the programming can nevertheless cause problems in synchronisation issues. Thus, a tolerance of about one second is automatically allowed in those moments, for example, when the computer must record into its memory a note or a passage played live by the performer, which will be used later in the piece. The SMPTE has been already programmed and included on the *MAX/MSP* patches (in these particular two cases, an audio file containing the bi-phased modulated square waves for each bit[41] of the frames), so that the performer can read it either on the computer screen directly or through a SMPTE display, whichever will be placed on the stage. In my own experience, and after talking to several performers of my music, I can categorically state, that to follow the SMPTE on a display or computer screen does not require further skills from a musician than those following a conductor's baton. Everyone found it a very pleasant experience, which required minimal efforts to become acquainted to.

(iv) Due to the automation of the patches, the performer is only required to follow the score looking at the SMPTE display. No further activity rather than playing the instrument is required, as the electronics play automatically. This has a big positive and relaxing impact on the concentration of both players and audience. In the case of pieces for several instruments, there are two ways of dealing with the issue: if it is an ensemble, the SMPTE display will be generally read by the conductor, thus making no difference at all to the ensemble's members; if it is a small group, such as duos or trios, and so on, several displays can be placed on the stage, all following however the same synchronised SMPTE time.

As it can be gathered by now, all features described in section 4.2 heavily depend on the electronics for the pieces working automatically. So, is automation the answer to all problems? The next section tries to answer this question by enumerating advantages and disadvantages related to the usage of complete automation in live-electronics situations.

---

[41] Each frame of the SMPTE (for example, 00:00:00:00, which is the first frame, with hours at the left and then minutes, seconds and frames to the right) is digitally encoded as an 80-bit word. The bi-phased modulated square wave is used to codify each of those bits during equal periods of time, where a 0 is coded if the square wave will not change its phase in each period, and a 1 if it does.

5 Advantages and Disadvantages of Using Automation in Live-Electronics

As usual, when composers make their decisions during the composition of a piece of music, some choices must be left aside. This implies the awareness about the limitations that choices bring with themselves. In the case of complete automation (or, at least, a high degree of partial automation), it is indeed a choice, which will have consequences for the piece, for the musicians involved in its performance and also for the degree of success the piece can achieve. These consequences must be weighted at the time of 'decision-making' while the piece is being composed. In any case, regardless of which decision will finally be made, each of them will present advantages and disadvantages. So, in my experience, these decisions should be made only based on a rather overwhelming percentage of advantages against disadvantages. These are explained in detail below, relating to full-automation.

5.1. Advantages

Among the advantages that full-automation of electronics can confer to the performance of a piece of music, we can identify:

- *Concentration and reduction of unnecessary activities.*

The most obvious advantage is that the performer can concentrate exclusively on the pure musical aspects of the performance. This has a further implication (even though it may be rather subjective): if a performer is involved in multiple activities that are required for the realisation of the electronics (for example, pressing pedals, touching the screen or the mouse, or making movements specially required for triggering the electronics, which might not suit the musical context they are required for), these can be a distraction factor not only for the player, but also for the audience. Even though it is true, that the nature of the piece may or may not diminish the impact regarding the audience's perception, these extra activities, if not part of another event involved in the piece (i.e. some type of acting), will always be to some extent a factor of distraction.

- *Relative independence of the electronics from the composer's presence during the performance.*

Regardless of the composer's presence or absence during the actual performance, the live-electronics should not need further manipulation, as everything is already programmed. This allows for more frequent performances of each piece and facilitates its transportation, as the patch itself (given the correct version of the software) needs only to be copied on any computer available capable of running that particular version of the software required for the performance. Once installed in the right environment, all events of the piece should run automatically after pressing the initial button

- *Better combination of processes and less risk.*

This method allows, in addition, for a more accurate and complex combination, crossover, fade-in-and-out, and so on, of different real-time processes. The risk of improvising in a live situation with the combination of processes cannot avoid sometimes the risk of accidentally exceeding the limits of the CPU performance on the computer. With automation however, as everything should have been tested beforehand, this rarely happens.

- *Ideal method for composer/performer in one person.*

As in the case of the viola piece described above, in which the composer is intended to be the performer as well, automation allows for shorter, safer and more efficient rehearsal times, as nobody else is needed to use or run the patch.

- *Additional solution for synchronization of events and processes related to performing time.*

For all the pieces mentioned in section 4.1, and as a consequence of the automation of all processes, following a time-code is a secure way to read a score and perform an interactive piece: it guarantees a very accurate interaction with the electronics, mostly in those cases, for example, when the recording of live samples for further use in the piece is required. Even though methods such as 'score-following' (through, for example, pitch recognition) are sometimes very successful (such as Lippe's *Music for clarinet and ISPW*), this is however not always the case, depending mostly on how they are applied (i.e. what type of technology is used to recognise the parameter followed). Negative effects such as some processes not triggered in time (or at all) seem to be frequent using score-following algorithms.

5.2 Disadvantages

In spite of all the advantages pointed above, some problems *do* appear though, which can be summarised in an overall lack of flexibility, imposing a variable degree of limitation to the performance/rehearsal situation.

From an aesthetical perspective, the following problematic becomes apparent: can processes, which repeat themselves invariably by each and every performance, be identified as 'authentic live-electronics'? I am aware that some composers and performers regard the very essence of live-electronics as relying mainly on the possibility of live manipulation, adaptation, interaction and variation during performance time. However, if we keep in mind the definition of live-electronics in the introduction of this paper, the answer to the former question should be definitely affirmative, as this should still be a clear case of "*performance involving electronic instruments which can be performed in real-time*". There is no contradiction between the concept of automation and this definition, as much as there is none regarding the definition of *interactivity* too. Moreover, the patches of my pieces using complete automation mentioned in this paper would not work if there would not be a 'live input' provided.

An obvious disadvantage is that the work is rather fixed in itself: if changes need to be introduced to the music in future revisions, the electronics must be therefore substantially reprogrammed. This includes the accurate re-programming of, for example, all amplitude levels within the patch, as much as all timing issues and all connections. It also requires an intensive testing during the re-programming/re-composing period.

Another disadvantage connected to the latter is, that the patch generally runs from the beginning to the end of the piece, without any break in the middle. This impacts on the rehearsal methods, as partial rehearsal of some parts of the pieces unfortunately cannot be achieved without waiting until the precise SMPTE location arrives. This said, the only way to rehearse pieces composed using this method is from the start to the end. Even though some performers may find this situation not ideal, my experience is that they feel rather comfortable after getting acquainted to the method.

6 Summary

The discussion of whether or not complete automation is an original method to compose, programme and perform interactive pieces is not the main intention of this article. The aim is simply to arise awareness about the fact, that applying complete automation to the real-time processes of the electronics can be far more beneficial than detrimental to the actual performance of interactive works, the achievement of their dramaturgical intentions, their rehearsals and their circulation in different venues. This is not only my view as a composer, but I also include here my rather long experience performing electroacoustic music.

Historically, automation was not quite possible in the first thirty years of the existence of live-electronics, simply because the technologies available could not afford such tasks.[42] But since 1997 at the latest, the development and evolution of hardware and software platforms have allowed for automation to be not only feasible, but a real, tangible option. We have observed in the former sections, that, in examples such as Tutschku's or in the system at Tempo Reale, automation has been introduced in a rather high degree, and in the case of the latter, even at the core of the system's conception. Manual operation of partial automation is indeed frequent, but the use of complete automation, at least as presented in section 4, does not seem to be so popular yet.

Programming live-electronics entirely timed on some type of 'electronic-score' (for example, a *MAX* sub-patch that automatically activates the start and end each process), allows performers for more artistic freedom during performances, as much as for concentration on basic activities such as reading the score, playing and interacting, without the need of extra activities such as switching pedals, etc. It also guarantees, that even if the composer does not attend the performances and/or rehearsals, the piece can be likewise easily activated. Environments such as *MAX/MSP* or *PD* allow for a simple and effective installation of patches on different computers/platforms, with just only minimal adjustments required. Making the patches an independent application is also a very elegant and efficient option (as in the case of Tutschku's *Zellen-Linien*).

If the synchronisation between performer and the electronics is achieved through simply following a time-code instead of more demanding activities, digital-signal-processes such as recording live to a buffer[43] as well as other interaction processes,

---

[42] One of the few exceptions could be drum machines, which appeared already in the 1930s with the *Rhythmicon* designed by Léon Theremin in 1932 as a commission from composer Henry Cowell. In 1959, the firm Wurlitzer released the *Sideman*, which was an electro-mechanical drum machine. The 1960s saw some others, such as the *Rhythm Synthesizer* or the *FR1 Rhythm Ace* by Ace Tone (later Roland). The list of devices that followed is too long for the frame of this article. However, it must be said that the automation was given in each case by some type of sequencing (the particular type depending on the technologies available at their own release times) and that they have never been relevant to pieces including live-electronics [36].

[43] For example: recording a few seconds of a specific part of the piece in order to use that sample later as, for example, a pick-up-sample for a convolution process.

can be easily achieved on a extremely accurate timing; further requirements such as additional programming, pitch detection or even other devices are therefore needless for any synchronisation purpose. Moreover, benefits shall increase, if the different DSP functions run steadily at the exact same set-times, as they can be entirely tested already while the work is being programmed and composed. This can result in less or simply no danger of exceeding the CPU's limits, as the entire patch can be completely monitored beforehand.

All the positive consequences, which complete automation offers thanks to the evolution of computing technology since the 1990s, result in an easier, costless, more effective and more frequent distribution of interactive pieces, as the only requirements for their performance are the score and the patch/software. It must also be observed that a better distribution should allow also for easier, perhaps even more artistic and definitely more frequent performances of these pieces. Hence, the minor disadvantages of complete automation do not, in my view, seriously impact on the overall musical end-results and can be considered minor inconveniences or limitations, which cannot be eluded, but however do not invalidate the usage of full automation.

References

[1] Live-electronics, EARS – English Glossary, http://www.ears.dmu.ac.uk (site accessed: November 2007)
[2] Interactivity, EARS – English Glossary, http://www.ears.dmu.ac.uk (site accessed: November 2007)
[3] UPIC system, http://en.wikipedia.org/wiki/UPIC (site accessed: September 2009)
[4] Lippe, C.: Real-Time Interactive Digital Signal Processing: A View of Computer Music. Computer Music Journal 20(4), 21–24 (Winter1996)
[5] Roads, C.: Microsound, pp. 27–28. MIT Press, Cambridge (2001)
[6] Truax, B.: Real-Time Granular Synthesis with a Digital Signal Processor. Computer Music Journal 12(2), 14 (1988)
[7] Microchips, http://en.wikipedia.org/wiki/Integrated_circuit (site accessed: August 2009)
[8] Personal Computer, http://en.wikipedia.org/wiki/Personal_computer (site accessed: August 2009)
[9] Fairlight, http://kmi9000.tripod.com/kmi_cmi.htm#models (site accessed: August 2009)
[10] Synclavier, http://en.wikipedia.org (site accessed: August 2009)
[11] ARP 2500, ARP 2500, http://en.wikipedia.org/wiki/ARP_2500 (site accessed: August 2009)
[12] SYNLAB, http://icem.folkwang-hochschule.de/ and http://www.klangbureau.de (sites accessed: October 2009)
[13] Vogt, H.: Neue Musik seit 1945. Philipp Reclam jun. Stuttgart, p. 94 (1982)
[14] Häusler, J.: Interview: Pierre Boulez on Répons, http://www.andante.com (site accessed: September 2009); This article was originally published in Teilton, Schriftenreihe der Heinrich-Strobel-Stiftung des Südwestfunks, Kassel, Bärenreiter, vol. IV, pp. 7–14 (1985)
[15] Boulez, P., Gerzso, A.: Computers in music. Scientific American 258(4), 44–50 (1988)
[16] Bohn, R.: Inszenierung und Ereignis: Beiträge zur Theorie und Praxis der Szenografie. Szenografie & Szenologie, vol. 1, pp. 149–150. Transcript Verlag, Bielefeld (2009)

[17] Lippe, C.: Real-Time Interactive Digital Signal Processing: A View of Computer Music. Computer Music Journal 20(4), 21–24 (1997)

[18] Roads, C.: The Computer Music Tutorial, pp. 330–335. MIT Press, Cambridge (1996)

[19] Andrenacci, P., Rosati, C., Paladin, A., Armani, F., Bessegato, R., Pisani, P., Prestigiacomo, A.: The new MARS workstation. In: Proceedings of the International Computer Music Conference 1997, Hellas, Thessaloniki, pp. 215–219 (1997)

[20] Garavaglia, J.: The necessity of composing with live-electronics. A short account of the piece "Gegensaetze (gegenseitig)" and of the hardware (AUDIACSystem) used to produce the real-time processes on it. In: Proceedings of the II Brazilian Symposium on Computer Music (SBC-UFRGS) Canela, Brazil Edited by Eduardo Reck Miranda, Gráfica Editora Pallotti, Brazil, pp. 65–71 (1995)

[21] Kyma System, http://www.symbolicsound.com/cgi-bin/bin/view/Products/FAQ (site accessed: September 2009)

[22] Nextstep, http://en.wikipedia.org/wiki/Nextstep (site accessed: September 2009)

[23] Mac OS X, http://en.wikipedia.org/wiki/Mac_OS_X (site accessed: September 2009)

[24] MAX/MSP, http://en.wikipedia.org/wiki/Max_software (site accessed: September 2009)

[25] PD, http://freesoftware.ircam.fr/article.php3?id_article=5 (site accessed: September 2009)

[26] Supercollider, http://en.wikipedia.org/wiki/SuperCollider (site accessed: September 2009)

[27] Spatialisateur Project (Spat) – IRCAM, http://support.ircam.fr/forum-ol-doc/spat/3.0/spat-3-intro/co/overview.html (site accessed: September 2009)

[28] Giomi, F., Meacci, D., Schwoon, K.: Live Electronics in Luciano Berio's Music. Computer Music Journal 27(2), 30–46 (Summer 2003)

[29] MMEG – Tempo Reale, http://www.temporeale.it (Under: Research – Development) (site accessed: September 2009)

[30] Game console control & electronics, http://jamespaulsain.com (site accessed: September 2009)

[31] FFTease, http://www.sarc.qub.ac.uk/~elyon/LyonSoftware/MaxMSP/FFTease (site accessed: September 2009)

[32] Puckette, M., Lippe, C.: Score Following in Practice. In: Proceedings of the International Computer Music Conference 1992, pp. 182–185 (1992)

[33] Score Following - IRCAM, http://imtr.ircam.fr (site accessed: September 2009)

[34] Lippe, C.: A Composition for Clarinet and Real-Time Signal Processing: Using Max on the IRCAM Signal Processing Workstation. In: Proceedings of 10th Italian Colloquium on Computer Music, Milan, pp. 428–432 (1993)

[35] Izmirli, O., Zahler, N., Seward, R.: Rewriting the Score: Score following in MAX/MSP and the Demands of Composers. In: Proceedings Book of the 9th Biennial Arts & Technology Symposium at Connecticut College Transparent Technologies, pp. 87–91 (2003)

[36] Drum Machines, http://en.wikipedia.org/wiki/Drum_Machine (site accessed: December 2009)

Polyphonic Alignment Algorithms for Symbolic Music Retrieval*

Julien Allali[1], Pascal Ferraro[2], Pierre Hanna[3], and Matthias Robine[3]**

[1] LaBRI - Université de Bordeaux, 33405 Talence cedex, France and Pacific Institute For the Mathematical Sciences - Simon Fraser University, Canada
julien.allali@labri.fr

[2] LaBRI - Université de Bordeaux, 33405 Talence cedex, France and Pacific Institute For the Mathematical Sciences - University of Calgary, Canada
pascal.ferraro@labri.fr

[3] LaBRI - Université de Bordeaux, 33405 Talence cedex, France
name.surname@labri.fr

Abstract. Melody is an important property for the perceptual description of Western musical pieces. A lot of applications rely on the evaluation of similarity between two melodies. While several existing techniques assume a monophonic context or extract a monophonic melody from polyphonic pieces, in this paper, we propose to consider the whole polyphonic context to evaluate the similarity without reducing to a monophonic melody. We thus propose a new model and a corresponding methodology that takes into account all the notes, even if they sound at the same time or if they overlap. Our model relies on a quotiented sequence representation of music. A quotiented sequence is a sequence graph defined with an additional equivalent relation on its vertices and such that the quotient graph is also a sequence graph. The core of the comparison method is based on an adaptation of edit-distance metrics, regularly applied in bio-informatic context. This algorithm is currently being used to evaluate the similarity between a monophonic or polyphonic query and a database of polyphonic musical pieces. First experiments show that the adaptation to polyphony does not degrade the quality of the algorithm with monophonic musical pieces. Furthermore, the results of experiments with polyphonic pieces are promising, even if they show some limitations.

1 Introduction

Searching for similarities in large musical databases has become a common procedure. One of the main goal of music retrieval systems is to find musical pieces in large databases given a description or an example. These systems compute a numeric score on how well a query matches each piece of the database and rank the music pieces according to this score. Computing such a degree of similarity

* This work has been partially sponsored by the French ANR SIMBALS (JC07-188930) and ANR Brasero (ANR-06-BLAN-0045) projects.

** Corresponding author.

S. Ystad et al. (Eds.): CMMR/ICAD 2009, LNCS 5954, pp. 466–482, 2010.

between two pieces of music is a difficult problem. Although, different methods have been proposed to evaluate this degree of similarity, techniques based on string matching [1] remain generally more accurate as they can take into account errors in the query or in the pieces of music of the database. This is a important property in the context of music retrieval systems since audio analysis always induces approximations. In particular, some music retrieval applications require specific robustness. Query by humming, a music retrieval system where the input query is a user-hummed melody, is a very good example. Since the sung query can be transposed, played faster or slower, without degrading the melody, retrieval systems have to be both transposition and tempo invariant. Edit distance algorithms, mainly developed in the context of DNA sequence recognition, have thus been adapted in the context of music similarity [2]. These algorithms, based on the dynamic programming principle, are generalizations of a local sequence alignment method proposed by Smith and Waterman [3] in the early 80's. Applications relying on local alignment are numerous and include cover detection [4], melody retrieval [2], query by humming [5], structural analysis, comparison of chord progressions [6], etc. Local alignment approaches usually provide very accurate results as shown at the recent editions of the Music Information Retrieval Evaluation eXchange (MIREX) [7].

Most existing techniques for evaluating music similarity consider a monophonic context. However, applications of this similarity measure may concern more complex musical information. For example, one might want to retrieve audio tracks similar to an audio query from a polyphonic audio database. In this case, without information about voice lines, most existing approaches consider transforming a polyphonic music into a monophonic music [8,9]. *Monophonic music* is assumed to be composed of only one melody. In a stricter sense, it implies that no more than one note is sounded at any given time. *Polyphony* is a texture consisting of two or more independent voices. Thus, in the polyphonic context, more than one note can sound at a given time. A monophonic sequence may be derived from a polyphonic source by selecting at most one note at every time step (this process is called monophonic reduction). In [8], a few existing methods are proposed for monophonic reduction from a polyphonic source. Experiments lead to the conclusion that the perfect technique does not exist, and that each algorithm produces significant errors. Another conclusion is that choosing the highest note of chords yields better results.

We propose here an alternative by considering directly polyphonic musical sequences instead of reducing a polyphonic source into a monophonic music. Such an approach requires a data structure adapted to the constraints induced by the polyphonic context and some adaptations of the algorithms.

In this paper, we propose to extend the edit-distance technique introduced in the musical retrieval context by Mongeau and Sankoff [2]. We detail the extension of this approach to the polyphonic context, in order to compute a similarity measure between a symbolic query and elements from a polyphonic symbolic database. After recalling the method to align two musical monophonic pieces in Section 2, an adequate representation is proposed for the polyphony in Section 3. Extensions of the operations proposed by Mongeau and Sankoff [2] are presented for taking into account the problems of overlapped notes and the transposition invariance. We propose finally in Section 4 a general framework for polyphonic music using the substitution score scheme set for monophonic music, which allows new operations. Experiments show the improvements induced by these adaptations in Section 5.

2 Aligning Two Pieces of Monophonic Music

In this section, we briefly recall the different representations and methods to compare two monophonic musics. Following Mongeau and Sankoff [2], any monophonic piece can be represented by a sequence of notes, each given as a pair *(pitch, length)*. Several alphabets and sets of numbers have been proposed to represent pitches and durations [10]. In the following, we are using the interval relative representation, *i.e.* the number of semitones between two successive notes reduced modulo 12. This representation presents the huge advantage to be transposition invariant.

2.1 General Sequence Alignment

Sequence alignment algorithms are widely used to compare strings. They evaluate the similarity between two strings t and q given on an alphabet A, and of respective sizes $|t|$ and $|q|$. Formally an alignment between t and q is a string z on the alphabet of pairs of letters, more precisely on $(A \cup \{\epsilon\}) \times (A \cup \{\epsilon\})$, whose projection on the first component is t and the projection on the second component is q. The letter ϵ does not belong to the alphabet A. It is often substituted by the symbol "-" and is called a *gap*. An aligned pair of z of type (a, b) with $a, b \in A$ denotes the *substitution* of the letter a by the letter b. A pair of type $(a, \text{-})$ denotes a *deletion* of the letter a. Finally, an aligned pair of type $(\text{-}, b)$ denotes the *insertion* of the letter b. A score $\sigma(t_i, q_j)$ is assigned to each pair (t_i, q_j) of the alignment. We suppose the score function respects the following conditions:

$$s(a, a) > 0 \quad \forall a \in A,$$
$$s(a, b) < 0 \quad \forall a \neq b, a, b \in A \cup \{\epsilon\}.$$

This means that the score between two symbols a and b become higher with their similarity.

The score S of an alignment is then defined as the sum of the costs of its aligned pairs. Computational approaches to sequence alignment generally fall into two categories: *global alignments* and *local alignments*. Computing a global alignment is a form of global optimization that constrains the alignment to span the entire length of all query sequences. By contrast, local alignments identify regions of similarity within long sequences that are often widely divergent overall. In query-by-humming applications, since the query is generally much shorter than the reference, one favours local alignment methods.

Both alignment techniques are based on dynamic programming [3,11,12]. Given two strings t and q, alignment algorithms compute a $(|t|+1) \times (|q|+1)$ matrix T such that:

$$T[i,j] = S(t[0 \dots i], q[0 \dots j]),$$

where $S(t[0 \dots i], q[0 \dots j])$ is the optimal score between the sub-strings of t and q ending respectively in position $0 \leq i \leq |t|$ and $0 \leq j \leq |q|$. $T[i,j]$ is recursively computed as follows:

$$T[i,j] = \min \left\{ \begin{array}{l} T[i-1,j] + \sigma(i,-) \\ T[i,j-1) + \sigma(-,j) \\ T[i-1,j-1] + \sigma(i,j) \end{array} \right\}$$

Dynamic programming algorithms can compute the optimal alignment (either global or local) and the corresponding score in time complexity $\mathcal{O}(|t| \times |q|)$ and in memory complexity $\mathcal{O}(\min\{|t|, |q|\})$ (see [3] for details).

2.2 Local Transposition

Queries produced by human beings can, not only be totally transposed, but can also be composed of several parts that are independently transposed. For example, if an original musical piece is composed of different harmonic voices, one may sing different successive parts with different keys to retrieve it. In the same way, pieces of popular music are sometimes composed of successive transposed choruses. A sung query may imitate these characteristics. Moreover, errors in singing or humming may occur, especially for users that are not trained to perfectly control their voice like professional singers. From a musical point of view, sudden tonal changes are disturbing. However, if these changes last during a long period, they may not disturb listeners. Figure 1 shows an example of query having two local transpositions.

Fig. 1. Example of a monophonic query not transposed (top) and a monophonic query with two local transpositions (bottom)

The two pieces in Figure 1 sound very similar, although the two resulting sequences are very different. This problem has been addressed in [13] by defining a local transposition algorithm. It requires to compute multiple score matrices simultaneously, one for each possible transposition value. The time complexity is $\mathcal{O}(\Delta \times |q| \times |t|)$, where Δ is the number of local transposition allowed during the comparison (for practical applications, Δ is set up to 12).

2.3 Pitch/Duration Scoring Scheme

The quality of an alignment-based algorithm heavily depends on the scoring function. Results may differ significantly whether one uses a basic scoring scheme or a more sophisticated scoring function [1]. For our experiments, we use the scoring schemes introduced in [1] and [2], where the score between two notes depends on the pitch, the duration and the consonance of both notes. For example, the fifth (7 semitones) and the third major or minor (3 or 4 semitones) are the most consonant intervals in Western music [14]. The score function between two notes is then defined as a linear combination of a function σ_p on pitches (its values are coded into a matrix) and a function σ_d on durations as:

$$\sigma(a, b) = \alpha \cdot \sigma_p(a, b) + \beta \cdot \sigma_d(a, b).$$

The cost associated to a gap only depends on the note duration. Finally a penalty (a negative score) is also applied to each local transposition.

3 Representation of Polyphony

3.1 Quotiented Sequence Model

In order to compare polyphonic music, existing works generally require reduction of a polyphonic piece [10]. Some methods [15] reduce polyphonic music as a set of separate tracks. We chose to not use any information about the voice lines, since they could be missing in the case of polyphonic music obtained by transcription from audio for example. Some monophonic reductions assume that the melody is only defined by the highest pitches, but in many cases it is not true. In order to avoid this kind of strong assumption, we propose to study representations using all the notes of a polyphonic musical piece. One way would be to consider all the distinct monophonic lines induced by the polyphony, but this naive approach may imply a combinatoric explosion in the cases of large scores [16].

To take into account the polyphonic nature of musical sequences, we then introduce the notion of quotiented sequence representation. Formally, a quotiented sequence is a sequence graph with an equivalence relation defined on the set of vertices, and such that the resulting quotient graph is also a sequence.

Definition 1. *A* quotiented sequence *is a 3-tuple $Q = (t, W, \pi)$ where t is a sequence called the support of Q, W is a set of vertices and π a surjective application from V, the set of vertices of t to W.*

The quotient graph $\pi(t)$ associated with Q is (W, E_π) such that:

$$\forall (t_i, t_j) \in E, (\pi(t_i), \pi(t_j)) \in E_\pi \Leftrightarrow \pi(t_i) \neq \pi(t_j)$$

By definition, in a quotiented sequence, quotient graph and support sequence are both sequences (Figure 2). A quotiented sequence can thus be considered as a self-similar structure represented by sequences on two different scales. Note that a quotiented structure can also be viewed as a tree graph.

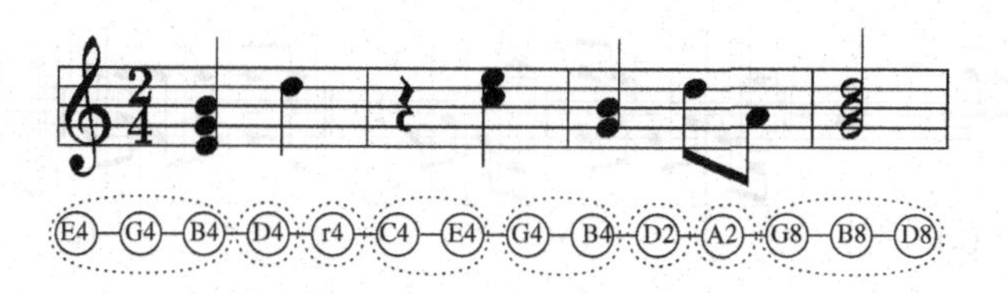

Fig. 2. Example of polyphonic musical score and its related quotiented sequence

This approach leads us to consider any polyphonic sequence as a series of ordered pairs. As for monophonic sequence [2], each pair is defined by the pitch of the note and its length (coded in sixteenth note values here). The pitch of the note is coded as the difference (in semitones) with the tonic. This difference is determined modulo an octave and is thus defined in the range $[0, 11]$ semitones.

3.2 Overlapped Notes

An important problem is induced by the representation of polyphonic music when considering note durations. Let us consider the polyphonic excerpt of *Carnaval des animaux (The swan)* by Camille Saint-Saëns (Figure 3). Some notes in the upper staff (first note of the third bar for example) still sound whereas other notes start in the lower staff. The representation of the third bar thus results in six triplets. This representation is not satisfactory since, for example, the second triplet contains only one eighth note (E, lower staff) and not the half note that still sounds at the same time (A, upper staff).

In order to take into account the possibility that one note still sounds at the same time another note starts, we propose to introduce the notion of *tied notes*. A new note n_2 starting whereas another note n_1 still sounds is represented by a chord composed of n_2 and a tied note n_1', where n_1' has onset time $o(n_1') = o(n_1)$, pitch $p(n_1') = p(n_1)$ but with a smaller duration $d(n_1')$:

$$d(n_1') = d(n_1) - [o(n_2) - o(n_1)] \quad (1)$$

The duration d_{n_1} of the note n_1 is also modified:

$$d(n_1) = o(n_2) - o(n_1) \quad (2)$$

Fig. 3. Polyphonic excerpt of *Carnaval des animaux (The swan)* by Camille Saint-Saëns

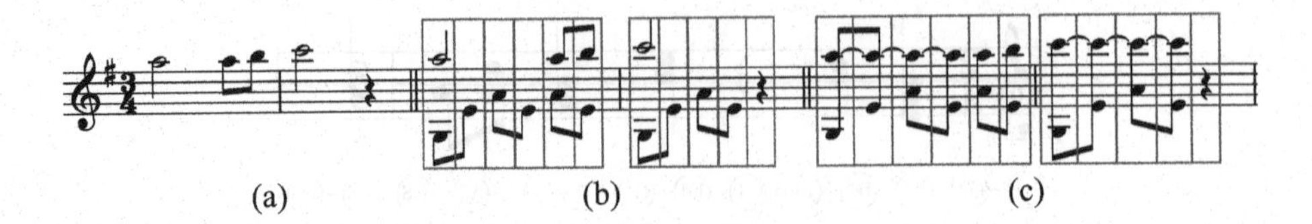

Fig. 4. Illustration of the representation of polyphonic music with tied notes: a monophonic query corresponding to one voice (a), a first polyphonic representation based on onsets (b), and the representation chosen that considers tied notes (c)

This representation does not consider only one tied note, but also consecutive tied notes. For example, Figure 4 illustrates a polyphonic representation with tied notes. The first bar of the excerpt is represented by six triplets in (c), four of which are composed of tied notes A.

The idea of considering tied notes is somehow similar to the concept of *notebits* introduced in [17]. However, one of the main difference is the algorithm related to these tied notes: we propose here to consider polyphonic music without taking into account informations about tracks or voices. The problem is thus more difficult. The representation of polyphonic music with tied notes allows alignment algorithm to take into account overlapped notes by defining one specific operation.

4 Polyphonic Comparison

As we have already seen, a lot of problems arise, when dealing with polyphonic music alignment. Actually, the definition of an alignment in the polyphonic case is not a straightforward application of the monophonic comparison. We propose hereafter a general method to align two polyphonic musical pieces.

Setting up a scoring scheme in the polyphonic case (*i.e.* fixing a score for two chords) is a difficult problem. Indeed, in one octave there are 12 possible pitch values for a chord made of a single note (in practical applications, it is common to work only on one octave), then 12×11 for two note chords ... $\binom{12}{p}$ for p note chords, which means the scoring scheme will be represented by a matrix of size $2^{12} \times 2^{12}$.

Fig. 5. Arpeggiated chord above, with its interpretation below. The notes are played successively from the lowest to the highest pitch, and not simultaneously.

Fig. 6. Similarity despite permutations (a) Main motif of the 14th string quartet in C# minor opus 131 by Beethoven (1st movement, bar 1). The motif is composed by 4 notes (sequence (1 2 3 4)). (b) First theme of the 7th movement, bar 1. The 4 last notes of the two groups are permuted notes of the main motif, sequence (1 4 3 2) and (1 4 2 3) (c) Second theme of the 7th movement, bar 21. The 4 notes are again a permutation of the main motif, sequence (3 2 4 1).

Moreover, complex note rearrangements may occur between two similar polyphonic pieces, and temporal deviations may appear between a score and its interpretation. Figure 5 shows such an example: in the notation score, the chord has to be arpeggiated, and the related interpretation is transcribed as successive notes. In this case, a comparison system has to detect the similarity between the chord indicated in the musical score and the successive notes interpreted.

More generally, we have to deal with note/chord merging and local rearrangements. For example, composers may choose to change the order of the notes in a melodic motif during a musical piece. Figure 6 shows three excerpts from a piece by Beethoven: the second (b) and third (c) excerpts correspond to a rearrangement of the main motif (a) with swapped notes.

4.1 Chord Comparison

In many cases, an arbitrary order is given to the notes composing the chords of a musical sequence. To avoid this arbitrary choice, one can consider chords as sets. The cost for substituting one chord by another one leads to the problem of computing the best permutation between both chords. Figure 7 shows an example of two cadences that sound similar, but that can be estimated as very dissimilar because of the different order of the notes in the chords. To avoid this sort of problem, we suggest that chords should be considered as unordered sets and the best permutation should be found. This optimization method allows the estimation of a high similarity between these two sequences of chords.

This optimization problem is actually a maximum score maximum bipartite matching problem and can be modeled as a weighted maximum matching

Fig. 7. Similarity between inverted chords. These successive two perfect cadences in C major are similar despite the different order of the notes in the chords composing the cadences.

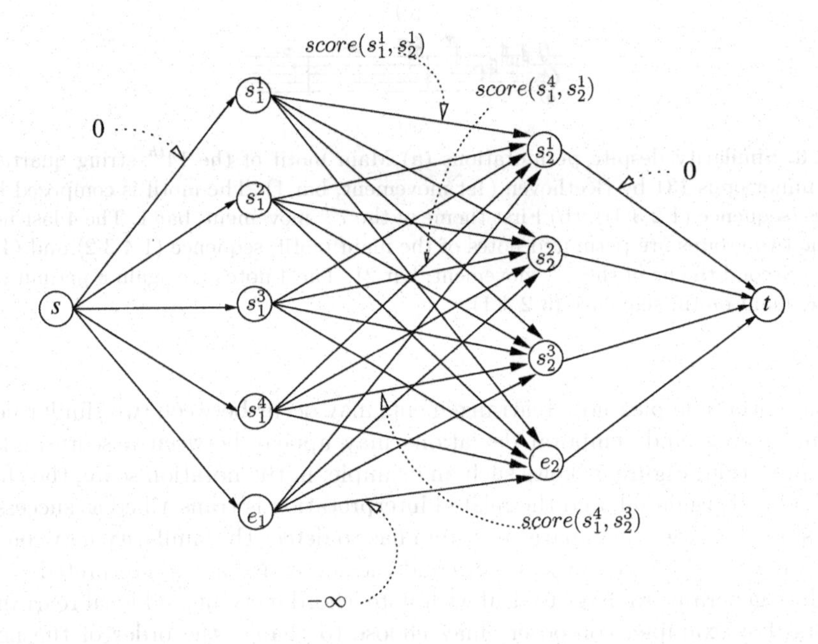

Fig. 8. Resolution of the optimal permutation as a maximum score flow problem

algorithm [18]. A similar approach has been proposed in the case of a geometric representation of musical pieces [19].

Let C_1 and C_2 be two chords of size n and m. We denote by $score_{bpg}(C_1, C_2)$ the score between these two chords. To compute $score_{bpg}(C_1, C_2)$ as the maximum score maximum bipartite matching problem, we consider the following graph $G(v, w) = (V, E)$ (Figure 8):

1. *vertex set* : $V = \{s, t, e_1, e_2\} \cup \{s_1^1, s_1^2, \dots s_1^n\} \cup \{s_2^1, s_2^2, \dots s_2^m\}$, where s is the source, t is the sink, $\{s_1^1, s_1^2, \dots s_1^n\}$ and $\{s_2^1, s_2^2, \dots s_2^m\}$ are the notes of the chords C_1 and C_2 and e_1, e_2 represent ϵ;
2. *edge set* : (s, s_1^k), (s, e_1), (e_2, t), (s_2^l, t) with a score 0, (s_1^k, s_2^l) with score $score(s_1^k, s_2^l)$, and (s_1^k, e) with score $score(s_1^k, \epsilon)$. All the edges have a capacity of 1 except (e, t) which capacity is $n - m$.

G is then a graph whose edges are labeled with integer capacities, non-negative scores in $\mathbb{R}$, and the maximum flow $f^* = n+m$. The score of the maximum flow is actually the score $score_{bpg}(C_1, C_2)$ and the complexity of computing local score is due to this maximum score maximum flow computation. This problem can be solved by the Edmonds and Karp's algorithm [18] improved by Tarjan [20] whose complexity is $O(|E||f^*|\log_2(|V|))$. For our graph, the maximum flow is $f^* = n+m$, the number of edges is $|E| = n \times m+2n+2m+3$ and the number of vertices is $|V| = n+m+4$. Finally the complexity of the score computation between two chords is bounded by $O(n^3 \times \log_2(n))$ where n represents the maximum number of notes in a chord.

In conclusion, computing alignment between two strings t and q leads to a total time complexity of $O(|t| \times |q| \times C^3 \times \log_2(C))$ where C is the maximum number of notes in a chord in t or q. In practical applications the parameter C is generally bounded by 4.

4.2 Extending Mongeau-Sankoff Operations

As previously stated, an accurate algorithm for music alignment must take into account both local rearrangements and merging operations. We thus propose allowing the merging of sub-sequences in both q and t simultaneously. Music alignment between two sequences is then extended as follows:

Definition 2 (Extended Music Alignment). *Given two sequences t and q of sets of symbols over Σ. A valid extended alignment X^a between t and q is a set of pairs of sub-sequences of t and q, that is $X^a = \{((i_s, i_e), (j_s, j_e))\}$ and respects the following for all $((i_s, i_e), (j_s, j_e)) \in X^a$:*

1. $1 \leq i_s < i_e \leq |t|$ and $1 \leq j_s < j_e \leq |q|$,
2. $\forall ((i'_s, i'_e), (j'_s, j'_e)), ((i'_s, i'_e), (j'_s, j'_e)) \in X^a$ such that $i'_s \neq i_s$:
(a) if $i'_s < i_s$ then $i_e \leq i_s$ and $j'_e \leq j_s$,
(b) if $i'_s > i_s$ then $i'_s \geq i_e$ and $j'_s \geq j_e$.

We define the set G_t (resp. G_q) as the set of positions of t (resp. q) not involved in X^a, that is

$$G_t = \{1, \ldots, |t|\} \setminus \bigcup_{((i_s,i_e),(j_s,j_e)) \in X^a} \{i_s \ldots i_e\}.$$

The score associated with X^a is defined as follows:

$$S(X^a) = \sum_{((i_s,i_e),(j_s,j_e)) \in X^a} score(t_{i_s \ldots i_e}, q_{j_s \ldots j_e}) + \sum_{i \in G_t} score(t_i, \epsilon) + \sum_{j \in G_q} score(\epsilon, q_j).$$

The adaptation of the alignment algorithm is straightforward. The recurrence 1 is modified by adding the following cases:

$$\forall (k, l) \in \{2..i\} \times \{2..j\} : T[k, l] + score(t_{i-k \ldots i}, q_{j-l \ldots j}). \tag{3}$$

In the dynamic programming table, the value of a given position $T[i, j]$ is obtained from any case $T[k, l]$ for $1 \leq k \leq i$, $1 \leq l \leq j$. We thus need to consider the computation of the value $score(t_{i-k...i}, q_{j-l...j})$ (*i.e.* the cost of aligning the sequence of chords $t_{i-k...i}$ and the sequence of chords $q_{j-l...j}$).

For the comparison of two polyphonic sequences t and q, we propose to define a scoring scheme based on the scoring scheme presented in Section 2. However, in this case, we are dealing with sequences of chords instead of single chords and we thus need a way to encode several chords into a single one.

Let us consider the sequence of chords $t_{i_s...i_e}$ that must be encoded (and that will be compared to a sequence of chords $q_{j_s...j_e}$ in q). The pitch of $t_{i_s...i_e}$ is then defined as the set T_p of all the different pitches that are present in this sequence. The duration of this sequence of chords is equal to T_d the duration elapsed from the beginning of t_{i_s} to the end of t_{i_e}.

For example, in Figure 2, the sequence of 4 chords:

$$(\{G4, B4\}, \{D2\}, \{A2\}, \{G8, B8, D8\})$$

is encoded by the chord:

$$(\{A, B, D, G\})$$

with a duration of 16.

Finally, a sequence of chords is only represented by a single set of unordered pitches and a single duration. To compare these chord representations the approach described in Section 4.1 and $score_p$ to weight the edges of the bipartite graph are used. Then, the score between two sequences of chords is given by:

$$score(t_{i_s...i_e}, q_{j_s...j_e}) = \alpha \times score_{bpg}(T_p, Q_p) + \beta \times score_d(T_p, Q_d)$$

Now let us consider the computation of the dynamic programming table M. We propose to illustrate the computation of a case $M[i][j]$ of this table through the example in Table 1. The score in position X is obtained either from f which implies the computation of scores $score_{bpg}(\{A\}, \{C, G\})$ and $score_d(2, 2)$ or:

- from e which implies the computation of $score_{bpg}(\{A\}, \{A, C, G\})$ and $score_d(2, 3)$
- from d which implies the computation of $score_{bpg}(\{A, D\}, \{C, G\})$ and $score_d(4, 2)$
- from c which implies the computation of $score_{bpg}(\{A, D\}, \{A, C, G\})$ and $score_d(4, 3)$

Table 1. Example of merged notes that must be considered for one step (X) of the alignment algorithm

		{A1}	{C2,G2}	{C1,E1,B1}
	...	...	...	...
{G4,B4 }	...	*a*	*b*	
{D2}	...	*c*	*d*	
{A2}	...	*e*	*f*	
{G8,D8,B8}	...			*X*

- from b which implies the computation of $score_{bpg}(\{A, B, D, G\}, \{C, G\})$ and $score_d(8, 2)$
- from a which implies the computation of $score_{bpg}(\{A, B, D, G\}, \{A, C, G\})$ and $score_d(8, 3)$.

One can observe that from one computation of $score_{bpg}$ to another one, we just add vertices in the bipartite graph. So, it is not necessary to recompute $score_{bpg}$ from scratch. Toroslu and Üçoluk give in [21] an incremental algorithm to compute the assignment problem in $O(V^2)$ where V is the number of vertices in the bipartite graph. Using this algorithm in our approach, the time complexity of the computation of all possible merges for the case i, j is bounded by $O(\sum_{i=1}^{C} i^2) = O(C^3)$ where C is number of different pitches in $t_{1...i}$ and $q_{1...j}$. The time complexity of the alignment becomes $O(|t|^2 \times |q|^2 \times C^3)$ where C is the number of different pitches in t and q.

4.3 Mongeau-Sankoff Operations and Tied Notes

Consolidation and fragmentation operations are particularly useful in the case of tied notes and rests. The main difficulty is to set up the score for these operations. In our process, the score associated to either a consolidation or a fragmentation is highly negative except in the case of a perfect match between a chord and a sequence of tied notes. Actually, one note matches with a sequence of tied notes only if the pitches and the onsets of all the tied notes are exactly the same. If these conditions are satisfied, the score of the consolidation/fragmentation operations depends on the sum of the durations of the tied notes concerned.

Following the definition proposed in Section 3.2, tied notes can have an unlimited size. However, from a practical point of view, the number of tied notes that could be involved in an edit operation (in particular a consolidation or a fragmentation) has to be limited in order to reduce the time computation of the retrieval algorithm. Results of experiments lead us to fix this limit to 6 notes.

4.4 Local Transposition

A recent interesting analysis of existing approaches for retrieval of polyphonic music compares the geometric and the alignment algorithms [16]. One of the conclusion is that alignment algorithms *face severe issues in combining polyphony and transposition invariance* [16]. Actually, the problem of the transposition invariance property is induced by the representation of polyphonic music. Interval or contour representations of pitches allows retrieval systems to be transposition invariant in the monophonic context, but such representations cannot be applied in polyphonic context.

However, the local transposition algorithm presented previously in a monophonic context [13] can be directly applied to our polyphonic representations. Experiments show that applying this algorithm lead to a transposition invariant polyphonic music retrieval system. However, since we need to compute 12 transposition matrices (in addition of the dynamic programming table) the main disadvantage of this extension increases drastically time computations.

5 Experiments

We present hereafter the results of the experiments we performed to evaluate the accuracy of the polyphonic music retrieval system presented in the previous sections. The Music Information Retrieval Evaluation eXchange (MIREX) [22] is a contest whose goal is to compare state-of-the-art algorithms and systems relevant for Music Information Retrieval. During the first contest, an evaluation topic about symbolic melodic similarity has been performed. Participants have discussed the process of evaluation and proposed an evaluation procedure. The experiments presented in this paper are based on these procedures.

During MIREX 2006[1], the second task of the symbolic melodic similarity contest consisted in retrieving the most similar pieces from mostly polyphonic collections given a monophonic query. Note in our approach we allow comparing two polyphonic musical pieces. Two collections were considered. The *karaoke* collection is composed of about 1000 `.kar` files (Karaoke MIDI files) with mostly Western popular music. A few short (less than 20 notes) monophonic queries have been defined from polyphonic musical pieces. Some of them are exact but some of them are interpreted and thus have slight time or pitch deviations. Some queries have been transposed in order to test the property of transposition invariance. The considered queries are excerpts of:

1. *Carnaval des animaux* by Saint-Saëns (19 notes)
2. *The Winner Takes It All* by ABBA (18 notes, inexact)
3. *Silent Night*, popular song (11 notes, inexact)
4. *Für Elise* by Beethoven (17 notes, transposed)
5. *Ah, vous dirai-je, Maman* by Mozart (14 notes, transposed)
6. *Angels We Have Heard On High*, popular song (14 notes)
7. *The Final Countdown* by Europe, (9 notes, inexact)

Experiments consist in computing the similarity score between one query and each piece of the collection. The similarity score is normalized, by dividing it by the maximum similarity score. This maximum score is obtained by computing the similarity score between the query and itself. In the following results, normalized scores are then positive values lower than 1. Table 2 shows the score, the rank (obtained according to the similarity score) of this corresponding piece and the score obtained by the piece that has been estimated as the most similar (first rank). In order to evaluate the improvements induced by the adaptations presented in the paper, we run two different algorithms. The first algorithm is the basic alignment algorithm, and is denoted *normal* in tables. The other algorithm consider tied notes and is denoted *improved*. This is the only difference between the two algorithms.

Without improvement, the alignment algorithm does not allow the retrieval of the correct musical piece. Many other pieces obtain a better similarity score (>50 indicates that the correct piece has not been ranked in the first 50 pieces). By introducing and considering tied notes, the results are far better. For the

[1] `http://www.music-ir.org/mirex2006/index.php/MIREX2006_Results`

Table 2. Results of polyphonic music retrieval from a monophonic query using either a basic alignment algorithm or an improved algorithm. The rank and the score of the correct piece are indicated. The correct piece is always retrieved by applying the improved alignment algorithm.

	Normal			Improved	
Query	**Score**	**Rank**	**Rank 1 Score**	**Score**	**Rank**
1	0.55	> 50	0.85	1	**1**
2	0.52	> 50	0.83	0.93	**1**
3	0.68	> 50	0.95	0.98	**1**
4	0.69	4	0.72	1	**1**
5	0.59	> 50	0.86	1	**1**
6	0.58	> 50	0.83	1	**1**
7	0.79	> 50	0.87	0.98	**1**

Table 3. Average Precision (AP) and Precision at N Documents (PND) obtained by edit-distance based retrieval systems for MIREX 2006 databases and queries. The Classical Alignment column presents the results obtained by Uitdenbogerd during (MIREX).

Collection		**Proposed System**	Classical Alignment
Karaoke	AP	**0.78**	0.36
	PND	**0.83**	0.33
Mixed	AP	**0.67**	0.52
	PND	**0.66**	0.55

proposed queries, the correct piece is always estimated as the most similar piece. The similarity score is always 1, except for queries that was not an exact excerpt of the polyphonic piece. The score remains very important in these cases.

The second collection - *mixed* collection - used in MIREX 2006 is composed of 10000 randomly picked MIDI files that were harvested from the Web and which include different genres. Eleven queries (hummed or whistled) were proposed in the contest. Table 3 presents the results obtained with *karaoke* and *mixed* collections using our general polyphonic alignment algorithm and a classical alignment algorithm. Algorithms have been evaluated according to two measures: the *average precision* and the *precision at N documents* (N is the number of relevant documents).

The results[2] presented in Table 3 show that the proposed method improves retrieval systems based on the alignment principle. Concerning the *karaoke* collection, the average precision is near 0.80 whereas it is only 0.36 when considering a monophonic alignment. This difference (although less pronounced) is also observed

[2] The complete results obtained by the different methods proposed during MIREX 2006 can be found at
`http://www.music-ir.org/mirex/2006/index.php/Symbolic_Melodic_Similarity_Results`

for the *mixed* collection. The average precision is 0.67 instead of 0.52. Although voices are not separated in the musical pieces, this improvement remains notable.

6 Conclusion

In this paper, we have proposed a general alignment algorithm for the comparison of two polyphonic musics. This method is based on an extension proposed by Mongeau and Sankoff [2], in order to compare two monophonic musical pieces. In particular, this algorithm only requires a scoring scheme between notes (and not between chords). Although new edit operations have been introduced, the algorithm remains quadratic, if the number of different pitches in a chord and the number of merged chords are bounded by constants.

The string-to-string alignment problem consists of determining the score between two strings as measured by the maximum cost sequence of edit-operations needed to change one string into the other. The edit operations generally allow changing *one symbol* of a string into *another single symbol*, or inserting a *single symbol* or deleting *a single symbol*. Due to our domain of application, we have firstly generalized this notion to the comparison of sequences of sets of symbols: an edit operation allows modifying a set of symbols. While the cost of symbol comparison is stored in an *alignment table*, now to compare two sets of symbols, a maximum score maximum matching problem must be solved.

Another key element introduced in this paper is the notion of merge also known as consolidation and fragmentation operations [2]. This operation is different from the deletions and insertions familiar in sequence comparison and from the compression and expansions of time warping in automatic speech recognition [23]. This new transformation involves the replacement of several elements of the initial sequence by several elements of the final sequence. However, although this new definition differs from the one proposed by Mongeau and Sankoff [2] and increases the combinatorics, it allows us to compute a more accurate similarity measure between two *polyphonic musical pieces* in polynomial time.

Finally, our methodology can also take into account local transpositions and tied notes in the melody.

Applications of this algorithm could be extended to any kind of sequences in which time is essentially a quantitative dimension (*e.g.* rhythm in music) and not simply an ordinal parameter. However, further studies might be carried out in a more general framework in order to evaluate the proper accuracy of this approach.

References

1. Hanna, P., Ferraro, P., Robine, M.: On Optimizing the Editing Algorithms for Evaluating Similarity Between Monophonic Musical Sequences. Journal of New Music Research 36(4), 267–279 (2007)
2. Mongeau, M., Sankoff, D.: Comparison of Musical Sequences. Computers and the Humanities 24(3), 161–175 (1990)

3. Smith, T., Waterman, M.: Identification of Common Molecular Subsequences. Journal of Molecular Biology 147, 195–197 (1981)
4. Serrà, J., Gómez, E., Herrera, P., Serra, X.: Chroma Binary Similarity and Local Alignment Applied to Cover Song Identification. IEEE Transactions on Audio, Speech and Language Processing 16, 1138–1151 (2008)
5. Dannenberg, R.B., Birmingham, W.P., Pardo, B., Hu, N., Meek, C., Tzanetakis, G.: A Comparative Evaluation of Search Techniques for Query-by-Humming Using the MUSART Testbed. Journal of the American Society for Information Science and Technology (JASIST) 58(5), 687–701 (2007)
6. Bello, J.: Audio-based Cover Song Retrieval using Approximate Chord Sequences: Testing Shifts, Gaps, Swaps and Beats. In: Proceedings of the 8th International Conference on Music Information Retrieval (ISMIR), Vienna, Austria, September 2007, pp. 239–244 (2007)
7. Downie, J.S., Bay, M., Ehmann, A.F., Jones, M.C.: Audio Cover Song Identification: MIREX 2006-2007 Results and Analyses. In: Proceedings of the 9th International Conference on Music Information Retrieval (ISMIR), September 14-18, pp. 51–56 (2008)
8. Uitdenbogerd, A.L., Zobel, J.: Manipulation of Music for Melody Matching. In: Proceedings of the Sixth ACM International Conference on Multimedia, Bristol, England, pp. 235–240 (1998)
9. Paiva, R.P., Mendes, T., Cardoso, A.: On the Detection of Melody Notes in Polyphonic Audio. In: Proceedings of the 6th International Conference on Music Information Retrieval (ISMIR), London, UK (September 2005)
10. Uitdenbogerd, A.L.: Music Information Retrieval Technology. PhD thesis, RMIT University, Melbourne, Australia (July 2002)
11. Needleman, S., Wunsch, C.: A General Method Applicable to the Search for Similarities in the Amino Acid Sequences of Two Proteins. Journal of Molecular Biology 48, 443–453 (1970)
12. Gusfield, D.: Algorithms on Strings, Trees and Sequences: Computer Science and Computational Biology. Cambridge University Press, Cambridge (1997)
13. Allali, J., Hanna, P., Ferraro, P., Iliopoulos, C.: Local Transpositions in Alignment of Polyphonic Musical Sequences. In: Ziviani, N., Baeza-Yates, R. (eds.) SPIRE 2007. LNCS, vol. 4726, pp. 26–38. Springer, Heidelberg (2007)
14. Horwood, F.J.: The Basis of Music. Gordon V. Thompson Limited, Toronto (1944)
15. Madsen, S., Typke, R., Widmer, G.: Automatic Reduction of MIDI Files Preserving Relevant Musical Content. In: Proceedings of the 6th International Workshop on Adaptive Multimedia Retrieval (AMR), Berlin, Germany (2008)
16. Lemström, K., Pienimäki, A.: Approaches for Content-Based Retrieval of Symbolically Encoded Polyphonic Music. In: Proceedings of the 9th International Conference on Music Perception and Cognition (ICMPC), Bologna, Italy, pp. 1028–1035 (2006)
17. Pardo, B., Sanghi, M.: Polyphonic Musical Sequence Alignment for Database Search. In: Proceedings of the 6th International Conference on Music Information Retrieval (ISMIR), London, UK, pp. 215–222 (2005)
18. Edmonds, J., Karp, R.M.: Theoretical improvements in algorithmic efficiency for network flow problems. Journal of the Association for Computing Machinery 19, 248–264 (1972)
19. Typke, R.: Music Retrieval based on Melodic Similarity. PhD thesis, Utrecht University (2007)
20. Tarjan, R.E.: Data Structures and Network Algorithms. CBMS-NFS - Regional Conference Series In Applied Mathematics (1983)

21. Toroslu, I.H., Üçoluk, G.: Incremental assignment problem. Information Sciences 177(6), 1523–1529 (2007)
22. Downie, J.S., West, K., Ehmann, A.F., Vincent, E.: The 2005 Music Information retrieval Evaluation Exchange (MIREX 2005): Preliminary Overview. In: Proceedings of the 6th International Conference on Music Information Retrieval (ISMIR), London, UK, pp. 320–323 (2005)
23. Kruskal, J.B.: An orverview of sequence comparison. In: Sankoff, D., Kruskal, J.B. (eds.) Time Wraps, Strings Edits, and Macromolecules: the theory and practice of sequence comparison, pp. 1–44. Addison-Wesley Publishing Company Inc., University of Montreal, Montreal (1983)

AllThatSounds: Associative Semantic Categorization of Audio Data

Julian Rubisch, Matthias Husinsky, and Hannes Raffaseder

University of Applied Sciences, St. Pölten
Institute for Media Production
3100 St. Pölten, Austria
{julian.rubisch,matthias.husinsky,hannes.raffaseder}@fhstp.ac.at

Abstract. Finding appropriate and high-quality audio files for the creation of a sound track nowadays presents a serious hurdle to many media producers. As most digital sound archives restrict the categorization of audio data to verbal taxonomies, this process of retrieving suitable sounds often becomes a tedious and time-consuming part of their work. The research project AllThatSounds tries to enhance the search procedure by supplying additional, associative and semantic classifications of the audio files. This is achieved by annotating these files with suitable metadata according to a customized systematic categorization scheme. Moreover, additional data is collected by the evaluation of user profiles and by analyzing the sounds with signal processing methods. Using artificial intelligence techniques, similarity distances are calculated between all the audio files in the database, so as to devise a different, highly efficient search algorithm by browsing across similar sounds. The project's result is a tool for structuring sound databases with an efficient search component, which means to guide users to suitable sounds for their sound track of media productions.

1 Introduction

Supply and demand for digitally stored audio increased rapidly in the recent years. Number and diversity, as well as quality of available sound files reached an unmanageable amount. Efficient processes for the retrieval of audio data from large digital sound libraries play a pivotal role in the process of media production. In many cases, the user is required to know important features of the sound he is seeking, such as its source or excitation, in advance. On the other hand it is hardly possible to search for semantic features of a sound which are closer to human perception rather than technical parameters. Another obstacle towards an adequate verbal description of audio data emanates from the medium's inherent volatility, which makes it difficult to uptake and formally subsume acoustic events. Thus, not sounds themselves, much more the causing events are described.

The research project AllThatSounds aimed at facilitating the process of finding suitable sounds for media productions. For this purpose many different possibilities to categorize and describe sonic events were analyzed, evaluated, and

S. Ystad et al. (Eds.): CMMR/ICAD 2009, LNCS 5954, pp. 483–491, 2010.

linked. Apart from the applicatory use of the tool, the research questions raised by the work on these topics trigger a discussion process about perception, meaning, function and effect of the sound track of a media product.

2 Present Situation

As indicated above, sound designers and media producers often face the difficulty of retrieving appropriate sounds from huge digital audio libraries. The indexing of such databases is mostly restricted to verbal descriptions of the items' sources and excitations, limiting the formulation of adequate search requests to these criteria. In addition, solely verbally tagged sound libraries generally display poor accessibility, since users often have to browse through huge flat text files, which also points at the need for a standardized sound categorization scheme and vocabulary. Hence, even though it might seem natural and sufficient to describe a sound by its source or the event it is caused by, these categorizations usually do not carry any semantic or sound-related information.

Still, the semantic information included in a sonic event, or its signal-related parameters present a useful search criterion for many possible applications. For example, in the field of movie sound design, for the dubbing of objects often sounds are used which in fact have nothing in common with the sonified object, apart from the transported meaning or certain signal characteristics. Even more apparent are use cases that deal with objects that do not even exist in reality, such as spacecraft engines or lightsabers. Since human auditory perception tends to accept differences between what is perceived visually and aurally as long as it occurs simultaneously, and the differences do not exceed a certain tolerable limit – an effect known as *synchresis* [2] – it is most sound designers' primary target to retrieve the sound that expresses a certain meaning best, which may not necessarily be accomplishable by the actual sound of the object in question.

Time is another factor that has to be considered when designing a search interface for sound databases. Since in many cases it is unfeasible to spend a large amount of time for the retrieval of an optimal sound, it was a primary objective of the prototypical development to devise an optimized user interface which is capable of assisting the user in his search process.

3 Description and Categorization of Audio Files

A central objective of the research project was to design a sound database with enhanced retrieval possibilities. A design issue that had to be tackled results from the fact that sound is a carrier of information in many different dimensions. Aside from technical signal parameters, many acoustic events are tagged with a multitude of meanings and messages that originate e.g. from sociocultural contexts associated with them.

Therefore a primary obstacle in designing a taxonomy for sounds results from the abundance of possible categories that have to be considered. Furthermore, these categories are subject to transformation relating to cultural differences and temporal developments. For example, the encoded meaning of a mechanical typewriter has altered from *modern*, 50 years ago, to *nostalgic* or *anachronistic* nowadays.

In order to address these design issues, a multi-dimensional approach was taken, consisting of four different approaches:

1. Descriptive Analysis
2. Listeners' Analysis
3. Machine Analysis
4. Semantic Analysis

3.1 Descriptive Analysis

The descriptive analysis enables for the uniform description of acoustical events from the sound designer's perspective. The aim is that already at the time the upload of a sound into the database takes place, it can be sufficiently and distinctly categorized. Based on relevant literature [5] [6] [7] [8] [9], a general classification scheme of audio events was developed, which allows for a differentiated description in the following categories:

1. *Type of Sound*
   (music, speech, sound effect, ambience)
2. *Source*
   (technical/mechanical, musical instruments, objects/materials, ambience, synthetical, human, animal, nature)
3. *Stimulation*
   (time structure, intensity, type of stimulation)
4. *Room*
   (reverberation, inside/outside)
5. *Timbre*
   (harmonicity, brightness, volume)
6. *Pitch*
7. *Semantic Parameters*
   (personal distance, familiarity, emotion/mood)
8. *Other Parameters*
   (source velocity, complexity)

A discrepancy that has to be addressed here concerns the amount of detail that is necessary for a satisfactory categorization of audio data. While it becomes evident from the considerations mentioned above, that in order to provide an optimal search accuracy, sounds also have to tagged with high precision and complexity, such a process also requires a lot of time. To keep time short when describing the sound at upload, full description cannot be reached.

3.2 Listeners' Analysis

To overcome the drawback that a single user's description of a sound may lead to an unwanted result in another context, collaborative methods using Web 2.0

technologies are employed. Tags, comments and descriptions of possible usages of a sound provide further assistance when trying to retrieve appropriate audio files.

Moreover, users are given the possibility to contribute alternative categorizations of sounds to the system and thus expressing their own interpretation of the sound in question. A large data pool also represents a basis for further analysis which is explained in Sect. 3.4.

3.3 Machine Analysis

Contemporary off-the-shelf computers offer enough computational power to enable for an automated processing and categorization of audio signals. In particular, such methods seem to be a promising approach regarding the categorization of raw audio data from existing archives or large databases, which are to be integrated in the system. To accomplish a technical analysis of audio files, two different methods were employed.

MPEG-7 Features. The MPEG-7 standard[1] defines a variety of audio descriptors which are usable for the annotation of multimedia files. Some of these features are suitable for the estimation of psychoacoustic measures, such as sharpness, volume, harmonicity and loudness of a sound. In a preliminary listening experiment, the AudioSpectrumCentroid, AudioSpectrumSpread, AudioHarmonicity and AudioPower descriptors respectively were determined to model the mentioned characteristics best.

Based on these descriptors, the users are provided suggestions of these factors, with the intention to accelerate the process of entering data into the system. The accuracy of these estimations was investigated during a quantitative analysis of 1273 sounds which were additionally subdivided according to their respective length. The measured correlation coefficients are depicted in the following tables 1, 2, 3 and 4.

As may have been anticipated, there is no significant correlation for AudioSpectrumSpread and the volume of a sound, since aside from bandwidth, volume is also commonly associated with level and frequency of a sound. However, the AudioPower, AudioSpectrumCentroid and AudioHarmonicity descriptors seem to show medium correlation values for perceived loudness, sharpness and harmonicity, respectively. Taking into account the relatively large sample, a significant connection between the technical and psychoacoustical parameters can be suspected nonetheless. Lower correlation values for shorter sounds can be primarily explained by temporal masking effects.

Similarity Analysis. As described in Sect. 2, a considerable acceleration of the search process can be anticipated by computing similarity measures of all the sounds in the database. To achieve this, a similarity model based on Mel Frequency Cepstral Coefficients (MFCCs) is used to compare all the sounds with

[1] http://www.chiariglione.org/mpeg/standards/mpeg-7/mpeg-7.htm

Table 1. Correlation Coefficients for MPEG7-AudioPower / Loudness

	$\Delta t \leq 0.5s$	$\Delta t \leq 1s$	$\Delta t \leq 10s$	Σ
N	70	178	902	1273
corr	0,3	0,28	0,43	0,44

Table 2. Correlation Coefficients for MPEG7-AudioSpectrumCentroid / Sharpness

	$\Delta t \leq 0.5s$	$\Delta t \leq 1s$	$\Delta t \leq 10s$	Σ
N	70	178	902	1273
corr	0,34	0,41	0,46	0,44

Table 3. Correlation Coefficients for MPEG7-AudioSpectrumSpread / Volume

	$\Delta t \leq 0.5s$	$\Delta t \leq 1s$	$\Delta t \leq 10s$	Σ
N	70	178	902	1273
corr	-0,09	0,07	0,03	0,04

Table 4. Correlation Coefficients for MPEG7-AudioHarmonicity / Harmonicity

	$\Delta t \leq 0.5s$	$\Delta t \leq 1s$	$\Delta t \leq 10s$	Σ
N	70	178	902	1273
corr	0,23	0,31	0,45	0,46

one another. This model proved to be very useful in comparing and classifying sounds and music [3].

Similarity measures are obtained by computing the mean values as well as covariances and inverted covariances of the first five MFCC coefficients on a single-frame basis, and calculating a Kullback Leibler divergence afterwards [4].

For every sound in the database, the 20 most related sounds are stored (see Fig. 1) which allows the user to browse for viable sounds in a novel way. Early tests have proved this feature to be of great value to the users, because thereby sounds that wouldn't have been found by pure metadata search can be discovered.

3.4 Semantic Analysis

In media production the semantic content of acoustic events plays a crucial role to achieve a certain perception at the consumer. Conclusions regarding semantic content of a sound obtained through signal analysis methods would present a rewarding facilitation of the search process. However, until today it is almost

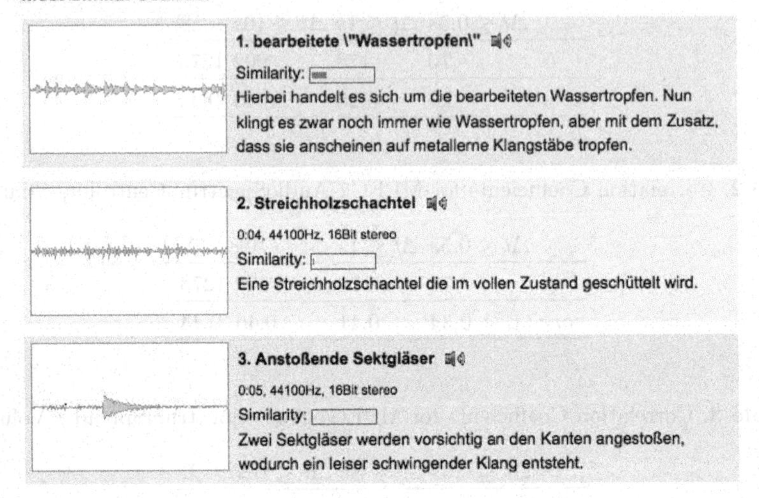
Most Similar Sounds:
1. bearbeitete \"Wassertropfen\"
Similarity:
Hierbei handelt es sich um die bearbeiteten Wassertropfen. Nun klingt es zwar noch immer wie Wassertropfen, aber mit dem Zusatz, dass sie anscheinen auf metallerne Klangstäbe tropfen.
2. Streichholzschachtel
0:04, 44100Hz, 16Bit stereo
Similarity:
Eine Streichholzschachtel die im vollen Zustand geschüttelt wird.
3. Anstoßende Sektgläser
0:05, 44100Hz, 16Bit stereo
Similarity:
Zwei Sektgläser werden vorsichtig an den Kanten angestoßen, wodurch ein leiser schwingender Klang entsteht.

Fig. 1. Similar sounds view: Every sound in the database is connected to its 20 most similar siblings; this figure shows the three most similar sounds for a certain database entry, along with an estimated measure of similarity. By clicking on one of the sounds, the database can be browsed in an efficient, more sound-related way.

impossible to automatically extract semantic meaning from audio signals, since not only signal parameters, but to a wide extent also cultural and social experience of listeners play a role.

The project aims at combining the collected user metadata and calculated sound features so as to explore possible correlations between signal parameters and semantic denotations. At present, such an approach is still unfeasible due to a lack of a sufficient number of user annotations. We intend to further investigate this field when numerous listeners have contributed to the database for a longer period of time.

4 Prototype

The first prototype with an improved search tool was released to the public on May 30th 2008 as a web application[2]. As of January 30, 2009, about 3700 sounds primarily recorded by students of the University of Applied Sciences St. Pölten's Media Engineering course. Sounds can be up- and downloaded in uncompressed wave-format and used under the Creative Commons Sampling Plus license[3].

[2] http://www.allthatsounds.net

[3] http://creativecommons.org/licenses/sampling+/1.0/

4.1 Usage and Architecture

To use the application and contribute to it, a user is required to register. Afterwards he or she is entitled to up- and download sounds to and from the database. Upon upload, the user is asked to provide a short name and description of the sound, as well as keywords that can be used for search purposes afterwards. The mentioned MPEG-7 descriptors are computed and employed to supply recommendation values for the sound's loudness, sharpness, harmonicity and volume. Additionally, the user is required to provide a first categorization according to the mentioned taxonomy. At the same time, the 20 most similar sounds are calculated and stored in the database. Furthermore, similarity distances for all sounds are renewed on a daily basis.

For retrieval purposes, a primitive text-based search can be used as well as the extended feature-based search (see Fig. 2). The user is furthermore enabled to obtain a sound's most similar counterparts from its detail page.

To improve architectural flexibility, the prototype's core functionality is exposed as a SOAP web service, enabling the implementation of a wide range of clients, including e.g. standalone applications as well as Digital Audio Workstation (DAW) plugins.

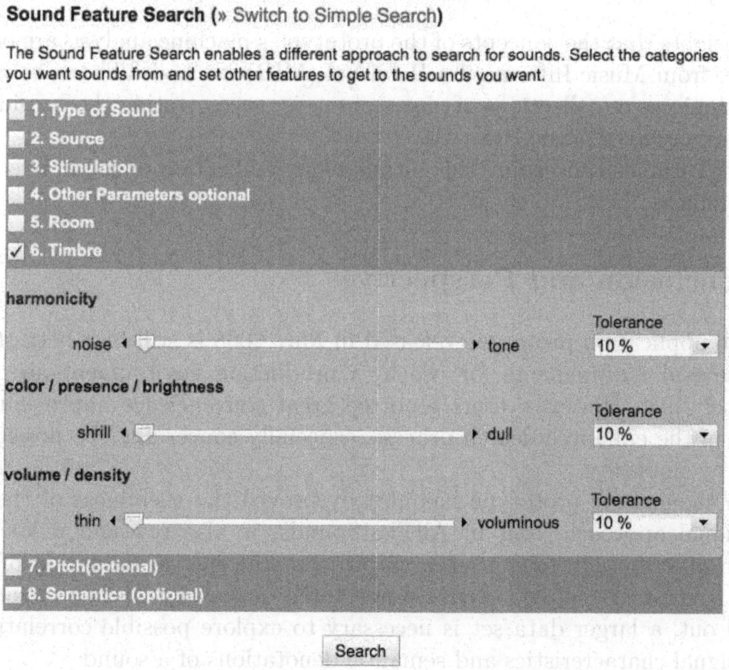

Fig. 2. Feature-based search form

4.2 Contents

This first prototype focuses on media sounds (jingles, music beds etc.), environmental or ambient sounds as well as sound effects, or foley sounds. While it is of course possible to upload music files, the musical features are not separately treated by the implemented algorithms. Durations range from approx. 30 milliseconds to about 9 minutes, with an average of 12.8 seconds.

Most of the sounds currently in the database have been provided by students of the University of Applied Sciences St. Pölten who were asked to record and upload sounds of a certain timbre, context or envelope as part of a course assessment.

4.3 Evaluation

At the present stage of the prototype, the evaluation process is limitied to internal reports by university employees and students, but we also encourage external users to provide feedback and report bugs.

Furthermore, several efforts are underway to form a cooperation with Austrian free radio stations to obtain feedback also from professional media producers.

5 Related Work

Many insights that the concepts of the prototype's machine analysis are based on originate from Music Information Retrieval (MIR) methods. There, researchers deal with pieces or collections of pieces of music, necessetating the consideration of the large area of music theory.

Related studies concerning the automated classification of sound effects have been conducted by Cano et. al. [1][4].

6 Conclusion and Perspective

The web application prototype released in May 2008 is still in a beta stadium, hence several requirements for use in a production environment are not yet completely met. However, users reported great convencience improvements by the use of the enhanced search process, especially concerning the possibility to browse by similarity.

Even though the prototype has already proved the usefulness of the multi-dimensional approach used in AllThatSounds, it also revealed a lot of new research questions. In future research projects a deeper examination of the semantic content of acoustic events is absolutely necessary. As has already been pointed out, a larger data set is necessary to explore possible correlations between signal characteristics and semantic denotations of a sound.

Moreover, concerning the usability and effectivity of the prototype, further studies have to be conducted. Enhancements of workflow could be reached by

[4] http://audioclas.iua.upf.edu/

the implementation of a text classificator (e.g. Naïve Bayes) for the estimation of probable sound sources and excitation types from the initial set of keywords. Also, pattern recognition methods which can be employed to classify and label sounds solely by taking into account their signal characteristics will have to be evaluated. Unsupervised machine learning methods, such as a k-means clustering approach using the mentioned MFCC features at upload time, could prove to be a valuable addition to the system.

Furthermore, when a larger data pool has been reached, it will be inevitable to revise the current categorization taxonomy, as it sometimes produces redundant information, and it still takes too much time for the user to complete the sound classification. Eventually, using the above-mentioned machine learning algorithms the user should not be forced to interfere with the upload process in most of the cases. In combination with an automated batch import routine which is yet to be designed, this should result in a drastic acceleration of database maintenance.

Acknowledgement

Starting in October 2005 a research team at the *University of Applied Sciences St.Pölten* worked together under project leader Hannes Raffaseder with the Vienna based companies *Audite / Interactive Media Solutions, Team Teichenberg* and, since April 2007 also the *University of Applied Sciences Vorarlberg*. The works were supported by the *Österreichische Forschungsförderungsgesellschaft (FFG)* in their FHplus funding programme until May 2008.

References

1. Cano, P., Koppenberger, M., Le Groux, S., Ricard, J., Herrera, P., Wack, N.: Nearest-neighbor generic sound classification with a wordnet-based taxonomy. In: Proceedings of AES 116th Convention, Berlin, Germany (2004)
2. Chion, M.: Audio-Vision - Sound on Screen. Columbia University Press (1994)
3. Feng, D., Siu, W.C., Zhang, H.J.: Multimedia Information Retrieval and Management. Springer, Berlin (2003)
4. Mandel, M.M., Ellis, D.P.W.: Song-Level Features and Support Vector Machines for Music Classification. In: Proceedings of the 6th International Conference on Music Information Retrieval, London, pp. 594–599 (2005)
5. Raffaseder, H.: Audiodesign. Hanser-Fachbuchverlag. Hanser-Fachbuchverlag, Leipzig (2002)
6. Schafer, M.R.: The Soundscape - Our Sonic Environment and the Tuning of the World. Destiny Books, Rochester (1994)
7. Sonnenschein, D.: Sound Design - The Expressive Power of Music, Voice and Sound Effects in Cinema. Michael Wiese Productions, Studio City (2001)
8. Truax, B.: Acoustic Communication, 2nd edn. Ablex Publishing, Westport (2001)
9. van Leeuwen, T.: Speech, Music, Sound. MacMillan Press Ltd., London (1999)

Author Index